Lecture Notes in Artificial Intelligence 4285

Edited by J. G. Carbonell and J. Siekmann

Subseries of Lecture Notes in Computer Science

Yuji Matsumoto Richard Sproat
Kam-Fai Wong Min Zhang (Eds.)

Computer Processing of Oriental Languages

Beyond the Orient: The Research Challenges Ahead

21st International Conference, ICCPOL 2006
Singapore, December 17-19, 2006
Proceedings

Springer

Volume Editors

Yuji Matsumoto
Nara Institute of Science and Technology, Japan
E-mail: matsu@is.naist.jp

Richard Sproat
University of Illinois at Urbana-Champaign
Dept. of Linguistics, Dept. of Electrical Engineering, USA
E-mail: rws@xoba.com

Kam-Fai Wong
The Chinese University of Hong Kong
Department of Systems Engineering and Engineering Management
Shatin, N.T., Hong Kong
E-mail: kfwong@se.cuhk.edu.hk

Min Zhang
Institute for Infocomm Research
21 Heng Mui Keng Terrace, Singapore 119613
E-mail: mzhang@i2r.a-star.edu.sg

Library of Congress Control Number: 2006937162

CR Subject Classification (1998): I.2.6-7, F.4.2-3, I.2, H.3, I.7, I.5

LNCS Sublibrary: SL 7 – Artificial Intelligence

ISSN 0302-9743
ISBN-10 3-540-49667-X Springer Berlin Heidelberg New York
ISBN-13 978-3-540-49667-0 Springer Berlin Heidelberg New York

Springer is a part of Springer Science+Business Media

springer.com

Printed in Germany

Typesetting: Camera-ready by author, data conversion by Scientific Publishing Services, Chennai, India
Printed on acid-free paper SPIN: 11940098 06/3142 5 4 3 2 1 0

Message from the President

"Beyond the Orient: The Research Challenges Ahead"

The International Conference on Computer Processing of Oriental Languages (ICCPOL) is a regular conference series of the Chinese and Oriental Languages Computer Society, COCLS (formerly known as the Chinese Language Computer Society, CLCS), which was established 30 years ago, to be exact on June 9, 1976. The society's name change was made in the winter of 2005 in response to the growing international demand in Chinese and oriental languages research and applications. The new vision of the society was also launched at the same time. COLCS was set "to be the international computer society driving the advancement and globalization of the science and technology in Chinese and oriental languages processing." On this front, the society's conference, ICCPOL, and journal, namely, the *International Journal on Computer Processing of Oriental Languages* (IJCPOL) provide two effective platforms.

This year marked the 21st meeting of the ICCPOL conference. I was delighted that despite his heavy workload, Kim-Teng Lua kindly accepted my invitation to host ICCPOL 2006 in Singapore. He put together one of the most energetic organizing committees: Minghui Dong, who looked after the local organization including the conference Website, Hui Wang the registration and Min Zhang the publication. Without their dedication and professionalism, ICCPOL 2006 would not have been so successful.

I am grateful to the Department of Systems Engineering and Engineering Management at The Chinese University of Hong Kong not only for allowing me to take up the conference chairmanship but, even more importantly, for providing financial aid for students in need. I am thankful to my colleague, Chris Yang, for working closely with me to assess every application in detail.

I would also like to thank the program Co-chairs Yuji Matsumoto and Richard Sproats, who jointly worked out an inspiring program. The combination of Asian and American scientists supported our theme of "Beyond the Orient." Many high-quality papers from all around the world were received and unfortunately due to limited space, only a few were accepted for publication in this year's proceedings. The accepted papers truly highlighted "The Research Challenges Ahead" in Chinese and Oriental language processing.

Kam-Fai Wong
Conference Chair ICCPOL 2006
and
President Colcs

Message from the Program Co-chairs

As the Co-chairs of the technical program of the 21st International Conference on Computer Processing of Oriental Languages (December 17-19, Singapore) we are delighted and honored to write the introduction to these proceedings.

The subtitle of this year's convocation was "Beyond the Orient: The Research Challenges Ahead," the goal being to broaden the scope of ICCPOL beyond its origins in East Asia, to become a truly international event. We believe that we made a very successful first step in this direction both in the composition of the Program Committee, which is made up of members from countries around the world and the accepted papers, which include a large number of contributions from outside ICCPOL's traditional home base.

We received 169 submissions from a variety of countries. We initially accepted only 38 full papers (30 oral presentations and 8 poster presentations determined by the authors' preference), but since this was a fairly small set, we decided to accept another 20 papers as short papers, all of which were presented as posters. Thus, the acceptance rate of full papers is 23% (38/169), and that of all accepted papers is 34% (58/169). Since two papers were withdrawn after the notification, this volume includes 56 papers (36 full papers and 20 short papers) presented at the conference.

As Technical Co-chairs we can claim but a small amount of credit for the success of the technical program. Our main thanks go to the Program Committee members who worked diligently to give fair reviews of the submitted papers, and most of whom spent additional time coming to a consensus on papers where there was a wide amount of disagreement.

We also thank the many authors who submitted their work for inclusion in the conference. Needless to say, the conference would not exist were it not for the technical presentations. We are mindful of the fact that there are many computational linguistics conferences and workshops available, and we are therefore happy that so many papers were submitted to ICCPOL 2006.

We would also like to thank our invited keynote speakers Gerald Penn, Claire Cardie and Hwee-Tou Ng for agreeing to present their work at ICCPOL.

In this year's conference, we used the START Conference Manager System for most of the paper handling process, that is, paper submission, paper reviews and discussion, notification of acceptance/rejection of papers, and uploading of final manuscripts, all of which went very smoothly. Thanks are especially due to Rich Gerber, the maintainer of the system, who was always quick to answer our queries, and even modified the system to handle the specific needs of our conference. We would also like to thank the committee members of ICCPOL, especially Kam-Fai Wong for his continuous support and timely advice, Minghui Dong for preparing very beautiful Web pages, and Min Zhang and Wai Lam for handling all the final manuscripts that are included in this volume.

December 2006

Yuji Matsumoto and Richard Sproat
ICCPOL 2006 Program Co-chairs

Organization

Conference Committee

Honorary Conference Co-chairs

Shi-Kuo Chang, University of Pittsburgh, USA (Co-founder, COLCS)
Benjamin Tsou, City University of Hong Kong, China (President, AFNLP)
Jun'ichi Tsujii, University of Tokyo, Japan (President, ACL)

Conference Chair

Kam-Fai Wong, The Chinese University of Hong Kong, China

Conference Co-chair

Jong-Hyeok Lee, POSTECH, Korea

Organization Chair

Kim-Teng Lua, COLIPS, Singapore

Program Co-chairs

Yuji Matsumoto, Nara Institute of Science and Technology, Japan
Richard Sproat, University of Illinois at Urbana-Champaign, USA

General Secretary

Minghui Dong, Institute for Infocomm Research, Singapore

Publication Co-chairs

Min Zhang, Institute for Infocomm Research, Singapore
Wai Lam, Chinese University of Hong Kong, China

Finance Co-chairs

Chris Yang, Chinese University of Hong Kong, China
Hui Wang, National University of Singapore, Singapore

Program Committee

Galen Andrew, Microsoft Research, USA
Masayuki Asahara, Nara Institute of Science and Technology, Japan

Hsin-Hsi Chen, National Taiwan University, Taiwan
Keh-Jiann Chen, Academia Sinica, Taiwan
David Chiang, ISI, USA
Lee-Feng Chien, Academia Sinica, Taiwan
Key-Sun Choi, KAIST, Korea
Susan Converse, University of Pennsylvania, USA
Robert Dale, Macquarie University, Australia
Guohong Fu, Hong Kong University, China
Pascale Fung, Hong Kong University of Science and Technology, China
Niyu Ge, IBM T. J. Watson Research Center, USA
Julia Hockenmaier, University of Pennsylvania, USA
Liang Huang, University of Pennsylvania, USA
Kenji Imamura, NTT, Japan
Kentaro Inui, Nara Institute of Science and Technology, Japan
Martin Jansche, Columbia University, USA
Donghong Ji, Institute for Infocomm Research, Singapore
Gareth Jones, Dublin City University, Ireland
Genichiro Kikui, NTT, Japan
Sadao Kurohashi, University of Tokyo, Japan
Kui-Lam Kwok, City University of New York, USA
Olivia Oi Yee Kwong, City University of Hong Kong, China
Gary Geunbae Lee, POSTECH, Korea
Gina-Anne Levow, University of Chicago, USA
Roger Levy, University of Edinburgh, UK
Haizhou Li, Institute for Infocomm Research, Singapore
Hang Li, Microsoft Research Asia, China
Mu Li, Microsoft Research Asia, China
Wenjie Li, Polytechnic University of Hong Kong, China
Chin-Yew Lin, ISI, USA
Qin Lu, Polytechnic University of Hong Kong, China
Bin Ma, Institute for Infocomm Research, Singapore
Qing Ma, Ryukoku University, Japan
Helen Meng, Chinese University of Hong Kong, China
Tatsunori Mori, Yokohama National University, Japan
Hwee Tou Ng, National University of Singapore, Singapore
Cheng Niu, Microsoft
Douglas Oard, University of Maryland, USA
Kemal Oflazer, Sabanci University, Turkey
Manabu Okumura, Tokyo Institute of Technology, Japan
Martha Palmer, University of Colorado, USA
Hae-Chang Rim, Korea University, Korea
Laurent Romary, LORIA, France
Tetsuya Sakai, Toshiba, Japan
Rajeev Sangal, International Institute of Information Technology, India
Jungyun Seo, Sogang University, Korea

Kiyoaki Shirai, Japan Advanced Institute of Science and Technology, Japan
Dawei Song, Open University, UK
Virach Sornlertlamvanich, Thai Computational Linguistics Lab., Thailand
Keh-Yih Su, Behavior Design Corporation, Taiwan
Jian Su, Institute for Infocomm Research, Singapore
Maosong Sun, Tsinghua University, China
Kumiko Tanaka-Ishii, University of Tokyo, Japan
Takenobu Tokunaga, Tokyo Institute of Technology, Japan
Kiyotaka Uchimoto, NICT, Japan
Takehito Utsuro, Tsukuba University, Japan
Hui Wang, National University of Singapore, Singapore
Patrick Wang, Northeastern University, USA
Andi Wu, Microsoft, GrapeCity Inc., USA
Fei Xia, University of Washington, USA
Yunqing Xia, The Chinese University of Hong Kong, China
Bo Xu, Chinese Academy of Sciences, China
Jie Xu, National University of Singapore, Singapore and Henan University, China
Nianwen (Bert) Xue, University of Colorado, USA
Tianshun Yao, Northeastern University, China
Zaharin Yusoff, Malaysian Institute of Micro-Electronics, Malaysia
Min Zhang, Institute for Infocomm Research, Singapore
Guodong Zhou, Institute for Infocomm Research, Singapore
Ming Zhou, Microsoft Research Asia, China

Hosted by the Chinese and Oriental Languages Computer Society (COLCS)

Organized by the Chinese and Oriental Languages Information Processing Society (COLIPS)

Supported by
Asian Federation of Natural Language Processing (AFNLP)
Department of Systems Engineering and Engineering Management (SEEM), The Chinese University of Hong Kong, China

Publisher Springer

[illegible] Korea Advanced Institute of Science and Technology, Korea
[illegible], USA
[illegible], Institute of Computational Linguistics, [illegible], Ireland
[illegible] Joo [illegible], Korea
[illegible] Institute for Infocomm Research, Singapore
[illegible] Microsoft Research Asia, China
[illegible] University, Hong Kong
Takenobu Tokunaga, Tokyo Institute of Technology, Japan
Kiyotaka Uchimoto, NICT, Japan
[illegible] University, China
[illegible] National University of Singapore, Singapore
[illegible] University, USA
[illegible], USA
[illegible] University of [illegible], Japan
[illegible] Harbin Institute of Technology, China
[illegible] Academy of Sciences, China
[illegible] National University of Singapore, Singapore and [illegible], China
[illegible] University of Colorado, USA
[illegible] University, China
[illegible] Institute of [illegible] Academy of Sciences, China
[illegible] Institute for Infocomm Research, Singapore
[illegible] Microsoft Research Asia, China
[illegible]

Hosted by Chinese and Oriental Languages Computer Society (COLCS)

Organized by [illegible] and Oriental Languages Information Processing Society (COLIPS)

Supported by

Asian Federation of Natural Language Processing (AFNLP)

Department of Systems Engineering and Engineering Management (SEEM), The Chinese University of Hong Kong, China

Publisher: Springer

Table of Contents

Information Retrieval/Document Classification/QA/ Summarization I

Machine Translation I

Information Retrieval/Document Classification/QA/Summarization II

Poster Session 1

Machine Translation II

Poster Session 2

Parsing/Semantics/Lexical Resources

Answering Contextual Questions Based on the Cohesion with Knowledge *

Tatsunori Mori, Shinpei Kawaguchi, and Madoka Ishioroshi

Graduate School of Environment and Information Sciences
Yokohama National University
79-7 Tokiwadai, Hodogaya, Yokohama 240-8501, Japan
{mori, kawaguchi, ishioroshi}@forest.eis.ynu.ac.jp

Abstract. In this paper, we propose a Japanese question-answering (QA) system to answer contextual questions using a Japanese non-contextual QA system. The contextual questions usually contain reference expressions to refer to previous questions and their answers. We address the reference resolution in contextual questions by finding the interpretation of references so as to maximize the cohesion with knowledge. We utilize the appropriateness of the answer candidate obtained from the non-contextual QA system as the degree of the cohesion. The experimental results show that the proposed method is effective to disambiguate the interpretation of contextual questions.

1 Introduction

In recent years, *contextual question-answering (QA) systems* have gained attention as a new technology to access information. In this paper, we propose a method to construct a Japanese contextual QA system using an existing Japanese non-contextual QA system[1]. Although a contextual question generally contains reference expressions[2], we expect that the non-contextual QA system will be able to find answers for such a question if reference expressions are appropriately completed along with their antecedents.

The completion of a question may be performed in the following steps: (1) detect reference expressions, and then (2) find an antecedent for each reference expression. However, there are ambiguities in these steps, and we may have multiple interpretations, namely, multiple *completed question candidates*, for one question. In the research area of discourse understanding, there exist many studies of the reference resolution in terms of *the cohesion with the context*. The

* This study was partially supported by Grant-in-Aid for Scientific Research on Priority Areas (No.18049031) and Scientific Research (C) (No.17500062) from the Ministry of Education, Culture, Sports, Science and Technology, Japan.

[1] We define *a contextual question* as "a question that may have references to the context, i.e., previously asked questions and their answers."

[2] In this paper, we use the term "reference expressions" to refer to not only reference expressions themselves, such as demonstratives and pronouns, but also ellipses like zero pronouns. Zero pronouns are ellipses of obligatory case elements.

Y. Matsumoto et al. (Eds.): ICCPOL 2006, LNAI 4285, pp. 1–12, 2006.

centering theory is one of the most widely used methods [1]. This type of reference resolution attempts to find an optimal interpretation so as to maximize the cohesion between a newly introduced sentence and the context. Such a method would definitely work in many cases, but it does not provide the method to resolve the ambiguity in the step (1).

In this paper, we propose another approach. It is the reference resolution in terms of *the cohesion with knowledge*. It is based on the fact that the QA system can refer to not only the context of the dialogue but also the knowledge base (i.e. a large text document collection). It is noteworthy that "answering a question" can be regarded as finding an object, i.e., an answer, whose context in the knowledge base is coherent with the question. Therefore, the cohesion with knowledge may be one of the promising criteria for finding the best interpretation of the question. Here, we hypothesize that the degree of cohesion with knowledge is analogous to *the appropriateness of the answer candidate* for each completed question candidate. The completed question candidate with the most appropriate answer can be accordingly considered as the best interpretation of the (original) question.

2 Related Work

2.1 Contextual Question Answering

The contextual QA was introduced as a subtask of the QA track in TREC 2001. However, Voorhees [2] summed up the evaluation as follows: "the first question in a series defined a small enough subset of documents such that the results were dominated by whether the system could answer the particular type of the current question, rather than by the system's ability to track context." For this reason, this task was excluded from all subsequent TRECs. On the other hand, a context task has been employed as a subtask of QAC in NTCIR, which is a series of evaluation workshops organized by the National Institute of Informatics, Japan. Kato et al. [3] summarized the lessons from the context task of TREC QA as follows: (1) the number of questions in a series is relatively small, and (2) there is no topic shift in a series. They prepared the test sets for NTCIR QAC according to the lessons, that is, (1) a series is relatively long, about seven questions (QAC3), and (2) two types of series are introduced, namely, *the gathering type* and *the browsing type*. A question series of the gathering type contains questions that are related to one topic. On the other hand, in a series of the browsing type, the topic varies as the dialogue progresses.

The approaches of the systems participating in the context tasks in the NTCIR QAC are mainly based on the cohesion with the context. In general, the approaches are classified into two types. The first type is based on the effort involved in the document/passage retrieval. It expands the query submitted to the IR system with the words/phrases that appeared in the previously asked questions[4]. The second type of approach is based on the completion of questions by resolving reference expressions[5]. One completed question is submitted to a non-contextual QA system. The method that we propose in this paper is similar to the second approach. However, our approach is based on the cohesion with knowledge as well as the cohesion with the context.

2.2 Reference Resolution

Detection of reference expressions: The detection of zero pronouns is particularly very important and is studied from various viewpoints. One of the most widely used methods is detection using a case-frame dictionary. A case-frame dictionary is used to find unoccupied cases in a sentence.

Identification of antecedents: Nariyama [6] proposed a modified version of the centering theory[1] in order to resolve Japanese zero pronouns. It utilizes a "*salient referent list (SRL)*" that pools all overt case elements that have appeared up to the sentence in question. If a new case element appears with a case marker identical to that of another case element already existing in the SRL, the new case element takes its place because of recency. In an SRL, the case elements are listed in the following order of salience: Topic > Nominative > Dative > Accusative > Others. A zero pronoun is resolved by selecting the most salient case element in the SRL.

3 Proposed Method

Figure 1 shows the overview of the proposed method. The method generates an answer list for each question in a given series by using the following procedure. It should be noted that the non-contextual QA system can perform list-type question answering, as described in Section 3.3. It is the task in which a system is requested to enumerate all correct answers.

Input. A new question and a list of antecedent candidates. The list is initialized to an empty list for the first question.
Output. An answer list and an updated antecedent candidate list.
Procedure

1. Detect reference expressions including zero pronouns in the new question using a case frame dictionary, and then generate question candidates with the zero pronouns. Generally we obtain multiple candidates because a single verb may have multiple case-frame entries of case frame.
2. Find antecedent candidates for reference expressions according to a selected strategy for completing reference expressions. We proposed three strategies: *CRE-C*, a strategy based on a modified version of the SRL-based centering theory; *CRE-H*, a heuristic strategy in which the characteristics of a series of questions are taken into account; and *CRE-A*, a strategy that adopts all possible noun phrases.
3. Generate all possible completed question candidates using the results of Step 2. Then, select the M-best completed question candidates according to the semantic consistency in the reference resolution.

4. Submit completed question candidates to the non-contextual QA system, and obtain a list of answer candidates for each question candidate. Each answer candidate in the list is associated with its answer score. From the answer scores, calculate the appropriateness of the list. We propose two measures of appropriateness: *AM-D*, a measure defined in terms of the distribution of scores of answer candidates; and *AM-M*, the maximum score of the answer candidates in the list.
5. Provide the most appropriate answer list as the final output.
6. Using the completed question candidate that provides the most appropriate answer list, update the list of antecedent candidates according to a selected strategy for completing the reference expressions, and output the list for the next question.

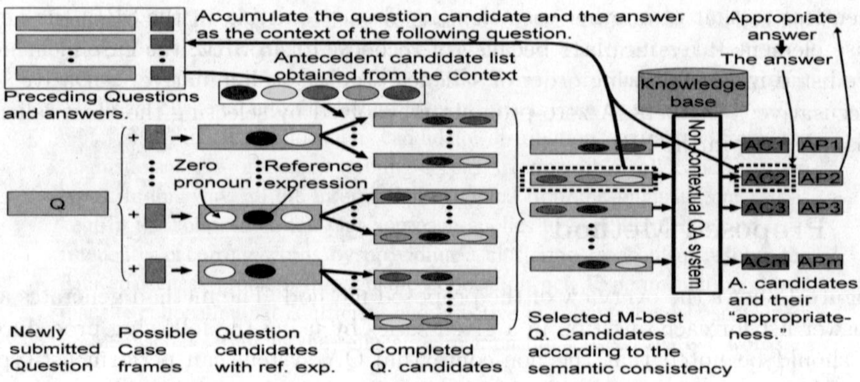

Fig. 1. Overview of the proposed method

3.1 Example

Before we explain the details of each step, we describe the flow of the procedure using the following example series of questions. In this example, we adopt Strategy CRE-C and Measure AM-D as the strategy for completing reference expressions and the appropriateness measure of the answer lists, respectively.

(1) a. Shizuoka-sutajiamu "ECOPA"-no kokeraotoshi-de Shimizu-S-Pulse-no
Shizuoka-studium "ECOPA"-REL opening-AT Shimizu-S-Pulse-REL
taisen-aite-wa doko de-su-ka
matched-team-TOP where/what BE-POL-INTERROG
Which team played a match against the Shimizu S-Pulse at the opening game of Shizuoka Stadium "ECOPA"?

b. sono-shuryoku-senshu-ni dare-ga i-masu-ka
its-leading-player-AS who-NOM exist-POL-INTERROG
Who is its leading player?

Since Question (1a) is the first in the series, it is submitted to a non-contextual QA system without any modification, and obtained the following answer list. The system also assigns semantic categories[3] to each answer in the list according to a thesaurus.

(2) a. {"Iwata" (CITY, STATION)}[4].

In Step 6 of the procedure, the system updates the list of antecedent candidates. When CRE-C is adopted, the list would be the following SRL obtained from the first question and its answer list[5]:

(3) [top:{"Iwata" (CITY, STATION), "Shimizu-S-Pulse-no taisen-aite" (OPPONENT, COMPANION, OTHER PERSON)}, nom:{}, dat:{}, acc:{}, other:{}].

As shown in this example, if the interrogative is the expression to be stored in the SRL, we replace it with the answer list. Each expression in the SRL is also assigned semantic categories.

Next, the system receives the second question (1b). Since the question has a context, the system tries to detect the reference expressions, as described in Section 2.2. First, the system checks whether the question contains demonstratives or pronouns, and it finds a demonstrative adjective called "sono." Next, the system tries to detect zero pronouns by using the following steps: (i) look up the verb of the question in a case-frame dictionary in order to obtain case frames, and (ii) detect unoccupied cases in the question as zero pronouns. The system also obtains the information of semantic categories for unoccupied cases from the case frames. In the case of Question (1b), the system obtains the following case frame for the verb "iru" (exist)[6], which satisfies the selectional restriction in terms of the semantic category, and detects that there are no zero pronouns.

(4) NP1(PLACE, PERSON)-ni NP2(PERSON)-ga iru

The antecedent candidates for the demonstrative adjective are obtained from SRL according to the order in the list. Since there are two candidates at the top of the list, we have the following two completed question candidates:

(5) a. Shimizu-S-Pulse-no taisen-aite-no shuryoku-senshu-ni
Shimizu-S-Pulse-REL matched-team-REL leading-player-AS
dare-ga i-masu-ka
who-NOM exist-POL-INTERROG
Who is the leading player of the team that played against the Shimizu S-Pulse?

b. Iwata-no shuryoku-senshu-ni dare-ga i-masu-ka
Iwata-REL leading-player-AS who-NOM exist-POL-INTERROG
Who is the leading player of Iwata?

The system selects the M-best completed question candidates according to the semantic consistency in the reference resolution, as described in Section 3.7, and submits them to the non-contextual QA system in order to obtain an answer list and a value of the appropriateness of the list for each question candidate. In

[3] Semantic categories are shown in parentheses.

[4] The full name of the team is "Jubilo Iwata," and "Iwata" is the name of city.

[5] In the Japanese sentence "NP1 wa NP2 da," the copula "da" represents that NP2 is equivalent to NP1. We accordingly treat not only NP1 but also NP2 as the topic.

[6] The expression "iru" is the root form of the verb "i" in Question (1b).

the example, we have the following results: {"NANAMI Hiroshi[7]","YAMADA Kosuke[8]"} and 0.019 for Question candidate (5a), and {"NANAMI Hiroshi"} and 0.031 for (5b). Since we hypothesize that the question candidate whose answer list has the highest value of appropriateness is the best interpretation in terms of the cohesion with knowledge, the system outputs the latter answer list as the answer for Question (1b).

After the process of the second question, the system updates the SRL and proceeds to the processing of the next question.

3.2 Non-contextual Japanese QA System

The non-contextual Japanese QA system used is a Japanese real-time QA system based on the study by Mori[7]. Mori reported that the MRR (mean reciprocal rank) of the system is 0.516 for the test set of NTCIR-3 QAC2. It treats each morpheme in retrieved documents as a seed of an answer candidate and assigns it a score that represents the appropriateness for an answer. Under the assumption that the morpheme is the answer, the score is calculated as the degree of matching between the question and the sentence where the morpheme appears. In his current implementation, the score is a linear combination of the following sub-scores: the number of shared character bigrams, the number of shared words, the degree of case matching, the degree of matching between dependency structures, and the degree of matching between the NE type of the morpheme and the type of question.

3.3 List-Type QA Processing in Non-contextual Japanese QA

We also introduce a list-type QA processing proposed by Ishioroshi et al. [8]. They assume that the distribution of answer scores contains a mixture of two normal distributions, $\phi_p(x; \mu_p, \sigma_p)$ and $\phi_n(x; \mu_n, \sigma_n)$, i.e., those of the correct answers and incorrect answers, where μ and σ are the average and the standard deviation, respectively. Under these assumptions, the correct answers may be separated from the mixture of the distributions by using the EM algorithm. Figure 2 shows an example of the score distribution in the case that the score distribution of the correct answers is separable from that of the wrong answers.

3.4 Detecting of Reference Expressions

Our method treats the three types of reference expressions — (i) demonstratives and pronouns, (ii) zero pronouns, and (iii) ellipsis of the adnominal modifier "NP_1-NO" in a noun phrase "NP_1-NO NP_2 (NP_2 of NP_1)."

The detection of the reference expressions of type (i) is not difficult because they appear explicitly. With regard to type (ii), we employ an existing method based on a case-frame dictionary as described in Section 2.2. We use "nihon-go

[7] A correct answer.

[8] A wrong answer.

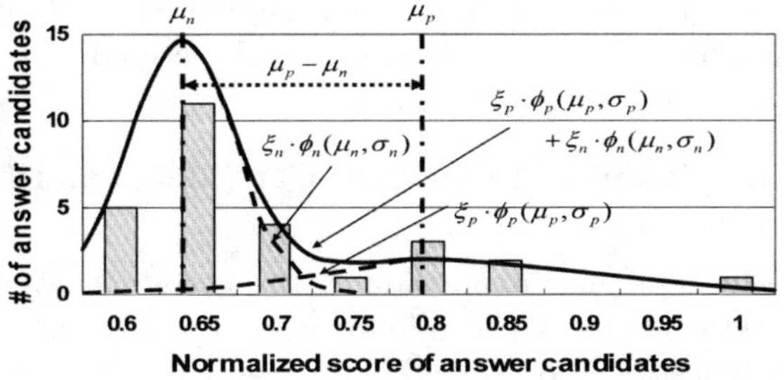

Fig. 2. An example distribution of answer scores

goi taikei" (a Japanese lexicon) as the case-frame dictionary. For type (iii), we adopt Takei's method [9].

If no reference expression is detected in a question, the system assumes that a topic phrase is omitted in the question and introduces a zero topic phrase in order to force the question to have a relation with the context.

3.5 Finding Antecedent Candidates of Reference Expressions

Strategy CRE-C: This strategy is based on a modified version of Nariyama's SRL-based centering theory. The list of antecedent candidates is SRL itself. This method is different from Nariyama's method in the following manner: (a) demonstratives and pronouns are resolved before zero pronouns, and (b) a zero topic phrase may refer to all possible case elements in SRL.

Strategy CRE-H and Strategy CRE-A: For these strategies, the list of antecedent candidates is maintained as described in Section 3.9. The system may select any element from the list for each reference expression.

3.6 Narrowing Down Antecedent Candidates Using Selectional Restriction

According to the selectional restriction, for each reference expression in the current question, the system filters out inappropriate candidates in the antecedent candidates that are obtained by one of the strategies described in Section 3.5. With respect to Strategy CRE-C, the system selects the candidate that is at the highest rank in SRL among the appropriate candidates. The selectional restriction is based on the similarity $sim(r, a)$ between the semantic categories of the antecedent candidates and reference expressions, a and r, defined in the thesaurus("nihon-go goi taikei"). The similarity is calculated by the following equation [10]:

$$sim(r, a) = \begin{cases} \frac{2 \times L_{ra}}{l_r + l_a} & \text{if } a \not\prec r \\ 1 & \text{if } a \prec r \end{cases} \tag{1}$$

where l_r and l_a are the depths of the categories r and a in the thesaurus respectively, and L_{ra} is the depth of the lowest common ancestor of r and a.

The symbol "$\prec$" represents the subsumption relation. We determine a threshold value of the similarity Th_{sim}, and filter out each antecedent candidate whose similarity is less than the threshold value.

3.7 Generating Completed Question Candidates and Narrowing Them Down

By completing each reference expression in the current question with all the possible antecedent candidates, the system generates all the possible candidates of the completed question. However, this process may generate many question candidates, and the non-contextual QA systems may take a very long time to process them. Therefore, we introduce a measure for a completed sentence in terms of *the degree of consistency in the reference resolution*, and select the M-best question candidates by using the measure. We defined the degree $C(Q)$ of a question candidate Q in Equation (2).

$$C(Q) = \sum_{<r_i, a_i> \in resolv(Q)} \frac{c_1(r_i, a_i)}{|resolv(Q)|} \tag{2}$$

$$c_1(r, a) = \begin{cases} 1 & \text{if } a \prec r \wedge NE(a) \\ 1.5 & \text{if } a \prec r \wedge \neg NE(a) \\ sim(r, a) & \text{if } a \not\prec r \end{cases}$$

where $resolv(Q)$ is the set of pairs of a reference expression and its antecedent in the question Q, and $NE(a)$ is true if a is a named entity.

3.8 Finding the Most Appropriate Answer List by Using the Non-contextual QA System

The selected question candidates are submitted to the non-contextual QA system. The completed question candidate with the most appropriate answer may be considered as the best interpretation of the original question. We propose the following two methods as the appropriateness measure.

Measure AM-D: The appropriateness of an answer candidate list is assumed to be measured by $\mu_p - \mu_n$ in Figure 2. Some of the candidates of question completion may not be consistent with the knowledge base. In such cases, the scores of highly ranked answer candidates are not very high and have almost the same distribution as that of the lower ranked candidates. Conversely, if the value $\mu_p - \mu_n$ is relatively large, we can expect that an appropriate list is obtained.

Measure AM-M: The appropriateness of an answer candidate list is assumed to be measured by the maximum score of the answer candidates in the list. It is based on the fact that the score of an answer is calculated according to the coherence between the question and the context of the answer.

3.9 Updating the List of Antecedent Candidates

By using the completed question candidate that provides the most appropriate list of answer candidates, the system updates the list of antecedent candidates according to a selected strategy for completing the reference expressions.

Strategy CRE-C: The list is updated in the same manner as the SRL. The difference is the treatment of the interrogatives. An interrogative is replaced with its answer list before the SRL is updated.

Strategy CRE-H: The list of antecedent candidates is maintained so as to have the following elements: (a) all the case elements in the current completed question, (b) all the phrases in the answer list of the current completed question, and (c) topic phrases, that is, all the case elements in the first question. Here, it should be noted that the system extracts the case elements not from the original current question but from the *completed* current question. The case elements in the questions before the current question may be retained in the completed current question if the question continues to refer to them.

Strategy CRE-A: We just adopt all the case elements in all the completed questions thus far.

4 Experimental Results

We evaluate the proposed systems in terms of the accuracy of the question answering by using the test set of NTCIR-5 QAC3 [11]. The test set comprises 50 series and 360 questions. In these series, 35 series (253 questions) are of the gathering type and 15 series (107 questions) are of the browsing type. It should be noted that the systems are not allowed to use the series type for answering the questions. The document collection as the knowledge source consists of all (Japanese) articles in the Mainichi Shimbun newspaper and Yomiuri Shimbun newspaper published in 2000 and 2001. The parameters are tuned with the test set of NTCIR-4 QAC2. The threshold value for the selectional restriction Th_{sim} is 0.54, and the number M of completed question candidates to be selected is 20. For measuring the accuracy of the list-type question answering, we use the mean of the modified F measure $MMF1$ defined by Kato et al. [11].

We also prepare systems that do not use cohesion with knowledge ("No-CK" in the following table) as baseline systems. These systems adopt the completed answer candidate that has the largest value of the degree of consistency in the reference resolution, which was described in Section 3.7. As an upper limit of accuracy, we also evaluate the non-contextual QA system with questions whose reference expressions are manually resolved.

The experimental results are shown in Table 1.

Table 1. Evaluation of Question Answering in MMF1

	(a) All Series			(b) Gathering Type			(c) Browsing Type		
	AM-D	AM-M	No-CK	AM-D	AM-M	No-CK	AM-D	AM-M	No-CK
CRE-C	**0.174**	0.166	0.164	0.177	0.178	0.168	**0.166**	0.137	0.146
CRE-H	0.147	0.147	0.136	0.142	0.168	0.141	0.134	0.097	0.123
CRE-A	0.169	0.157	0.133	**0.180**	0.166	0.146	0.146	0.135	0.105
Manually resolved	0.242								

5 Discussion

5.1 Overall Performance of Question Answering

With regard to the effect of the introduction of cohesion with knowledge, as shown in Table 1, both AM-D and AM-M, which utilize the cohesion with knowledge, outperform the baseline No-CK, which is only based on the degree of consistency in the reference resolution. In particular, the appropriateness measure AM-D works well for all strategies for completing the reference expressions.

With regard to the differences between the strategies for completing the reference expressions, Strategy CRE-C exhibits a better performance than the others. However, the difference between CRE-C and CRE-A is not significant when the measure AM-D is used.

In order to investigate the situation in further detail, let us compare (b) and (c) in Table 1. The systems based on Strategy CRE-C are stable over both types of series. The combination of CRE-C and AM-D is more robust even when topic shifts occur. The systems with the measure AM-M seem to have a better performance in the gathering type series. The reason for this is as follows. Because of the composition of the answer score, the score of a longer question tends to be larger than that of a shorter question. Consequently, the use of the measure promotes the selection of the case frames that have a larger number of case elements. As a result, the system tends to select question candidates that are more cohesive with the context. However, the systems easily fail to track the shift of topic, as shown in Table 1 (c). The strategies CRE-H and CRE-A have good performance for the series of the gathering type, but are not good at the series of the browsing type because they could not track topic shifts properly.

5.2 Failure Analysis

A detailed analysis of success and failure is summarized in Table 2. In this table, "Success" implies that the generated answer list contains at least one correct answer. The other cases are "Failure." Among the success cases, there are many cases where the reference resolution fails but the system successfully finds the answers for the question. This implies that the introduction of expressions of the context into the current question has a positive effect on the performance of question answering even if the accuracy of the reference resolution is insufficient. One of the reasons for this is that these newly introduced expressions may work well in the early stages of question answering, such as document/passage retrieval.

The main reason for the failure lies in the stage of updating the list of antecedent candidates in the CRE-C. The failure is caused by, at least, the following reasons: (1) failure in finding correct answers for some previous questions, (2) failure in finding the appropriate antecedents for the reference expressions in the list of antecedent candidates.

The number of failures in the updating stage in CRE-A is relatively low because the restriction on the list of antecedent candidates is relatively weak

Table 2. Detailed analysis of success and failure

Method	Success		Failure				
	Res. OK Ans. OK	Res. NG Ans. OK	Ante. NG	Q. Gen. NG	Q. Sel. NG	Ans. NG	Others
CRE-C + AM-D	11.0% (34)	13.5% (42)	26.5% (82)	21.0% (65)	9.7% (30)	18.1% (56)	0.3% (1)
CRE-C + AM-M	9.4% (29)	13.2% (41)	29.4% (91)	20.3% (63)	11.0% (34)	16.5% (51)	0.3% (1)
CRE-C + No-CK	9.4% (29)	11.9% (37)	29.4% (91)	21.0% (65)	10.6% (33)	17.4% (54)	0.3% (1)
CRE-A + AM-D	11.3% (35)	11.3% (35)	17.4% (54)	14.5% (45)	29.4% (91)	15.2% (47)	1.0% (3)
CRE-A + AM-M	10.6% (33)	8.7% (27)	18.1% (56)	13.5% (42)	37.1% (115)	11.0% (34)	1.0% (3)
CRE-A + No-CK	6.8% (21)	9.0% (28)	19.4% (60)	15.2% (47)	35.2% (109)	13.5% (42)	1.0% (3)

Res.: reference resolution
Ans.: question answering by the non-contextual system
Ante.: appropriate antecedent
Q. Gen.: generation of completed question candidates
Q. Sel.: selection of an appropriate question candidate

and the list may have more candidates than CRE-C. On the other hand, there are many failures in the stage involving the selection of an appropriate question candidate. Since the cohesion with knowledge is taken into account in this stage, the ratio of failures in this stage is closely related to the effectiveness of the cohesion with knowledge. However, we can *not* jump to the conclusion that the proposed method is not effective because the method based on the cohesion with knowledge works correctly only when the non-contextual QA system can find at least one correct answer in the knowledge base.

In order to estimate the ability of disambiguation based on the cohesion with knowledge more accurately, we investigate the accuracy of answering the questions that satisfy the following conditions: (a) each of them has multiple completed question candidates, and at least one candidate is a proper interpretation in the given context, and (b) the non-contextual QA system can find at least one correct answer for at least one of the completed answer candidates with the correct interpretation. The result shown in Table 3 implies that the use of cohesion with knowledge significantly improves the accuracy. By comparing Table 3 with Table 1, we also find that the accuracy may be improved when the non-contextual QA system is able to find appropriate answers.

Table 3. MMF1 values when the cohesion with knowledge may work correctly

	AM-D	AM-M	No-CK
CRE-C	0.478	0.437	0.399
CRE-A	0.436	0.374	0.334

6 Conclusion

In this paper, we introduced the notion of "*cohesion with knowledge*" and on its basis, proposed a question-answering system to answer contextual questions using a non-contextual QA system. Experimental results showed that the system works effectively under the combination of the strategy based on the SRL-based centering theory, i.e., CRE-C and the appropriateness measure of an answer that is defined in terms of the score distribution of the answer candidates, i.e., AM-D.

According to the failure analysis, the main reason for the failure was that the appropriate antecedents of the reference expressions in the current question do not appear in the list of antecedent candidates. Therefore, further improvement in the non-contextual QA system is required.

References

1. Walker, M., Iida, M., Cote, S.: Japanese discourse and the process of centering. Computational Linguistics **20**(2) (1994) 193–232
2. Voorhees, E.M.: Overview of the TREC 2001 question answering track. In: Proceedings of the tenth Text Retrieval Conferene (TREC 2001). (2001)
3. Kato, T., Fukumoto, J., Masui, F., Kando, N.: Are open-domain question answering technologies useful for information access dialogues? ACM Transactions on Asian Language Information Processing (TALIP) **4**(3) (2005) 243–262
4. Murata, Y., Akiba, T., Fujii, A., Itou, K.: Question answering experiments at NTCIR-5: Acquisition of answer evaluation patterns and context processing using passage retrieval. In: Proceedings of the Fifth NTCIR Workshop Meeting. (2005)
5. Matsuda, M., Fukumoto, J.: Answering questions of IAD task using reference resolution of follow-up questions. In: Proceedings of the Fifth NTCIR Workshop Meeting. (2005)
6. Nariyama, S.: Grammar for ellipsis resolution in Japanese. In: Proceedings of the 9th International Conference on Theoretical and Methodological Issues in Machine Translation. (2002) 135–145
7. Mori, T.: Japanese question-answering system using A* search and its improvement. ACM Transactions on Asian Language Information Processing (TALIP) **4**(3) (2005) 280–304
8. Ishioroshi, M., Mori, T.: A method of list-type question-answering based on the distribution of answer score generated by a ranking-type Q/A system. SIG Technical Reports 2005-NL-169, Information Processing Society of Japan (2005) (in Japanese).
9. Yamura-Takei, M.: Approaches to zero adnominal recognition. In: Proceedings of the 41st Annual Meeting of the Association for Computational Linguistics (ACL2003) Student Research Workshop, ACL (2003) 87–94
10. Kawahara, D., Kurohashi, S.: Zero pronoun resolution based on automatically constructed case frames and structural preference of antecedents. In: Proceedings of the 1st International Joint Conference on Natural Language Processing. (2004) 334–341
11. Kato, T., Fukumoto, J., Masui, F.: An Overview of NTCIR-5 QAC3. In: Proceedings of the Fifth NTCIR Workshop Meeting. (2005)

Segmentation of Mixed Chinese/English Document Including Scattered Italic Characters

Yong Xia, Chun-Heng Wang, and Ru-Wei Dai

Laboratory of Complex System and Intelligence Science
Institute of Automation, Chinese Academy of Sciences, Beijing, 100080, China
{yong.xia, chunheng.wang, ruwei.dai}@ia.ac.cn

Abstract. It is difficult to segment mixed Chinese/English documents when there are many italic characters scattered in documents. Most contributions attach more attention to English documents. However, mixed document is different from English document and some special features should be considered. This paper gives a new way to solve the problem. At first, an appropriate character area is chosen to detect italic. Next, a two-step strategy is adopted. Italic determination is done first and then if the character pattern is identified as italic, the estimation of slant angle will be done. Finally the italic character pattern is corrected by shear transform. A method of adopting two-step weighted projection profile histogram for italic determination is introduced. And a fast algorithm to estimate slant angle is also introduced. Three large sample collections, including character and character-pair and document respectively, are provided to evaluate our method and encouraging results are achieved.

1 Introduction

We have been engaging in the recognition of mixed Chinese/English document for years, and good results have been reported in [15] and [16], but determination and correction of italic characters haven't been considered before, which is important to find a general way to the recognition of all kinds of documents. In this paper, we will report our further research on this special problem. Many methods have been introduced to detect italic characters in documents. In [1]-[4], a method of adopting direction chain code of pattern is discussed in detail. The experimental results reported in the paper are very good, but there are some limitations that the ground truth is not absolute because the experimental objects are handwritten characters and the ground truth is constructed by people's subjective identification. A comparison of positive and negative direction elements between sheared pattern and original pattern is presented to determinate italic in [8]. But the experimental results are not provided. The projection profile histogram is very useful to determinate italic according to [5], [7], [12]. In [12], the distribution of projection profile histogram is adopted. A simple method is introduced in [9] that bounding box of every connected component is adopted to estimate the slant angle and the bounding box is a parallelogram rather than a rectangle. It seems that this method is simple and efficient. However how to get

Y. Matsumoto et al. (Eds.): ICCPOL 2006, LNAI 4285, pp. 13–21, 2006.

the parallelogram is the key and it is not easy. The authors do not mention this point in the paper. In [11], a method based on several special rules is presented and these rules are from some features of printed English characters. However it is not robust very much when there are noises in the pattern and it is only suitable for printed English documents. Besides, some complicated methods are considered in order to get more accurate estimation of slant angle. In [6], [13] and [14], strokes are adopted to estimate the slant angle. And slant estimation of handwritten characters by means of Zernike moments is also introduced in [10].

So far, most contributions mentioned above discuss English documents. But how about mixed Chinese/English documents? In general, most algorithms of estimating slant angle of pattern are language independent. However how to adopt these algorithms is not still easy to the segmentation of mixed documents. Besides, some italic characters are scattered in the document, which will increase the difficulties. As for English documents, it is very prevalent that a word rather than a character is used to detect italic. But How about mixed documents? All these problems will be discussed and a new method of italic determination and estimation of slant angle will be presented.

In the next section, segmentation of mixed documents including scattered italic characters will be discussed. Italic determination and estimation of slant angle are introduced in detail in this section. Finally, some experimental results are given in section 3 and conclusion is drawn in section 4.

2 Segmentation of Mixed Chinese/English Document Including Scattered Italic Characters

In general, if a document includes italic characters, we should find these italic areas at first and then correct them by shear transform according to the estimation of slant angle. So the key of segmentation is to find those italic characters scattered in the document and estimate the slant angle accurately.

In this paper, we only discuss printed mixed Chinese/English documents. There are some features in printed documents that can improve the performance of italic determination. First, the slant angle of italic characters in a document is constant. Second, although the slant angle may be different in different documents, the differences of the slant angles are not very big and in general the slant angle ranges from 12 to18 which is dependent on the typography software. Finally, it is easy to evaluate the performance because the ground truth is reliable and absolute, which is not easy to be done in handwritten documents.

Obviously, it is difficult or even impossible to estimate the slant angle accurately that only the original pattern of scattered character is adopted. The reason is that the number of Chinese character classes is very large and various fonts also give some troubles. So comparison of features of patterns between the original pattern and the sheared pattern is a good way to italic determination. Some researchers adopted this method and good results are achieved. But how to get the shear angle? And how many sheared patterns are needed? These problems haven't been discussed thoroughly so far and it is very important.

Some researchers shear the pattern by a step of 1 degree, many patterns are achieved and comparison of features will be done and then the slant angle can be estimated. This method gets the slant angle firstly and then italic determination is done according to the slant angle. It is good if speed is not considered. But as for scattered italic characters, the speed of italic detection is very important. Other researchers assume the slant angle is a constant angle such as 12 degree. So they set the shear angle as 12 and give italic determination by comparing the features of the original pattern and sheared pattern. However, as mention above, the slant angle in different documents may be different. So this method is not a general way.

Next we will give a new way to italic determination and estimation of slant angle. The strategy of two-step is adopted. Italic determination will be done firstly and then if the character pattern is identified as italic, we will estimate the slant angle.

2.1 Preprocess

As for scattered italic characters, the method of italic determination should be simple and efficient. Otherwise, the speed of the whole OCR system is very slow. Therefore some complicated methods are not considered in this paper, no matter how accurate they can be. Furthermore, some methods may be efficient if the detected pattern is very large such as a document page or a large text block, but they can fail in detecting scattered italic characters. So a series of detection areas should be obtained before italic determination, and these areas should be large as possible because it is obvious that larger the area is, more accurate the result of detection is. In English document, the area is often the pattern of a word. But how to get the detection area in mixed Chinese/English document? If the area includes only one Chinese character, the result of italic determination is not very good whatever methods are adopted and some experimental results are shown in section 3. In fact, it is rare that an italic area only includes one Chinese character. In general, the number of Chinese characters in the italic area is no less than two. Besides, some obvious gaps exist between italic characters and normal characters, which is a premise to this paper and it can be established in most applications. These features are very important and can give some help to solve the segmentation of mixed documents including italic characters. Character-pair rather than character is used as the basic unit while detecting italic characters.

We assume that the document image has been split into text lines. The procedure of getting character-pair area in a text line is as follows.

First, connected component analysis is conducted.

Second, conservative mergence is done based on the dependency of position of connected components. The set of merged components is defined as BC.

Third, the width of Chinese character is evaluated. The width is defined as W_H.

Fourth, as for BC, a procedure of recursive mergence is done and the new created set from BC is defined as MC.

The italic determination and estimation of slant angle will be done based on the set of MC.

2.2 Italic Determination

In this section, we will give a constant shear angle and the angle is 15 degree that is the median of general slant angles. Although the shear angle can not be just the real slant angle, it is enough to determinate whether the character is italic or not. The comparison of features of patterns between the original pattern and the sheared pattern will be done.

The extracted features of pattern can be gratitude, projection profile histogram or its distribution, direction elements and so on. We prefer the projection profile histogram. This method is very simple and efficient. We call it as PP.

The feature is extracted according to the formula as:

$$f(x) = \frac{1}{w}\sum_{i=0}^{w-1} p_i(x) \quad (1)$$

where x is a character area pattern, w is the width of pattern and $p_i(x)$ is the ith column projection histogram of pattern. We presume I0 is the original pattern and I1 is the sheared pattern. In general, the result of italic determination is given as:

$$font = \begin{cases} italic & f(I0) < f(I1) \\ normal & f(I0) > f(I1) \\ confusion & f(I0) = f(I1) \end{cases} \quad (2)$$

Furthermore, in [5], a method of weighted projection profile histogram is introduced and we find it very useful to improve the performance. This method give weights to those connected black runs and bigger the length of the black run is, higher the weight is. The difference between different patterns is enlarged. We call this method as WPP. Comparison of projection profile histogram between a italic Chinese character-pair "北京" and its sheared pattern by PP and WPP is shown in Fig. 1. In the figure, the left part shows the features of the original italic character-pair and the right part shows the features of the sheared character-pair.

Finally, a new way, which is similar to the motivation of [5], is presented in this paper to amplify the difference of feature between original pattern and sheared pattern. After analyzing the patterns of both italic and normal characters, we find a feature in the projection profile histogram that there are more blank columns in normal character pattern than in italic character pattern. According to the formula (1) and (2), these blank columns can cause error italic determination. We assume that the width of pattern is w_o and the number of blank columns is w_b, then the normalization width is $w = w_o - w_b$. As a result of fact, the difference between $f(I0)$ and $f(I1)$ is enlarged. We call this method as WWPP. Comparison of italic determination to the patterns in Fig. 1 is shown in table 1. Although the results of italic determination are all right by different methods, but obviously WWPP is the most robust.

Table 1. Comparison of italic determination by PP, WPP and WWPP

Methods	f(I0)	f(I1)	f(I1) - f(I0)	Determination
PP	7.69	8.18	0.49	Italic
WPP	30.31	59.45	29.14	Italic
WWPP	33.19	75.51	42.32	Italic

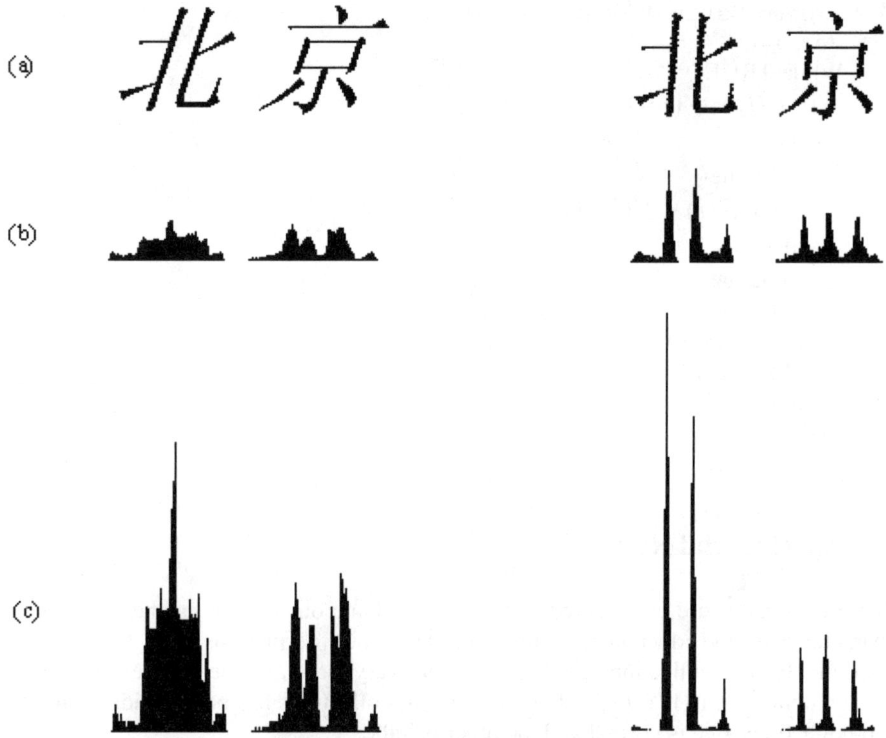

Fig. 1. Comparison of projection profile histogram between a Chinese character-pair "北京" and its sheared pattern by PP and WPP. (a) The original and sheared patterns. (b) Histogram by PP. (c) Histogram by WPP. The left part is the features of the original pattern and the right part is that of sheared pattern. The histogram in the right of (c) is dwindled to one-fifth of the original histogram.

2.3 Estimation of Slant Angle

If a pattern area is identified as italic, estimation of slant angle should be done. It is very important to correct italic character pattern because it will be relevant closely to character recognition. The valid range of slant angle in this paper is from 12 to 18 degree. Because the slant angle is constant in a text line in general, we can merge all italic-like patterns in a line into only one pattern and then estimate the slant angle of the pattern.

In order to get the slant angle quickly, a fast algorithm is used. The procedure of estimation of slant angle is as follows.

First, we assume that MIN is the minimum of valid slant angle, here it is 12, and similarly, MAX is the maximum and the value is 18.

Second, F(x) is the value of f(I1) when the shear angle is x.

Finally, the following program will be executed. In the program, the variant of i is the estimation of slant angle.

```
Program EstimateSlantAngle
  var i=15
  While(TRUE)
    If F(i) < F(i-1)
      then i--
      continue
    else if F(i) < F(i+1)
      then i++
      continue
    Endif
    if i=min or i=max
      then break
    Endif
  Endwhile
  End.
```

3 Experimental Results

Three collections are constructed in order to evaluate our method, namely character, character-pair and document collection. They are defined as C1, C2 and C3 respectively. The collections of C1 and C2 are very large and the number of samples in each collection is 17926056. Each of the two collections has been divided into two sub-collections, one is normal and the other is italic.

The number of normal character samples in C1 is 6729×333, where 6729 is the number of Chinese character classes and 333 is the number of samples in each class. The italic collection is constructed from the normal collection by artificial shear transform of normal samples, and the shear angle is 12, 13, 14, 15, 16, 17 and 18. So the number of samples in italic collection is $6729 \times 333 \times 7$.

The construction of C2 is similar to C1 and the number of samples in C2 is the same as that of C1.

The third collection is made up of 200 documents which are from books, journals and newspapers. There are many mixed Chinese/English italic characters scattered in the documents.

3.1 Evaluation of Italic Determination

The experimental results by adopting the methods of PP, WPP, WWPP are shown respectively in table 2-4.

Table 2. Experimental results of italic determ by PP

Libs	Error rate of italic determination in various shear angles (%)							Average error rate(%)	
	12	13	14	15	16	17	18	Italic	Normal
C1	14.09	11.21	8.98	5.48	5.43	4.64	3.40	7.60	10.03
C2	18.71	15.12	11.48	7.23	6.95	5.43	3.96	9.84	7.97

Table 3. Experimental results of italic determination by WPP

Libs	Error rate of italic determination in various shear angles (%)							Average error rate(%)	
	12	13	14	15	16	17	18	Italic	Normal
C1	5.43	5.53	4.21	0.36	1.84	1.49	6.74	3.66	3.93
C2	2.79	3.43	2.16	0.06	0.89	0.54	0.19	1.44	1.20

Table 4. Experimental results of font determination by WWPP

Libs	Error rate of italic determination in various shear angles (%)							Average error rate(%)	
	12	13	14	15	16	17	18	Italic	Normal
C1	3.53	3.62	3.15	0.31	1.75	1.26	3.52	2.45	3.25
C2	1.84	2.05	1.22	0.03	0.44	0.25	0.10	0.85	0.79
C3	—	—	—	—	—	—	—	0.92	0.21

Based on the above results, we can find that the method of WWPP is the best. In table 4, this method is tested in C3. The result is good enough to application.

3.2 Evaluation of Estimation of Slant Angle

When an italic pattern is confirmed, the estimation of the slant angle will be done. The results are shown in table 5.

Table 5. Experimental results of estimation of slant angle by WWPP

Libs	Absolute error of estimation of slant angle in various shear angles							
	12	13	14	15	16	17	18	Average
C1	2.21	2.08	1.33	1.54	2.85	2.12	3.05	2.17
C2	1.07	0.91	0.96	0.88	1.25	0.72	1.56	1.05

Based on all above results, we can see that the results in C2 are better than C1 whether to determinate italic or estimate the slant angle.

4 Conclusion

After analyzing some popular methods of italic detection, a new method of italic determination and estimation of slant angle is introduced to solve the segmentation of mixed documents including scattered italic characters. The unique features in printed Chinese document are discussed and adopted. Three collections are constructed to evaluate our method. So large collections haven't been reported in other contributions and we get good results in the collections. Italic determination is difficult, especially for scattered Chinese italic characters, some further researches are needed to improve the accuracy.

References

1. Ding, Y.M., Okada, M., Kimura, F., Miyake, Y.: Application of Slant Correction to Handwritten Japanese Address Recognition. Proceedings of the Sixth International Conference on Document Analysis and Recognition, (2001) 670-674
2. Ding, Y.M., Kimura, F., Miyake, Y., Shridhar, M.: Slant estimation for handwritten words by directionally refined chain code. Proceedings of the Seventh International Workshop on Frontiers in Handwritten Recognition, (2000) 53-62
3. Ding, Y.M., Ohyama, W., Kimura, F., Shridhar, M.: Local slant estimation for handwritten English words. Proceedings of the Ninth International Workshop on Frontiers in Handwritten Recognition, Kokubunji, Tokyo, Japan (2004) 328-333
4. Simoncini, L., Kovacs-V, Zs. M.: A system for reading USA census '90 hand-written fields. Proceedings of the Third International Conference on Document Analysis and Recognition, Vol.1, Montreal (1995) 86-91
5. Nicchiotti, G., Scagliola, C.: Generalised projections: a tool for cursive character normalization. Proceedings of Fifth International Conference on Document Analysis and Recognition, Bangalore (1999)
6. Fan, K.C., Huang, C.H., Chuang, T.C.: Italic Detection and Rectification. Proceedings of 2005 International Conference on Image Processing, Vol.2, (2005) 530-533
7. Li, Y., Naoi, S., Cheriet, M., Suen, C.Y.: A segmentation method for touching italic characters. Proceedings of Seventeenth International Conference on Pattern Recognition, Vol.2, (2004) 594-597
8. Su L., Restoration and segmentation of machine printed documents, Ph.D dissertation, University of Windsor, Canada (1996) 92-95
9. Sun C.M., Si, D.: Skew and slant correction for document images using gradient direction. Proceedings of the Fourth International Conference on Document Analysis and Recognition, Vol.1, (1997) 142-146
10. Ballesteros, J., Travieso, C.M., Alonso, J.B., Ferrer, M.A.: Slant estimation of handwritten characters by means of Zernike moments. Electronics Letters, 41(20), (2005) 1110-1112
11. Chaudhuri, B.B., Garain, U.: Automatic detection of italic bold and all-capital words in document images. Proceedings of Fourteenth International Conference on Pattern Recognition, Vol.1, (1998) 610-612
12. Kavallieratou, E., Fakotakis, N., Kokkinakis, G.: Slant estimation algorithm for OCR system. Pattern Recognition, 34(12), (2001) 2515-2522
13. Zhang, L., Lu, Y., Tan C.L.: Italic font recognition using stroke pattern analysis on wavelet decomposed word images. Proceedings of Seventeenth International Conference on Pattern Recognition, Vol.4, (2004) 835-838

14. Bozinovic, R.M., Srihari, S.N.: Off-line cursive script word recognition. IEEE Transactions on Pattern Analysis and Machine Intelligence, 11(1), (1989) 68-83
15. Xia, Y., Wang C.H., Dai, R.W.: Segmentation of mixed Chinese/English document based on AFMPF model. Acta Automatica Sinica, 32(3), (2006) 353-359
16. Xia, Y., Xiao B.H., Wang, C.H., Li, Y.D.: Segmentation of mixed Chinese/English documents based on Chinese Radicals recognition and complexity analysis in local segment pattern. Lecture Notes in Control and Information Sciences, Vol.345, Springer-Verlag, (2006) 497-506

Using Pointwise Mutual Information to Identify Implicit Features in Customer Reviews*

Qi Su, Kun Xiang, Houfeng Wang, Bin Sun, and Shiwen Yu

Institute of Computational Linguistics, School of Electronic Engineering and Computer Science, Peking University, 100871, Beijing, China
{sukia, xk, wanghf, bswen, yusw}@pku.edu.cn

Abstract. This paper is concerned with automatic identification of implicit product features expressed in product reviews in the context of opinion question answering. Utilizing a polarity lexicon, we map each adjectives in the lexicon to a set of predefined product features. According to the relationship between those opinion-oriented words and product features, we could identify what feature a review is regarding without the appearance of explicit feature nouns or phrases. The results of our experiments proved the validity of this method.

Keywords: pointwise mutual information, customer review, implicit feature.

1 Introduction

In recent years a large amount of research has been done on the identification of semantic orientation in text. The task is also known as opinion mining, opinion extraction, sentiment analysis, polarity detection and so on. It has focused on classifying reviews as positive or negative and ranking them according to the degree of polarity. The technology has proved to have many potential applications. For example, it could be integrated with search engines to provide quick statistical summary of whether a product was recommended or not in the World Wide Web, thus help consumers in making their purchasing decision and assist manufacturers in performing market analysis.

Our work studies the problem of answering natural language questions on customer reviews of products. The research is based on our previous projects of personalized information retrieval and natural language question answering(QA) system. Some similar works include multi-perspective question answering [1] [2] [3] and opinion question answering [4]. For a QA system, the questions from users could be of two types: objective questions and subjective questions. Subjective questions are those related to opinion, attitude, review, and etc. The identification of semantic orientation from corpora provides a possible way to answer user's subjective questions. According to the observation, those opinion

* This work was supported by National Natural Science Foundation of China (No. 60435020, No.60675035).

Y. Matsumoto et al. (Eds.): ICCPOL 2006, LNAI 4285, pp. 22–30, 2006.

related questions asked by users could be either general or feature-based. The following are such two query instances with opinions:

(1)"宝马汽车好不好？"(*How is BMW?*)
(2)"宝马汽车的动力性怎样？"(*How is the power of BMW?*)

For sentence (1), users want to get a general opinion about the automobile "BMW", while for sentence (2), users only want to get those opinions about the power of BMW and don't care other features. So, for our task, we need to first identify product features on which customers express their opinions.

Given a set of product reviews, the task of identifying semantic orientation could be divided into four subtasks [5]: (1)Identify product features expressed in the reviews; (2)Identify opinions regarding product features(including those general opinions); (3)Calculate the semantic orientation of opinions; (4)Rank the reviews based on their strength of semantic orientation.

Most of existing researches focused on subtask (3). In general, there are two approaches to semantic orientation calculation: namely, the rule based approach and the machine learning based approach. The former calculates the semantic orientation of each opinion-oriented word/phrase appeared in reviews and uses the average semantic orientation to represent the sentiment orientation of reviews. According to the value of average semantic orientation, it could rank reviews (subtask 4) easily. The latter, which is more widely employed, applies supervised automatic classification technology to classify reviews into bipolar orientation of positive or negative. Utilizing the semantic orientation of opinion-oriented word/phrase, this approach is proved to get an improved accuracy than using usual bag-of-word or n-gram features.

Although important, researches on the identification of product features are relatively few. Hu [6] classified product features that customers have expressed their opinion on as explicit features and implicit features. In their definition, the features which appear explicitly in opinion sentences are explicit features; the features which do not appear explicit in sentences are implicit features. But they did not propose a method for identifying implicit features. In this paper we address the issue of automatically identifying implicit features from domain dependent product reviews. To the best of our knowledge, no previous work has been conducted on exactly this problem. The existing researches on feature identification such as [5] [6] [7]mainly focus on finding features that appear explicitly as nouns or noun phrases in product reviews. The identification of implicit feature is a harder task than identifying explicit one. According to the observation, opinion-oriented word/phrase could usually be a good indicator to the feature which it modifies. So the key issues to the task are to define a feature set related to specific domain and map those indicators to the set of predefined features.

We take a mutual information approach to address the problem. In this paper we first present our approaches in the identification of opinion-oriented words/phrases, and map the adjective ones to the corresponding implicit product features in reviews. By identifying implicit features, we could undoubtedly get more reasonable semantic orientation scores on different product features.

2 Definition of Features in Product Reviews

2.1 Explicit Features and Implicit Features

Product features include attributes, such as "加速性"(*acceleration*) for automobile products, and parts, such as "刹车"(*brake*). A feature can appear either explicitly or implicitly in product reviews. Features that can be extracted directly from the reviews are Explicit Features, e.g., 动力性(*power*) in "宝马汽车的动力性不错"(*BMW power is not bad*). Sometimes users express their opinions without explicit feature words. But we could still deduce the features which their opinions toward from the opinion sentences. Those kinds of features are Implicit Features. For example, in the sentence of "宝马汽车很漂亮"(*BMW is beautiful*), we could judge that the feature which users talk about is BMW's *exterior* although the word doesn't appear explicitly.

The identification of explicit features is relatively easy. Using frequent nouns and noun phrases appeared in reviews with some pruning strategies, the experiment in [6] could reach the best average precision of 79%. Evaluating each noun phrase by computing mutual information between the phrase and meronymy discriminators associated with the product class, the experiment in [5] achieved 86% in precision and 89% in recall.

In the research of [8], they represented a method to identify implicit features by tagging the mapping of specific feature values to their actual feature. Our work is close to their research in the sense that we also use mapping rules, but it is also different in that we propose an automatic method to generate the mapping rules and use opinion-oriented words.

2.2 Feature Structures in the Domain of Automobile Review

Our work in this paper is focused on identifying implicit features in product reviews. Since there are no feature words appeared in reviews, we need firstly to define a set of review features for automatic opinion mining. The definition of features are domain dependent. In our experiment, we consider the domain of automobile review.

According to the investigation of several automobile review websites, we found that people usually evaluate a car from several aspects, such as 动力性(Power), 操控性(Handling), 外观(Exterior) , 内饰(Interior), 经济性(Economy), 工艺性(Craft), 市场性(Marketability) and etc. So in this paper we define the product feature sets of automobile according to the 7 items above. This is a very detailed review feature proposal. In fact, many automobile review websites classify automobile review features on a rough level. For example, they combine the features of 动力性(Power) and 操控性(Handling) as 性能(Performance), combine the features of 外观(Exterior) and 内饰(Interior) as 设计(Design). From the rough classification schemes, we may have an impression that the feature of 动力性(Power) and 操控性(Handling) may be hard to divide, and so as the feature of 外观(Exterior) and 内饰(Interior). For our task of implicit feature identification, they are also problems. For example, when users speak of *an exquisite automobile*, they may consider both exterior design and interior design.

So, we propose that a feature indicator should be mapped to several implicit features. It also seems to accord with our instinct.

Product features which appear explicitly in automobile reviews can also be classified into the feature structure. For example, *acceleration* belongs to the feature of Power, *brake* belongs to the feature of Handling.

3 Feature Identification

3.1 The Method of Pointwise Mutual Information

Pointwise Mutual Information (PMI) is an ideal measure of word association norms based on information theory [9]. Researchers have applied this measurement to many natural language processing problems such as word clustering. PMI compares the probability of observing two items together with the probabilities of observing two items independently. So it can be used to estimate whether the two items have a genuine association or just be observed by chance.

If two words $word_1$ and $word_2$ have probabilities $P(word_1)$ and $P(word_2)$, then their mutual information $PMI(word_1,word_2)$ is defined as [10]:

$$PMI(word_1, word_2) = \log\left(\frac{P(word_1 \& word_2)}{P(word_1)P(word_2)}\right) \tag{1}$$

Usually, word probabilities $P(word_1)$, $P(word_2)$ and joint probabilities $P(word_1 \& word_2)$ can be estimated by counting the number of observations of $word_1$, $word_2$ and the co-occurrence of $word_1$ and $word_2$ in a corpus normalizing by the size of the corpus. The co-occurrence range of $word_1$ and $word_2$ is usually limited in a window of w words.

The quality of the PMI algorithm largely depends on the size of training data. If there is no co-occurrence of $word_1$ and $word_2$ in the corpus, the accuracy of PMI becomes an issue. The PMI-IR algorithm introduced by Turney in [11] used PMI to analyze statistical data returned by the query of Information Retrieval(IR). So, the corpus used by PMI-IR algorithm is the document collection which is indexed by IR system. The PMI-IR algorithm is proposed originally for recognizing synonyms in TOEFL test. Then Turney [12] used this method in their sentiment classification experiments, where the sentiment orientation $\hat{\sigma}(w)$ of word/phrase w is estimated as follows.

$$\hat{\sigma}(w) = PMI(w, positive) - PMI(w, negative) \tag{2}$$

In equation (2), *positive* and *negative* means a set of Reference Words Pair (RWP)[13] with the sentiment orientation of positive or negative respectively. Choosing the RWP of *excellent* and *poor*, the equation above can be written as:

$$\hat{\sigma}(w) = \log \frac{hits(w, excellent)/\ hits(excellent)}{hits(w, poor)/\ hits(poor)}, \tag{3}$$

Where $hits(query)$ is the number of hits (the number of documents retrieved) when the query *query* is given to IR system. And $hits(query1, query2)$ is the number of hits "*query1* NEAR *query2*".

3.2 Feature-Based PMI Algorithm

In Turney's research, semantic orientation was calculated by the word association with a positive paradigm word minus a negative paradigm word. For our research, we expand the RWP of two words to a set of features. In our case, we have a set $S_{feature}$ which holds different features.

$$S_{feature} = \{feature_1, feature_2, ..., feature_n\} \tag{4}$$

For each selected word w, we calculate the PMI between w and $feature_i$ ($feature_i \in S_{feature}$). Using the PMI method, we get the word association between w and different features, then map w to one or several features according to the probability.

The PMI-IR algorithm use a NEAR operator to simulate the co-occurrence window limit of two words in PMI algorithm. But using NEAR operator has its limitation. The two NEAR words may be distributed on the different sentences or clauses, or even two paragraphs. Thus their semantics may be non-sequential and not represent the correct co-occurrence relationship. For keeping the two words in a semantic sequence, we suggest an improvement of PMI-IR algorithms for our task.

We use conjunction operator to construct queries. Because we have limit *word* to be adjectives in the polarity lexicon. As features are always nouns/noun phrases, so the query of *"word feature"* would form a normal collocation relation.

$$Score_{conjunction}(word, feature) = \log \frac{p\,(word\ feature)}{p\,(word)\ p\,(feature)} \tag{5}$$

Since we calculate the association between w and the feature set $S_{feature}$, we can drop $p(word)$. Using World Wide Web as the corpus (using the query results returned by search engine), the equation(5) can be simplified as follows:

$$Score_{conjunction}(word, feature) = \log \frac{hits(\text{“}word\ feature\text{”}) + \varepsilon}{hits(feature)} \tag{6}$$

Where ε is a parameter to prevent the numerator from getting zero when there is no hits returned for the query of *"word feature"*.

4 Experiment

4.1 Opinion Words Collection

For collecting opinion words, we first download automobile review webpages from the internet, and then extract the opinion words from them manually. Thus we get a small/basic polarity lexicon. Using these words as seeds, we enlarge our polarity lexicon utilizing synonym and antonym sets in the Chinese Concept Dictionary(CCD), a Chinese version of the online electronic dictionary Wordnet. The consideration of utilizing CCD is, in general, words share the same orientation as their synonyms and opposite orientation as their antonyms [6]. For example, the positive word 可爱(*lovely*) has the synonym set of {优美 伶俐 可爱 喜人 柔情 深情 漂亮 甜蜜 痛快 秀丽 绝妙 讨人喜欢} and the antonym set of{可恨 可恶 可憎 讨厌}. The two sets take the opposite orientation of positive and negative accordingly. In our research, according to the structural characteristic of CCD and the expansion results, we only use the result of synonym expansion, and enlarge the polarity lexicon more than 5 times larger.

Most of the opinion words in our lexicon are adjectives. Being good modifiers for product features and carriers of sentiment, in our experiment, only the adjectives in polarity lexicon were chosen to identify implicit product features. We map each adjective to the set of product features which it could modified.

4.2 Mapping Opinion Words to Features

The IR system used in our experiments is Google. Google provides API to refer queries conveniently and rapidly. Google API returns estimated hit counts rather than actual values. This would add some noise into our model, but still has the necessary stability for our estimation.

For each product feature in the above mentioned feature set $S_{feature}$, we choose one or several representative nouns as feature words to calculate the PMI between them and opinion words. Such as, for the feature of 动力性(Power), we choose the words of 动力(power),引擎(engine),马力(horsepower) and etc.

Table 1 shows PMI scores for some word examples.

Table 1. PMI scores for some word examples

Words	Power	Handling	Exterior	Interior	Economy	Craft	Marketability
漂亮(beautiful)	-12.01	-14.15	-4.79	-8.67	-12.87	-9.41	-12.76
强劲(powerful)	-4.26	-13.20	-9.91	-7.96	-13.04	-10.46	-8.76
一般(ordinary)	-7.86	-9.14	-9.51	-7.47	-6.24	-5.20	-6.15
昂贵(expensive)	-15.50	-15.76	-13.40	-9.86	-16.09	-11.78	-5.58
考究(exquisite)	-17.70	-15.76	-10.00	-8.23	-16.09	-3.94	-16.21
灵活(flexible)	-10.27	-4.13	-11.00	-8.35	-10.16	-14.65	-12.15
别致(unique)	-17.70	-15.76	-9.92	-9.60	-16.09	-9.49	-16.78
方便(convenient)	-12.29	-5.07	-9.51	-9.47	-12.37	-11.44	-12.40
不足(inadequate)	-4.37	-10.15	-11.88	-10.27	-8.99	-9.75	-8.75
完美(perfect)	-9.88	-9.64	-10.03	-7.95	-4.86	-8.41	-11.88

For each opinion word, we have got its PMI scores on each $feature_i \in S_{feature}$. We need to further map them to one or several most suitable features according to their PMI values.

For this purpose, we first define a function to describe the difference between $\hat{S}(\hat{S} \subset S_{feature})$ and $S_{feature}$-$\hat{S}$ for a word w.

$$score_{diff}(\hat{S}, S_{feature}, w) = \frac{\sum_{f \in \hat{S}}(f, w)}{|\hat{s}|} - \frac{\sum_{f \in (S_{feature} - \hat{S})}(f, w)}{|S_{feature} - \hat{s}|} \tag{7}$$

Then we calculate a series of $score_{diff}$ for each word w using an algorithm as described in the following pseudo code:

Algorithm 1. The calculation of $score_{diff}$

```
Set S1 = ∅, S2 = S_feature
while S2 ≠ ∅ do
    f_i = arg max_{f∈S2} Score(f_i, w)
    Set S1 = S1 + {f_i}, S2 = S2 − {f_i}
    Calculate Score_diff(S1, w)
end
```

Table 2 is the $score_{diff}$ of some word examples.

Table 2. Value of $score_{diff}$ of some word examples

考究(exquisite)		强劲(powerful)	一般(ordinary)	灵活(flexible)
Craft	10.06	Power 6.30	Craft 2.53	Handling 6.97
Interior	9.07	Interior 4.96	Marketability 2.37	Interior 5.41
Exterior	9.05	Marketability 4.66	Economy 2.63	Economy 4.47
Handling	7.18	Exterior 4.51	Interior 2.57	Power 4.37
Economy	6.15	Craft 4.85	Power 2.74	Exterior 4.62
Marketability	6.00	Handling 4.14	Handling 2.5	Marketability 5.31

We use value *gap* to describe the margin between two $diff_{score}$s and set a experiential threshold $\xi = 1.0$. When the margin is greater than ξ, we say that the two adjacent $diff_{score}$ have a *gap*. With each word and its $diff_{score}$, we find every gap value from high to low of a set of $diff_{score}$, and judge the feature class which the word should belong to according to the sequence.

From Table 2, the biggest *gap* for 考究(exquisite) should between the feature of Exterior and Handling. So we map the word 考究(exquisite) to feature sets of Craft,Interior,Exterior. For the word of 强劲(powerful), the biggest *gap* should between Power and Interior. So we map the word 强劲(powerful) to the feature Power. As for the word 一般(ordinary), we find that there is no *gap* between every two features. So we consider that the word 一般(ordinary) is a general descriptive adjective.

Based on this method, we map the adjective words in our polarity lexicon to the predefined product feature set. Table 3 gives some experimental example results.

Table 3. Mapping results of some example words

words	feature sets
漂亮(beautiful)	Exterior
强劲(powerful)	Power
一般(ordinary)	-
昂贵(expensive)	Marketability
考究(exquisite)	Craft,Interior,Exterior
灵活(flexible)	Handling
别致(unique)	Craft,Interior,Exterior
方便(convenient)	Handling
不足(inadequate)	Power
完美(perfect)	Economy

5 Conclusion

Feature identification in product reviews is the first step of opinion QA and other opinion mining tasks. In product reviews, the appearance of feature nouns or noun phrases is usually an important clue which aids in mapping to predefined product features. But according to our observation of product review webpages, in many cases, those explicit feature nouns or noun phrases do not appeared in the context. Here comes the importance of the identification of implicit product features. According to the mapping between opinion words and product features, we could judge what features a review is regarding to, without the explicit appearance of feature nouns or noun. In this paper, we describe a method which uses PMI to calculate the association between opinion-oriented adjectives and a set of product review features. We do not conduct a quantitative evaluation of our method, because the relationship between some adjectives, especially those general ones and product features are likely to be highly contextual. But the results of our experiments prove the validity of this method intuitionistically.

Using the method proposed in this paper, we supply our polarity lexicon with the corresponding product feature information. With this resource, we could score a product review from different aspect of features. Future work should include the weighting of explicit features and implicit features when both of the feature types appear in a review.

References

1. Cardie, C., Wiebe, J.,Wilson, T. and et al.: Combining Low-Level and Summary Representations of Opinions for Multi-Perspective Question Answering. Proceedings of the AAAI Spring Symposium on New Directions in Question Answering(2003)
2. Wiebe, J., Breck, E., Buckley, C and et al.:Recognizing and Organizing Opinions Expressed in the World Press.Proceedings of the 2003 AAAI Spring Symposium on New Directions in Question Answering(2003)

3. Stoyanov, V., Cardie, C., Wiebe, J.:Multi-Perspective Question Answering Using the OpQA Corpus.Proceedings of HLT-EMNLP 2005(2005)
4. Yu, H., Hatzivassiloglou, V.: Towards Answering Opinion Questions: Separating Facts from Opinions and Identifying the Polarity of Opinion Sentences.Proceedings of EMNLP-03, 8th Conference on Empirical Methods in Natural Language Processing(2003)129–136
5. Popescu, A. M. and Etzioni, O.: Extracting Product Features and Opinions from Reviews. Proceedings of HLT-EMNLP(2005)
6. Hu, M. Q. and Liu, B.: Mining and Summarizing Customer Reviews. In KDD(2004)168-177
7. Kobayashi, N., Inui, K., Tateishi, K and et al.:Collecting Evaluative Expressions for Opinion Extraction. In IJCNLP(2004)596-605.
8. Liu, Bing., Hu, M. Q., and Cheng, J. S.:Opinion observer: analyzing and comparing opinions on the Web. Proceedings of WWW '05, the 14th international conference on World Wide Web(2005)342–351
9. Thomas, M. Cover. and Joy A. Thomas.: Elements of Information Theory. John Wiley(1991)
10. Church, K. W. and Hanks, P.: Word Association Norms, Mutual Information and Lexicography. Proceedings of the 26th Annual Conference of the Association for Computational Linguistics(1989)
11. Turney, P.D.: Mining the Web for synonyms: PMI-IR versus LSA on TOEFL. Proceedings of the Twelfth European Conference on Machine Learning. Berlin: Springer-Verlag(1991) 491-502
12. Turney, P. D.: Thumbs Up or Thumbs Down? Semantic Orientation Applied to Unsupervised Classification of Reviews. In Proc. of the 40th Annual Meeting of the Associationfor Computational Linguistics(2002) 417-424
13. Ye, Q., Shi, W., Li, Y. J.: Sentiment Classification for Movie Reviews in Chinese by Improved Semantic Oriented Approach. Proceedings of the 39th Annual Hawaii International Conference on System Sciences(2006)
14. Bethard, S., Yu, H., Thornton, A. and et al.:Automatic extraction of opinion propositions and their holders. In 2004 AAAI Spring Symposium on Exploring Attitude and Affect in Text(2004)

Using Semi-supervised Learning for Question Classification

Nguyen Thanh Tri, Nguyen Minh Le, and Akira Shimazu

School of Information Science
Japan Advanced Institute of Science and Technology
1-1 Asahidai, Nomi, Ishikawa, 923-1292, Japan
{t-thanh, nguyenml, shimazu}@jaist.ac.jp

Abstract. This paper tries to use unlabelled in combination with labelled questions for semi-supervised learning to improve the performance of question classification task. We also give two proposals to modify the Tri-training which is a simple but efficient co-training style algorithm to make it more suitable for question data type. In order to avoid bootstrap-sampling the training set to get different sets for training the three classifiers, the first proposal is to use multiple algorithms for classifiers in Tri-training, the second one is to use multiple algorithms for classifiers in combination with multiple views. The modification prevents the error rate at the initial step from being increased and our experiments show promising results.

1 Introduction

Question classification is the task of identifying the type of a given question among a predefined set of question types. The type of a question can be used as the clue to narrow down the search space to extract the answer, and used for query generation in a question answering (QA) system. Therefore, it has a significant impact on the overall performance of QA systems. There have been several studies to solve this problem focusing on supervised learning [4,10,13]. Zhang and Lee [4] and Li and Roth [13] explore different types of features for improving the accuracy. Zhang and Lee consider *bag-of-words*, *bag-of-ngrams* (all continuous word sequences in a question) features. Especially, they propose a kernel function called *tree kernel* to enable support vector machine (SVM)[3] to take advantage of the syntactic structures of questions. Li and Roth focus on several features: *words*, *pos tags*, *chunks* (non overlapping phrases), *named entities*, *head chunks* (e.g., the first noun chunk in a question) and *semantically related words* (words that often occur in a specific question type). They also use hierarchical classifiers, in which a question is classified by two classifiers: the first one classifies it into a coarse category; the second determines the fine category from the result of the first classifier. Kadri and Wayne [10] employ error correcting codes in combination with support vector machine to improve the results of classification.

Y. Matsumoto et al. (Eds.): ICCPOL 2006, LNAI 4285, pp. 31–41, 2006.

However, above methods do not use unlabelled questions which are available with high volume to improve the performance of classification. In order to utilize both labelled and unlabelled data, we propose to use semi-supervised learning. For semi-supervised learning algorithm, we consider the Tri-training of Zhou and Li [14] which uses three classifiers of the same algorithm. Tri-training has a simple but efficient way of deciding to label an unlabelled instance, that is, if any two classifiers of the three classifies agree to label an unlabelled instance, while the confidence of the labelling of the classifiers are not needed to be explicitly measured, that instance is used for further training the other classifier. Such simplicity makes Tri-training have advantages over other Co-training algorithms, such as the Co-training algorithm presented by Goldman and Zhou [11], which frequently uses 10-fold cross validation on the labelled set to determine how to label the unlabelled instances and how to produce final hypothesis. If the original labelled set is rather small, cross validation will give high variance and is not useful for model selection. Additionally, the simplicity makes Tri-training faster than the algorithm of Goldman with the frequently use of cross validation making the learning process time-consuming. At the beginning, Tri-training bootstrap-samples the labelled data to generate different training sets for three classifiers in order to make the three classifiers diverse enough so that the Tri-training algorithm does not degenerate into *self-training* with a single classifier.

However, question data is sparse and imbalanced. A class may include a few questions, so if the bootstrap-sampling procedure duplicates some questions while omitting some questions in the few question class, then classifiers being trained on these bootstrap-sampled sets have higher error rates than those of classifiers being trained on the labelled set. In order to avoid this drawback while still keeping classifiers diverse, we propose to use more than one classifier with different algorithms. The original training set is initially used by the three classifiers without bootstrap-sampling. Another proposal is to apply more than one *views* (feature spaces) in the learning process. This allows the three classifiers to initially be trained from the labelled set with different feature spaces and still have diversity. For the sake of simplicity, in the experiments, we used two different algorithms and two views: *bag-of-words* and *bag-of-pos-tags*. Two classifiers which use the first algorithm are assigned different views, i.e., the first classifier gets bag-of-words and the other gets pos-tags. The third classifier uses the second algorithm with bag-of-words features. Our experiments show promising results.

The remainder of the paper is organized as follows: Section 2 gives detail about the Tri-training algorithm and our modification. The experimental results are given in Section 3 and conclusions are given in Section 4.

2 Tri-training Semi-supervised Learning and Its Modifications

In this section, we describe the original Tri-training algorithm and give two proposals to improve it.

a) Original Tri-training algorithm

```
tri-training(L,U,Learn)
 for i ∈ {1..3} do
   S_i ← BootstrapSample(L)
   h_i ← Learn(S_i)
   e'_i ← 0.5; l'_i ← 0
 end for
 repeat until none of h_i (i ∈ {1..3}) changes
   for i ∈ {1..3} do
     L_i ← ∅;update_i ← FALSE
     e_i ← MeasureError(h_j&h_k) (j,k≠i)
     if (e_i<e'_i) then
       for every x ∈ U do
         if h_j(x)=h_k(x) (j,k ≠ i)
         then L_i ← L_i ∪ {(x,h_j(x))}
       end for
       if (l'_i=0) then l'_i ← [e_i/e'_i + 1]
       if (l'_i<|L_i|) then
         if(e_i|L_i|<e'_i l'_i) then update_i ← TRUE
         else if l'_i>e_i/e'_i
         then L_i ← Subsample(L_i,[e'_i l'_i/e_i − 1]);
                 update_i ← TRUE
   end for
   for i ∈ {1..3} do
     if update_i = TRUE then
       h_i ← Learn(L ∪ L_i);e'_i ← e_i;l'_i ← |L_i|
   end for
 end repeat
 Output:h(x) ← arg max_{y∈label} Σ_{i:h_i(x)=y} 1
```

b) Tri-training with multiple learning algorithms

```
tri-training(L,U,Learn_1,
        Learn_2, Learn_3)
 for i ∈ {1..3} do

   h_i ← Learn_i(L)
   e'_i ← 0.5; l'_i ← 0
 end for
...
h_i ← Learn_i(L ∪ L_i);
   e'_i ← e_i;l'_i ← |L_i|
...
```

c) Tri-training with multiple learning algorithms and views

```
tri-training(L,U,Learn_1,
        Learn_2, Learn_3)
 for i ∈ {1..3} do

   h_i ← Learn_i(view_i(L))
   e'_i ← 0.5; l'_i ← 0
 end for
...
h_i ← Learn_i(view_i(L ∪ L_i));
   e'_i ← e_i;l'_i ← |L_i|
...
```

Fig. 1. Original and modified versions of Tri-training

2.1 Semi-supervised Tri-training Algorithm

In Tri-training algorithm [14], three classifiers: h_1, h_2 and h_3 are initially trained from a set by bootstrap-sampling the labelled set L. For any classifier, an unlabelled instance can be labelled as long as the other two classifiers agree on the labelling this instance, while the confidence of the labelling of the classifiers are not needed to be explicitly measured. For example, if h_1 and h_2 agree on the labelling of an instance x in the unlabelled set U, then x can be labelled for h_3. Obviously, in this scheme, if the prediction of h_1 and h_2 on x is correct, then h_3 will receive a valid new instance for further training; otherwise, h_3 will get an instance with a noisy label. Nonetheless, as claimed in [14], even in the worse case, the increase in the classification noise rate can be compensated for, if the number of newly labelled instances is sufficient.

Also in the algorithm, each classifier is initially trained from a data set generated via bootstrap-sampling from the original labelled training set in order to make classifiers diverse. If all the classifiers are identical, then for any of three classifiers, the unlabelled instances labelled by the other two classifiers will be the same as those labelled by itself, thus, Tri-training becomes *self-training* with a single classifier. The pseudo-code of the algorithm is described in Fig. 1a, where *Learn* is a classification algorithm; S_i is a training set bootstrap-sampled from original set L; e'_i is the error rate of h_i in the $(t\text{-}1)^{th}$ round. With the assumption that the beginning error rate is less than 0.5, therefore e'_i is initially set to 0.5; e_i is the error rate of h_i in the tth round; L_i is the set of instances that are labelled for h_i in the tth round; l'_i is the size of L_i at $(t\text{-}1)^{th}$ round and in the first round it is estimated by $[\frac{e_i}{e'_i} + 1]$; $Subsample(L_i, s)$ function randomly removes $|L_i| - s$ number of instances from L_i in order to make current round have better performance than that of previous round as proved in [14]; $MeasureError(h_j \& h_k)$ function attempts to estimate the classification error rate of the hypothesis derived from the combination of h_j and h_k. Because it is difficult to estimate the classification error rate on the unlabelled instances, the algorithm only estimates on the labelled training set, with the assumption that both the labelled and unlabelled instance sets have the same distribution.

The interesting point in Tri-training algorithm is that in order to ensure current round of training to have better performance than that of previous round, the size of each newly labelled set L_i must not be greater than $[\frac{e'_i l'_i}{e_i} - 1]$. If it is greater than this value, the function $Subsample(L_i, s)$ is used to randomly remove redundant instances. The three classifiers are refined in the training process, and the final hypothesis is produced via *majority voting*. For the sake of saving space, other details of the Tri-training algorithm can be seen in [14].

2.2 Modified Versions of Tri-training

Due to the nature of question data type, which is very sparse and imbalanced as given in Table 1. As stated in [12], text data type, when represented in a vector space model, is very sparse. For each document, the corresponding document vector contains only a few entries which are non zero. A question contains very few words in comparison with a document, so question data is even more sparse than text data. Because of the imbalance, after bootstrap- sampling, each newly created training set misses a number of questions as compared to the original labelled set. If the missed questions are in a class which contains a few questions, then it makes the initial error rate of each classifier increase when being trained from these data sets. The final improvement after learning sometimes does not compensate for this problem. In order to avoid this drawback, we propose to use more than one algorithm for the three classifiers. Each classifier is initially trained on the labelled training set. Our experiments showed that, if the performance of one of the three classifiers is far better (or worse) than that of the others, the final result is not improved. For this reason, a constraint on three classifiers is that their performances be approximate. The modified version is depicted in Fig. 1b, where $Learn_i$ stands for different algorithms. We omit other lines that are identical to those of the original algorithm in Fig. 1a.

Table 1. Question distribution. #Tr and #Te are the number of training and testing questions.

Class	#Tr	#Te	Class	#Tr	#Te	Class	#Tr	#Te
ABBREV.	**86**	**9**	letter	9	0	country	155	3
abb	16	1	other	217	12	mountain	21	3
exp	70	8	plant	13	5	other	464	50
DESC.	**1162**	**138**	product	42	4	state	66	7
definition	421	123	religion	4	0	NUMERIC	**896**	**113**
description	274	7	sport	62	1	code	9	0
manner	276	2	substance	41	15	count	363	9
reason	191	6	symbol	11	0	date	218	47
ENTITY	**1250**	**94**	technique	38	1	distance	34	16
animal	112	16	term	93	7	money	71	3
body	16	2	vehicle	27	4	order	6	0
color	40	10	word	26	0	other	52	12
creative	207	0	HUMAN	**1223**	**65**	period	27	8
currency	4	6	group	47	6	percent	75	3
dis.med.	103	2	individual	189	55	speed	9	6
event	56	2	title	962	1	temp	8	5
food	103	4	description	25	3	size	13	0
instrument	10	1	LOCATION	**835**	**81**	weight	11	4
lang	16	2	city	129	18			

Another proposal is to use more than one view, such as two or three views in the learning process, so that each classifier can be trained from the original labelled set with different feature spaces while still making sure that they are diverse enough. The modified algorithm seems to have the standard Co-training style in the framework of Tri-training. The modified version according to this proposal is given in Fig. 1c, where $view_i(L)$ is the ith view of the data set L. Other lines that are the same as those in Fig. 1a are ignored.

3 Experiments

This section gives details about our implementation as well as the evaluation.

3.1 Question Data Sets and Feature Selection

The Question Answering Track in Text Retrieval Conference (TREC) [5,6,7] defines six question classes, namely, *abbreviation*, *description*, *entity*, *human*, *location* and *numeric*. However, for a question answering system in an open domain, six classes are not sufficient enough. The larger the number of question classes, the better a system locates and extracts answers to questions. Hence, from six coarse classes defined by TREC, Li and Roth [13] proposed to divide questions into 50 fine-grained classes. We follow this proposal to classify questions into these finer-grained classes. Our data set[1] is the same as that used in [13] with a total of about

[1] The annotated question sets are available at http://L2R.cs.uiuc.edu/~cogcomp/

6000 questions (the exact number is 5952). We keep 500 questions from TREC 10 [7] as the test set while the other 5500 questions are divided into non-overlapping labelled and unlabelled sets. We employ two strategies for dividing the file containing 5500 questions into labelled and unlabelled sets. The first strategy is manually partitioning, in which a number of consecutive questions (such as 1000, 2000, 3000 or 4000) from the beginning of the file is used as a labelled set, while the rest of the file serves as an unlabelled set. The purpose of this strategy is to compare the performance of all algorithms on the same labelled and unlabelled sets. The second one is random partitioning, in which we randomly pick up a number of questions (such as 1000, 2000, 3000 or 4000) from the file to generate a labelled set, and let other questions form a unlabelled set. The distribution of training and testing data is shown in Table 1, where the coarse classes are in capital followed by its fine classes. As listed in the table, some classes consists of a few questions, such as 4 questions in *currency* and *religion* class.

In experiments, we used two primitive feature types automatically extracted for each question, namely, *bag-of-words* and *bag-of-pos-tags* (or *pos-tags* for short).

Question classification is a little different from text classification. Because a question contains a small number of words, while a document can have a large number of words. In text classification, common words like 'what', 'is', etc. are considered to be "*stop-words*" and omitted as a dimension reduction step in the process of creating features. This is an important step in improving the performance of classification as proven in [12]. However, these words are very important for question classification. Also, word frequencies play an important role in document classification while those frequencies are usually equal to 1 in a question, thus, they do not significantly contribute to the performance of classification. In order to keep these words while still reducing the dimension space, we use a preprocessing step for bag-of-words features: all verbs are restored into infinitive forms, such as 'is', 'were', 'was', 'are' and 'am' are converted to 'be'; plural nouns are changed to singular forms, such as 'children' is converted to 'child'; words having the CD (*cardinal number*) part-of-speech are made the same value. For example, 23000, 19, 1995 are changed into 100. Interestingly, this dimension reduction step makes SVM [3] reach the precision of 81.4% training on 5500 questions while the same feature with SVM used in [4] gives the precision of 80.2% training on the same data set and with the same *linear* kernel. For pos-tags, each *word* in a question is converted into the form of *POS-word*, where *POS* is the part-of-speech tag of the *word*. We also used the preprocessing step similarly to what applied to the process to generate bag-of-words features, for example 'how' is transformed into 'WRB-how', 'who' is converted to 'WP-who', 'are', 'is', 'am', 'were' and 'was' are converted to 'AUX-be', etc.

3.2 Experiments with Multiple Classifiers

In the first experiment, we develop our programs based on the Sparse Network of Winnows (SNoW) learning architecture[2] [2], which implements three learning

[2] The software is freely available at http://L2R.cs.uiuc.edu/~cogcomp/software.php

Table 2. The precision (%) of Tri-training with single algorithm and Tri-training with Bayes, Perceptron and Winnow

Manually partitioning		TBayes		TPerceptron		TWinnow		TBPW	
Labelled	Unlab.	Super.	Final	Super.	Final	Super.	Final	Super.	Final
1000	4452	59.8	57.0	60.2	60.4	58.0	60.0	60.2	**65.8**
2000	3452	58.4	58.0	67.2	66.6	67.0	64.8	67.2	**68.8**
3000	2452	57.2	56.4	68.4	**70.0**	49.4	65.4	68.4	**70.0**
4000	1452	51.8	51.8	66.4	65.2	71.6	70.8	71.6	**72.0**
Randomly partitioning		TBayes		TPerceptron		TWinnow		TBPW	
Labelled	Unlab.	Super.	Final	Super.	Final	Super.	Final	Super.	Final
1000	4452	56.2	51.4	57.6	62.8	37.2	40.0	61.8	**63.0**
2000	3452	52.8	49.8	61.6	64.0	67.0	61.8	63.8	**65.4**
3000	2452	54.2	53.8	67.4	68.4	46.2	47.6	65.2	**71.2**
4000	1452	52.2	52.2	70.2	**72.0**	43.8	45.0	70.6	**72.0**

algorithms: Perceptron, Bayes and Winnow. We used these three learning algorithms to follow the Tri-training algorithm. Besides, we implemented an individual algorithm following the original Tri-training algorithm. All the parameters of the these algorithms, such as the learning rate α, threshold and initial weight of Perceptron and Winnow are default values. The bag-of-words feature is used in this experiment.

The comparison of experimental results is listed in Table 2, where TBayes, TPerceptron and TWinnow, respectively, stand for original Tri-training with Bayes, Perceptron and Winnow algorithm; TBPW stands for the modified Tri-training algorithm with Bayes, Perceptron and Winnow following the algorithm depicted in Fig. 1b. For Bayes, Perceptron and Winnow Tri-training algorithm, we compare the final result (*Final* column) with the result produced by the correspondingly supervised learning algorithm being trained on the same labelled set (*Super.* column). For TBPW, we compared the final result (*Final* column) with the best of individually supervised learning of classifiers (*Super.* column).

The results show that the precision of supervised learning of Bayes, Perceptron and Winnow is not sensitive to the size of training sets. Concretely, when the size of the training set increases, the corresponding precision does not increase. Maybe, the question data type and word features are not suitable for these learning algorithms. Moreover, manually and randomly partitioning may form training sets with different distributions. As the results, the corresponding precision is different. Fortunately, the final precision of our algorithm is improved. The sign test [9] shows that our algorithm is significant at the level of 95% (p=0.05).

In the second experiment, we use two algorithms: the first is Maximum Entropy Model[3](MEM) [1], the second one is Support Vector Machine[4](SVM) [3]

[3] We use a free open source implementation of Maximum Entropy Model available at http://homepages.inf.ed.ac.uk/s0450736/pmwiki/pmwiki.php

[4] We use a free implementation of SVM available at http://www.csie.ntu.edu.tw/~cjlin/libsvmtools/

Table 3. The precision (%) of original Tri-training with single MEM, SVM algorithm and modified Tri-training with both MEM and SVM

Manually		TMEM (%)			TSVM (%)			TSSM (%)	
Label.	Unlab.	Super.	Initial	Final	Super.	Initial	Final	Super.	Final
1000	4452	67.6	65.4	67.0	68.4	64.8	67.4	68.4	68.4
2000	3452	74.8	72.2	75.2	75.6	72.8	75.2	75.6	**76.0**
3000	2452	76.8	74.8	76.4	78.2	76.6	78.0	78.2	**78.6**
4000	1452	77.2	76.4	77.4	78.6	77.4	78.6	78.6	**78.8**
Randomly		TMEM (%)			TSVM (%)			TSSM (%)	
Label.	Unlab.	Super.	Initial	Final	Super.	Initial	Final	Super.	Final
1000	4452	69.8	67.0	**71.0**	68.0	66.6	68.6	66.8	69.0
2000	3452	76.8	74.8	78.0	75.4	70.6	74.4	77.8	**78.2**
3000	2452	79.0	73.2	79.8	80.6	77.6	79.0	78.2	**79.0**
4000	1452	80.0	74.4	**80.0**	79.8	77.6	**80.0**	79.8	**80.0**

which has been proven to perform well for text classification [12]. Because, in this domain, SVM classifier has better performance than that of MEM classifier, thus, we use two SVM classifiers and one MEM classifier in the implementation, with the expectation of making two SVM classifiers to have high degree of decision on final hypothesis. With SVM classifiers, we set them to use *linear* kernel and other parameters (e.g., parameter C) are default. In this domain, other kernels of SVM, such as *polynomial*, *radial basic function* or *sigmoid*, give poor performance. For MEM classifier, we use all default values of parameters (e.g., L-BFGS parameter estimation). Bag-of-words features are used for the experiment. Table 3 shows the results of different algorithms, where TSVM and TMEM, respectively, stand for original Tri-training algorithm with SVM and MEM algorithms; TSSM stands for modified Tri-training with two different SVM classifiers and a MEM classifier following the algorithm described in Fig. 1b. As shown in the table, MEM and SVM are sensitive to the size of the training sets. The precision is increased when the size of training set increases. In other words, MEM and SVM are more suitable for this type of data.

For original Tri-training algorithm with MEM and SVM, we also give the worst initial precision of a classifier in Step 4 of the original Tri-training in Fig. 1a (after being trained on a bootstrap-sampled set) in the column *Initial*. This column shows that the initial precision of a classifier may be decreased (or its error rate may increase).

Because the precision of manually partitioned test with the size of training set of 1000 is not improved, we did not carry out the sign test. Except for the manually partitioning test of training set size 4000, our other tests are significant at the level of 95%.

3.3 Experiments with Two Different Algorithms and Two Views

In the third experiment, we implement the second proposal of using more than one view following the algorithm described in Fig. 1c. In theory, we can use three different algorithms with distinct views, however, our primary purpose is to make the three classifiers diverse at the initial step, so two different algorithms, two views and a suitable assignment of views to classifiers are sufficient. Concretely, among the three classifiers, two of them are SVM classifiers and the third one is a MEM classifier. The first view (feature space) is *bag-of-words* the same as that used in previous experiments, and the second view is *pos-tags*. We set two SVM classifiers to use two different views, while the MEM classifier can use either of them, for example, the first SVM classifier uses bag-of-words features, the second SVM classifier uses pos-tags features and the final MEM classifier uses bag-of-words features. Let TMEM-word and TMEM-pos be the original Tri-training algorithm with MEM using bag-of-words and pos-tags features, respectively; Let TSVM-word and TSVM-pos respectively be the original Tri-training with SVM using bag-of-words and pos-tags features; Let TSSM-2views be a modified Tri-training with two SVM classifiers and a MEM classifier following the algorithm described in Fig. 1c using two views: bag-of-words and pos-tags. The results of the experiment are given in Table 4. Interestingly, the test on randomly partitioning with the training set size of 4000 reaches the precision of 81.4% which is also the precision of SVM being trained on the whole training set of the size 5500. Except for the two tests with the size of 1000 (both manually and randomly partitioning), our other tests are significant at the level of 95%.

Table 4. The precision (%) of Tri-training with single algorithm and Tri-training with MEM and SVM with two views

Manually		TMEM-word		TMEM-pos		TSVM-word		TSVM-pos		TSSM-2views	
Lab.	Unlab.	Super.	Final	Super.	Final	Super.	Final	Super.	Final	Super.	Final
1000	4452	67.6	67.0	68.8	**69.0**	68.4	67.4	69.2	65.6	69.2	68.4
2000	3452	74.8	75.2	75.4	73.2	75.6	75.2	75.2	74.4	75.6	**76.0**
3000	2452	76.8	76.4	76.8	76.2	78.2	78.0	77.0	76.6	78.2	**79.0**
4000	1452	77.2	77.4	77.8	77.8	78.6	78.6	79.0	78.4	79.0	**80.0**
Randomly		TMEM-word		TMEM-pos		TSVM-word		TSVM-pos		TSSM-2views	
Lab.	Unlab.	Super.	Final	Super.	Final	Super.	Final	Super.	Final	Super.	Final
1000	4452	69.8	71.0	71.0	69.0	68.0	68.6	69.2	67.2	71.0	**71.2**
2000	3452	76.8	78.0	73.4	73.8	75.4	74.4	75.2	75.0	76.4	**78.2**
3000	2452	79.8	77.6	75.2	75.6	80.6	79.0	77.0	77.4	79.0	**79.2**
4000	1452	80.0	80.0	76.2	76.8	79.7	80.0	79.0	78.8	80.8	**81.4**

4 Conclusion

This paper applied semi-supervised learning to explore unlabelled question to improve the performance of question classification task, and proposed two ways of modifying the Tri-training algorithm presented by Zhou and Li [14] to make it more suitable for question data type. The proposals dealt with a problem at the initial step of Tri-training, where the original training set is bootstrap-sampled to generate three different training sets, which can make the initial error rate of each classifier increase. With the purpose of using the original training set for all classifiers, while ensuring that they are still diverse, in the first proposal, we used more than one learning algorithm for the three classifiers, and the second proposal is to use multiple learning algorithms in combination with more than one view. Our experiments indicate that the performance is improved.

Our proposed algorithms have the same property as that of semi-supervised learning algorithms, that is instability. Which means the results of different running are not the same as each other. The reason is that the unlabelled instances may be wrongly labelled during the learning process. Although Tri-training has the function $Subsample(L_i, s)$ to limit the number of newly labelled instances for re-training the classifiers, it does not have the ability to throw away those mislabelled instances by random selection. A possible solution, in the future, for this problem is to eliminate the mislabelled instances as presented in [8] before calling the $Subsample(L_i, s)$ function.

The random selection of the function $Subsample(L_i, s)$ is not a good method. Because, the randomness may select many newly labelled questions of one class while omitting questions of other classes. A possible solution is to select roughly equal numbers of questions for each class, or select a number of questions for each class according to its distribution in the training set.

Our second proposal of applying two different algorithms of classification with two views may be extended to the *fourth-training* case, in which there are four classifiers. The first two classifiers use the same learning algorithm with different views. The other two classifiers use the second algorithm with two different views. This is also a possible study in the future.

In the current implementation, we have not considered to select other better feature types, such as those used in [13]. This is one interesting issue to explore in future to achieve higher precision.

Our modified versions of the Tri-training algorithm do not have any constraints on data types, therefore, one more issue which is worth studying in the future is to apply these algorithms in other domains, such as text classification.

References

1. Adam Berger, Stephen Della Pietra, and Vincent Della Pietra, "A maximum entropy approach to natural language processing," *Computational Linguistics*, Vol. 22, No. 1, (1996).
2. Andrew Carlson, Chad Cumby and Dan Roth, "The SNoW learning architecture," *Technical Report UIUC-DCS-R-99-2101, UIUC Computer Science Department* (1999).
3. Corinna Cortes and Vladimir Vapnik, "Support vector networks," *Machine Learning*, Vol. 20, No. 3, pp. 273-297, (1995).
4. Dell Zhang and Wee Sun Lee, "Question classification using Support vector machine," *In Proceedings of the 26th Annual International ACM SIGIR Conference*, pp. 26-32, (2003).
5. Ellen Voorhees, "The TREC-8 Question Answering Track Report," *In Proceedings of the 8th Text Retrieval Conference (TREC8)*, pp. 77-82, (1999).
6. Ellen Voorhees, "The TREC-9 Question Answering Track," *In Proceedings of the 9th Text Retrieval Conference (TREC9)*, pp. 71-80, (2000).

7. Ellen Voorhees, "Overview of the TREC 2001 Question Answering Track," *In Proceedings of the 10th Text Retrieval Conference (TREC10)*, pp. 157-165, (2001).
8. Fabrice Mulenbach, et al, "Identifying and handling mislabelled Instances," *Journal of Intelligent Information Systems*, Vol. 22, No. 1, pp. 89-109, (2004).
9. Gopal K Kanji, "100 Statistical Tests," *SAGE Publications*, (1994).
10. Hacioglu Kadri and Ward Wayne, "Question classification with Support vector machines and error correcting codes," *In Proceedings of NAACL/Human Language Technology Conference*, pp. 28-30, (2003).
11. Sally Goldman and Yan Zhou, "Enhancing supervised learning with unlabeled data," *In Proceedings of the 17th International Conference on Machine Learning*, pp. 327-334, (2000).
12. Thorsten Joachims, "Text categorization with Support vector machines: Learning with many relevant features," *In Proceedings of ECML-98, the 10th European Conference on Machine Learning*, pp. 137-142, (1998).
13. Xin Li and Dan Roth, "Learning question classifiers," *In Proceedings of the 19th International Conference on Computational Linguistics*, pp. 556-562, (2002).
14. Zhi-Hua Zhou and Ming Li, "Tri-training: Exploiting unlabeled data using three classifiers," *IEEE Transactions on Knowledge and Data Engineering*, Vol 17, No. 11, (2005).

Query Similarity Computing Based on System Similarity Measurement

Chengzhi Zhang, Xiaoqin Xu, and Xinning Su

Department of Information Management of Nanjing University,
Nanjing 210093, China
zcz51@citiz.net

Abstract. Query similarity computation is one of important factors in the process of query clustering. It has been used widely in the field of information processing. In this paper, a unified model for query similarity computation is presented based on system similarity. The novel approach of similarity computation uses the literal, semantic and statistical relative features of query. The method can take advantage of the normal approaches to improve the computation accuracy. Experiments show that the proposed method is an effective solution to the query similarity computation problem, and it can be generalized to measure the similarity of other components of text, such as sentences, paragraphs etc.

Keywords: query similarity, query clustering, similarity unit, system similarity measuring, literal similarity, semantic similarity.

1 Introduction

In the field of information processing, the similarity computation between strings or queries, such as words, phases, etc, plays an important part in dictionary compilation, machine translation based on examples, information retrieval, automatic question answering, information filtering and so on. Strings or queries similarity computation is one of important factors in the process of query clustering. It has been used widely in the field of information processing.

This paper builds a unified model to compute the similarity between queries by integrated using the advantages of the three methods, e.g. literal similarity measurement, statistical relevant similarity measurement, semantic similarity measurement, and overcoming their shortcomings. Namely, a unified model of similarity computation is built, which is based on similarity system theory[1] and the measurement of multiple features. It takes similar cell as the queries basic processing unit and considers the similar cell's literal, semantic and statistical relevant features synthetically. At the same time, the model amends the position information missing problem in the processing of sorting similar unit.

Y. Matsumoto et al. (Eds.): ICCPOL 2006, LNAI 4285, pp. 42 – 50, 2006.

2 Related Work

According to the different features, the existing methods of queries similarity computation could be classified into three types: methods based on literal similarity, methods based on statistical relevant similarity, methods based on semantic similarity. Of which, the computation methods based on literal similarity are mainly computed based on edit distance[2] and based on common words or phrases[3]. Methods based on statistical relevant similarity mainly compute words co-occurrence [4], vector space model[5], grammatical analysis[6] and so on. The improved methods based on large-scale corpus such as PMI-IR[7] and various smoothing algorithms[8] is used to resolve the problem of data sparseness in corpus. Methods based on semantic similarity would mainly make use of paraphrase dictionary[9] or some large-scale Ontology[10][11] to do semantic similarity computation.

Method based on literal similarity is simple, and be easy to achieve. But it is not flexible enough and doesn't consider the synonym substitution. Methods based on statistical relevant similarity could get much efficient relevancy between the strings, which could not be observed by people only. But this method depends on the training corpus, and is largely affected by the problem of data sparseness and data noise. Sometimes, methods based on semantic similarity may compute similarity between the strings, which are visual to be literal dissimilarity and statistical to be weak relevancy. But the Ontology are usually built by hand, which need to spend a lot of time.

3 Unified Modeling of Queries Similarity Computation

3.1 Mathematical Description of Queries Similarity Computation

Traditional approaches mostly compute similarity from a certain feature of queries. The similarity computation methods, which combine the literal, semantic and statistical relevant features of queries, have not yet been reflected in any report. Before unified modeling to the similarity computation of queries, we will give several related notes and definitions.

$\boldsymbol{\Omega}$: a set of Chinese strings or queries ;

$\boldsymbol{S}$: $\boldsymbol{\Omega}$'s subset, that is $\boldsymbol{S} \subseteq \boldsymbol{\Omega}$;

$\boldsymbol{\Psi}$: semantic dictionary; the authors use it to segment the Chinese text and each listing has a corresponding semantic codes; $\boldsymbol{\Psi} \subset \boldsymbol{S}$;

$\boldsymbol{S}_1$、$\boldsymbol{S}_2$: two given queries, including:

$\boldsymbol{S}_1 = \{a_1, a_2, \ldots, a_i, \ldots, a_M\}$,$i \in [1, M]$, the element quantity of $\boldsymbol{S}_1$ is M;

$\boldsymbol{S}_2 = \{b_1, b_2, \ldots, b_j, \ldots, b_N\}$,$j \in [1, N]$, the element quantity of $\boldsymbol{S}_2$ is N;

The element of $\boldsymbol{S}_1$、$\boldsymbol{S}_2$ could be single character, semantic words segmented by semantic dictionary (or Ontology) or its corresponding semantic codes[11]. Take query '计算机控制' for example, when the element is single character, the query may be expressed as follows: { 计, 算, 机, 控, 制}; if it is segmented by semantic dictionary (including the words without semantic codes, this paper takes *Tongyici* Cilin[12] as semantic system), the query may be expressed as {Bo010127, Je090101}.

s_i: similar unit; to identify the similar features between S_1 and S_2, the elements having similar features are known as similar unit. Similar elements which become similar units between S_1 and S_2, are called similar units, notated as $s(a_i, b_j)$, abridged notated as s_{ij}. Element a_i of S_1 is similar to element b_j of S_2. Element a_i and b_j are similar elements, which constitute the element cell $s(a_i, b_j)$. According to the similarity priority, we order the character string S_1 and S_2. And we could get:

$S_1' = \{a_1, a_2, \ldots, a_i \ldots, a_M\}$,

$S_2' = \{b_1, b_2, \ldots, b_j, \ldots, b_N\}$,

At this point, the similar elements between the strings are a_i, and b_i. The similar unit is $s(a_i, b_i)$, abridged notates as s_i.

Definition 1: Quantity of the similar units is the similar degree between element a_i of S_1 and the corresponding element b_i of S_2, which is notated as $q(s_i)$.

Definition 2: Similarity of the strings is the similar degree between queries S_1 and S_2, which is notated as $\mathbf{Sim}(S_1, S_2)$.

The common mathematical description of the queries similarity is as follows:

$$\mathbf{Sim}(S_1, S_2) = f(M, N, K, q(s_i)), \quad (i \in [1, K]) \tag{1}$$

Namely, similarity $\mathbf{Sim}(S_1, S_2)$ is a multiple function, whose variables are the quantity of element M in S_1 , the quantity of element N in S_2, the quantity of similar units K between S_1 and S_2 and $q(s_i)$ which reflected the similar degree between each similar elements.

According to the primary method of similarity measurement between the similar systems in the similarity system theory[1], we should consider two aspects when we do the similarity computation between the queries. That is, the quantity of the similar units and the similar units' numerical value of the similar units. The formula is as follows:

$$\mathbf{Sim}(S_1, S_2) = \mathbf{Qn} \cdot \mathbf{Qs} = \frac{K}{M+N-K} \sum_{i=1}^{K} \lambda_i q(s_i) \quad (i \in [1, K]) \tag{2}$$

And, λ_i is the weight which reflected the influence degree of similar cell s_i makes to the strings' similarity, $\lambda_i \in [0, 1]$, $\sum_{i=1}^{K} \lambda_i = 1$.

Considering the similarity degree by the quantity of the similar units, i.e. $\mathbf{Q}_n$, and the the similarity degree by the similar units' numerical value of the similar units, i.e. $\mathbf{Q}_s$, has a co-complementary function to compute the whole similarity of the similar units, we could assign different weights to $\mathbf{Q}_n$ and $\mathbf{Q}_s$, respectively α and β. And that $\alpha, \beta \in [0,1]$, $\alpha+\beta=1$. So there is:

$$\mathbf{Sim}(S_1, S_2) = \alpha \cdot \mathbf{Q}_n + \beta \cdot \mathbf{Q}_s = \alpha \cdot \frac{K}{M+N-K} + \beta \cdot \sum_{i=1}^{K} \lambda_i q(s_i) \quad (i \in [1, K]) \tag{3}$$

3.2 Improvement of Literal Similarity Computation

If we just consider the literal feature of the queries, namely element a_i of $\boldsymbol{S}_1$ and element b_i of $\boldsymbol{S}_2$ are completely literal matching, a_i and b_i could be regarded as similar elements. And the influence degree of similar unit makes to the queries is equal, namely $\boldsymbol{q}(\boldsymbol{s}_i)=1$ and $\lambda_i=1/K$. According to formula (2), we could get:

$$\mathbf{Sim}(S_1,\ S_2) = \frac{K}{M+N-K} \tag{4}$$

Formula (4) is common and is a simple similarity computation method based on literalness. For example, according to formula (4), the similarity between '计算机' and '微机' is **Sim**（'计算机', '微机'） =0.25. Because formula (4) computes the queries similarity excluding the similar elements' position information, the computation result would not be reliable. For instance, **Sim**('计算机', '机计算')=1.

According to formula (3), set α=0.6, β=0.4. Because Chinese character string has the feature that the topic kernel lies often back of it, we define λ_i as formula (5).

$$\lambda_i = \left[i / \sum_{k=1}^{K} k + j / \sum_{k=1}^{K} k \right] \Big/ 2 \tag{5}$$

Where, i and j respectively expresses that element unit s_{ij} is the number i element of $\boldsymbol{S}_1$ and the number j of $\boldsymbol{S}_2$. So formula (3) could be transformed as follows.

$$\mathbf{Sim}(S_1,\ S_2) = 0.6 * \frac{K}{M+N-K} + 0.4 * \sum_{i=1}^{K} \left[i / \sum_{k=1}^{M} k + j / \sum_{k=1}^{M} k \right] \Big/ 2 \tag{6}$$

We could see that, when computing the similarity of queries, formula(6) haven't considered the similar units' position information in the queries completely. For example, when computing the similarity between string '机微' and '微机' by formula (6), we could get **Sim**('机微', '微机')= 1. The reason for this result is that, the assumption of $\boldsymbol{q}(\boldsymbol{s}_i)$ equal to '1' is improper. Since, in addition to completely matching with the similar element's literalness, the computation of similar units' quantity is also correlated to the different positions of similar elements in different queries.

This paper introduces the moving distance (Distance(s_{ij})=| i-j |) of similar elements to optimize the value of $\boldsymbol{q}(\boldsymbol{s}_i)$. | i-j | expresses the absolute value of the distance between the similar element s_{ij}'s position in $\boldsymbol{S}_1$ and in $\boldsymbol{S}_2$. Taking the moving cost factor into account, $\boldsymbol{q}(\boldsymbol{s}_i)$ could be computed by formula (7):

$$\boldsymbol{q}(\boldsymbol{s}_i) = \frac{1}{1+|i-j|} \tag{7}$$

And, formula (3) would be transformed into:

$$\mathbf{Sim}(S_1,\ S_2) = 0.6 * \frac{K}{M+N-K} + 0.4 * \sum_{i=1}^{K} \left[\frac{i / \sum_{k=1}^{M} k + j / \sum_{k=1}^{M} k}{2} \cdot \frac{1}{1+|i-j|} \right] \tag{8}$$

According to formula (8), the similarity between query '机微' and '微机' is **Sim**('机微','微机')= 0.8.

Furthermore, the difference between literal similarity and word's similarity is that the similar elements' granularity is different. Considering the methods of similarity computation, they are both based on the literal feature. Therefore, in essence they are the same.

3.3 Quantity Computation of Multi-feature Similar Units

A key step of seeking the queries similarity is the computation of similar units' quantity $q(s_i)$. Obtaining similar units is the necessarily previous step of computing similar units. But in fact, because the similarity of reviewed object is very complex, the similar units are difficult to obtain. This paper takes a simple strategy, which takes a certain feature of the elements as the foundation to judge weather they are similar units. That is, giving a threshold quantity δ, taking a certain feature as object, and without taking into account the position difference of elements in different queries. If the similarity of this feature between element a_i and b_i, i.e. $q(s_i)$, is exceed δ, a_i and b_i would be similar elements. When judging weather two elements are similar elements or not, this paper is based on the literal feature, semantic feature and statistic relevant feature.

For semantic feature, we could set $\delta_1=0.25$. The queries are segmented by semantic dictionary, i.e. Ψ, and the segmented results are represented as semantic codes: Code[i], Code[j]. Without taking into account the position difference of elements in different queries, if the similarity between Code[i] and Code[j] is greater than 1/4, the two elements could be viewed as similar. $q(s_i)_1$ could be computed as follows.

$$q(s_i)=\begin{cases} 1 & \text{if } \mathit{strcomp}\,(\text{Code[i]}\,,\ \text{Code[j]}) = 0 \\ 1/[2 * (6 - n)] \cdot \dfrac{1}{1+|i-j|} & \text{elsewise} \end{cases} \tag{9}$$

Where, n stands for the first different layer number in the processing of comparing the two semantic codes from root nod, n∈[1, 5].

For literal feature, without taking into account the position difference of elements in different strings, we set $\delta_2=1$. That is, $q(s_i)_2$ could be got by formula (7). If we just consider the literal feature, the similarity computation of strings would be degenerated to the literal similarity computation, which is the situation of formula (6).

For statistical relevant features, without taking into account the position difference of elements in different queries, we set $\delta_3=0.5$. Statistic relevant degree is measuring the similarity between elements with a view to statistic distribution. Through the training corpus resources, we compute the mutual information between queries. For words in queries, i.e. a_i and b_j, if their mutual information $MI(a_i, b_j)$ is greater than the threshold quantity δ_3, we could consider that they are similar and could save the computation result into the statistical relevant table. It could be noted as Tab_Relation. When computing the statistical relevant similarity between elements a_i and b_j, we could find it in Tab_Relation directly. If elements a_i and b_j belong to Tab_Relation, it would mean that the two elements are similar. $q(s_i)_3$ could be computed as follows.

$$q(s_i)_3 = \mathrm{MI}(a_i - b_j) \Big/ Max(MI) \tag{10}$$

And, Max(MI) is the maximal value of the words' mutual information in Tab_ Relation.

After considering the multiple features, it's hard to estimate the influence weight λ_i of each similar unit s_i gives to queries' similarity. This paper takes it as equal weight, that is $\lambda_i = 1/K$. And K is the quantity of similar unit.

3.4 Description of Similarity Computation Algorithm of the Queries

For the given queries S_1 and S_2, the similarity between them could be computed by formula (3). If query $S_1=\{a_1, a_2, \ldots, a_i, \ldots, a_m\}$ and $S_2=\{b_1, b_2, \ldots, b_j, \ldots, b_n\}$ are completely different, we could use Ψ to segment S_1 and S_2 by the maximal matching segment method. The result would be $S_1' = \{A_1, A_2, \ldots, A_i, \ldots, A_M\}$, $S_2' = \{B_1, B_2, \ldots, B_j, \ldots, B_N\}$. Clearly for A_i (or B_j), if A_i is not belong to Ψ, A_i would be single character. If A_i and B_j are belong to Ψ, the similar unit's quantity $q(s_i)$ could be computed according to the priory of 'semantic > statistical relevance > literalness'. The detailed algorithm of the queries' similarity computation could be described as follows.

Algorithm: Similarity_Query compute the similarity between character query S_1 and S_2

Input: character string S_1, S_2

Output: **Sim,** the similarity of the queries: S_1, S_2

Process:

- Initialize: **Sim**=Q_s=0.0, M= N=K=Num=0, α=0.6, β=0.4, δ_1=0.25, δ_2=1, δ_3=0.5
- Segment S_1 and S_2 by Ψ, get A[i], [j], and create the Corresponding semantic codes
- For each A[i]
  - For each B[j]
    - If A[i], B[j] $\in \Psi$ then
      - If $q(s_i)_1 \geq \delta_1$ then
        K=K+1, M=M+1, N=N+1
        Compute $q(s_i)_1$
      - Else if $q(s_i)_3 \geq \delta_3$ then
        K=K+1, M=M+1, N=N+1
        Compute $q(s_i)_3$
- Segment the element A[i] or B[j]which has no corresponding similar element by single character
- Num = 'the number of the same single character between S_1 and S_2', K=K+Num, M = M+ 'number of the single characters of which couldn't match on literalness *of* S_1', N=N+ 'number of the single characters which couldn't match on literalness *of* S_2',compute $q(s_i)_2$
- λ_i =1/K, $Q_s = \sum_{i=1}^{K} \lambda_i q(s_i)$
- **Sim**=$\alpha \cdot Q_n + \beta \cdot Q_s = \alpha \cdot K/(M+N-K) + \beta \cdot Q_s$
- Return **Sim,** the similarity of the queries.

4 Experiments and Results

This paper has done two experiments, and each experiment computes the queries' similarity based on literal, semantic and multiple features. The first group experiment is a close testing, whose objects are Chinese queries. It tests the synonym search of Chinese words and phrases. The test processing is: first, draw out 100 pairs of economy and military queries from the search log database, which are viewed as highly similar to each other. Then, make them into disorder and generate nearly 40,000 pairs of synonym automatically. Then compute the similarity by the methods based on literal similarity, semantic similarity and multiple features respectively. After that, select those words which similarity is greater than 0.66 to compare with the first 100 pairs of words. The testing result shows as table 1.

The second group experiment is an open testing. Testing set is made up of unordered queries. And the search testing of synonyms are doing based on the open set. The test processing is: select 891queries of politics and 200 queries of economy, and make each pair of these words out of order, then compute their similarity and choose those words whose threshold quantity is greater than 0.66, then identify these words by hand. According to the similarity, we could divide them into synonym, quasi -synonym, hypogynous word, relevant words and irrelevant words. The first three kinds could be viewed as synonym, while the other two kinds are no- synonym. The testing result shows as table 2.

From the statistic data of table 1, we could see that through the experiment of searching for synonymic compound word, the recall of the pairs of synonym could

Table 1. The search result of synonymic compound word corresponding to the Chinese compound word

Domain	Testing words	word pairs	Recall（%）		
			A	B	C
economy	196	38,213	40	87	93
military affairs	200	39,800	49	87	92
Total	396	78,013	44.5	87	92.5

Notes：A,B,C stands for the，literal similarity measurement, semantic similarity measurement and multi-feature-based similarity measurement.

Table 2. The synonym extraction result on open testing set

Domain	Word pairs (Sim≥0.66)	Precision（%）		
		A	B	C
politics	347	11.24	24.78	27.64
economy	4,730	10.34	23.70	29.34
Total	5,077	10.40	25.52	29.23

achieve 92.5 % when the computation is based on multiple features, and 87% when based on semantic similarity, and just 44.5% when based on literal similarity. All this showed that, judging from the angle of the recall, the method based on multiple features is much better than the method just based on semantic or literal similarity.

From the data in table 2, we could see that, the precision is 29.23 % when using the method based on multiple features to recognize the synonym, and 25.52 % when based on semantic similarity, and just 10.40 % when based on the literal similarity. It shows that, on the aspect of synonym identification, the method based on multiple features is better than the method based on just literal or semantic similarity. This indicates that using the method based on multiple features is much more effective than the method based on just literal or semantic similarity. Besides, by the method based on multiple features, the rate of the searched relevant words whose similarity is greater threshold quantity is 80.08%, which is higher than the searching result by the other two methods, whose rate is respective 72.15% and 60.78%. It indicates that, the method has obvious advantage when it is used on the queries clustering.

5 Conclusion and Future Work

This paper builds a unified model to compute the queries' similarity. It takes similar unit as the queries' basic processing unit and considers the similar unit's literal, semantic and statistical relevant features synthetically. At the same time, the model amends the position information missing problem in the processing of sorting similar unit. The result of the experiments shows that, the method based on multiple features is efficient and it also has heuristic significance in the similarity computation between sentences and paragraphs. For the research of queries' similarity is also related to the knowledge of semantics, system theory and so on. For there are some questions existed in the present semantic system, the setting of similar unit's weight still needs farther research, i.e. estimate the combination coefficients from the data instead of using predefined value. Furthermore, the future work includes also testifying whether the method is applicable to other oriental languages, especially the languages in which Chinese characters are not used. It will be interesting to see the application of the proposed algorithm in English queries, running in a larger text corpus.

Acknowledgments. We thank the reviewers for the excellent and professional revision of our manuscript.

References

1. Zhou ML. Some concepts and mathematical consideration of similarity system theory. Journal of System Science and System Engineering 1(1)(1992)84-92.
2. Monge AE, Elkan CP. The field-matching problem: algorithm and applications. Proceedings of the Second Internet Conference on Knowledge Discovery and Data Mining, Oregon, Portland(1996)267-270.

3. Nirenburg S, Domashnev C, Grannes DJ. Two approaches to matching in example-based machine translation. Proceedings of TMI-93, Kyoto, Japan(1993)47-57.
4. http://metadata.sims.berkeley.edu/index.html, accessed: 2003.Dec.1.
5. Crouch CJ. An approach to the automatic construction of global thesauri. Information Processing and Management 26(5)(1990)629-640.
6. Lin DK. Automatic retrieval and clustering of similar words. Proceedings of the 17th International Conference on Computational Linguistics and 36th Annual Meeting of the Association for Computational Linguistics, Montreal(1998)768-774.
7. Peter D. Turney. Mining the web for synonyms: PMI-IR versus LSA on TOEFL. Proceedings of the 12th European Conference on Machine Learning, Freiburg(2001) 491-502.
8. Weeds J. The Reliability of a similarity measure. Proceedings of the Fifth UK Special Interest Group for Computational Linguistics, Leeds(2002)33-42.
9. Pierre P. Senellart. Extraction of information in large graphs: Automaitc search for synonyms. Masters Intership Reports. University catholique de Louvam, Louvain-la-Neuve, Belgium(2001)1-17.
10. Resnik P. Semantic similarity in a taxonomy: an information-based measure and its application to problems of ambiguity in natural language, Journal of Artificial Intelligence research 11(1999)95-130.
11. Li SJ, Zhang J, Huang X, Bai S. Semantic computation in Chinese question-answering system, Journal of Computer Science and Technology 17(6)(2002)933-939.
12. Mei Jiaju. *Tongyici Cilin*. Shanghai Lexicographical Publishing House(1983).

An Improved Method for Finding Bilingual Collocation Correspondences from Monolingual Corpora

Ruifeng Xu[1], Kam-Fai Wong[1], Qin Lu[2], and Wenjie Li[2]

[1] Dept. of Systems Engineering and Engineering Management, The Chinese University of Hong Kong, N.T., Hong Kong
[2] Dept. of Computing, The Hong Kong Polytechnic University, Kowloon, Hong Kong
{rfxu, kfwong}@se.cuhk.edu.hk,
{csluqin, cswjli}@comp.polyu.edu.hk

Abstract. Bilingual collocation correspondence is helpful to machine translation and second language learning. Existing techniques for identifying Chinese-English collocation correspondence suffer from two major problems. They are sensitive to the coverage of the bilingual dictionary and the insensitive to semantic and contextual information. This paper presents the *ICT* (Improved Collocation Translation) method to overcome these problems. For a given Chinese collocation, the word translation candidates extracted from a bilingual dictionary are expanded to improve the coverage. A new translation model, which incorporates statistics extracted from monolingual corpora, word semantic similarities from monolingual thesaurus and bilingual context similarities, is employed to estimate and rank the probabilities of the collocation correspondence candidates. Experiments show that *ICT* is robust to the coverage of bilingual dictionary. It achieves 50.1% accuracy for the first candidate and 73.1% accuracy for the top-3 candidates.

Keywords: Bilingual collocation correspondence, Monolingual corpora.

1 Introduction

A collocation is defined as an expression consisting of two or more words that correspond to some conventional way of saying things [11]. Collocations are popular word combinations widely used in natural habitual language. It is however hard to find the correct bilingual collocation correspondences merely using a bilingual dictionary for many collocation translations are idiosyncratic in the sense that they are unpredictable by syntactic or semantic features [9]. Normally, there are several translations for each word in the bilingual lexicon. However, the translations of collocations are not simply the combination of the word translations with the highest frequencies. For example, a English verb *raise* is the top frequency word translation for a Chinese verb *提出*. However, when collocated with *证明* (*proof*), *提出* should be translated to *produce*. Similar problems are existed when translate English collocations to Chinese. Therefore, the knowledge of bilingual collocation correspondence is helpful to improve

Y. Matsumoto et al. (Eds.): ICCPOL 2006, LNAI 4285, pp. 51 – 62, 2006.

machine translation [21]. Furthermore, a word may have more than one sense and the sense in a given context is determined by its collocated words [18]. Thus, the knowledge of bilingual collocation correspondence is helpful to both monolingual and bilingual word sense disambiguation [7]. Finally, bilingual collocation correspondence knowledge is important to second language learning. Normally, collocation learning is a barrier to non-native speakers. This is because collocation correspondence on a word-by-word basis is ineffective. Instead, a speaker must consider a collocation habitually in the source language and learn the common correspondence in the target language [16].

The knowledge of bilingual ccollocation correspondence cannot be compiled manually. Several research works were reported on automatic acquisition of collocation correspondences. Generally speaking, these techniques follow two approaches. The first approach is based on parallel corpora. Collocation correspondences are identified from aligned sentences by using dictionary and co-occurrence statistics [16]. Enlightened by the observation that collocations within different languages have a strong direct dependency correspondence and more than 80% of them can be mapped directly, the approach based on dictionary, syntactic dependency and monolingual corpora has been investigated [9,21].

There are two problems in the existing techniques. Firstly, these techniques are sensitive to the coverage of the underlying dictionary, where coverage is defined as the percentage of the collocations directly found from the dictionary. Secondly, most of these techniques only utilize the dictionary and syntax information, and they ignore the useful semantic and contextual information. An improved method for finding Chinese-English collocation correspondences from monolingual corpora, *ICT*, is proposed in this paper to overcome the problems. In this study, *ICT* is designed to handle verb-object (VO) collocation translation, which is most important and yet difficult. For a given Chinese VO collocation, the translation candidates for the verb and the object are generated by bilingual dictionary look-up. They are then expanded by using different strategies. A new model is employed to estimate and rank the correspondence probabilities of the collocation candidates. To do this, three types of information are used: (1) word translation probability learned from monolingual corpora; (2) word semantic similarities based on monolingual thesaurus; and (3) bilingual context word correspondences. The candidate with highest probabilities is identified as the collocation correspondence. Experiments show that the proposed candidate expansion method improves the coverage of the collocation correspondences and the accuracy of the new correspondence identification model outperforms the existing techniques.

The rest of this paper is organized as follows. Section 2 presents related work. Section 3 describes the method for candidate expansion and Section 4 presents the detail design of the model for identifying bilingual collocation correspondences. Section 5 outlines the evaluations and Section 6 concludes this paper.

2 Related Work

Most previous techniques extract collocation correspondences from parallel corpus. The basic idea of this approach is that in a parallel corpus, the bilingual collocation correspondences have similar occurrence frequencies and word and sentence order as well as the word distributions are also similar. These techniques identify correspondence candidates with similar statistical metrics from aligned sentences. Common statistical metrics include dice coefficient [16], mutual information [4], χ^2-test and log likelihood [1]. Note that these techniques are also applicable to translation of noun phrases [6] and multi-word unit [14]. The extensive use of statistical features renders this approach effective. However, despite of different efforts to compile parallel corpora, the size of current parallel corpora are still not enough to achieve high quality collocation correspondence extraction. It has been shown that at least 100M words are required [17]. Meanwhile, manual or semi-automatic preparation of parallel corpora is costly and impractical. Finally, the bilingual corpora preparation for arbitrary domains is not flexible.

Based on the observation that about 80% bilingual collocation correspondences have direct semantic and syntactic consistency and they can be compositional translated [9]. Degan et al. [2] investigated an approach to identify collocation correspondences from monolingual corpora [2]. Most existing works following this approach are at word level including probability estimation of word translation [5], identification of word translations [3, 15, 20] and word translation disambiguation [7]. Based on these, Zhou et al. proposed a method to extract collocation correspondences by using statistical cross-language similarity learned from monolingual corpora [21]. Lv and Zhou then proposed an improved method by using EM algorithm based on dependency correspondence [9]. Since the sizes of available monolingual corpora are generally much larger than parallel ones, data sparseness is not serious. Moreover, this approach is flexible to handle new texts. Therefore, it attracts much interest in recent years.

There are two problems faced by the existing collocation correspondence identification techniques. Firstly, these techniques cannot overcome the out-of-vocabulary (OOV) problems, i.e. they are sensitive to the coverage of the underlying bilingual dictionary. Experiments in [9] showed that the coverage of bilingual dictionary has only 83.98%. It means that there is an upper accuracy of collocation correspondences identification. Secondly, the existing techniques identify bilingual collocation correspondences mainly using statistical information and syntactic dependency and ignore the semantic and contextual information, which are crucial.

3 Candidates Expansion

From a linguistic point of view, less than 20% collocations are non-compositional and their meaning cannot be predicted by their components [21]. The correspondence identification of these collocations mainly uses a bilingual dictionary or parallel corpus and thus it is not the focus of this study. This study addresses the correspondence identification of compositional VO collocations [21]. Let a Chinese VO collocation labeled as $C_{col}= c_v c_o$, where c_v is a Chinese verb or a verbal phrase and c_o is a Chinese noun or a nominal phrase, the objective of our research is to find out the corresponding English VO collocation $E_{col} = e_v e_o$,where E_{col} consists of an English verbal component e_v and an English nominal component e_o. Most existing techniques assume that $e_v \in Trans(c_v)$ and $e_o \in Trans(c_o)$, where *Trans*(*x*) is a function for determining word translations from a Chinese-English dictionary. These techniques rank all possible combinations of $e_v e_o$ according to pre-defined statistical correspondence and dependency features. The one with the maximal correspondence probability, labeled as $E_{col\text{-}max}$, is identified as the output. However, as shown in the previous research [21], there are many cases that e_v or e_o are beyond $Trans(c_v)$ and $Trans(c_o)$. For these cases, the existing techniques based on dictionary-lookup fail. Therefore, the performance of these techniques has an upper limit which is sensitive to the coverage of underlying dictionary coverage.

We studied 1000 VO bilingual collocation correspondences collected from "*Everyday English word collocations (with Chinese explanation)*" [19] and measured the dictionary coverage for these collocation correspondences. Our dictionary, which consisted of 153,515 entries, was provided by Harbin Institute of Technology, China. It was observed that the coverage was 97.2% for e_o and 87.5% for e_v. Theoretically, the coverage for collocation correspondences was estimated to be 97.2%*87.5%=85.1%. The candidate coverage should be enhanced in order to improve the accuracy of the collocation correspondence identification model. Intensively, one would expand the bilingual dictionary to improve the coverage. However, natural languages are too dynamic rendering a comprehensive to bilingual dictionary costly to establish and too large to operate. Furthermore, from our observation, in fact a larger dictionary contributes little improvement on coverage (See Section 5).

Considering the different characteristics of e_v and e_o, we propose an effective method to expand the candidates of e_v and e_o to $Trans(c_v)^*$ and $Trans(c_o)^*$ using different strategies. It is observed that most OOV objects are the synonyms of $Trans(c_o)$; thus we perform object expansion using the synonyms in WordNet. For $Trans(c_o)$, we find the "predominate synset" for an object, labeled as *pso*, covers most words in $Trans(c_o)$. Similarly, the predominate synset for a verb is defined as the synset which covers most words in $Trans(c_v)$, labeled as *psv*. If *pso* has entries beyond $Trans(c_o)$, these entries are appended to $Trans(c_o)$ leading to $Trans(c_o)^*$. To avoid over expansion, only the synset containing more than 3 entries that belong to $Trans(c_o)$ are considered. The coverage of $Trans(c_v)$ is only 87.5% which is much lower than $Trans(c_o)$ because people use verbs flexible and thus verbs always have many senses. Therefore, the confidence of *psv* is lower than *pso*. The expansion for $Trans(c_v)$ following a different way based on monolingual collocation extraction. To extract monolingual collocations, we used a window-based algorithm [17] to collect context words co-occurring within the observing headwords. It estimates the collocation confidence of each bi-gram by using bi-directional strength and bi-directional spread. The former measures the co-occurrence frequency significance and the latter measures the co-occurrence distribution significance. As a result of this algorithm, we obtained a VO collocation list for $Trans(c_o)^*$. All of the verbs collocated with $Trans(C_o)^*$ led to $Trans(c_v)^*$. By expanding the candidates, coverage of collocation correspondence candidates is enhanced.

4 Collocation Correspondence Identification

4.1 Collocation Correspondence Identification Model

Our collocation correspondence identification model is based on two assumptions. At the outset, it is observed that there is a strong correlation between the co-occurrences of word patterns, which are translations of each other [15].

> **Assumption 1:** If an English translation E_{col} of a Chinese collocation C_{col} has the highest probability out of all possible translations, C_{col} will be the corresponding translation of E_{col} with the highest probability.

This assumption indicates that the bilingual collocation correspondences are symmetrical. Zhang proposed and verified another assumption that, if a word w has strong correlations with a set of context words, labeled as $w_1, w_2, .., w_n$, there will be similar correlations between w' and $w_1', w_2', .., w_n'$ where w' and $w_1', w_2', .., w_n'$ are the corresponding translations of w and $w_1, w_2, .., w_n$ in another language [20]. This assumption is extended as follows.

> **Assumption 2:** If a Chinese collocation C_{col} is associated with has different correlations with a set of context words $w_1, w_2, .., w_n$ to form different collocations, there will be similar correlations between E_{col} and $w_1', w_2', .., w_n'$.

Based on these assumptions, the collocation correspondence should have high bi-directional translation probability and strong context words consistency. This is used to maximize the following equation:

$$E_{col-\max} = \arg\max(P(E_{col} \mid C_{col}) + P(C_{col} \mid E_{col}) + Sim(CContext(C_{col}), EContext(E_{col}))) \quad (1)$$

where $E_{col}=(e_v,e_o)$ in which $e_v \in Trans(c_v)^*$ and $e_o \in Trans(c_o)^*$. In Equation 1, $P(E_{col}|C_{Col})$ is the Chinese-English translation probability, $P(C_{col}|E_{Col})$ is the English-Chinese translation probability, and $Sim(CContext(C_{col}), EContext(E_{col}))$ is the bilingual similarity between $CContext(C_{col})$ which is the context words of C_{col} in Chinese and $EContext(E_{col})$ which is the English context words of E_{col}. Our collocation correspondence identification model estimates the overall correspondence probabilities of all candidates and identifies the one with the highest overall probability as the final collocation correspondence.

4.2 Estimation of Chinese-English Translation Probability

Given $C_{col}=(c_v,c_o)$. We have correspondence candidates $E_{col}=(e_v, e_o)$. Following Bayes's theorem, the Chinese-English translation probability is calculate as follows,

$$P(E_{col} \mid C_{col}) = P(E_{col}) \cdot P(C_{col} \mid E_{col}) = P(E_{col}) \cdot P(c_v c_o \mid e_v e_o) \quad (2)$$

where $e_v \in Trans(c_v)^*$ and $e_o \in Trans(c_o)^*$. We then simplify Equation 2 by assuming that the translation operations for c_v and c_o are independent.

$$P(E_{col} \mid C_{col}) = P(E_{col}) \cdot P(c_v c_o \mid e_v e_o) = P(E_{col}) \cdot P(c_v \mid e_v) \cdot P(c_o \mid e_o) \quad (3)$$

The existing methods estimate the word translation probability of $P(e_v|c_v)$ and $P(e_o|c_o)$ based on statistical correspondence between monolingual corpora [5,9]. Considering that a collocation correspondence should have both strong statistic consistency and close semantic consistency, we further incorporate word semantic similarity between the candidate and the predominate synsets to estimate the Chinese-English word translation probability.

$$\begin{aligned} P(E_{col} \mid C_{col}) &= P(E_{col}) \cdot P(e_v \mid c_v) \cdot P(e_o \mid c_o) \\ &= P(E_{col}) \cdot (P_s(c_v \mid e_v) + Sim(e_v, psv)) \cdot (P_s(c_o \mid e_o) + Sim(e_o, pso)) \end{aligned} \tag{4}$$

where $P_s(c_v|e_v)$ and $P_s(c_o|e_o)$ are the word translation probabilities based on statistical correspondence. $Sim(e_v, psv)$ and $Sim(e_o, pso)$ are word similarities between e_v and psv, and e_o.and pso, respectively. The values of word similarities are from 0 to 1.

A concordance is performed to collect all of the sentences containing $Ecol=e_ve_o$, where $e_v \in Trans(c_v)^*$ and $e_o \in Trans(c_o)^*$. Suppose that the number of matched sentences is N_s, and the occurrence of e_v and e_o in the matched sentences are $N(e_v)$ and $N(e_o)$, respectively. The word translation probability based on statistical correspondence can be calculated by,

$$P_s(c_v \mid e_v) = N(e_v)/N_s \qquad P_s(c_o \mid e_o) = N(e_o)/N_s \tag{5}$$

We employ a well-developed tool package based on WordNet [13] to calculate the word similarities of $Sim(e_v, psv)$ and $Sim(e_o, pso)$. It is noteworthy that, if e_v or e_o has more than one sense in WordNet, similarity values between each sense and the predominate synset are calculated and the one with the highest value is adopted.

4.3 Estimation of English-Chinese Translation Probability

The estimation of English-Chinese translation probability is similar as the Chinese-English translation probability estimation. It is formulized as,

$$\begin{aligned} &P(C_{col} \mid E_{col}) = \\ &P(C_{col}) \cdot (P_s(e_v \mid Trans(e_v)) + Sim(Trans(e_v), c_v)) \cdot (P_s(e_o \mid Trans(e_o)) + Sim(Trans(e_o), c_o)) \end{aligned} \tag{6}$$

In Equation 6, the similarity between $Trans(e_v)$ and c_v, and $Trans(e_o)$ and c_o are measured because c_v and c_o are known. Furthermore, word similarity estimation is based on similarity distance between two observing words in TongYiCi CiLin [12], a Chinese thesaurus. The algorithm is similar to the one adopted in [17].

4.4 Estimation of Bilingual Context Similarity

We rank the context words surrounding C_{col} and E_{col} according to their co-occurrence frequencies, respectively. In practice, we believe that nouns and adjectives within a context window are helpful to distinguish the sense of VO collocations. 20 top-frequent nouns and adjectives surrounding C_{col} and E_{col} are identified and defined as "predominate context words" which are respectively labeled as pcc_i for Chinese and pce_i for English (i=1 to 20). The percentage of pcc_i among 20 predominate context words is calculated by,

$$p(pcc_i) = f(pcc_i) / \sum_{j=1}^{20} f(pcc_j) \tag{7}$$

This percentage is used to determine the significance in collocation sense determination. A larger value indicates that this word is more important for determining the collocations sense. Similarly, $p(pce_i)$ (i=1 to 20) are also obtained. Bilingual context similarity between C_{col} and E_{col} is then calculated by,

$$Sim(\,CContext(C_{col}), EContext(E_{col})) = \sum_{i=1,20\ j=1,20} sim(pcc_i, pce_j) \qquad (8)$$

where, $sim(pcc_i, pce_j) = p(pcc_i) \cdot p(pce_j)$, if $pce_i \in Trans(pcc_i)$ and $p(pce_i)$ is the highest one among all $p(Trans(pcc_i))$, otherwise $sim(pcc_i, pce_j)$ equals 0. The value of bilingual context similarity is from 0 to 1. A larger value indicates C_{col} and E_{col} are more similar.

Following Equation 1, 4, 6 and 8, the overall collocation correspondence probability for each candidate is calculated. The candidate with the highest overall probabilities is identified as the final collocation correspondence.

5 Experiments and Evaluation

5.1 Experiment Data Preparation

The British National Corpus which contains nearly 100 million English words and the Chinese corpus proposed in [17] were used. The latter contained about 90 million segmented and tagged Chinese words derived from several newspapers. Their sizes were much larger than any existing parallel corpus [4, 16]. Three bilingual dictionaries were prepared. They were (1) Sundict developed by Harbin Institute of Technology; (2) the dictionary extracted from HowNet; and (3) LDC Chinese-English dictionary V2.0. Furthermore, a standard test set containing 1000 Chinese-English VO collocation correspondences was built based on a bilingual collocation dictionary [19].

5.2 Improvement on the Coverage

The first experiment evaluated and compared the coverage improvement of three different dictionaries and our proposed expansion method. The coverage values for verbs, objects and collocations (labeled as $c(v)$, $c(o)$ and $c(col)$, respectively), as well as the number of entries in each dictionary are listed in Table 1.

Table 1. Improvement in collocation translation coverage

	no. entries	$c(v)$ (%)	$c(o)$ (%)	$c(col)$(%)
HowNet	109,441	87.3	96.4	84.2
LDC CE 2.0	128,366	87.0	96.8	84.2
SunDict	153,515	87.5	97.2	85.1
Candidate Expansion	153,515	97.2	98.3	95.5

It is observed that, while the number of entries in the dictionary increases from 109,441 (HowNet) to 128,366 (LDC C-E 2.0), $c(col)$ roughly remains constant. Also, when the number increases to 153,515 (SunDict), the coverage of collocation correspondences slightly increases (0.8%). This indicates that individually expanded dictionary cannot ensure higher coverage. It is also shown that using SunDict as bilingual dictionary, the proposed candidate expansion method increases $c(v)$, $c(o)$ and $c(col)$ to 97.2%, 98.3% and 95.5%. This result shows that this method is effective to improve coverage.

5.3 Evaluations of Existing Techniques

This experiment evaluated the three existing collocation correspondences identification techniques. The performances achieved by these models with the proposed candidate expansion method were also evaluated. The baseline system (*Model A*) selects the top-frequent translation for each word within the collocation as the collocation correspondence.

$$\text{Model A}: \quad E_{col\_max} = \arg\max_{e_v \in Trans(c_v)} (freq(e_v), \arg\max_{e_o \in Trans(c_o)} (freq(e_o)) \tag{9}$$

Dagan proposed another model (*Model B*) which selected collocation candidates with the maximal probability in the object language [2].

$$\text{Model B}: \quad E_{col\_max} = \arg\max p(E_{col}) = \arg\max_{e_v \in Trans(c_v), e_o \in Trans(c_o)} p(e_v e_o) \tag{10}$$

Based on dependency parsing result, Lv proposed a translation model (*Model C*) which estimate dependency triple translation probabilities by using EM algorithm and bilingual dictionary [9]. Here, a dependency triple as a collocation is $C_{col}=(c_v, \textit{verb-object}, c_o)$.

$$\begin{aligned} \text{Model C}: \quad E_{col\_max} &= \arg\max p(E_{col}) \cdot p(C_{col} \mid E_{col}) \\ &= \arg\max_{e_v \in Trans(c_v), e_o \in Trans(c_o)} (p(e_v e_o) \cdot p_{head}(c_v \mid e_v) \cdot p_{dep}(c_o \mid e_o) \cdot p(r_c \mid r_e)) \end{aligned} \tag{11}$$

These three models assumed that $e_v \in Trans(c_v)$ and $e_o \in Trans(c_o)$. The coverage and translation accuracy for the standard test set achieved by these models are given in Table 2. Note that NLPWin parser adopted in [9] is not a public resource. In this experiment, we used another two dependency parser, namely Minipar for English [8] and the Chinese parser provided by Harbin Institute of Technology [10] to simulate the NLPWin parser.

Table 2. Translation performances by three existing models

	Without expansion $c(col)$=85.1%		Candidate expansion $c(col)$=95.5%	
	Top-1	*Top-3*	*Top-1*	Top-3
Model A	20.2	33.1	16.8	23.4
Model B	36.2	54.1	23.3	30.7
Model C	42.5	65.2	31.5	36.3

In this table, *Top-1* and *Top-3* are the accuracies of the highest and the top three ranked collocation correspondence candidates identified by the three models, respectively. In the standard test set, some Chinese collocations have more than one recorded correspondences. Thus, if the resulting correspondence candidate matches any of the recorded correspondence answers, such a correspondence identification is regarded successful.

It is shown that *Model C* achieves the best result. It reflects the efficiency of dependency parsing for collocation correspondence identification and that the incorporation of syntactic dependency information improves accuracy. It is also shown that after expanding the word translation candidates, the coverage for these models increase to 95.5%. However, the achieved accuracy decreases significantly. This means that the enlarged word translation set brinks the risk to decrease collocation correspondence identification accuracy if no additional discriminative features are incorporated. Our proposed method further incorporated word translation probability, word semantic similarities and contextual similarity into estimation of correspondence probability.

5.4 Evaluations of *ICT*

This experiment evaluated the accuracies of *ICT* as well as each of its main component. *ICT* estimates the correspondence probability by three components. They are (1) collocation translation probability from Chinese to English (labeled as $ICT_{C\text{-}E}$); (2) collocation translation probability from English to Chinese (labeled as $ICT_{E\text{-}C}$); and (3) bilingual context similarity (labeled as $ICT_{context}$). The correspondence accuracies achieved by incrementally adding these components are given in Table 3. For comparison, the corresponding translation accuracies achieved with/without candidate expansion are also shown in Table 3.

Table 3. Translation accuracy of *ICT*

	Without candidate expansion $c(col)$=85.1% (%)		With candidate expansion $c(col)$=95.5% (%)	
	Top-1	*Top-3*	*Top-1*	*Top-3*
$ICT_{C\text{-}E}$	41.2	60.8	38.4	54.0
$ICT_{C\text{-}E}+ICT_{E\text{-}C}$	46.5	62.7	45.4	63.8
$ICT_{C\text{-}E}+ICT_{E\text{-}C}\,ICT_{context}$	48.7	66.4	50.1	73.1

Observe that with candidate expansion, the top-1 accuracy achieved by $ICT_{C\text{-}E}$+$ACE_{E\text{-}C}$ decreases 1.1% from 46.5% to 45.4%, while the top-3 accuracy increases 1.1%. On one hand, this result indicates that candidate expansion improves the coverage so that the top-3 accuracy increases. On the other hand, this result also shows that without more discriminative features, an enlarged candidate candidates set may have negative impact. By incorporating $ICT_{context}$, the final *ICT* achieves the best accuracy. Without candidate expansion, *ICT* achieves 48.8% top-1 accuracy which is 6.2% higher than *Model C* and the top-3 accuracy improvement is 1.2%. This result indicates that the proposed correspondence identification model in *ICT* outperforms Model C. With candidate expansion, the final *ICT* achieves 50.1% top-1 accuracy which is 7.6% improvement on *Model C*. Furthermore, *ICT* achieves 73.1% top-3 accuracy which is 7.9% improvement on Model C. This result means that candidate expansion can improve both coverage and accuracy of collocation translation if effective discriminative features are employed. Finally, we find that there is 73.1-50.1=23.4% correct collocation correspondences not ranked first. This indicates that there is room for further improvement.

5.5 Discussions

Although our approach achieves promising results, its performance can be further improved by ameliorating the following problems.

(1) The collocation candidates ranking

As shown in Section 5.4, the difference between the accuracy of the top-1 candidate and the top-3 candidates is obvious. Meanwhile, many correct collocation correspondence are ranked even latter positions. This indicates that the correspondence ranking algorithm must be further improved. Firstly, the parameters in Equation 1 are not weighted, i.e. the contributions of the parameters from different sources are regarded equal which is not accurate. The algorithm for finding optimal weights is required. Secondly, there are several default values given in *ICT* which is determined empirically. Such values influence the final ranking result. The optimization of such default values should be considered.

(2) The noise filtering of monolingual collocation extraction

The monolingual collocation extraction is one of the key components of *ICT*. Its performance significantly influences final *ICT*. The close observation on the output of *ICT* has shown that about 34% errors of collocation correspondence identification attribute to the monolingual collocation extraction errors. Especially, the collocation extraction accuracy for Chinese should be further improved.

(3) The out of vocabulary (OOV) problem

The OOV problem is relatively not serious in *ICT*. Most OOV cases in Chinese-English candidate generation can be recovered by *ICT*. However, if a word is OOV in both Chinese-English and English-Chinese candidate generation, it cannot be recovered.

(4) Non-compositional collocation translation.

Our model is based on the correspondence assumption, which assumes that most collocations can be compositional translated. But there are still many collocations that can't be translated word by word. *ICT* cannot deal with this kind of problem. Normally, the identification of such kind of collocations relies on the parallel corpora.

6 Conclusions

This paper presents an improved method for identifying Chinese-English collocation correspondences from non-parallel corpora. This method first expands word translation candidates to reduce the influence of OOV problem. For this reason, this method is relatively robust to dictionary coverage. A new model incorporates three features is employed to refine the estimation of collocation correspondence probability. These features come from (1) statistical word translation probability learned from monolingual corpora; (2) word semantic similarity from monolingual thesauruses; and (3) bilingual context similarity. Experiments on 1000 collocation correspondence samples show that *ICT* outperforms three existing techniques. The results also show the effectiveness of word semantic similarity and context similarity in collocation correspondence identification. The developed system, *ICT*, can be used to extract bilingual collocation translations from monolingual corpora with good quality and good efficiency which outperforms the time-consuming manually compiling. The acquired knowledge of bilingual collocation correspondence is valuable for monolingual and bilingual word sense disambiguation and machine translation.

Acknowledgements

This research is partially supported by The Chinese University of Hong Kong under the Direct Grant Scheme project (2050330) and Strategic Grant Scheme project (4410001), and Hong Kong Polytechnic University (A-P203) and a CERG Grant (5087/01E).

References

1. Chang, B.B.: Translation Equivalent Pairs Extraction Based on Statistical Measures. Chinese Journal of Computers. 26. 1. (2003) 616-621
2. Dagan, I., Itai, A.: Word Sense Disambiguation Using a Second Language Monolingual Corpus. Computational Linguistics. 20. 4 (1994) 563-596
3. Fung, P., Yuen, Y.L.: An IR Approach for Translating New Words from Nonparallel, Comparable Texts. Proc. of ACL'1998. (1998) 414-420
4. Haruno, M., Ikehara, S., Yamazaki, T.: Learning Bilingual Collocations by Word-level Sorting. Proc. 16th COLING (1996) 525-530
5. Koehn P., Knight, K.: Estimating Word Translation Probabilities from Unrelated Monolingual Corpora using the EM Algorithm. Proc. of NCAI'2000. (2000) 711-715
6. Kupiec, J.: An Algorithm for Finding Noun Phrase Correspondences in Bilingual Corpora. Proc. of ACL'1993 (1993) 23-30
7. Li, H., Li, C.: Word Translation Disambiguation Using Bilingual Bootstrapping, Computational Linguistics, 30. 1. (2004)
8. Lin, D.K.: Principar – An Efficient, Broad-coverage, Principle-based Parser. Proc. of 12th COLING. (1994) 482-488
9. Lv, Y.J., Zhou, M.: Collocation Translation Acquisition Using Monolingual Corpora. Proc. of ACL'2004. (2004) 167-174
10. Ma, J. S., Zhang, Y., Liu, T., Li, S.: A Statistical Dependency Parser of Chinese under Small Training Data, Proc. of 1st IJCNLP, (2004)
11. Manning, C.D., Schütze, H.: Foundations of Statistical Natural Language Processing, MIT Press (1999)
12. Mei, J. J. et al. (eds.): TongYiCiCiLin, Shanghai Dictionary Press (1996)
13. Patwardhan.: Incorporating Dictionary and Corpus Information into a Context Vector Measure of Semantic Relatedness, MSc. Thesis, University of Minnesota, U.S (2003)
14. Piao, S.L., McEnery, T.: Multi-word Unit Alignment in English-Chinese Parallel Corpora. Proceedings of Corpus Linguistic 2001 (2001) 466-475
15. Rapp, R.: Automatic Identification of Word Translations from Unrelated English and German Corpora. Proc. of ACL' 1999. (1999) 519-526
16. Smadja, F., Mckeown K.F., Hatzivassiloglou V.: Translation Collocations for Bilingual Lexicons: A Statistical Approach. Computational Linguistics. 22 (1996) 1-38

17. Xu, R.F., Lu, Q. 2005: A Multi-stage Chinese Collocation Extraction System. Lecture Notes in Computer Science, Vol. 3930, Springer-Verlag. (2006) 740-749
18. Yarowsky, D.: Unsupervised Word Sense Disambiguation Rivaling Supervised Methods. Proc. of ACL'1995 (1995) 189-196
19. Zhang, X.Z., Dai, W., P., Gao, P., Chen, S. B.: Everyday English Word Collocations, Dalian University of Technology Press (2003)
20. Zhang, Y.C., Sun, L., et al.: Bilingual Dictionary Extraction for Special Domain Based on Web Data. Journal of Chinese Information Processing. 20. 2. (2006) 16-23
21. Zhou, M., Yuan, M., Huang, C.N.: Improving Translation Selection with a New Translation Model Trained by Independent Monolingual Corpora. Computational Linguistics and Chinese Language Processing. 6. 1. (2001) 1-26

A Syntactic Transformation Model for Statistical Machine Translation

Thai Phuong Nguyen and Akira Shimazu

Japan Advanced Institute of Science and Technology
School of Information Science
{thai, shimazu}@jaist.ac.jp

Abstract. We present a phrase-based SMT approach in which the word-order problem is solved using syntactic transformation in the preprocessing phase (There is no reordering in the decoding phase.) We describe a syntactic transformation model based on the probabilistic context-free grammar. This model is trained by using bilingual corpus and a broad coverage parser of the source language. This phrase-based SMT approach is applicable to language pairs in which the target language is poor in resources. We considered translation from English to Vietnamese and from English to French. Our experiments showed significant BLEU-score improvements in comparison with Pharaoh, a state-of-the-art phrase-based SMT system.

1 Introduction

1.1 Phrase-Based SMT

In the field of statistical machine translation (SMT), several phrase-based SMT models [17,12,9] have achieved the state-of-the-art performance. These models have a number of advantages in comparison with the original IBM SMT models [2] such as word choice, idiomatic expression recognition, and local restructuring. These advantages are the result of the moving from words to phrases as the basic unit of translation.

Although phrase-based SMT systems have been successful, they have some potential limitations when it comes to modeling word-order differences between languages. The reason is that the phrase-based systems make little or only indirect use of syntactic information. In other words, they are still "non-linguistic". That is, in phrase-based systems tokens are treated as words, phrases can be any sequence of tokens (and are not necessarily phrases in any syntactic theory), and reordering models are based solely on movement distance [17,9] but not on the phrase content.

1.2 Approaches to Exploiting Syntactic Information for SMT

Several previous studies have proposed translation models which incorporate syntax representations of the source and/or target languages. Yamada and Knight [25] proposed a new SMT model that uses syntax information in the

Y. Matsumoto et al. (Eds.): ICCPOL 2006, LNAI 4285, pp. 63–74, 2006.

target language alone. The model is based on a tree-to-string noisy channel model and the translation task is transformed into a parsing problem. Melamed [14] used synchronous context free grammars (CFGs) for parsing both languages simultaneously. This study showed that syntax-based SMT systems could be built using synchronous parsers.

Charniak et al. [4] proposed an alternative approach to using syntactic information for SMT. The method employs an existing statistical parsing model as a language model within a SMT system. Experimental results showed improvements in accuracy over a baseline syntax-based SMT system.

A third approach to the use of syntactic knowledge is to focus on the preprocessing phase. Xia and McCord [24] proposed a preprocessing method to deal with the word-order problem. During the training of a SMT system, rewrite patterns were learned from bitext by employing a source language parser and a target language parser. Then at testing time, the patterns were used to reorder the source sentences in order to make their word order to that of the target language. The method achieved improvements over a baseline French-English SMT system. Collins et al. [6] proposed reordering rules for restructuring German clauses. The rules were applied in the preprocessing phase of a German-English phrase-based SMT system. Their experiments showed that this method could also improve translation quality significantly. Our study differs from those of [24] and [6] in several important respects. First, our transformational model is based on statistical decisions, while neither of the previous studies used probability in their reordering method. Second, the transformational model is trained by using bitext and only a source language parser, while Xia and McCord [24] employed parsers of both source and target languages. Third, we consider translation from English to Vietnamese and from English to French.

Reranking [20,18] is a frequently-used postprocessing technique in SMT. However, most of the improvement in translation quality has come from the reranking of non-syntactic features, while the syntactic features have produced very small gains [18].

1.3 Our Work

In our previous work [23], we studied about improving phrase-based SMT using morphological and syntactic transformation in preprocessing phase. We proposed a syntactic transformation model based on the probabilistic context free grammar. We considered translation from English to Vietnamese on small corpora. Our various experiments showed improvements in translation quality. However, there were several open questions. First, the corpora are small, which leads to suspicion that the improvements (made by syntactic transformation) over Pharaoh will vanish as the corpora scales up.[1] Second, word-order problem is considered in both preprocessing and decoding phases. It is not clear whether the translation quality is improved if decoding is carried out without reordering. Third, when the syntactic transformation is used, does the SMT system need

[1] Works which use morphological transformation for SMT have the property of vanishing improvement [7].

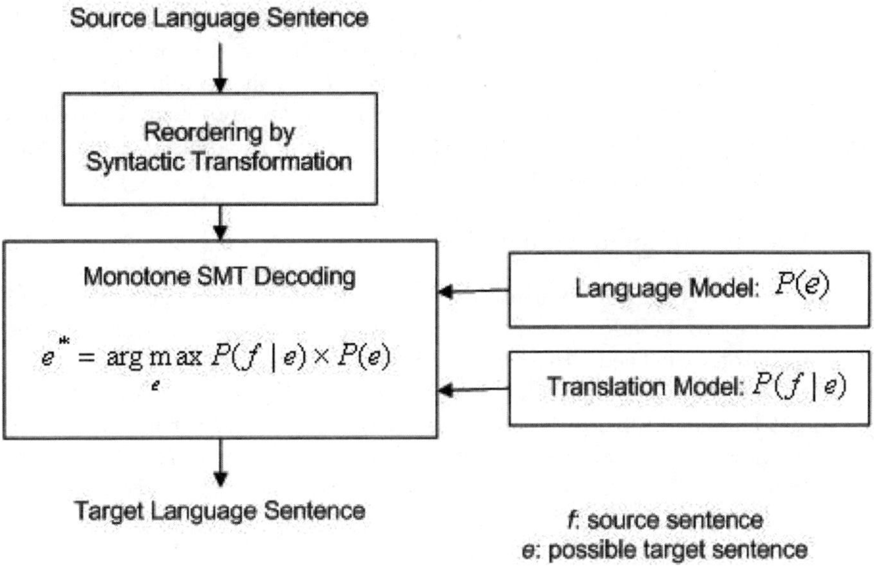

Fig. 1. The architecture of our SMT system

long phrases to achieve high translation quality? In this paper, we aim to find the answers to these questions.

Our approach is shown in Fig. 1. In the syntactic-transformation phase, first a source sentence is parsed, resulting in a syntactic tree. Next the tree is transformed into the target language structure. Then the surface string is extracted from the transformed tree. The resulting source sentence has a word order of the target language. In the decoding phase, the decoder searches for the best target sentence without reordering source phrases. The syntactic analysis and transformation are applied in both training and testing.

For syntactic transformation, we describe a transformational model [23] based on the probabilistic context free grammar. The knowledge of the model is learned from bitext in which the source text has been parsed. In order to demonstrate the effectiveness of the proposed method, we carried out experiments for two language pairs: English-Vietnamese and English-French. We used Pharaoh [10] as a baseline phrase-based SMT system. The experiments showed significant improvements of BLEU score.

The rest of this paper is organized as follows: In section 2, syntactic transformation is presented, including the transformational model, training, and applying the model. Section 3 describes experimental results.

2 Syntactic Transformation

One major difficulty in the syntactic transformation task is ambiguity. There can be many different ways to reorder a CFG rule. For example, the rule[2] $NP \rightarrow$

[2] NP: noun phrase, DT: determiner, JJ: adjective, NN: noun.

$DTJJNN$ in English can become $NP \to DTNNJJ$ or $NP \to NNJJDT$ in Vietnamese. For the phrase "a nice weather", the first reordering is most appropriate, while for the phrase "this nice weather", the second one is correct. Lexicalization of CFG rules is one way to deal with this problem. Therefore we propose a transformational model which is based on probabilistic decisions and also exploits lexical information.

2.1 Transformational Model

Suppose that S is a given lexicalized tree of the source language (whose nodes are augmented to include a word and a part of speech (POS) label). S contains n applications of lexicalized CFG rules $LHS_i \to RHS_i$, $1 \le i \le n$, (LHS stands for left-hand-side and RHS stands for right-hand-side). We want to transform S into the target language word order by applying transformational rules to the CFG rules. A transformational rule is represented as $(LHS \to RHS, RS)$ which is a pair consisting of an unlexicalized CFG rule and a reordering sequence (RS). For example, the rule $(NP \to JJNN, 10)$ implies that the CFG rule $NP \to JJNN$ in source language can be transformed into the rule $NP \to NNJJ$ in target language. Since the possible transformational rule for each CFG rule is not unique, there can be many transformed trees. The problem is how to choose the best one. Suppose that T is a possible transformed tree whose CFG rules are annotated as $LHS_i \to RHS_i'$ which is the result of reordering $LHS_i \to RHS_i$ using a transformational rule $(LHS_i \to RHS_i, RS_i)$. Using the Bayes formula, we have:

$$P(T|S) = \frac{P(S|T) \times P(T)}{P(S)} \tag{1}$$

The transformed tree T^* which maximizes the probability $P(T|S)$ will be chosen. Since $P(S)$ is the same for every T, and T is created by applying a sequence Q of n transformational rules to S, we can write:

$$Q^* = \arg\max_Q [P(S|T) \times P(T)] \tag{2}$$

The probability $P(S|T)$ can be decomposed into:

$$P(S|T) = \prod_{i=1}^{n} P(LHS_i \to RHS_i | LHS_i \to RHS_i') \tag{3}$$

where the conditional probability $P(LHS_i \to RHS_i | LHS_i \to RHS_i')$ is computed with the unlexicalized form of the CFG rules. Moreover, we constraint:

$$\sum_{RHS_i} P(LHS_i \to RHS_i | LHS_i \to RHS_i') = 1 \tag{4}$$

To compute $P(T)$, a lexicalized probabilistic context free grammar (LPCFG) can be used. LPCFGs are sensitive with both structural and lexical information. Under a LPCFG, the probability of T is:

$$P(T) = \prod_{i=1}^{n} P(LHS_i \rightarrow RHS'_i) \tag{5}$$

Since application of a transformational rule only reorders the right-hand-side symbols of a CFG rule, we can rewrite (2):

$$Q^* = \{RS_i^* : RS_i^* = \arg\max_{RS_i}[P(LHS_i \rightarrow RHS_i | LHS_i \rightarrow RHS'_i) \times P(LHS_i \rightarrow RHS'_i)], \\ i = 1, .., n\} \tag{6}$$

Suppose that a lexicalized CFG rule has the following form:

$$F(h) \rightarrow L_m(l_m)...L_1(l_1)H(h)R_1(r_1)...R_k(r_k) \tag{7}$$

where $F(h)$, $H(h)$, $R_i(r_i)$, and $L_i(l_i)$ are all lexicalized non-terminal symbols; $F(h)$ is the left-hand-side symbol or parent symbol, h is the pair of head word and its POS label; H is a head child symbol; and $R_i(r_i)$ and $L_i(l_i)$ are right and left modifiers of H. Either k or m may be 0, k and m are 0 in unary rules. Since the number of possible lexicalized rules is huge, direct estimation of $P(LHS \rightarrow RHS')$ is not feasible. Fortunately, some LPCFG models [5,3] can compute the lexicalized rule's probability efficiently by using the rule-markovization technique [5,3,8]. Given the left hand side, the generation process of the right hand side can be decomposed into three steps: Generate the head constituent label, generate the right modifiers, and generate the left modifiers. This is zeroth order markovization (the generation of a modifier does not depend on previous generations). Higher orders can be used if necessary.

The LPCFG which we used in our experiments is Collins' Grammar Model 1 [5]. We implemented this grammar model with some linguistically-motivated refinements for non-recursive noun phrases, coordination, and punctuation [5,1]. We trained this grammar model on a treebank whose syntactic trees resulted from transforming source language trees. In the next section, we will show how we induced this kind of data.

2.2 Training

The required resources and tools include a bilingual corpus, a broad-coverage statistical parser of the source language, and a word alignment program such as GIZA++ [16]. First, the source text is parsed by the statistical parser. Then the source text and the target text are aligned in both directions using GIZA++. Next, for each sentence pair, source syntactic constituents and target phrases (which are sequences of target words) are aligned. From this hierarchical alignment information, transformational rules and transformed syntactic tree are induced. Then the probabilities of transformational rules are computed. Finally, the transformed syntactic trees are used to train the LPCFG.

Fig. 2 shows an example of inducing transformational rules for English-Vietnamese translation. Source sentence and target sentence are in the middle

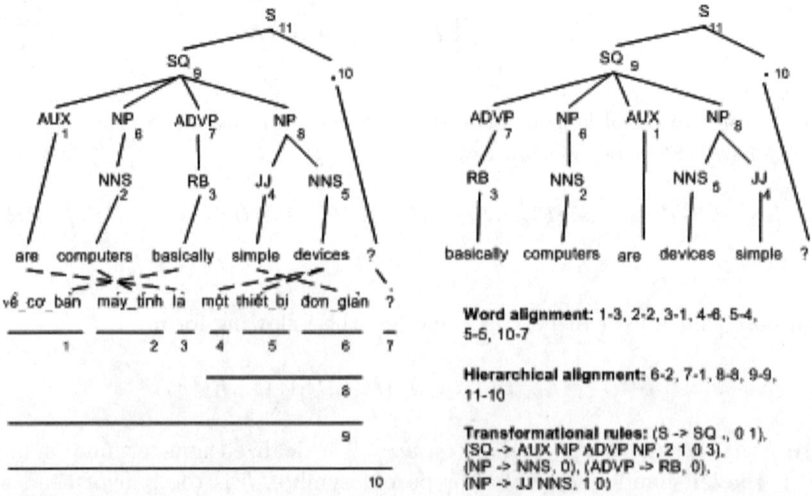

Fig. 2. Inducing transformational rules

part of the figure, on the left. The source syntactic tree is in the upper left part of the figure. The source constituents are numbered. Word links are represented by dotted lines. Words and aligned phrases of the target sentence are represented by lines (in the left lower part of the figure) and are also numbered. Word alignment results, hierarchical alignment results, and induced transformational rules are in the lower right part of the figure. The transformed tree is in the upper right.

To determine the alignment of a source constituent, link scores between its span and all of the target phrases are computed using the following formula [24]:

$$score(s,t) = \frac{links(s,t)}{words(s) + words(t)} \tag{8}$$

where s is a source phrase, t is a target phrase; links(s,t) is the total number of source words in s and target words in t that are aligned together; words(s) and words(t) are, respectively, the number of words in s and t. A threshold is used to filter bad alignment possibilities. After the link scores have been calculated, the target phrase, with the highest link score, and which does not conflict with the chosen phrases will be selected. Two target phrases do not conflict if they are separate or if they contain each other.

We supposed that there are only one-to-one links between source constituents and target phrases. We used a number of heuristics to deal with ambiguity. For source constituents whose span contains only a word which is aligned to many target words, we choose the best link based on the intersection of directional alignments and on word link score. When applying formula (8) in determining alignment of a source constituent, if there were several target phrases having the highest link score, we used an additional criterion:

- for every word outside s, there is no link to any word of t
- for every word outside t, there is no link to any word of s

Given a hierarchical alignment, a transformational rule can be computed for each constituent of the source syntactic tree. For a source constituent X with children $X_0, ..., X_n$ and their aligned target phrases $Y, Y_0, ..., Y_n$ (in which Y_i are sorted increasingly according to the index of their first word), the conditions for inducing the transformational rule are as follows:

- Y_i are adjacent to each other.
- Y contains $Y_0, ..., Y_n$ but not any other target phrases.

Suppose that f is a function in which $f(j) = i$ if X_i is aligned to Y_j. If the conditions are satisfied, a transformational rule $(X \rightarrow X_0...X_n, f(0)...f(n))$ can be inferred. For example, in Fig. 2, the constituent SQ (9) has four children AUX_0 (1), NP_1 (6), $ADVP_2$ (7), and NP_3 (8). Their aligned target phrases are[3]: Y (9), Y_0 (1), Y_1 (2), Y_2 (3), and Y_3 (8). From the alignment information: 1-3, 6-2, 7-1, and 8-8, the function f is determined: f(0)=2, f(1)=1, f(2)=0, and f(3)=3. Since the target phrases satisfy the previous conditions, the transformation rule $(SQ \rightarrow AUX\,NP\,ADVP\,NP$, 2 1 0 3) is induced.

For a sentence pair, after transformational rules have been induced, the source syntactic tree will be transformed. The constituents which do not have a transformational rule remain unchanged (all constituents of the source syntactic tree in Fig. 2 have a transformational rule). Their corresponding CFG rule applications are marked as untransformed and are not used in training the LPCFG.

The conditional probability for a pair of rules is computed using the maximum likelihood estimate:

$$P(LHS \rightarrow RHS|LHS \rightarrow RHS') = \frac{Count(LHS \rightarrow RHS, LHS \rightarrow RHS')}{Count(LHS \rightarrow RHS')} \quad (9)$$

In training the LPCFG, a larger number of parameter classes have to be estimated such as head parameter class, modifying nonterminal parameter class, and modifying terminal parameter class. Very useful details for implementing Collins' Grammar Model 1 were described in [1].

2.3 Applying

After it has been trained, the transformational model is used in the preprocessing phase of a SMT system. Given a source sentence, first it is parsed. Next the resulting syntactic tree is lexicalized by associating each non-terminal node with a word and a part of speech (computed bottom-up, through head child). Then the best sequence of transformational rules is computed by formula (6). Finally, by applying transformational rules on the source tree, the best transformed tree is generated. Finally, the surface string is extracted from the transformed syntactic tree.

[3] For clarity, we use Y symbol instead of Vietnamese phrases.

Table 1. Corpora and data sets

Corpus	Sentence pairs	Training set	Dev test set	Test set
Computer	8718	8118	251	349
Conversation	16809	15734	403	672
Europarl	740000	95924	2000	1122

Table 2. The numbers of unlexicalized CFG rules (UCFGRs), transformational rule groups (TRGs), and ambiguous groups (AGs)

Corpus	UCFGRs	TRGs	AGs
Computer	4779	3702	951
Conversation	3634	2642	669
Europarl	14462	10738	3706

3 Experiments

3.1 Experimental Settings

We carried out some experiments of translation from English to Vietnamese and from English to French. For the first language pair, we used two small corpora: one collected from some computer text books (named Computer) and the other collected from some grammar books (named Conversation). For the second language pair, we used the freely available Europarl corpus [9]. Data sets are described in Table 1. For the quick experimental turn around, we used only a part of the Europarl corpus for training. We created test set by choosing sentences randomly from the common test part [9] of this corpus.

A number of tools were used in our experiments. Vietnamese sentences were segmented using a word-segmentation program [22]. For learning phrase translations and decoding, we used Pharaoh [10], a state-of-the-art phrase-based system which is available for research purpose. For word alignment, we used GIZA++ tool [16]. For learning language models, we used SRILM toolkit [21]. For MT evaluation, we used BLEU measure [19] calculated by the NIST script version 11b. For the parsing task, we used the Charniak's parser [3].

3.2 Training the Transformational Model

On each corpus, the transformational model was trained resulting in a large number of transformational rules and an instance of Collins' Grammar Model 1. We restricted the maximum number of syntactic trees used for training the transformational model to 40000. Table 2 shows the statistics which resulted from learning transformational rules. On three corpora, the number of transformational rule groups which have been learned is smaller than the corresponding number of CFG rules. The reason is that there were many CFG rules which

Table 3. BLEU scores

Corpus	Baseline	Syntactic transformation
Computer	45.12	47.62
Conversation	33.85	36.26
Europarl	26.41	28.02

appear several times, however their hierarchical alignments did not satisfy the condition of inducing transformational rule. Another reason is that there were CFG rules which required nonlocal transformation.[4]

3.3 BLEU Scores

Table 3 shows BLEU scores of the Pharaoh system (baseline) and the Pharaoh decoder with preprocessing using syntactic transformation and monotone setting. On the Vietnamese corpora, the improvements are 2.5% and 2.4%. On the Europarl corpus, the improvement is smaller, only 1.61%. The difference of those values can be explained in some ways: First, we are considering word order problem, so the improvement is higher with language pairs which are more different in word order. According to our knowledge, Vietnamese and English are more different in word order than that of French and English. Second, by using phrases as the basic unit of translation, phrase-based SMT captures local reordering quite well if there is a large amount of training data.

In order to test the statistical significance of our results, we chose the sign test[5] [11]. We selected a significance level of 0.05. The Computer test set was divided into 23 subsets (15 sentences per subset), and the BLEU metric was computed on these subsets individually. The translation system with syntactic transformation was then compared with the baseline system over these subsets. We found that the system with preprocessing had a higher score than the baseline system on 20 subsets, and the baseline system had a higher score on 3 subsets. With the chosen significance level of 0.05 and the number of subsets 23, the critical value is 7. So we can state that the improvement made by the system with syntactic transformation was statistically significant. The same experiments were carried out on the other test sets (see Table 4). All the improvements were statistically significant.

3.4 Maximum Phrase Length

Table 5 displays the performance of the baseline SMT system and the syntactic-transformation SMT system with various maximum phrase lengths.[6] Obviously, the translation quality of both systems changes up when the maximum phrase length increases. The second system can achieve a high performance with a

[4] That is carried out by reordering subtrees instead of CFG rules.

[5] Sign test was also used in [6].

[6] We used Europarl data sets.

Table 4. Sign tests

Test set	Subsets	Test result	Critical value
Computer	23	20/3	7
Conversation	22	20/2	6
Europarl	22	17/5	6

Table 5. Effect of maximum phrase length on translation quality (BLEU score)

Maximum phrase size	2	3	4	5	6
Pharaoh	21.71	24.84	25.74	26.19	26.41
Syntactic transformation	24.1	27.01	27.74	27.88	28.02

short maximum phrase length, while the first system requires a longer maximum phrase length to achieve a similar performance. The improvement of the SMT system with syntactic transformation over the baseline SMT system decreases slightly when the maximum phrase length increases. This experiment gives us two suggestions. First, a maximum phrase length of three or four is enough for the SMT system with syntactic transformation. Second, the baseline SMT system relies on long phrases to solve the word order problem while the other SMT system is based on syntactic transformation to do that.

3.5 Training-Set Size

In this section, we report BLEU scores and decoding times corresponding to various sizes of the training set (in terms of sentence pairs). In this experiment, we used Europarl data sets and we chose a maximum phrase length of four. Table 6 shows an improvement in BLEU score of about 2% for all training sets. It means the improvement over Pharaoh does not decrease as the training set scales up. Note that studies which use morphological analysis for SMT have a contrary property of vanishing improvement [7]. Table 7 shows that, for all training sets, the decoding time of the SMT system with syntactic transformation is about 5-6% that of the Pharaoh system. This is an advantage of monotone decoding. Therefore we save time for syntactic analysis and transformation.

Table 6. Effect of training-set size on translation quality (BLEU score)

Training-set size	10K	20K	40K	80K	94K
Pharaoh	21.84	23.35	24.43	25.43	25.74
Syntactic transformation	23.65	25.67	26.86	27.52	27.74

Table 7. Effect of training-set size on decoding time (seconds/sent)

Training-set size	10K	20K	40K	80K	94K
Pharaoh	1.98	2.52	2.93	3.45	3.67
Syntactic transformation	0.1	0.13	0.16	0.19	0.22

4 Conclusion

We have demonstrated that solving the word-order problem in the preprocessing phase using syntactic transformation can improve phrase-based SMT significantly. For syntactic transformation, we have described a transformational model based on the probabilistic context free grammar and a technique of inducing transformational rules from source-parsed bitext. Our method can be applied to other language pairs, especially when the target language is poor in resources.

By experiments, we have found out the answers for the questions mentioned in Section 1.3. First, by using syntactic transformation in preprocessing and monotone decoding, the translation quality is improved and the decoding time is reduced. Second, the improvement does not vanish as the training-set size increases. Third, in order to achieve the same performance, the maximum phrase length is shorter than that of the baseline system.

In the future, we would like to apply this approach to other language pairs in which the difference in word order is greater than that of English-Vietnamese and English-French. We also would like to extend the transformational model to dealing with non-local transformations.

References

1. D. M. Bikel. 2004. Intricacies of Collins' Parsing Model. Computational Linguistics. 30(4): 479-511.
2. P. F. Brown, S. A. D. Pietra, V. J. D. Pietra, R. L. Mercer. 1993. The mathematics of statistical machine transla-tion. Computational Linguistics. 22(1): 39-69.
3. E. Charniak. 2000. A maximum entropy inspired parser. In Proceedings of HLT-NAACL 2000.
4. E. Charniak, K. Knight, and K. Yamada. 2003. Syntax-based language models for statistical machine translation. In Proceedings of the MT Summit IX.
5. M. Collins. 1999. Head-Driven Statistical Models for Natural Language Parsing. PhD Thesis, University of Pennsylvania.
6. M. Collins, P. Koehn, amd I. Kucerova. 2005. Clause restructuring for statistical machine translation. In Proceedings of ACL 2005.
7. S. Goldwater and D. McClosky. 2005. Improving statistical MT through morphological analysis. In Proceedings of EMNLP 2005.
8. D. Klein and C. D. Manning. 2003. Accurate unlexicalized parsing. In Proceedings of ACL 2003.
9. P. Koehn, F. J. Och, and D. Marcu. 2003. Statistical phrase-based translation. In Proceedings of HLT-NAACL 2003.

10. P. Koehn. 2004. Pharaoh: a beam search decoder for phrase-based statistical machine translation models. AMTA 2004.
11. E. L. Lehmann. 1986. Testing Statistical Hypotheses (Second Edition). Springer-Verlag.
12. D. Marcu, W. Wong. 2002. A phrase-based, joint probability model for statistical machine translation. In Proceedings of EMNLP 2002.
13. M. P. Marcus, B. Santorini, and M. A. Marcinkiewicz. 1993. Buildind a large annotated corpus of English: The Penn TreeBank. Computational Linguistics, 19: 313-330.
14. I. D. Melamed. 2004. Statistical machine translation by parsing. In Proceedings of ACL 2004.
15. S. Niessen and H. Ney. 2004. Statistical machine translation with scarce resources using morpho-syntactic information. Computational Linguistics, 30(2):181-204.
16. F. J. Och and H. Ney. Improved statistical alignment models. In Proceedings of ACL 2000.
17. F. J. Och, H. Ney. 2004. The alignment template approach to statistical machine translation. Computational Linguistics, 30: 417-449.
18. F. J. Och, D. Gildea, S. Khudanpur, A. Sarkar, K. Yamada, A. Fraser, S. Kumar, L. Shen, D. Smith, K. Eng, V. Jain, Z. Jin, and D. Radev. 2004. A smorgasbord of features for statistical machine translation. In Proceedings of HLT-NAACL 2004.
19. K. Papineni, S. Roukos, T. Ward, W.-J. Zhu. 2001. BLEU: a method for automatic evaluation of machine translation. Technical Report RC22176 (W0109-022), IBM Research Report.
20. L. Shen, A. Sarkar, F. J. Och. 2004. Discriminative reranking for machine translation. In Proceedings of HLT-NAACL 2004.
21. A. Stolcke. 2002. SRILM - An Extensible Language Modeling Toolkit", in Proc. Intl. Conf. Spoken Language Processing, Denver, Colorado, September 2002
22. T. P. Nguyen, Nguyen V. V. and Le A. C.. 2003. Vietnam-ese Word Segmentation Using Hidden Markov Model. International Workshop for Computer, Information, and Communication Technologies in Korea and Vietnam.
23. T. P. Nguyen and Akira Shimazu. 2006. Improving Phrase-Based SMT with Morpho-Syntactic Analysis and Transformation. In Proceedings of AMTA 2006.
24. F. Xia, M. McCord. 2004. Improving a statistical MT system with automatically learned rewrite patterns. In Proceedings of COLING 2004.
25. K. Yamada, K. Knight. 2001. A syntax-based statistical translation model. In Proceedings of ACL 2001.

Word Alignment Between Chinese and Japanese Using Maximum Weight Matching on Bipartite Graph

Honglin Wu[1] and Shaoming Liu[2]

[1] Natural Language Processing Lab, Institute of Software and Theory, Northeastern University, Shenyang, 110004, China
wuhl@mail.neu.edu.cn
[2] Corporate Research Group, Fuji Xerox, Co., Ltd., Kanagawa, Japan
Liu.Shaoming@fujixerox.co.jp

Abstract. The word-aligned bilingual corpus is an important knowledge source for many tasks in NLP especially in machine translation. Among the existing word alignment methods, the unknown word problem, the synonym problem and the global optimization problem are very important factors impacting the recall and precision of alignment results. In this paper, we proposed a word alignment model between Chinese and Japanese which measures similarity in terms of morphological similarity, semantic distance, part of speech and co-occurrence, and matches words by maximum weight matching on bipartite graph. The model can partly solve the problems mentioned above. The model was proved to be effective by experiments. It achieved 80% as F-Score than 72% of GIZA++.

Keywords: word alignment, matching, bipartite graph, similarity measure.

1 Introduction

Word alignment is an object for indicating the corresponding words in a parallel text [1]. Word alignment result tells which units in target text link to the units in source text. One main purpose of the word alignment is to provide data for machine translation, while another purpose can be to produce lexical data for bilingual dictionaries or/and terminology research for human translators.

There is a vast literature on word alignment and many approaches have been proposed to solve this problem. Concerning whether the linguistic knowledge has been used, these approaches can be divided into two kinds: statistical approaches and linguistic approaches.

Statistical approaches also refer to corpus-based approaches, which align word with statistical information gotten from the corpus. Most work in statistical word alignment has been inspired by the work on IBM SMT introduced by Brown [2]. IBM models can be improved using dependencies on word class, smoothing techniques for the estimation of probabilities, etc [3], [1]. Besides, heuristic models are usually used. In heuristic models, the word alignments are usually computed by analyzing some

Y. Matsumoto et al. (Eds.): ICCPOL 2006, LNAI 4285, pp. 75 – 84, 2006.

association score of a link between a source language word and a target language word. Heuristic methods often involve using a statistic to create a score to measure the strength of correlation between source and target words [4].

Linguistic approaches also refer to knowledge-based approaches, which align word with the help of linguistic knowledge such as cognate, dictionary, thesaurus, and so on. The alignment problem is the problem of bilingual word similarity calculation in substance. Bilingual dictionary is the most direct linguistic knowledge in getting bilingual word similarity. Besides, there are semantic class similarity (by thesaurus) and character similarity (by cognates) between some languages [5], [6], [7]. Stochastic inversion transduction grammars were also proposed [8]. For Chinese and Japanese, Zhang presented a method using English as an intermediate [9]. Ma proposed a method using self-organizing semantic maps [10].

Among the existing word alignment methods, the unknown word problem, the synonym problem and the global optimization problem are very important factors impacting the recall and precision of alignment results. a) In the real text, there are many unknown words. Most word alignment methods deal with unknown words statistically. These methods could hardly avoid the problem of data sparseness; b) For human translators, translations of words are not selected strictly from bilingual lexicons. There are many delicately translated words in real text. To these words, bilingual lexicons are useless for word aligning. We named this synonym problem; c) Many widely used matching methods could only get local optimal solution. We need improved matching methods to get global optimization.

In this paper, we propose a word alignment model between Chinese and Japanese which measures similarity in terms of morphological similarity, semantic distance, part of speech and co-occurrence, and matches words by maximum weight matching on bipartite graph. We introduce the morphological similarity to partly solve the unknown word problem; introduce the semantic similarity to partly solve the synonym problem; use the maximum weight matching on bipartite graph to partly solve the local optimization problem.

The paper is structured as follows: in section 2, we describe the details of our word alignment model; in section 3, the experiment design and results are shown, as well as analysis; conclusions are given in section 4.

2 Word Alignment Model

The word alignment model proposed in this paper represents sentence pair by weighted bipartite graph. Words in the sentence pair are represented by vertexes. Weights of links are equal to the similarity between the Chinese word and Japanese word. The similarity is measured in terms of morphology similarity (SimM), semantic distance (SimS), part of speech (SimP) and co-occurrence (Asso). And we get the alignment result by maximum weight matching (with preprocessing) on the bipartite graph.

2.1 Representation

In our alignment model, sentence is represented by set of words, which includes all the words in the sentence. Chinese sentence and Japanese sentence are represented by set $C = \{c_1, c_2, \cdots, c_m\}$ and set $J = \{j_1, j_2, \cdots, j_n\}$, respectively.

Sentence pair is represented by weighted bipartite graph $G = (V, E)$, where $V = C \cup J$ is the set of vertexes, and $E = \{e_{hk} \mid 1 \le h \le m, 1 \le k \le n\}$ is the set of links.

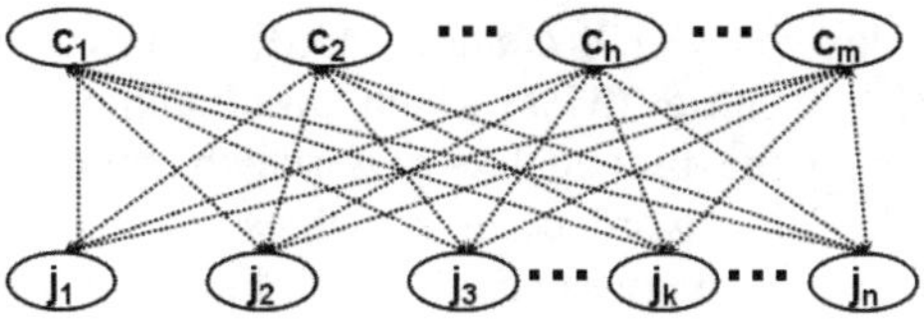

Fig. 1. Sentence pair represented by weighted bipartite graph

Vertices correspond to Chinese words and Japanese words in the given sentence. Each edge e_{hk} corresponds to a weighted link between Chinese word c_h and Japanese word j_k, whose weight w_{hk} is equal to the similarity between c_h and j_k. These weights can be calculated with the similarity metric described in the next subsection.

2.2 Similarity Measure

In this subsection, we measure the similarity between word pairs in terms of morphological similarity, semantic distance, part of speech and co-occurrence.

Sim(c,j) denotes the similarity between a Chinese word c and a Japanese word j, and it is calculated as follows:

$$Sim(c, j) = \begin{cases} 1 & SimD(c, j) = 1 \\ a_1 \times SimM(c, j) + a_2 \times SimS(c, j) + \\ a_3 \times SimP(c, j) + a_4 \times Asso(c, j) & SimD(c, j) = 0 \end{cases} \quad (1)$$

Bilingual lexicon is the most direct linguistic knowledge in getting bilingual word similarity. SimD(c,j) measures the similarity by bilingual lexicon, and it returns 1 when Japanese word j is a translation of Chinese word c, or Chinese word c is a translation of Japanese word j; else returns 0. When SimD(c,j) is equal to 0, we need more information to measure the similarity, such as morphological similarity (SimM), semantic distance (SimS), part of speech (SimP) and co-occurrence (Asso). And we can get Sim(c,j) by calculating SimM(c,j), SimS(c,j), SimP(c,j) and Asso(c,j) as follows.

a) Morphological Similarity (SimM)

Japanese was partly derived from Classical Chinese. Many Japanese words include kanji which was imported from Chinese characters. Kanji have the similarity of morphology with Chinese characters in high level and many of them are same to

Chinese characters. Over half of the Japanese words include kanji and 28% consist of kanji only [7]. Morphology may play a certain role in measuring the similarity between Chinese word and Japanese word. Modern Chinese has two main character sets: traditional character set and simplified character set. The Japanese character set shares many characters with these two Chinese character sets.

We assumed that a Chinese word and a Japanese word tend to be a translation pair if they have common characters. To prove our assumption, we made two examinations on a Chinese-Japanese lexicon and a bilingual corpus aligned in word level.

By examining the lexicon which contains 43,072 entries (translation pairs), we found that there are 15,970 entries (37%) which have common character(s).

Besides, we examined the bilingual corpus which contains 100 sentence pairs aligned in word level. For a word c_h in a Chinese sentence, we calculated Dice Coefficient (at character level) between c_h and each Japanese word j_k in corresponding Japanese sentence; made statistics of these data in the bilingual corpus; tested the correlation between "whether c_h aligns to j_k" and "whether the Dice Coefficient between c_h and j_k is larger than 0". Table.1 shows the data concerning correlation.

Table 1. Correlation between word alignment and morphological similarity

	$Dice(c_h, j_k) > 0$	$Dice(c_h, j_k) = 0$
aligned	306	363
not aligned	4	4916

We applied a χ^2 test upon Table 1, and calculated the Correlation Coefficient (r) as follows:

$$r = \sqrt{\chi^2 / (a+b+c+d)} \tag{2}$$

As a result, the Correlation Coefficient of 0.64 strongly supported our assumption of correlation between word alignment and morphological similarity.

We thus introduced morphological similarity to our alignment model. SimM(c,j) measures the similarity in terms of morphology, and it is calculated as follows:

$$SimM(c, j) = \frac{2 \times |(c \cap j) \cup (c^* \cap j)|}{|c| + |j|} \tag{3}$$

String c^* is acquired by converting simplified Chinese character of string c to traditional Chinese character. Our word alignment task is between simplified Chinese and Japanese. SimM(c,j) is a Dice-like statistic. Through formula 3, traditional Chinese characters can act as an intermediary that is able to discover the relationship between simplified Chinese and Japanese.

b) Semantic Similarity (SimS)

SimS(c,j) measures the similarity in terms of semantic distance, and it is calculated as follows:

$$SimS(c, j) = \underset{e \in Dict(c)}{Max} \{(1 - Distance(e, j))\} \tag{4}$$

Dict(c) is the set of translation words (in Japanese) of Chinese word c. The value of Distance(e,j) can be determined by any semantic distance algorithm, such as Jiang-Conrath's measure [11], [12].

$$Distance(e, j) = 2\log(p(lso(e, j))) - (\log(p(e)) + \log(p(j))) \tag{5}$$

c) Part of Speech (SimP)

SimP(c,j) measures the similarity in terms of part of speech, and it is calculated as follows:

$$SimP(c, j) = \begin{cases} 1 & POS(c) = POS(j) \\ 0 & POS(c) \neq POS(j) \end{cases} \tag{6}$$

d) Co-occurrence (Asso)

Asso(c,j) measures the similarity in terms of co-occurrence, and it can be calculated by many co-occurrence model, such as χ^2, Dice Coefficient, etc [4].

2.3 Matching

In subsection 2.2, we measured the similarity between every Chinese word c and Japanese word j by Sim(c,j). In other words, we got all the weights of edges (links) in bipartite graph which represented the given sentence pair. In this subsection, we proposed two matching algorithms which can select a set of links as word alignment result.

Definition 1. A matching in the bipartite graph G = (C ∪ J, E) is a subset of the set of edges E assigning vertices of set C to vertices of set J, such that no two edges in the matching share a common vertex [13]. M(G) denotes matching of bipartite graph G.

Definition 2. A maximum weight matching (MWM) is a matching such that the sum of the weights of the edges in the matching is maximized. $M^{Max}(G)$ denotes the maximum weight matching of bipartite graph G.

$$M^{Max}(G) = \underset{\sum_{h,k} w_{hk} x_{hk}}{ArgMax} \; M(G) \tag{7}$$

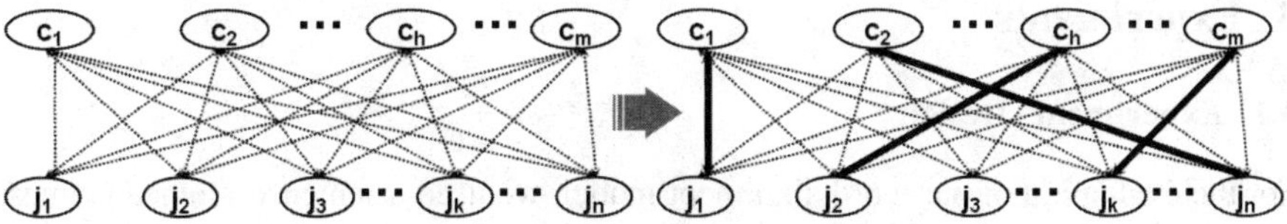

Fig. 2. Maximum weight matching on bipartite graph

The word alignment task can be the process of seeking the MWM on the bipartite graph which represents the input sentence pair. The algorithm is as follows:

Algorithm A.1: For a given sentence pair represented by bipartite graph $G=(C \cup J, E)$, discard the edges with the weight lower than θ^A. θ^A is a heuristically decided threshold to avoid statistical noise. We named the new set of edges $E^* = \{e_{hk} \mid e_{hk} \in E, w_{hk} \geq \theta^A\}$ and build a new bipartite graph $G^* = (C \cup J, E^*)$. Then obtain $M^{MAX}(G^*)$ by formula 7. Maximum weight matching can be solved by the Hungarian Algorithm in $O(n^3)$ time [13]. Finally output $M^{MAX}(G^*)$ as the alignment result.

However, the MWM matching boils down to finding 1-to-1 alignments, as a matching is restricted such that no two edges share the same vertex. This will limit the expected results of real text. We thus extended the MWM by giving additional attention to the inner component of link weights, and designed algorithm A.2 to enhance the ability for indicating m-to-n alignments.

Algorithm A.2: For a given sentence pair represented by bipartite graph $G=(C \cup J, E)$, obtain $M^{MAX}(G^*)$ by algorithm A.1. Obtain $M^M(G)$ and $M^S(G)$ as follows:

$$M^M(G) = \{e_{hk} \mid e_{hk} \in E, SimM(c_h, j_k) \geq \theta^M\} \tag{8}$$

$$M^S(G) = \{e_{hk} \mid e_{hk} \in E, SimS(c_h, j_k) \geq \theta^S\} \tag{9}$$

$M^M(G)$ is a subset of E. It consists of the edges with the morphological similarity (SimM) not lower than a heuristically decided threshold θ^M. These edges have very high confidence based on the analysis in subsection 2.2. We select these edges as a part of final alignment result. So does $M^S(G)$.

Then get $M^{A.2}(G)$ as follows:

$$M^{A.2}(G) = M^{Max}(G^*) \cup M^M(G) \cup M^S(G) \tag{10}$$

In formula 10, each subset that constitutes the union achieves high precision in selecting right links, and each subset selects links from a different aspect. Therefore, the union of the three subsets will lead to a large increase of recall with a few loss of precision.

Finally output $M^{A.2}(G)$ as the alignment result.

3 Experiments

3.1 Experiments Design

For the evaluation of our word alignment model, we used a sentence-aligned Chinese-Japanese parallel corpus (for Asso), a bilingual lexicon between Chinese and Japanese (for SimD) and the EDR thesaurus (for SimS).

A randomly chosen subset of the corpus is used as the test set which contains 100 sentence pairs, and the remaining sentence pairs are used as the training corpus which contains about 15,000 sentence pairs. The average length of sentence pairs is 33.2 words. There are 511,326 tokens and 24,304 types in the corpus. Chinese and Japanese sentences are segmented by NEUCSP (available in www.nlplab.com) and Chasen (available in chasen.aist-nara.ac.jp) respectively.

To evaluate our alignment model, we built seven alignment systems (WA_1a, b, c, d, e and WA_2a, b) by our alignment model, and built three more systems (Baseline1, Baseline2 and GIZA++[1]) by other methods as baseline.

Baseline1 is built by introducing bilingual lexicon similarity measure (SimD); WA_1a, b and c are built by introduced morphological similarity (SimM), semantic similarity (SimS) and part of speech (SimP) upon Baseline1 respectively; WA_1d is a corpus only system which use χ^2 as co-occurrence measure (Asso); WA_1e is built by introducing all of our similarity measure methods. In Expt.1, we compare these systems to evaluate the effect of our similarity measure methods.

Baseline2, WA_2a and WA_2b are built by introducing all of our similarity measure methods. The difference among these three systems lies in the matching method. Baseline2 used Competitive Linking matching; WA_2a and WA_2b used algorithm A.1 and A.2 we proposed in this paper respectively. (In fact, WA_2b is same to WA_1e.) Coefficients and thresholds are heuristically decided. The testing data are not used as the tuning data for these coefficients and thresholds. In Expt.2, we compare these three systems to evaluate the performance of our matching methods. Also, we compare our best system WA_2b with GIZA++.

The details of these experiments are shown in Table.2.

Table 2. Design of Experiments

Methods Used		for Similarity Measuring					for Matching		
		SimD	SimM	SimS	SimP	Asso	CompetitiveLinking	A1	A2
Expt.1	Baseline1	●							●
	WA_1a	●	●						●
	WA_1b	●		●					●
	WA_1c	●			●				●
	WA_1d					●			●
	WA_1e	●	●	●	●	●			●
Expt.2	Baseline2	●	●	●	●	●	●		
	WA_2a	●	●	●	●	●		●	
	WA_2b	●	●	●	●	●			●
	GIZA++	GIZA++							

For all the experiments, we manually create the correct alignment, and evaluate the recall, precision and F-Score.

3.2 Experimental Results

Table 3 shows the result of Expt.1:

Table 3. Result of Expt.1

	Recall	Precision	F-Score
Baseline1	0.38	0.93	0.54
WA_1a	0.56	0.90	0.69
WA_1b	0.43	0.90	0.58
WA_1c	0.44	0.85	0.58
WA_1d	0.55	0.93	0.69
WA_1e	0.73	0.89	0.80

From Table 3, we found that WA_1a achieved F-Score of 0.69, which is much higher than the F-Score of Baseline1 (0.54). This is due to the introduction of morphological similarity (SimM). Many unknown words have been aligned by SimM. Because the Chinese character set shares many characters with the Japanese character set, during translating Chinese unknown word to Japanese, people tend to use common characters between Chinese and Japanese, especially for named entity. SimM could indicate almost all such corresponding word pairs. It thus achieved a large increase of recall with a few loss of precision.

Also, we found that WA_1b achieved F-Score of 0.58, which is higher than Baseline1 (0.54). This is due to the introduction of semantic similarity (SimS). For real text, translations of words are not strict with bilingual lexicons. Using bilingual lexicon can only align a small part of words. Baseline1 thus get a low recall of 0.38. Introducing SimS, the system WA_1b can align many delicately translated word pairs by synonym chain.

WA_1c achieved F-Score of 0.58 which shows how much the SimP will do. WA_1d achieved F-Score of 0.69 which shows how much the corpus alone will do.

WA_1e is built by introducing all of our similarity measure methods. The improvement of F-Score (26% up) thus proved the effect of our similarity measure methods.

Table 4 shows the result of Expt.2:

Table 4. Result of Expt.2

	Recall	Precision	F-Score
Baseline2	0.75	0.78	0.77
WA_2a	0.61	0.92	0.74
WA_2b	0.73	0.89	0.80
GIZA++	0.74	0.70	0.72

From Table 4, we find that WA_2b improved the F-score than Baseline2 (3% up). This is due to introduction of our maximum weight matching on bipartite graph which can apply global optimal matching.

Baseline2 use Competitive Linking (CL) approach for matching [14]. In this approach, the alignment is done in a greedy "best-first" search manner. In this way, the method can only find local optimal alignment which may include wrong links. For example, in table 5, the similarity scores between word pairs in a sample sentence pair are listed.

Table 5. Association scores between word pairs in a sample sentence pair

	j1	j2	j3	j4	j5	j6
c1	143	8	0	2	1	15
c2	6	35	7	0	15056	337
c3	0	59155	538	22	8	3
c4	21	419	27573	6	3	69
c5	536	10	7	421	23	148

In table 5, under the matching strategy of the Competitive Linking approach, the links c3-j2, c4-j3, c2-j5, c5-j1 and c1-j6 will be selected. Link c5-j1: Duode(get)-Ginmedaru (silver medal) and link c1-j6: Yajvn(runner-up)-Ta(an auxiliary word) are false alignments. The false link of c5-j1 resulted in the false link of c1-j6 because the algorithm selects link one by one. If a link is wrongly selected, such mistake will influence the alignment of the remaining words. In contrast, with our algorithm, the links c3-j2, c4-j3, c2-j5, c5-j4 and c1-j1 are selected. The sum of the weights of these links is 102,348 which is the largest. Our matching method compares the weights of all the possible combination of links. It can find globally optimal solution. WA_2b thus performed better than Baseline2.

From Table 4, we also find that WA_2b improved the F-score than WA_2a does (6% up). This is due to the improvement of our matching algorithm A.2 based on algorithm A.1. A.1 restricts the alignment within 1-to-1 alignment. A.2 extend A.1 with the union of $M^{MAX}(G^{*})$, $M^{M}(G)$ and $M^{S}(G)$. These subsets achieved high precision in selecting right links, and each subset selects links from a different aspect. Therefore, A.2 led to a large increase of recall with a few loss of precision.

WA_2b is also the full model we proposed in this paper which achieved the best result in our experiments. It achieved F-Score of 80% than 72% of GIZA++. Our word alignment approach is proved to be effective by these experimental results.

4 Conclusion

In this paper, we described the word alignment model which measures similarity in terms of morphological similarity, semantic distance, part of speech and co-occurrence, and matches words by the maximum weight matching on bipartite graph. We made a word alignment system based on the model, and evaluated the system on the Chinese-Japanese corpus. As a result, it achieves 80% as F-Score than 72% of GIZA++.

The alignment model we proposed in this paper is our first step of attempt. To enhance the ability in aligning m-to-n alignments, the model can be improved by a) introducing more similarity measure methods (such as phonetic similarity), and b) enhancing the algorithm A.2 by adding more restricted conditions to formula 8, 9 and 10. Our final goal is to align parallel text in structure level, which can provide more help to machine translation.

Acknowledgments

This research was supported by The Document Company Fuji-Xerox under the VFP Program 7th. This research was supported in part by the National Natural Science Foundation of China (No.60473140), 985 project of Northeastern University (No.985-2-DB-C03) and Program for New Century Excellent Talents in University (No.NCET-05-0287).

References

1. F. Och, H. Ney. 2003. A Systematic Comparison of Various Statistical Alignment Models. *Computational Linguistics*, 29(1):19-51.
2. P. F. Brown, J. Cocke, S.A.D. Pietra, V.J.D. Pietra, F. Jelinek, J.D. Lafferty, R.L. Mercer and P.S. Roossin. 1990. A statistical approach to machine translation. *Computational Linguistics*, 16(2):79-85.
3. K. Toutanova, H. T. Ilhan, and C. D. Manning. 2002. Extensions to hmm-based statistical word alignment models. In *Proceedings of Conference on Empirical Methods for Natural Language Processing*, pp.87–94, Philadelphia, PA.
4. W. Gale, K. Church. 1991. Identifying Word Correspondances in Parallel Texts. In *Proceedings of DARPA Workshop on Speech and Natural Language*, pp.152-157. Pacific Grove, CA.
5. S. J. Ker, and J.S. Chang. 1997. A Class-based Approach to Word Alignment. *Computational Linguistics*, 23(2):313-343.
6. M. Simard, G. Foster and P. Isabelle. 1992. Using Cognates to Align Sentences in Bilingual Corpora. In *Proceedings of the Fourth International Conference on Theoretical and Methodological Issues in Machine translation* (TMI92), (Montreal), 67-81.
7. Y. Zhang, Q. Ma, H. Isahara. 2004. Use of Kanji Information in Constructing a Japanese-Chinese Bilingual Lexicon. In *Proceedings of The 4th workshop on ALR*, Hainan, China.
8. D. WU. 2000. Bracketing and aligning words and constituents in parallel text using Stochastic Inversion Transduction Grammars. In *Parallel Text Processing: Alignment and Use of Translation Corpora*. Dordrecht: Kluwer. ISBN 0-7923-6546-1.
9. Y. Zhang, Q. Ma, H. Isahara. 2003. Automatic acquisition of a Japanese-Chinese Bilingual lexicon Using English as an Intermediary. IEEE NLPKE-2003, Beijing.
10. Q. Ma, Y. Zhang, M. Masaki, H. Isahara. 2003. Semantic Maps for Word Alignment in Bilingual Parallel Corpora. *ACL2003 Workshop: Second SIGHAN Workshop on Chinese Language Processing*, pp. 98-103, Sapporo.
11. A. Budanitsky and G. Hirst. 2001. Semantic Distance in WordNet: An Experimental, Application-oriented Evaluation of Five Measures. In *the North American Chapter of the Association for Computational Linguistics*, Pittsburgh, PA.
12. J. Jiang and D. Conrath. 1997. Semantic similarity based on corpus statistics and lexical taxonomy. In *Proceedings of International Conference on Research in Computational Linguistics*, Taiwan.
13. H.W. Kuhn. 1955 The Hungarian Method for the assignment problem. *Naval Research Logistic Quarterly* 2, 83-97.
14. I. Melamed. 1996. Automatic construction of clean broad-coverage lexicons. In *Proceedings of the 2nd Conf. AMTA*, pp.125-134, Montreal, CA.

Improving Machine Transliteration Performance by Using Multiple Transliteration Models

Jong-Hoon Oh[1], Key-Sun Choi[2], and Hitoshi Isahara[1]

[1] Computational Linguistics Group, National Institute of Information and Communications Technology (NICT), 3-5 Hikaridai, Seika-cho, Soraku-gun, Kyoto 619-0289 Japan
{rovellia, isahara}@nict.go.jp

[2] Computer Science Division, EECS, KAIST, 373-1 Guseong-dong, Yuseong-gu, Daejeon 305-701 Republic of Korea
kschoi@cs.kaist.ac.kr

Abstract. Machine transliteration has received significant attention as a supporting tool for machine translation and cross-language information retrieval. During the last decade, four kinds of transliteration model have been studied — grapheme-based model, phoneme-based model, hybrid model, and correspondence-based model. These models are classified in terms of the information sources for transliteration or the units to be transliterated — source graphemes, source phonemes, both source graphemes and source phonemes, and the correspondence between source graphemes and phonemes, respectively. Although each transliteration model has shown relatively good performance, one model alone has limitations on handling complex transliteration behaviors. To address the problem, we combined different transliteration models with a "*generating transliterations followed by their validation*" strategy. The strategy makes it possible to consider complex transliteration behaviors using the strengths of each model and to improve transliteration performance by validating transliterations. Our method makes use of web-based and transliteration model-based validation for transliteration validation. Experiments showed that our method outperforms both the individual transliteration models and previous work.

1 Introduction

Machine transliteration has received significant attention as a supporting tool for machine translation (MT) [1,2] and cross-language information retrieval (CLIR) [3,4]. During the last decade, several transliteration models – grapheme[1]-based transliteration model (GTM) [5,6,7,8], phoneme[2]-based transliteration model (PTM) [1,9,10], hybrid transliteration model (HTM) [2,11], and correspondence-based transliteration model (CTM) [12,13,14] – have been proposed. These models

[1] Graphemes refer to the basic units (or the smallest contrastive units) of a written language: for example, English has 26 graphemes or letters.

[2] Phonemes are the simplest significant unit of sound. We used ARPAbet symbols to represent source phonemes (`http://www.cs.cmu.edu/~laura/pages/arpabet.ps`).

Y. Matsumoto et al. (Eds.): ICCPOL 2006, LNAI 4285, pp. 85–96, 2006.

are classified in terms of the information sources for transliteration or the units to be transliterated; GTM, PTM, HTM, and CTM make use of source graphemes, source phonemes, both source graphemes and source phonemes, and the correspondence between source graphemes and phonemes, respectively. Although each transliteration model has shown relatively good performance, it often produced transliterations with errors. The errors are mainly caused by complex transliteration behaviors, meaning that a transliteration process dynamically uses both source graphemes and source phonemes. Sometimes either source graphemes or source phonemes contribute to the transliteration process; while sometimes both contribute. Therefore, it is hard to consider the complex transliteration behaviors depending on one transliteration model because one model just concentrates on only one of the complex transliteration behaviors. To address this problem, we combined the different transliteration models with a "*generating transliterations followed by their validation*" strategy as shown in Fig. 1. First, we generate transliteration candidates (or a list of transliterations) using GTM, PTM, HTM, and CTM. Then, we validate the candidates using two measures — a transliteration model-based measure and a web-based measure.

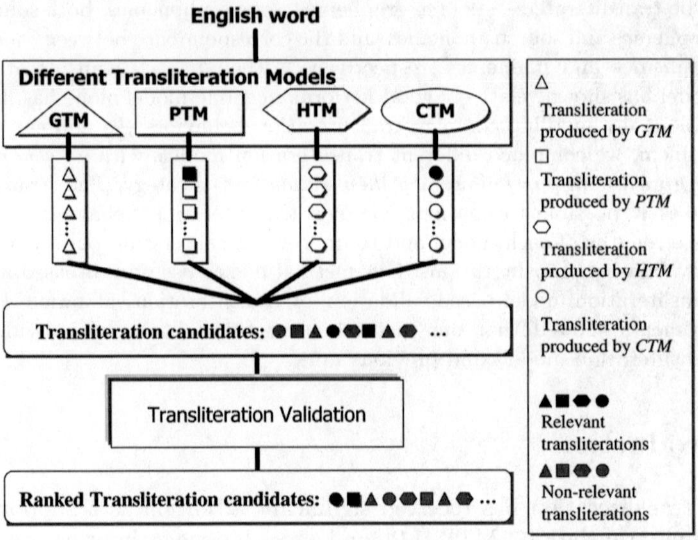

Fig. 1. System architecture

This paper is organized as follows. In section 2, we review previous work based on the four transliteration models. In section 3, we describe the framework of different transliteration models, and in section 4, we describe the transliteration validation. In section 5, we describe our experiments and results. We then conclude in section 6.

2 Previous Work

2.1 Grapheme-Based Transliteration Model

The grapheme-based transliteration model (GTM) is conceptually a direct orthographical mapping model from source graphemes to target graphemes. Several different transliteration methods have been proposed within this framework. Kang & Choi [5] proposed a decision tree-based transliteration method. Decision trees, which transform each source grapheme into target graphemes, are learned and then they are directly applied to machine transliteration. Kang & Kim [6] and Goto *et al.* [7] proposed a method based on a transliteration network. The transliteration network is composed of nodes and arcs. A node represents a chunk of source graphemes and its corresponding target grapheme. An arc represents a possible link between nodes and it has a weight showing its strength. Li *et al.* [8] used a joint source-channel model to simultaneously model both the source language and the target language contexts (bigram and trigram) for machine transliteration. Its main advantage is the use of bilingual contexts.

The main drawback of GTM is that it does not consider any phonetic aspect of transliteration.

2.2 Phoneme-Based Transliteration Model

Basically, the phoneme-based transliteration model (PTM) is composed of *source grapheme-to-source phoneme* transformation and *source phoneme-to-target grapheme* transformation. Knight & Graehl [1] modeled Japanese-to-English transliteration with weighted finite state transducers (WFSTs) by combining several parameters such as romaji-to-phoneme, phoneme-to-English, English word probability models, and so on. Meng *et al.* [10] proposed an English-to-Chinese transliteration model. It was based on English grapheme-to-phoneme conversion, cross-lingual phonological rules and mapping rules between English and Chinese phonemes, and Chinese syllable-based and character-based language models. Jung *et al.* [9] modeled English-to-Korean transliteration with extended Markov window. First, they transformed an English word into English pronunciation by using a pronunciation dictionary. Then they segmented the English phonemes into chunk of English phonemes, which corresponds to one Korean grapheme by using predefined handcrafted rules. Finally they automatically transformed each chunk of English phoneme into Korean graphemes by using extended Markov window.

The main drawback of PTM is error propagation caused by its two-step procedure – errors in *source grapheme-to-source phoneme* transformation make it difficult to generate correct transliterations in the next step.

2.3 Hybrid Transliteration Model and Correspondence-Based Transliteration Model

There have been attempts to use both source graphemes and source phonemes in machine transliteration. Such research falls into two categories, the correspondence-based transliteration model (CTM) [12,13,14] and the hybrid transliteration model (HTM) [2,11]. The CTM makes use of the correspondence between a source grapheme and a source phoneme when it produces target language graphemes; the HTM just combines GTM and PTM through linear interpolation. The hybrid transliteration model requires the grapheme-based transliteration probability ($Pr(GTM)$) and phoneme-based transliteration probability ($Pr(PTM)$), and then it combines the two probabilities through linear interpolation.

Oh & Choi [12] considered the contexts of a source grapheme and its corresponding source phoneme for English-to-Korean transliteration. It is based on semi-automatically constructed context-sensitive rewrite rules in a form, $A/X/B \rightarrow y$, meaning that X is rewritten as target grapheme y in the context A and B. Note that X, A, and B represent correspondence between English grapheme and phoneme like "$r : |R|$" – English grapheme r corresponding to English phoneme $|R|$. Oh & Choi [13,14] trained a generative model representing transliteration rules by using the correspondence between source grapheme and source phoneme, and machine learning algorithms. The correspondence makes it possible to model machine transliteration in a more sophisticated manner.

Several researchers [2,11] have proposed hybrid model-based transliteration methods. They modeled GTM and PTM with WFSTs or a source-channel model. Then they combined GTM and PTM through linear interpolation. In their PTM, several parameters are considered, such as the *source grapheme-to-source phoneme* probability, *source phoneme-to-target grapheme* probability, *target language word* probability, and so on. In their GTM, the *source grapheme-to-target grapheme* probability is mainly considered.

3 Framework of Different Transliteration Models

Let SW be a source word, P_{SW} be the pronunciation of SW, T_{SW} be a target word corresponding to SW, and C_{SW} be a correspondence between SW and P_{SW}. P_{SW} and T_{SW} can be segmented into a series of sub-strings, each of which corresponds to a source grapheme. Then, we can write $SW = s_1, \cdots, s_n = s_1^n$, $P_{SW} = p_1, \cdots, p_n = p_1^n$, $T_{SW} = t_1, \cdots, t_n = t_1^n$, and $C_{SW} = c_1, \cdots, c_n = c_1^n$, where s_i, p_i, t_i, and $c_i = < s_i, p_i >$ represent the ith source grapheme, source phonemes corresponding to s_i, target graphemes corresponding to s_i and p_i, and the correspondence between s_i and p_i, respectively. With this definition, GTM, PTM, CTM, and HTM can be represented as Eqs. (1), (2), (3), and (4), respectively.

$$Pr_g(T_{SW}|SW) = Pr(t_1^n|s_1^n) \approx \prod_i Pr(t_i|t_{i-k}^{i-1}, s_{i-k}^{i+k}) \quad (1)$$

$$\begin{aligned} Pr_p(T_{SW}|SW) &= Pr(p_1^n|s_1^n) \times Pr(t_1^n|p_1^n) \quad (2) \\ &\approx \prod_i Pr(p_i|p_{i-k}^{i-1}, s_{i-k}^{i+k}) \times Pr(t_i|t_{i-k}^{i-1}, p_{i-k}^{i+k}) \end{aligned}$$

$$\begin{aligned} Pr_c(T_{SW}|SW) &= Pr(p_1^n|s_1^n) \times Pr(t_1^n|c_1^n) \quad (3) \\ &\approx \prod_i Pr(p_i|p_{i-k}^{i-1}, s_{i-k}^{i+k}) \times Pr(t_i|t_{i-k}^{i-1}, c_{i-k}^{i+k}) \end{aligned}$$

$$Pr_h(T_{SW}|SW) = \alpha \times Pr_p(T_{SW}|SW) + (1-\alpha) \times Pr_g(T_{SW}|SW) \quad (4)$$

With the assumption that each transliteration model depends on the size of the contexts, k, Eqs. (1), (2), (3) and (4) can be simplified. To estimate the probabilities in Eqs. (1), (2), (3), and (4), we used the maximum entropy model, which can effectively incorporate heterogeneous information [15]. In the maximum entropy model, event ev is composed of a target event (te) and a history event (he), and it is represented by a bundle of feature functions ($f_i(he, te)$), which represent the existence of certain characteristics in the event ev. The feature function enables a model based on the maximum entropy model to estimate probability [15]. Therefore, designing the feature functions, which effectively support certain decisions made by the model, is important. Our basic philosophy for the feature function design for each transliteration model is that the context information collocated with the unit of interest is important. With this philosophy, we designed the feature functions with all possible combinations of (s_{i-k}^{i+k}, p_{i-k}^{i+k}, c_{i-k}^{i+k}, and t_{i-k}^{i-1}). Generally, a conditional maximum entropy model is an exponential log-linear model that gives the conditional probability of event $ev =< te, he >$, as described in Eq. (5), where λ_i is a parameter to be estimated, and $Z(he)$ is the normalizing factor [15].

$$Pr(te|he) = \frac{1}{Z(he)} exp(\sum_i \lambda_i f_i(he, te)) \quad (5)$$

$$Z(he) = \sum_{te} exp(\sum_i \lambda_i f_i(he, te))$$

With Eq. (5) and feature functions, conditional probabilities can be estimated in Eqs. (1), (2), (3), and (4). For example, we can write $Pr(t_i|t_{i-k}^{i-1}, c_{i-k}^{i+k}) = Pr(te_{CTM}|he_{CTM})$ because we can represent target events (te_{CTM}) and history events (he_{CTM}) of CTM as t_i and tuples $< t_{i-k}^{i-1}, c_{i-k}^{i+k} >$, respectively. In the same way, $Pr(t_i|t_{i-k}^{i-1}, s_{i-k}^{i+k})$, $Pr(t_i|t_{i-k}^{i-1}, p_{i-k}^{i+k})$, and $Pr(p_i|p_{i-k}^{i-1}, s_{i-k}^{i+k})$ can be represented as $Pr(te|he)$ with their target events and history events. We used a maximum entropy modeling tool [16] to estimate Eqs. (1), (2), (3), and (4).

4 Transliteration Validation

We validated transliterations by using web-based validation, $S_{web}(s, tc_i)$, and transliteration model-based validation, $S_{tm}(s, tc_i)$, like in Eq. (6). Using Eq. (6), we can validate transliterations in a more correct and robust manner because $S_{web}(s, tc_i)$ reflects real-world usage of the transliterations in web data and $S_{tm}(s, tc_i)$ ranks the transliterations independent of the web data.

$$S_{TV}(s, tc_i) = S_{tm}(s, tc_i) \times S_{web}(s, tc_i) \quad (6)$$

4.1 Transliteration Model-Based Validation: S_{tm}

Our transliteration model-based validation, S_{tm}, uses the rank assigned by each transliteration model. For a given source word (s), each transliteration model generates transliterations (tc_i in TC) and ranks them using the probability described in Eqs. (1), (2), (3), and (4). The underlying assumption in S_{tm} is that the rank of the correct transliterations tends to be higher, on average, than the wrong ones. With this assumption, we represented $S_{tm}(s, tc_i)$ as Eq. (7), where $Rank_g(tc_i)$, $Rank_p(tc_i)$, $Rank_h(tc_i)$, and $Rank_c(tc_i)$ represent the rank of tc_i assigned by GTM, PTM, HTM, and CTM, respectively.

$$S_{tm}(s, tc_i) = \frac{1}{4} \times \left(\frac{1}{Rank_g(tc_i)} + \frac{1}{Rank_p(tc_i)} + \frac{1}{Rank_h(tc_i)} + \frac{1}{Rank_c(tc_i)}\right) \quad (7)$$

4.2 Web-Based Validation: S_{web}

Korean or Japanese web pages are usually composed of rich texts in a mixture of Korean or Japanese (main language) and English (auxiliary language). Let s and t be a source language word and a target language word, respectively. We observed that s and t tend to be near each other in the text of Korean or Japanese web pages when the authors of the web pages describe s as translation of t, or vice versa. We retrieved such web pages for transliteration validation.

There have been several web-based validation methods for translation validation [17,18] or transliteration validation [2,19]. They usually rely on the web frequency (the number of web pages) derived from "BILINGUAL KEYWORD SEARCH (BKS)" [2,17,18] or "MONOLINGUAL KEYWORD SEARCH (MKS)" [2,19]. BKS retrieves web pages by using a query composed of two keywords, s and t; while MKS retrieves web pages by using a query composed of t. Qu & Grefenstette [17] and Wang *et al.* [18] proposed BKS-based translation validation methods, such as relative web frequency and chi-square (χ^2) test. Al-Onaizan & Knight [2] used both MKS and BKS and Grefenstette *et al.* [19] used only MKS for validating transliterations. However, web pages retrieved by MKS tend to show whether t is used in target language texts rather than whether t is a translation of s. BKS frequently retrieves web pages where s and t have little relation to each other because it does not consider distance between s and t in the web pages. To address these problems, we developed a validation method based on "BILINGUAL PHRASAL SEARCH (BPS)", where a phrase composed of s and t is used as a query for a search engine. Let '[s t]' or '[t s]', 's AND t', and 't', respectively, be queries for BPS, BKS, and MKS. The difference among BPS, BKS, and MKS is shown in Fig. 2. In Fig. 2, '[s t]' or '[t s]' retrieves web pages where '[s t]' or '[t s]' exists as phrases; while 's AND t' retrieves web pages where s and t simply exist in the same document. Therefore, the number of web pages retrieved by BPS is more reliable for validating transliterations, because s and t usually have high co-relation in the web pages retrieved by BPS. For example, web pages retrieved by BPS in Fig. 3 usually contain correct Korean and Japanese transliterations and their corresponding English word *amylase* as translation pairs in parentheses expression. For these reasons, BPS is more suitable for our transliteration validation.

Let TC be a set of transliterations (or transliteration candidates) produced by different transliteration models, tc_i be the ith transliteration candidate in TC, s be the source language word resulting in TC, and $WF(s, tc_i)$ be the web

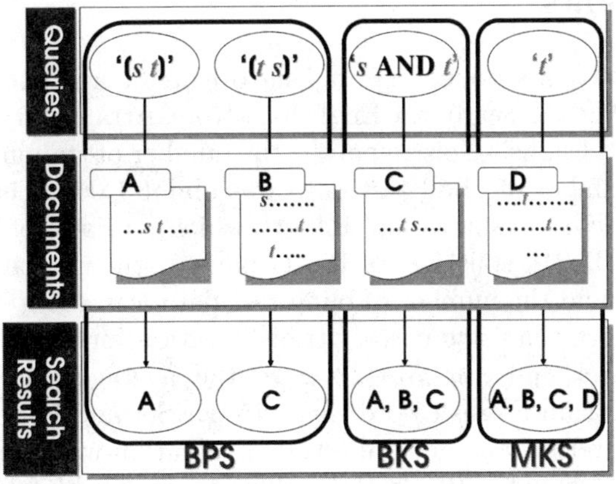

Fig. 2. Difference among BPS, BKS, and MKS

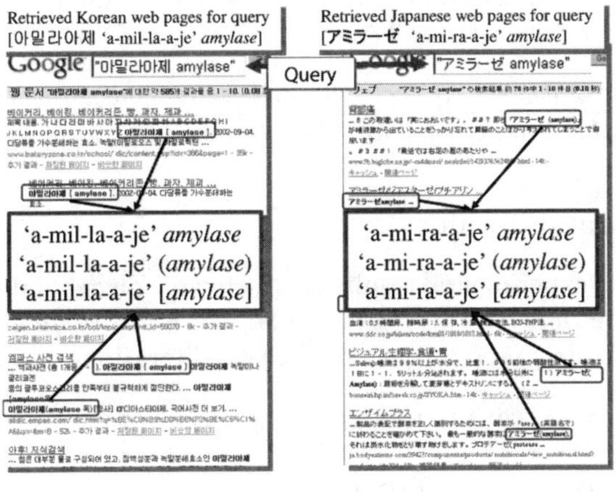

Fig. 3. Web pages retrieved by BPS

frequency for $[s\ tc_i]$. Our web-based validation method, S_{web}, can be represented as Eq. (8), which is the relative web frequency derived from BPS.

$$S_{web}(s,tc_i) = \frac{WF(s,tc_i) + WF(tc_i,s)}{\sum_{tc_k \in TC}(WF(s,tc_k) + WF(tc_k,s))} \tag{8}$$

Let TC for $s = data$ be $\{tc_1$=데이터, tc_2=데이타, tc_3=데타$\}$ and $WF(s,tc_1) + WF(tc_1,s)$, $WF(s,tc_2) + WF(tc_2,s)$, and $WF(s,tc_3) + WF(tc_3,s)$ be 94,100, 67,800, and 54, respectively. Then, S_{web} for each tc_i can be calculated as follows.

- $S_{web}(s,tc_1) = 94{,}100/161{,}954 = 0.5811$
- $S_{web}(s,tc_2) = 67{,}800/161{,}954 = 0.4186$
- $S_{web}(s,tc_3) = 54/161{,}954 = 0.0003$.

5 Experiments

Our experiments were done for English-to-Korean and English-to-Japanese transliteration. The test set for the English-to-Korean transliteration (EKSet) [20] consisted of 7,172 English-Korean pairs – the number of training data was about 6,000 and the number of blind test data was about 1,000. The test set for the English-to-Japanese transliteration (EJSet), which consisted of English-katakana pairs from EDICT [21], consisted of 10,417 pairs — the number of training data was about 9,000 and the number of blind test data was about 1,000. EJSet contained one or more than one correct transliteration for one English word, like <*micro*, マイクロ>, and <*micro*, ミクロ>; the average number of Japanese transliterations for an English word was 1.15. EKSet and EJSet covered proper names, technical terms, and general terms. Evaluation was done in terms of the word accuracy (WA) in Eq. (9). In the evaluation, we used k-fold cross-validation ($k = 7$ for the EKSet and $k = 10$ for the EJSet). The test set was divided into k subsets. Each one was used for testing, while the remainder was used for training. Then the average WA across all the k trials was computed. Through the cross-validation, we set α (0.4 for the EKSet and 0.5 for the EJSet) for HTM in Eq. (4).

$$WA = \frac{the\ number\ of\ correct\ transliterations\ output\ by\ the\ system}{the\ number\ of\ transliterations\ in\ the\ blind\ test\ data} \quad (9)$$

5.1 Experimental Results

Summaries of our experimental results conducted on EKSet and EJSet are shown in Table 1. In the table, GTM, PTM, HTM, and CTM represent the individual transliteration models used for generating transliterations. S_{tm}, S_{web}, and S_{TV}

Table 1. Summary of Results (%)

Methods		EKSet				EJSet			
		Top-1	Top-3	Top-5	Top-10	Top-1	Top-3	Top-5	Top-10
GTM		56.8	76.9	82.5	88.0	51.6	76.6	84.6	94.1
PTM		49.6	68.3	75.1	82.5	47.6	72.8	81.6	92.6
HTM		60.6	78.5	83.5	88.6	55.7	77.6	85.1	94.0
CTM		60.8	79.6	84.7	89.6	58.2	81.1	87.9	96.5
GPC [6]		55.1	N/A	N/A	N/A	53.2	N/A	N/A	N/A
GMEM [7]		55.9	N/A	N/A	N/A	56.2	N/A	N/A	N/A
HWFST [11]		58.3	N/A	N/A	N/A	62.5	N/A	N/A	N/A
S_{tm}		71.0	81.2	85.4	89.1	66.9	81.3	87.5	93.2
S_{web}	MKS	50.5	75.9	84.3	91.6	49.9	75.1	83.0	93.0
	BKS	74.4	88.8	91.2	92.1	76.1	94.2	96.0	96.8
	BPS	83.6	91.5	91.9	92.1	81.3	95.6	96.9	97.8
S_{TV}	MKS	59.5	81.3	87.8	92.1	56.7	77.3	83.5	92.8
	BKS	79.5	90.5	91.7	92.1	79.6	94.7	96.2	96.8
	BPS	**84.4**	**91.7**	**92.0**	**92.1**	**82.0**	**95.6**	**96.9**	**98.1**

represent experimental results validated by Eqs. (7), (8), and (6), respectively. Moreover, we tested S_{web} and S_{TV} according to web search methods (BPS, BKS, and MKS) to show the effect of BPS on transliteration validation. We compared our proposed method with the previous work, GPC [6], GMEM [7], and HWFST [11][3]. Note that only Top-1 was considered in the previous work because they, except for HWFST, focused only on the Top-1. The Top-n considers whether the correct transliteration is in the Top-n ranked transliterations[4].

Compared to individual transliteration models and previous work [6,7,11], S_{tm}, S_{web}(BPS), and S_{TV}(BPS) are more effective, especially in the Top-1[5]. Although S_{tm} by itself showed higher performance than individual transliteration models and previous work [6,7,11], S_{TV}(BPS) (the combination of S_{tm} and S_{web}(BPS)) shows much better performance. The Top-1 of S_{TV}(BPS) has the best performance[6]. The powerful transliteration validation ability of S_{TV}(BPS) enables our method to achieve the best result in the Top-1. More specifically, S_{web}(BPS) contributes highly to the performance improvement. This indicates that the web data used as the knowledge source for transliteration validation is very useful. Although S_{tm} makes a small contribution to the performance improvement of S_{TV}(BPS) because S_{web}(BPS) correctly validates transliterations whenever S_{tm} does, the errors of S_{web}(BKS) and S_{web}(MKS) are well compensated for by S_{tm} in S_{TV}(BKS) and S_{TV}(MKS). For example, Korean and Japanese transliterations for the English words *methoxyl* and *netware* were validated by each validation method, as shown in Tables 2 and 3 Note that the value coupled with each transliteration was assigned by each validation method. In Tables 2 and 3 S_{tm} causes the rank of correct transliterations to be higher in S_{TV}(BKS) and S_{TV}(MKS) than in S_{web}(BKS) and S_{web}(MKS).

When comparing BPS with BKS and MKS, BPS is the most effective web search method for transliteration validation. S_{web} based on MKS has the worst performance because it tends to validate whether tc_i is used in a target language rather than whether it is used as a translation of s. Actually, S_{web} based on BKS is effective because BKS considers both s and tc_i while it retrieves web pages. However, the more powerful retrieval ability of BPS protects S_{web} based on BPS from errors that S_{web} based on BKS causes as shown in Tables 2 and 3. So higher performance can be had with S_{web}(BPS) than with S_{web}(BKS) – about 12% improvement in Top-1 of EKSet and about 7% improvement in Top-1 of EJSet[7]. The

[3] We implemented the three previous methods [6,7,11] and then trained and tested them using the same data as our proposed method.

[4] For one English word, there are one or more than one correct transliterations in EJSet but there is only one correct transliteration in EKSet. Therefore, we had higher TOP-1 accuracies but lower TOP-10 accuracies in EKSet than in EJSet.

[5] A one-tail paired t-test showed that the results of S_{tm}, S_{web}(BPS), and S_{TV}(BPS) were always significantly better than those of individual transliteration models and previous work (level of significance = 0.001).

[6] A one-tail paired t-test showed that the results of S_{TV}(BPS) were always significantly better than those of the others (level of significance = 0.001).

[7] A one-tail paired t-test showed that the results of S_{web}(BPS) were always significantly better than those of S_{web}(BKS) (level of significance = 0.001).

Table 2. Korean transliterations for English word *methoxyl* and their validation (The underlined 메톡실 is the correct Korean transliteration)

S_{tm}	**메톡실 (0.772)**, 메타일 (0.081), 메토실 (0.253), ...
S_{web}(MKS)	메토실 (0.482), 메타실 (0.334), 메서일 (0.148), **메톡실 (0.029)**,...
S_{web}(BKS)	메타일 (0.564), **메톡실 (0.413)**, 메토실 (0.011),...
S_{web}(BPS)	**메톡실 (0.947)**, 메토실 (0.053),...
S_{TV}(MKS)	메토실 (0.122), **메톡실 (0.022)**, 메타일 (0.0003),...
S_{TV}(BKS)	**메톡실 (0.319)**, 메타일 (0.045), 메토실 (0.003),...
S_{TV}(BPS)	**메톡실 (0.742)**, 메토실 (0.013),...

Table 3. Japanese transliterations for English word *netware* and their validation (The underlined ネットウェア is the correct Japanese transliteration)

S_{tm}	**ネットウェア (0.875)**, ネットエア (0.458), ネットウエア (0.319), ...
S_{web}(MKS)	ネットワー (0.988), ネットエア (0.009), **ネットウェア (0.002)**,...
S_{web}(BKS)	ネットワー (0.626), **ネットウェア (0.274)**, ネットエア (0.100),...
S_{web}(BPS)	**ネットウェア (0.860)**, ネットワー (0.079), ネットウエア (0.061),...
S_{TV}(MKS)	ネットワー (0.189), ネットウエア (0.004), **ネットウェア (0.002)**, ...
S_{TV}(BKS)	**ネットウェア (0.240)**, ネットワー (0.120), ネットウエア (0.032),...
S_{TV}(BPS)	**ネットウェア (0.752)**, ネットウエア (0.019), ネットワー (0.015),...

higher performance of S_{web}(BPS) positively effects S_{TV}(BPS), thus the performance of S_{TV}(BPS) is higher than that of S_{TV}(BKS). Both S_{web} and S_{TV} based on BPS outperform those based on BKS or MKS.

The experimental results can be summarized as follows.

- S_{tm}, S_{web}(BPS), and S_{TV}(BPS) are more effective than individual transliteration models and previous work [6,7,11], especially in the Top-1.
- S_{TV}(BPS) shows the best performance.
- S_{web}(BPS) mainly contributes the high performance of S_{TV}(BPS).
- BPS is the most effective web search method for transliteration validation among BPS, BKS, and MKS.

6 Conclusion

We proposed a novel approach for improving machine transliteration performance by combining multiple transliteration models. We applied a "*generating transliterations followed by their validation*" strategy. We generated transliteration candidates using four different transliteration models and validated them using web-based validation and transliteration model-based validation. Experiments showed that combining multiple transliteration models was one way for considering complex transliteration behaviors and that transliteration validation was very important for improving machine transliteration performance. Our two transliteration validation methods were effective. The web-based validation method effectively filtered out wrong transliterations by using web data, which reflects real-world usage of transliterations, and the transliteration model-based validation method as a web-independent validation measure complemented the web-based validation method. Moreover, we showed that a web search method significantly affects the performance of the web-based validation method. Experiments showed that our "BILINGUAL PHRASAL SEARCH (BPS)" is more suitable than "BILINGUAL KEYWORD SEARCH (BKS)" and "MONOLINGUAL KEYWORD SEARCH (MKS)" in transliteration validation.

References

1. Knight, K., Graehl, J.: Machine transliteration. In: Proc. of the 35th Annual Meetings of the Association for Computational Linguistics. (1997) pp.128–135
2. Al-Onaizan, Y., Knight, K.: Translating named entities using monolingual and bilingual resources. In: Proc. of ACL 2002. (2002) 400–408
3. Fujii, A., Tetsuya, I.: Japanese/English cross-language information retrieval: Exploration of query translation and transliteration. Computers and the Humanities **35** (2001) 389–420
4. Lin, W.H., Chen, H.H.: Backward machine transliteration by learning phonetic similarity. In: Proc. of the Sixth Conference on Natural Language Learning (CoNLL). (2002) 139–145
5. Kang, B.J., Choi, K.S.: Automatic transliteration and back-transliteration by decision tree learning. In: Proc. of the 2nd International Conference on Language Resources and Evaluation. (2000) 1135–1411
6. Kang, I.H., Kim, G.C.: English-to-Korean transliteration using multiple unbounded overlapping phoneme chunks. In: Proc. of the 18th International Conference on Computational Linguistics. (2000) 418–424
7. Goto, I., Kato, N., Uratani, N., Ehara, T.: Transliteration considering context information based on the maximum entropy method. In: Proc. of MT-Summit IX. (2003) 125–132
8. Li, H., Zhang, M., Su, J.: A joint source-channel model for machine transliteration. In: Proc. of ACL 2004. (2004) 160–167
9. Jung, S.Y., Hong, S., Paek, E.: An English to Korean transliteration model of extended markov window. In: Proc. of the 18th conference on Computational linguistics. (2000) 383–389
10. Meng, H., Lo, W.K., Chen, B., Tang, K.: Generating phonetic cognates to handle named entities in English-Chinese cross-language spoken document retrieval. In: Proc. of Automatic Speech Recognition and Understanding, 2001. ASRU '01. (2001) 311–314
11. Bilac, S., Tanaka, H.: Improving back-transliteration by combining information sources. In: Proc. of IJCNLP2004. (2004) 542–547
12. Oh, J.H., Choi, K.S.: An English-Korean transliteration model using pronunciation and contextual rules. In: Proc. of COLING2002. (2002) 758–764
13. Oh, J.H., Choi, K.S.: An ensemble of grapheme and phoneme for machine transliteration. In: Proc. of IJCNLP05. (2005) 450–461
14. Oh, J.H., Choi, K.S.: Machine learning based English-to-Korean transliteration using grapheme and phoneme information. IEICE Transaction on Information & Systems **E88-D** (2005) 1737–1748

15. Berger, A.L., Pietra, S.D., Pietra, V.J.D.: A maximum entropy approach to natural language processing. Computational Linguistics **22** (1996) 39–71
16. Zhang, L.: Maximum entropy modeling toolkit for python and C++. `http://homepages.inf.ed.ac.uk/s0450736/software/maxent/manual.pdf` (2004)
17. Qu, Y., Grefenstette, G.: Finding ideographic representations of Japanese names written in Latin script via language identification and corpus validation. In: ACL. (2004) 183–190
18. Wang, J.H., Teng, J.W., Lu, W.H., Chien, L.F.: Exploiting the web as the multilingual corpus for unknown query translation. Journal of the American Society for Information Science and Technology **57** (2006) 660–670
19. Grefenstette, G., Qu, Y., Evans, D.A.: Mining the web to create a language model for mapping between English names and phrases and Japanese. In: Proc. of Web Intelligence. (2004) 110–116
20. Nam, Y.S.: Foreign dictionary. Sung An Dang (1997)
21. Breen, J.: EDICT Japanese/English dictionary .le. The Electronic Dictionary Research and Development Group, Monash University. `http://www.csse.monash.edu.au/~jwb/edict.html` (2003)

Clique Percolation Method for Finding Naturally Cohesive and Overlapping Document Clusters

Wei Gao, Kam-Fai Wong, Yunqing Xia, and Ruifeng Xu

Department of Systems Engineering and Engineering Management, The Chinese University of Hong Kong, Shatin, N.T., Hong Kong, China
{wgao, kfwong, yqxia, rfxu}@se.cuhk.edu.hk

Abstract. Techniques for find document clusters mostly depend on models that impose strong explicit and/or implicit priori assumptions. As a consequence, the clustering effects tend to be unnatural and stray away from the intrinsic grouping natures of a document collection. We apply a novel graph-theoretic technique called *Clique Percolation Method* (CPM) for document clustering. In this method, a process of enumerating highly cohesive maximal document cliques is performed in a random graph, where those strongly adjacent cliques are mingled to form naturally overlapping clusters. Our clustering results can unveil the inherent structural connections of the underlying data. Experiments show that CPM can outperform some typical algorithms on benchmark data sets, and shed light on its advantages on natural document clustering.

1 Introduction

Clustering is an important technique that facilitates the navigation, search and analysis of information in large unstructured document collections. It is an unsupervised process to identify inherent groupings of similar documents, where documents exhibit high intra-cluster similarity and low inter-cluster similarity.

Many existing clustering algorithms optimize criterion functions with respect to the employed similarity measures over all the documents assigned to each possible partition of the collection [10,19]. They always impose some explicit and/or implicit constraints as to the number, size, shape or disjoint characteristics of target clusters. For example, partitional algorithms like k-means assume cluster number k and do not allow one document belonging to multiple groups. Although fuzzy clustering, such as fuzzy C-means algorithm [2,12], does support overlapping clusters by a membership function and a fuzzifier parameter, they are still confined by cluster number and can find only spherical shape clusters. Some algorithms are model-based, e.g., Naive Bayes or Gaussian Mixture model [1,13]. They assume certain probabilistic distributions of the documents and try to find a model maximizing the likelihood of data. When data cannot fit the presumed distribution, poor cluster quality can result. k-way clustering or bisection algorithms [19] force clusters to be equally sized. Spectral clustering [7,8] has emerged as one of the most effective clustering tools based on max-flow/min-cut theorem [4]. However, they prohibit overlapping clusters.

Y. Matsumoto et al. (Eds.): ICCPOL 2006, LNAI 4285, pp. 97–108, 2006.

We define natural document clustering as a problem of finding unknown number of overlapping as well as cohesive document groups with varied sizes and arbitrary distributions of the data. We try to obtain the clustering results with these free characteristics by reducing as many external constraints as feasible and leaving things to the inherent grouping nature among documents. For this purpose, we propose a document clustering technique using a novel graph-theoretic algorithm, named Clique Percolation Method (CPM). The idea is to identify adjacent maximal complete subgraphs, which is referred to as Maximal Document Cliques (MDC), in the document similarity graph using a threshold clique, and then mingle those strongly adjacent MDCs to form naturally overlapping document clusters. Although it does introduce an explicit parameter k, which is the size of the threshold clique, our algorithm can automatically settle the critical point, at which the natural clustering can be achieved. We show that CPM outperforms representative clustering methods with experiments on the benchmark data.

The rest of this paper is organized as follows: Section 2 describes the proposed CPM; Section 3 presents the algorithmic implementation of this technique; Section 4 gives related work; Section 5 presents experimental evaluation results; Finally, we conclude this paper.

2 Document Clustering by Clique Percolation Method

2.1 Preliminaries

In general, suppose $V = \{d_1, d_2, \ldots, d_{|V|}\}$ is a collection of documents. We represent the collection by an undirect graph $G = (V, E)$, where V is the vertex set and E is the edge set such that each edge $\{i, j\}$ is a set of two adjacent vertices d_i, d_j in V. The adjacent matrix M of the graph is defined by $M = [m_{ij}]_{i,j=1}^{|V|}$, where each entry w_{ij} is the edge weight which is the value of similarity metric (in what follows we use Cosine coefficient) between d_i and d_j. The graph can also be unweighted where an edge exists indicating the distance of its two vertices smaller than some threshold, in which case w_{ij} is binary.

A *clique* in G is a subset $S \subseteq V$ of vertices, such that $\{i, j\} \in E$ for all distinct $\{d_i, d_j\} \in S$. Thus any two vertices are adjacent in a clique that constitutes a complete subgraph of G. A clique is said to be *maximal* if its vertices are not a subset of the vertices of a larger clique, which is referred to as a Maximal Document Clique (MDC) in a document similarity graph. MDC is considered the strictest definition of a cluster [15]. In graph theory, enumerating all maximal cliques (equivalently, all maximal independent sets or all minimal vertex covers) is believed NP-hard [3,18].

Suppose $|V|$ number of documents are given in a measure space with a similarity metric w_{ij}. We define a binary relation $\sim_t$ between documents on $G = \{V, E\}$ with respect to parameter t: $i \sim_t j := w_{ij} \leq t$, which is self-reflexive, symmetric and non-transitive. There is an edge $\{i, j\} \in E$ connecting vertices d_i and d_j whenever $i \sim_t j$ with respect to threshold t. Figure 1 illustrates that given a

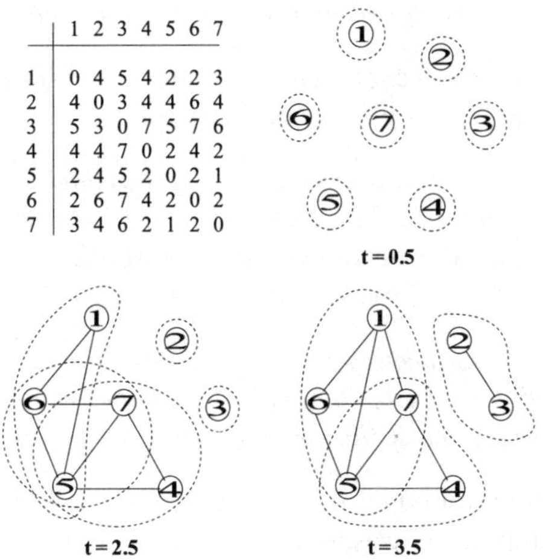

	1	2	3	4	5	6	7
1	0	4	5	4	2	2	3
2	4	0	3	4	4	6	4
3	5	3	0	7	5	7	6
4	4	4	7	0	2	4	2
5	2	4	5	2	0	2	1
6	2	6	7	4	2	0	2
7	3	4	6	2	1	2	0

Fig. 1. Graphs with respect to threshold level t and different cohesive MDC clusters (in dotted regions) resulted from it

matrix reflecting the distances between 7 documents and the t value, a series of graphs for the relation $i \sim_t j$ are produced with different connectivity densities. Clearly, if each MDC is considered as a cohesive form of cluster, we can discover different number of clusters from these graphs, where $t = 0.5$, 2.5 and 3.5 results in 7, 5 and 3 number of clusters, respectively. They display interesting properties of natural clusters except for excessive intra-cluster cohesiveness.

The series of graphs parameterized by t above can be seen as random graphs with constant set of vertices and a changing set of edges generated with probability p, the probability two vertices can be connected by an edge. Intuitively, tuning the value of t is somehow equivalent to adding or removing some edges according to p in monotonic manner. In order for an appropriate t, we first determine p_c, the critical value of p, and then derive t from p_c by making use of their interdependency relationship. The critical value p_c is defined as the probability, under which a giant k-clique percolation cluster will emerge in the graph, and is known as the percolation threshold for a random network [6]. At this threshold, the percolation transition takes place (see Section 2.2). For clustering, the assumption behind is that no cluster can be excessively larger than others, which can be achieved by commanding $p < p_c$.

2.2 Clique Percolation Method in Random Graphs

Concepts of k-Clique Percolation. The concept of k-clique percolation is fundamental for Clique Percolation Method (CPM) in random networks, which was studied in [6]. The successful applications of CPM for uncovering community structure of co-authorship networks, protein networks and word association graphs can be found in [14]. Hereby we briefly present some related notions.

Definition 1. *k-clique is defined as a complete subgraph of k vertices.*

Definition 2. *k-clique adjacency: Two k-cliques are adjacent if they share $k-1$ vertices, i.e., if they differ only in a single vertex.*

Definition 3. *k-clique percolation cluster is a maximal k-clique-connected subgraph, i.e., it is the union of all k-cliques that are k-clique adjacent. Obviously, a k-clique percolation cluster is unnecessarily a MDC, but it must be equivalent to the union of all MDCs adjacent by at least $k-1$ vertices.*

Definition 4. *k-clique adjacency graph is the compressed form of the original graph, where the vertices denote the k-cliques of the original graph and there is an edge between two vertices if the corresponding k-cliques are adjacent.*

Moving a particle along an edge on a k-clique adjacency graph is equivalent to rolling a k-clique template (threshold clique) from one k-clique on the original graph to an adjacent one. A k-clique template can be placed onto any k-clique of the original graph, and rolled to an adjacent k-clique by relocating one of its vertices and keeping other $k-1$ vertices fixed. Thus, the k-clique percolation clusters are all those subgraphs that can be fully explored by rolling a k-clique template in them [6]. Note that a k-clique percolation cluster consists of all MDCs adjacent by at least $k-1$ vertices. Thus, the cohesiveness of documents in a k-clique percolation cluster as well as the overlap degree between clusters can be tuned by the k value. The goal of CPM is to find all k-clique percolation clusters.

Percolation Threshold p_c. How to estimate the threshold probability p_c of k-clique percolation with respect to k ($k \geq 2$)? The clique percolation theory emphasizes that under such p_c (critical point), a giant k-clique percolation cluster that is excessively larger than other clusters will take place [9,6]. Intuitively, the greater the p ($p > p_c$) is, the more likely the giant cluster appears, and the larger its size (which includes most of graph nodes), as if using a k-clique can percolate the entire graph.

Consider the heuristic condition of template rolling at the percolation threshold: after rolling a k-clique template from a k-clique to an adjacent one by relocating one of its vertices, the expectation of the number of adjacent k-cliques, where the template can roll further by relocating another of its vertices, be equal to 1. The intuition behind is that a larger expectation value would allow an infinite series of bifurcations for the rolling, ensuring that a giant cluster is present in the graph. The expectation value can be estimated as $(k-1)(|V|-k)p_c^{k-1} = 1$, where $(k-1)$ is the number of template vertices that can be selected for the next relocation, $(|V|-k)$ is the number of potential destinations for this relocation, out of which only the fraction p^{k-1} is acceptable, because each of the new $k-1$ edges must exist in order to reach a new k-clique after relocation. Therefore, the percolation threshold function $p_c(k)$ with respect to k and $|V|$ is as follows:

$$p_c(k) = [(k-1)(|V|-k)]^{-\frac{1}{k-1}} \tag{1}$$

Generation of Random Graph. According to Eq. (1), we can obtain a series of critical values with regard to the threshold clique sizes provided, which are actually the threshold probabilities of connecting two document vertices by an edge at these critical points. How to generate a random graph with the exactly desirable connectivity is technically very challenging since the degree distribution of each vertex needs to be appropriately modeled. Some work on systematically modeling degree distribution has been done in the field of random networks [9]. In this study, we prefer to simplify our specific problem by using two heuristics.

First, for each vertex, we consider its N-Nearest Neighbors (NNB) instead of using a fixed similarity threshold value, where N is determined by the formula:

$$N(k) = p_c(k) \times \frac{|V|-1}{k-1} \tag{2}$$

where the factor $\frac{|V|-1}{k-1}$ actually scale the size of the original graph down to the level of k-cliques in the graph and $p_c(k)$ is considered as the average proportion of k-cliques are the nearest neighbors of a given clique in the k-clique adjacency graph. Here we actually use the connectivity of the k-clique adjacency graph to simulate the original graph. The reason we don't use $N(k) = p_c(k) \times (|V|-1)$ is because the generated graph tends to be over dense since $p_c(k)$ is not the proportion of NNB nodes, but in fact the probability of two vertices being connected by an edge.

Secondly, we examine the co-relation between p and the similarity threshold t. Given p_c, we can estimate the bound(s) of t_c so that the graph with the approximated connectivity as that under p_c could be generated. Because p-t are monotone, a graph could be produced with edge weights t greater than t_c. We derive t_c by a simple approximation:

$$t_c(k) = 0.5 + p_c(k) \times (w_{max} - w_{min}) \tag{3}$$

where w_{max} and w_{min} are the maximum and minimum values of document similarity in the collection, respectively. Intuitively, we deem that only edge weights somewhat larger than 0.5 are considered similar. We also observe $p_c(k)$ is well below 0.5 for the normal size of corpus as k is not too large (< 20), which can be shown in Fig. 2 and guarantees $t_c \leq 1$.

We then apply the two heuristics incrementally during the graph generation process, i.e. by generating connections between NNBs for each vertex at first place and then prune the edges with weights less then t_c. The intuition is that denser graphs are penalized more heavily by the combination of the two heuristics.

3 Algorithmic Implementation of CPM

The clustering process is turned out to be a problem of finding all MDCs and then merging those with at least $k-1$ common nodes into clusters. The proposed CPM clustering algorithm includes 5 major steps:

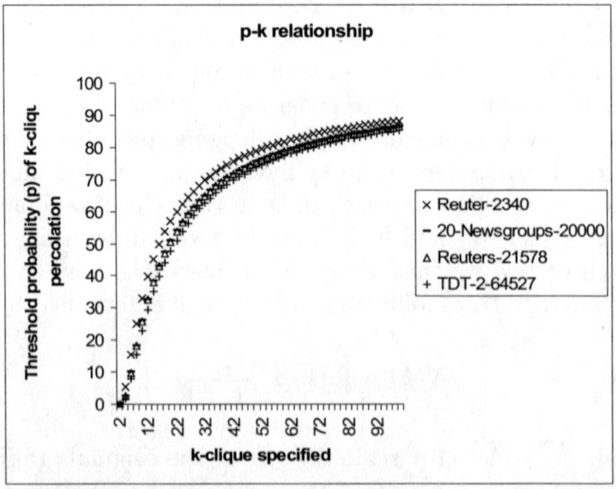

Fig. 2. The curve of p-k relationship given typical sizes of benchmark corpora indicates p_c is below 0.5 when $k < 20$

1. Preprocessing: Eliminate words in the stop list, use Porter's stemmer as the stemming algorithm, build document vectors, and create a $|V| \times |V|$ document similarity matrix $\mathbf{A}$;
2. Given k as parameter, compute Eq. (1) for $p_c(k)$, Eq. (2) for $N(k)$ and Eq. (3) for $t_c(k)$;
3. Create document similarity graph G from $\mathbf{A}$, where each vertex is connected with $N(k)$ NNBs. For each edge, prune it if its weight $w_{ij} < t_c(k)$;
4. Enumerate all MDCs in G using Algorithm 1;
5. Create a $M \times M$ adjacent matrix $\mathbf{B}$ (where M is the number of MDCs), find k-clique percolation clusters using Algorithm 2 on $\mathbf{B}$.

Enumerating Maximal Document Cliques. Algorithms for finding maximal cliques (step 4) were studied in [3] and achieved processing time bounded by $O(v^2)$ where v is the number of maximal cliques. Their algorithms are distinctive because they can be applied to a graph of comparatively large size. We implement an efficient counterpart of the algorithm using back-tracking method (see Algorithm 1). A MDC is output at each end of back-track. The running time is $O(v)$.

Finding *k*-Clique Percolation Clusters. When all the MDCs are enumerated, a clique-clique adjacent matrix is prepared. It is symmetric where each row and column represents a MDC and the entries are the number of common vertices between two cliques (the diagonal values are the sizes of MDCs). The k-clique percolation clusters are one-to-one correspondent to the connected components in the clique-clique adjacency graph represented by the matrix, which can be obtained using Algorithm 2 (step 5 above). The algorithm first

Algorithm 1. Enumerate All MDCs

input: Vertex set V and edge set E of graph G.
output: All MDCs of G into C.

```
procedure EnumMDC (C, U, E)
  if (U = φ) then
    output C
    return
  end if
  for every vertex u ∈ U do
    U := U − {u}
    EnumMC (C ∪ {u}, U ∩ {v|(v, u) ∈ E})
  end for
end procedure

C := φ
EnumMDC (C, V, E)
```

creates a clique-clique adjacent matrix **B**, in which every off-diagonal entry smaller than $k-1$ and every diagonal element smaller than k are erased (line 2–12), and then carrying out a depth-first-search (DFS) to find all the connected components.

4 Related Work

Traditional hierarchical agglomerative clustering (HAC) are intrinsically graph-based like CPM. HAC treats each data point as a singleton cluster and then successively merges pairs of clusters until all clusters have been merged into a single cluster that contains all documents. Single-link, complete-link and average-link are the most popular HAC algorithms.

In single-link algorithm [16], the similarity between clusters is measured by their most similar members (minimum dissimilarity). Generally, agglomerative process are rather computationally intensive because the minimum of inter-cluster distances must be found at each merging step. For single-link clustering, an efficient implementation of Minimum Spanning Tree (MST) algorithms of a weighted graph is often involved. Therefore, single-link produces clusters that are subgraphs of the MST of the data and are also connected components. It is capable of discovering clusters of varying shapes, but often suffers from the so-called chaining effect. Complete-link [11] measures the similarity between two clusters by their least similar members (maximum dissimilarity). From graph-theoretic perspective, complete-link clusters are non-overlapping cliques and are related to the node colorability of graphs. Complete-link is not vulnerable to chaining effect, but generates excessive compact clusters and is thus very sensitive to outliers. Average-link clustering [5] is a compromise between single-link

Algorithm 2. Find All k-Clique Percolation Clusters

input: A set of all MDCs C and k.
output: All k-clique percolation clusters into P.

```
procedure Find-k-CPC (P, C, k)
  B := 0 //Initialize B's elements as 0
  for i from 1 to M do
    for j from 1 to M do
      B[i][j] := |C_i ∩ C_j| //# of common nodes of two MDCs
      if (i = j) ∧ (B[i][j] < k) then
        B[i][j] := 0 //Off-diagonal element < k is replaced by 0
      end if
      if (i ≠ j) ∧ (B[i][j] < k − 1) then
        B[i][j] := 0 //Diagonal element < k − 1 replaced by 0
      end if
    end for
  end for
  P := φ; i := 1 //Initialize output container P and recursion counter i
  DFS(P,B,i) //DFS to output connected components in B into P
  output P
end procedure
Find-k-CPC(P, C, k)
```

and complete-link: the similarity between one cluster and another is the averaged similarity from any member of one cluster to any member of the other cluster; it is less susceptible to outliers and elongated chains.

5 Experimental Evaluations

5.1 Data Sets

We conduct the performance evaluations based on Reuters-21578[1] corpus, which is popular for document classification evaluation purpose. It contains 21,578 documents manually grouped into 135 topic classes. The size of classes is very unbalanced, ranging from 1 to 3945. Many documents have multiple category labels, and documents in each cluster have a broad scape of contents. In our experiments, we select documents that are assigned to one or more topics, and have the attribute LEWISSPLIT="TEST" with <BODY> and </BODY> tags. There are 2,745 such original documents, denoted by OC2745, from which we then extract 2,349 documents with unique class labels to form our data set UC2349, and the rest of 396 documents with multiple classes to form MC396. Table 1 shows the statistics of these three resulted data sets.

[1] http://www.daviddlewis.com/resources/testcollections/reuters21578

Table 1. Statistics of data sets OC2745, UC2349 and MC396 extracted from Reuter-21578 corpus

	OC2745	UC2349	MC396
# of documents	2745	2349	396
# of classes	92	58	87
max class size	1045	1041	127
min class size	1	1	1
avg. class size	38	41	13

5.2 Evaluation Metrics

We adopt two quality metrics widely used for document clustering [17], i.e., F-measure and Entropy. The F-measure of a class i is defined as $F(i) = \frac{2PR}{P+R}$. The precision and recall of a cluster j with respect to a class i are defined as: $P = Precision(i, j) = \frac{N_{ij}}{N_j}$ and $R = Recall(i, j) = \frac{N_{ij}}{N_i}$, where N_{ij} is the number of members of class i in cluster j, N_j is the size of cluster j, and N_i is the size of class i. The overall F-measure of the clustering result is the weighted average of $F(i)$:

$$F = \frac{\sum_i (|i| \times F(i))}{\sum_i |i|}$$

where $|i|$ is the number of documents in class i.

Entropy provides a measure of homogeneity of a cluster. The higher the homogeneity, the lower the entropy, and vice versa. For every cluster j in the clustering result, we compute p_{ij}, the probability that a member of cluster j belonging to class i. The entropy of each cluster j is calculated using $E_j = -\sum_i p_{ij} \log(p_{ij})$, where the sum is taken over all classes. The total entropy for a set of clusters is calculated as the sum of entropies of each cluster weighted by its size:

$$E = \sum_{j=1}^{m} (\frac{N_j}{N} \times E_j)$$

where N_j is the size of cluster j, m is the number of clusters, and N is the size of document collection.

5.3 Performance Evaluation

Experiment 1. Table 2 shows the performance of CPM given the size of threshold clique. Obviously CPM produces more clusters than the number of categories in the benchmark. This is because Reuters corpus are manually classified according to a set of pre-defined keywords (one for each class roughly). Thus the schema of categorization is rather unifarious. One document may belong to far more groups since the grouping criterion could be diverse. CPM is less limited by external constrains, which favors multifarious categorization schemes, and thus has more clusters. The least number of clusters are found at $k = 2$ where

CPM is degenerated to find connected components, which actually partitions the collections. With larger k, the cluster number increases as larger k allows for more overlapping clusters.

In terms of both F-measure and Entropy, CPM performance improves rapidly at the first few k augments, but worsens slowly with the further increases. Interestingly, there are some close optimal values of k on these data sets around 4–6. Unlike our expectation, the results on MC396, which contains documents all belonging to multiple classes, show inconsistencies on F-measure and Entropy. For Entropy, it is reasonable that CPM performs the best on MC396 since CPM favors overlapping clusters. But F-measure gives the worst results on it. F-measure seems very sensitive to outliers and penalizes their recalls heavily. As we found relatively larger proportion of outliers in MC396 clustering results, this may explain the low F-values.

We originally expected that the results on OC2745 would be far and few between UC2349 and MC396, but the worst Entropy results are observed on it. One possible reason is that Entropy favors small cluster number and cluster size. This may also explain the obviously low Entropy values on MC396 other than the advantages on overlapping clusters. Note that when $k = 2$, the performance is significantly poorer than other choices. This is also because at $k = 2$, CPM algorithm can only find connected components, which are the most relaxed criterion for clustering.

Table 2. Performance of CPM with respect to different sizes of the threshold clique

	# of clusters			F-measure			Entropy		
k	OC2745	UC2349	MC396	OC2745	UC2349	MC396	OC2745	UC2349	MC396
2	**1045**	**874**	**212**	0.093	0.083	0.234	0.596	0.366	0.749
3	1157	962	281	0.287	0.353	0.287	0.499	0.413	0.177
4	1455	1318	281	0.398	0.407	**0.302**	**0.481**	**0.376**	**0.140**
5	1964	1813	286	**0.525**	0.503	0.294	0.490	0.383	0.153
6	2390	2392	291	0.495	**0.594**	0.289	0.543	0.411	0.153
7	3177	3045	294	0.488	0.568	0.284	0.635	0.458	0.163
8	3782	3762	297	0.421	0.501	0.277	0.740	0.514	0.164
9	4503	4330	300	0.409	0.488	0.269	0.903	0.549	0.164
10	5261	4894	300	0.401	0.441	0.269	0.907	0.587	0.164

Experiment 2. In this experiment, we compare CPM with the other two representative clustering algorithms, k-means and complete-link. Because it is impossible to command CPM to produce exact number of clusters with the benchmark, we use $k = 4$, at which CPM reaches nearly optima based on Table 2. To make comparisons fair under this condition, we examine both k-means and complete-link twice: one uses the same number of clusters as the benchmark, and the other uses the same number of clusters as CPM, which are denoted by KM-B, KM-C, CL-B, and CL-C (suffixes B and C represent Benchmark and CPM, respectively). Furthermore, because k-means is well-known to be sensitive to local

optima, we repeat the algorithm 50 times with different initial centroids and average the outcomes achieved. The threshold for complete-link distance measure is set according to the computed values of t_c by CPM (see Section 3). This is to align with CPM.

Table 3 shows that CPM performs worse than k-means and complete-link if the standard number of clusters as the benchmark are produced. Because CPM generates far more clusters than the standard, this comparison is somewhat unfair to CPM. However, when the number of CPM clusters is used, its advantages can be clearly observed. Under this condition, k-means performs the worst among the three. Its poor performance on MC396 is very obvious because k-means can only produce partitioning of the corpus. Complete-link clusters are non-overlapping MDCs. The results show that CPM outperforms complete-link on all three test sets as well. This testifies the advantages of our method over the typical conventional clustering algorithms in terms of unrestraint cluster number.

Table 3. Comparisons of CPM and the other two representative algorithms, k-means (KM) and complete link (CL). We use $k = 4$.

	# of clusters			F-measure			Entropy		
	OC2745	UC2349	MC396	OC2745	UC2349	MC396	OC2745	UC2349	MC396
CPM	1455	1318	281	0.398	0.407	0.302	0.481	0.376	0.140
KM-B	92	58	87	0.510	0.503	0.391	0.319	0.270	0.117
KM-C	1455	1318	281	0.294	0.251	0.190	0.520	0.475	0.358
CL-B	92	58	87	0.531	0.511	0.437	0.276	0.203	0.105
CL-C	1455	1318	281	0.362	0.305	0.285	0.503	0.447	0.266

6 Conclusion and Future Work

We present a novel clustering algorithm CPM by applying clique percolation technique introduced from the area of biological physics. A more generalized framework related to it is the so-called "small-world network" describing many kinds of community structures in nature and society, which is extensively studied in random networks [9]. This is the pioneer work for the CPM being applied in document clustering. The preliminary results demonstrate it is feasible and promising for document clustering. We are confident that CPM is interesting and worth of further studies. There are still many issues left to be studied more deeply. So far, the heuristic relationship between p_c, N and t_c has not been well studied. To generate an appropriate random graph, an alternative is to make use of the degree distribution of graph vertices. For each vertex, some nearest neighbors associated with the precise degree distribution can be considered. This will lead to the further exploration on techniques to analyze complex networks. Furthermore, due to the NP-hardness of MDC enumeration algorithms, the CPM is time-consuming. Improvements on efficiency are required. In the future, we will also compare CPM to some more advanced clustering algorithms.

References

1. Baker, L., McCallum, A.: Distributional clustering of words for text classification. In Proc. of ACM SIGIR (1998):96–103
2. Bezdek, J.C.: Pattern recognition with fuzzy objective function algorithms. Plenum Press, New York
3. Bron, C., Kerbosch, J.: Finding all cliques of an undirected graph. Communications of the ACM **16** (1971):575–577
4. Cormen, T.H., Leiserson, C.E., Rivest, R.L., Stein C.: Introduction to algorithms, 2nd Edition. McGraw-Hill
5. Cutting, D., Karger, D., Pedersen, J., Tukey, J.W.: Scatter/Gather: A cluster-based approach to browsing large document collections. In Proc. of the 15th ACM SIGIR Conference (1992):318–329
6. Derenyi, I., Palla, G., Vicsek T.: Clique percolation in random networks. Physics Review Letters **95** (2005):160202
7. Dhillon, I.S.: Co-clustering documents and words using bipartite spectral grpah partitioning. In Proc. of the 7th ACM-KDD (2001): 269–274
8. Ding, C.H.Q., He, X.F., Zha, H.Y., Gu, M., Simon, H.D.: A min-max cut algorithm for graph partitioning and data clustering. In Proc. of IEEE ICDM (2001) 107–114
9. Dorogovtsev, S.N., Mendes, J.F.F.: Evolution of networks. Oxford Press, New York
10. Jain, A.K., Murty, M.N., Flynn, P.J.: Data clustering: a review. ACM Computing Surveys **31** (1999):264–323
11. King, B.: Step-wise clustering procedures. Journal of the American Statistical Association **69** (1967):86–101
12. Krishnapuram, R., Joshi, A., Nasraoui, O., Yi, L.Y.: Low-complexity fuzzy relational clustering algorithms for web mining. IEEE Transactions on Fuzzy Systems **9** (2001):595–607
13. Liu, X., Gong, Y.: Document clustering with clustering refinement and model selection capabilitities. In Proc. of ACM SIGIR (2002):191–198
14. Palla, G., Derenyi, I., Farkas, I., Vicsek, T.: Uncovering the overlapping community structure of complex netowrks in nature and society. Nature **435** (2005):814–818
15. Raghavan, V.V., Yu, C.T.: A comparison of the stability characteristics of some graph theoretic clustering methods. IEEE Transactions on Pattern Analysis and Machine Intelligence **3** (1981):393–402
16. Sneath, P.H.A., Sokal, R.R.: Numerical taxonomy: the principles and practice of numerical classification. Freeman, London, UK
17. Steinbach, M., Karypis, G., Kumar, V.: A comparison of doucment clustering techniques. In Proc. of KDD-2000 Workshop on Text Mining (2000)
18. Tsukiyama, S., Ide, M., Ariyoshi, H., Shirakawa, I.: A new algorithm for generating all the maximal independent sets. SIAM Journal on Computing **6** (1977):505–517
19. Zhao, Y., Karypis, G.: Criterion functions for document clustering. Technical Report #01-40, Department of Computer Science, University of Minnesota

Hybrid Approach to Extracting Information from Web-Tables

Sung-won Jung[1,2], Mi-young Kang[1,2], and Hyuk-chul Kwon[1,2]

[1] Pusan National University, Korean Language Processing Laboratory, Department of Computer Science Engineering,
[2] Pusan National University, Center for U-Port IT Research and Education, Jangjeon-dong, Geumjeong-gu, 609-735, Busan, Korea
{swjung, mykang, hckwon}@pusan.ac.kr

Abstract. This study concerns the extracting of information from tables in HTML documents. In our previous work, as a prerequisite for information extraction from tables in HTML, algorithms for separating meaningful tables and decorative tables were constructed, because only meaningful tables can be used to extract information and a preponderant proportion of decorative tables in training harms the learning result. In order to extract information, this study separated the head from the body in meaningful tables by extending the head extraction algorithm that was constructed in our previous work, using a machine learning algorithm, C4.5, and set up heuristics for table-schema extraction from meaningful tables by analyzing their head(s). In addition, table information in triples was extracted by determining the relation between the data and the extracted table schema. We obtained 71.2% accuracy in extracting table-schemata and information from the meaningful tables.

Keywords: Text mining; Information extraction; Table mining; Meaningful table.

1 Introduction

Table information extraction (i.e. table mining) is a sub-domain of information extraction. Yang [8] conceives the database as being constructed with attribute-value pairs, and *table mining* as a reverse process of table publishing from a database. That is, table mining is the process which converts a table into machine-understandable form and extracts information based on that form.

In order to extract information from web-tables, a preprocessing stage is necessary. As is well known, HTML does not distinguish between presentation and structure: tables on the Internet are used for not only the original purpose of the table but also for constructing HTML-document layouts. Our previous work's main focus was distinguishing meaningful from meaningless tables. In order to identify

Y. Matsumoto et al. (Eds.): ICCPOL 2006, LNAI 4285, pp. 109 – 119, 2006.

meaningful tables, we set 24 features which interact in defining the meaningfulness of a table; we built a separation model that utilized, with those features, a machine-learning algorithm.

If meaningful tables are extracted from raw HTML documents, we can extract information from those tables. Although the web-table is expressed by table tags, those tags do not express the web-table's semantic structure. However, when we interpret a table, we understand it based on the semantic structure. Therefore, if we conceive *the semantic structure of a table* as a *table-schema*, then we can declare that the data in that table is organized by a table-schema.

The analysis of the relation between a table-schema and table data corresponds to that between the traditional table components: the table head and body. The table body contains data that provides the main information, whereas the table head abstracts those data. Generally, the table head is just *an area* in the table, lacking semantic structure. Therefore, the table-schema is extracted by reorganizing the head in order that the reorganization reflects the semantic structure of the head. The table body is converted into a machine-understandable form based on the table-schema thus extracted. The relation between a table-schema and data corresponds to that of database schema and records, or that of ontology and triples in the semantic web.

This paper is organized as follows. Section 2 briefly summarizes several recent studies undertaken to develop information extraction from web-tables. Section 3 describes the method of extracting the head. Section 4 describes the method of extracting table-schema and that of extracting triples. Section 5 illustrates the experiments and, finally, concluding comments follow in Section 6.

2 Related Work

Research in the *table mining* field can be classified into two categories: (1) domain-specific research and (2) domain-independent research. As far as we are aware, most research on table mining has focused on extracting the meaningful table, and information extraction from web tables has been treated as a side issue.

Domain-specific research is based on wrapper induction [4], which performs particular information extraction using extraction rules. Using these extraction rules, several studies [1, 2, 4] have extracted table information according to a special tabular form. Because these studies dealt only with the special tabular form, the researchers experienced difficulty in coping with the various web-document formats.

We can find, among domain-independent approaches, information extraction from *web tables* Yang [8]. Yang's study conceives the database as being constructed with attribute-value pairs, and table mining as the reverse process of table publishing from a database. He extracts the attribute-value pairs using entity-patterns and extraction rules. However, Yang's method can hardly cope with new tables that contain unknown words. We need to repetitively and manually update those rules and patterns with linguistic bias, which is rather far from domain independence.

3 Extracting the Head

The objective of this study is to apply table mining to general HTML documents. In order to satisfy this objective, in constructing our model we used only structural information and avoided domain-specific information.

The aim of information extraction from web-tables is to establish machine-understandable information, and this can be achieved by converting a table to a table-schema and a triple. Because a table head abstracts related data in the body, the head can be a strong candidate for the table-schema. The triple can be extracted using this table-schema and body elements. Therefore, we should separate the head from the body in order to extract the semantic structure, which is the basis of our ultimate goal, that is, information extraction.

Once meaningful tables are extracted, we can rather easily extract their heads. Accordingly, our previous work proposed meaningful table- and head extraction methods. [3] Section 3.1 summarizes these methods briefly. Section 3.2 complements the head extraction method with supplementary heuristics.

3.1 Constructing Head Extraction Model

The table head is defined as a row(s) or a column(s). Most features for extracting heads were instituted based on two important factors of web-table editing:

- Specific techniques for separating the head from the data are used in order that readers understand a table more clearly.
- The row or column related to a head contains repetition, because the head abstracts related data in a table.

Both of these factors concern (1) rows' and columns' appearance characteristics and (2) their inter-relations. While analyzing rows and columns as parts of a head in meaningful tables and considering their appearance and relations, the following features and their values, which are used in machine-learning for extracting heads, were formulated[1].

For the construction of a head extraction model (hereafter, HEM), the meaningful tables are converted to an input data set for the machine-learning algorithm. As we mentioned above, the table head can be a row(s) or a column(s). Therefore, the rows and columns in the tables were converted into a 14-dimensional vector using the features in Table 1. (See Fig. 1) From this input data, we constructed the HEM using a decision tree classifier, C4.5 [7].

[1] For further discussion on features for extracting head see Ref. [3].

Table 1. Features for extracting head according to rows and columns' characteristics

• Appearance characteristics			
Criterion	No.	Feature	Value
Tag	1	If a row or column is expressed by <th> tags, the feature value is 1, otherwise 0.	0 or 1
	2	If a row or column contains a span tag, the feature value is 1, otherwise 0.	0 or 1
Table Structure	3	If a table has an empty cell in the first row, first column, the row and column that includes that empty cell is 1, otherwise 0.	0 or 1
	4	The index-number of a row or column in a table	Integer
No. of characters	5	The average number of characters in a cell of a row or a column	Prime number
	6	The standard deviation of the number of characters in a cell of a row or a column	Prime number

• Inter-relational characteristics			
Criterion	No.	Feature	Value
Cell-contents type	7	The fraction of a row's or column's representative CCT[a)] in a row or a column	Percentage
	8	The fraction of TRT[b)] in a row or a column	Percentage
	9	If a row's or column's CPT[c)] is different from its successive row's or column's CPT, the feature value is 1, otherwise 0. The lower-most row and the right-most column are always 0.	0 or 1
Cell-contents pattern	10	The fraction of a row's or a column's representative CCP[d)] in a row or a column	Percentage
	11	The fraction of TRP[e)] in a row or a column	Percentage
	12	If a row's or column's CPP[f)] is different from its successive row's or column's CPP, the feature value is 1, otherwise 0. The lower-most row and the right-most column are always 0.	0 or 1
Possibility of head presence	13	The degree of possibility of the presence of a head based on background color	Prime number
	14	The degree of possibility of the presence of a head based on font attributes	Prime number

a) CCT : cell-contents type such as link, image, digit, and words
b) TRT : a table's representative CCT
c) CPT : contents pattern based on CCT in a row or a column
d) CCP : cell-contents pattern[2]
e) TRP : a table's representative CCP
f) CPP : contents pattern based on CCP in a row or a column

[2] Cells have a particular sequence of token types. A token is the part of a sentence separated by specific delimiters such as space and punctuation marks, among others. We divide them into four types: word, digit, tag, and specific character, and cell content assumes a pattern according to these token types. We term it the cell-contents pattern (hereafter, CCP).

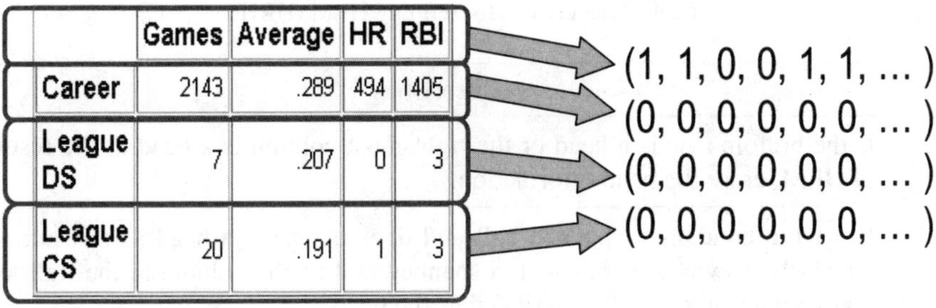

Fig. 1. Converting rows and columns to vectors

3.2 Supplementary Heuristics of Head Extraction Model

In Sections 3.1, we considered only whether a row or a column is a head or not. However, table-head extraction can proceed only when considering the semantics of a whole table and the structure of a table to which the table semantics are related. For example, Figure 2 is the result of erroneous extraction of a head from a meaningful table.

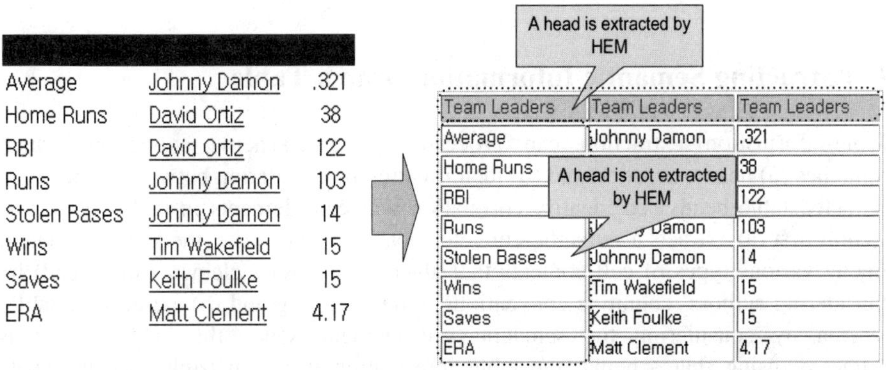

(a) Original (b) Result

Fig. 2. Example of erroneous head extraction

Therefore, this section complements the HEM reported in Sections 3.1 with heuristics constructed by extracting general factors from the results of a comparative analysis on correct table structures in the training data and those extracted by applying HEM. Accordingly, the following heuristics are formulated.

Table 2. Heuristics for refining HEM (HRH)

No.	Heuristics
1	If the bottom row is a head or the right-most column is a head as the result of HEM, erase the head information.
2	If a table contains a spanned cell and does not have a head as a result of HEM, the row below that with a spanned cell or the column to the right of that with a spanned cell is assigned as the head.
3	If a table contains only a row head and the left-most cell in the head row is empty, the left-most column is a head.
4	If a table contains only a column head and the upper-most cell in the head column is empty, the upper-most row is a head.
5	If a table contains both a head column and a head row and they are not present at the upper-most cell or the left-most column, erase the head information.
6	If a table contains a repetitive head column and head row and if the upper-most cell or the left-most column do not show the repetitiveness (in terms of contents, background color, font, among others), erase the head information.

4 Extracting Semantic Information from a Table

Table-information extraction can proceed via table-schema extraction, which combines (a) head extraction and (b) identification of correlation between elements in extracted table heads. To identify correlation between elements in table heads, this section first (a) extracts a semantic-core element from a head (i.e. the text-cell content among various types of cell contents that abstracts all other elements in a head) by considering authors' cognitive conventions in table editing and (b) extracts the table-schema by considering the semantic-core element. Once this table-schema is extracted, using that schema, the table information itself, in triple, can be easily extracted.

4.1 Extracting Semantic-Core Element from Extracted Head

Generally, a table provides a semantic-core element (hereafter, SCE) in a head, which assumes the role of the "pivot" in understanding table information. For example, in Figure 3a, intuitively, we first identify 'Team' as SCE, then reorganize the table structure as the hierarchical structure in Figure 3b. Accordingly, we interpret that 25 is W (the number of wins) and 4 is D (the number of defeats), in this case for Chelsea.

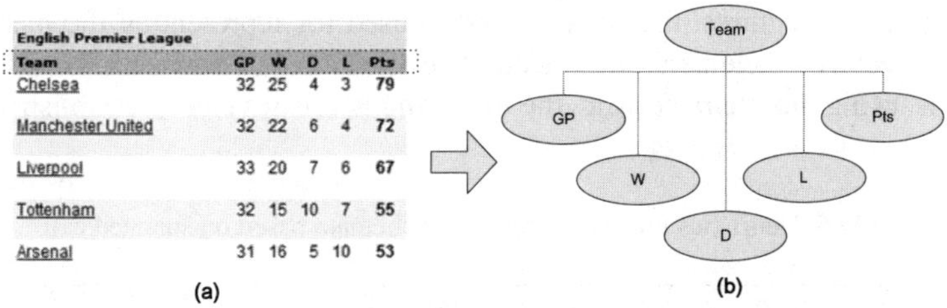

Fig. 3. Interpretation of a table

Considering this intuitive organization, we institute several heuristics for detecting an SCE, as below.

Table 3. Heuristics for detecting SCE

No.	Heuristics
1	An SCE in a head is located on the left-most side of a head row or on the upper-most side of a head column.
2	If a head cell has digit-type contents in its related body part, it has a weak possibility of being an SCE.
3	If all of the body contents in a table are the digit type, the left-most or upper-most element in a head is an SCE.
4	If a head cell is a spanned cell, it has a weak possibility of being an SCE.
5	If a head cell contains (a) word(s) which belong(s) to (an) attribute(s), it has a weak possibility of being an SCE.

4.2 Extracting Table-Schemata Using SCE and Table Structures

For the purpose of extracting table-schemata, we institute the following heuristics based on (a) the SCE that is detected by heuristics in Table 3 and (b) the characteristics of the table structure. As we mentioned above, the SCE is the core element of information and reorganizes the head by assuming the role of the pivot, as shown in Figure 3. Accordingly, the heuristics for extracting table-schemata are formulated as below.

Table 4. Heuristics for extracting table-schemata based on SCE

No.	Heuristics
1	If a $c_{m,*}$ is a head that is located in the m-th row, and if $c_{m,k}$ is the SCE, the $c_{m,k}$ abstracts all other $c_{m,j}$ elements in the head.
2	If a $c_{*,m}$ is a head that is located in the m-th column and if $c_{k,m}$ is the SCE, the $c_{i,m}$ abstracts all other $c_{i,m}$ elements in the head.

Among table structures, the spanned cell is used for representing the hierarchical semantic structure of elements in a head. The spanned cell abstracts its consecutive cells. Thus, using this characteristic, the following heuristics can be established.

Table 5. Heuristics for extracting table-schemata based on spanned cell

No.	Heuristics
1	If a $c_{i,j}$[a)] is spanned over n cells horizontally, the $c_{i,j}$ abstracts subcells from $c_{i+1,j}$ to $c_{i+1,\ j+n}$.
2	If a $c_{i,j}$ is spanned over n cells vertically, the ci,j abstracts subcells from $c_{i,j+1}$ to $c_{i+n,\ j+1}$.
a) $c_{i,j}$ is the cell that is located in the *i*-th row and the j-th cell.	

Based on the heuristics in Table 5, if a table contains a spanned cell in the head, it can be interpreted as the semantic hierarchy shown in Figure 4. This semantic hierarchy corresponds to a table-schema.

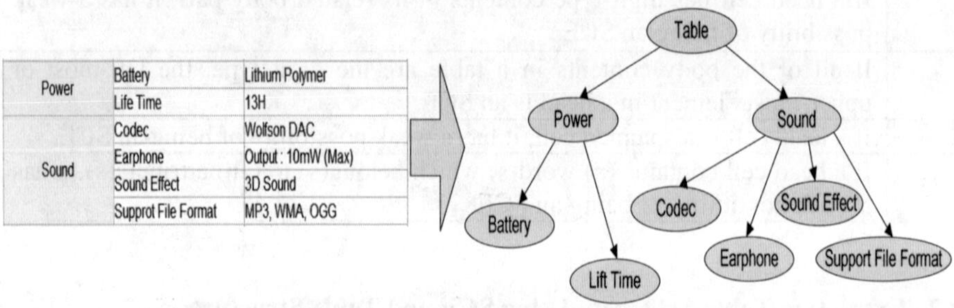

Fig. 4. Semantic-hierarchy extraction from spanned cell

4.3 Extracting Triple

Table information in a triple is extracted by applying the table-schema. In the semantic web, the fundamental ordering concept is the triple, which is composed of *object, attribute* and *value*. The *elements in a body* of a table correspond to *object* or *value* and the *elements in a head* of a table correspond to *attribute*. The following criteria are used to distinguish this triple.

Criterion 1

Detection of attribute: if an element in the head is an SCE, it is the class of object, and if not, is the attribute of the class.

Fig. 5. Attribute detection

The SCE can be a criterion for determining whether an element in the body is object or value.

Criterion 2

Distinction of elements in a body according to object and value: if an element in the body is related to the SCE, it is conceived as the object, and if not, as the value.

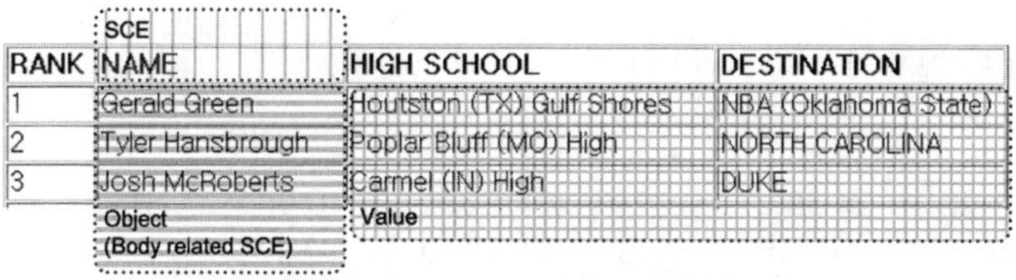

Fig. 6. Object and value distinction

Figure 7 shows triple extraction from the Table shown in Figures 5 and 6, using its table schema.

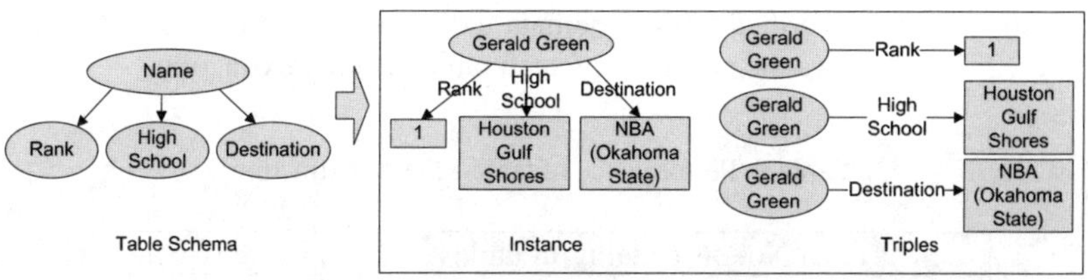

Fig. 7. Extraction of triple using table schema

5 Experimental Results

To test the performance of table-information extraction, we composed our test-data-suit by randomly selecting HTML documents from the Internet and from some of Wang's data [7].

Table 6. The characteristics of data sets

Items	Our training data	Wang's data	Total
No. of meaningful tables	964	969	1,933
No. of decorative tables	2,249	2,009	4,258
Total	3,213	2,978	6,191

In Section 3.1, we extracted the head using HEM based on the decision tree algorithm, C4.5. Based on those results, we refined head extraction by formulating supplementary heuristics, HRH in Table 2. For the performance test, 10-fold cross-validation was used. Table 7 shows the final performance of the head classifier: we obtained 82.2% accuracy in extracting the head.

Table 7. Performance of our head classifier

• Row- or Column-based performance

		Assigned class			Precision	Recall	F-Measure
		Body	Head				
True class	Body	24,587	329	Body	0.984	0.987	0.986
	Head	389	2,237	Head	0.872	0.852	0.862

• Table-based performance

Extraction model	No. of correct tables/ No. of total tables	Accuracy (%)
HEM	2,148/ 2,626	81.8
HEM with HRH	2,159/ 2,626	82.2

In Section 4, we described table schemata extraction based on table structure and heuristics. Table 8 shows the accuracy of such table-schema extraction.

Table 8. Performance of table-schema extraction

Total No. of meaningful tables	2,626
Total No. of correct extractions	1,870
Total No. of incorrect extractions	756
Accuracy	71.2%

6 Conclusions

The ultimate goal of information extraction from tables is to extract table-schemata and triples based on defining the relationships between the table components. Therefore, we suggested a method for extracting table-schemata based on table structure and heuristics. Using this method, a table is converted into a table-schema and a triple. In future work, using the results of this paper, we will expand the application domain to information retrieval systems, the construction of primary data for ontology, and to other areas.

Acknowledgements. This work was supported by the Regional Research Centers Program(Research Center for Logistics Information Technology), granted by the Korean Ministry of Education & Human Resources Development.

References

1. Chen, H.H., Tsai, S.C., Tsai, J.H.: Mining Tables from Large Scale HTML Texts. Proceedings of 18th International Conference on Computational Linguistics, Saabrucken, Germany, July (2000)
2. Hurst, M.: Layout and Language: Beyond Simple Text for Information Interaction - Modeling the Table. Proceedings of the 2nd International Conference on Multimodal Interfaces, Hong Kong, (1999)
3. Jung, S.W., Kwon, H.C.: A Scalable Hybrid Approach for Extracting Head Components from Web Tables, accepted and to be appeared in IEEE transaction on knowledge and data engineering, vol. 18. No. 2.
4. Kushmerick, N., Weld, D. S., Doorenbos, R.: Wrapper Induction for Information Extraction, 15th International Joint Conference on Artificial Intelligence(IJCAI-97), Nagoya, August (1997)
5. Ning, G., Guowen, W., Xiaoyuan, W., Baile, S.: Extracting web table information in cooperative learning activities based on abstract semantic model. Computer Supported Cooperative Work in Design, The Sixth International Conference (2001) 492-497
6. Wang, Y., Hu, J.: A Machine Learning Based Approach for Table Detection on The Web in Proceedings of The Eleventh International World Wide Web Conference WWW2002, Sheraton Wailili Honolulu, Hawaii, USA (2002) 7-11
7. Witten, I.H., Frank, E.: Data Mining: Practical Machine Learning Tools and Techniques with Java Implementations, Morgan Kaufmann Pub. (2000)
8. Yang, Y.: Web Table Mining and Database Discovery. M.Sc. thesis, Simon Fraser University, August (2002)

A Novel Hierarchical Document Clustering Algorithm Based on a kNN Connection Graph*

Qiaoming Zhu, Junhui Li, Guodong Zhou, Peifeng Li, and Peide Qian

School of Computer Science & Technology
Soochow University
215006, Suzhou, China
{qmzhu, jhli, gdzhou, pfli, pdqian}@suda.edu.cn

Abstract. Bottom-up hierarchical document clustering normally merges two most similar clusters in each step iteratively. This paper proposes a novel bottom-up hierarchical document clustering algorithm to merge several pairs of most similar clusters in each step. This is done via a concept of "kNN-connectedness", which measures the mutual connectedness of clusters in kNNs, and a kNN connection graph, which organizes given clusters into several sets of kNN-connected clusters. In such a graph, a connection between any two clusters only exists in the kNN-connected clusters of the same set. Moreover, a new kNN-based attraction function is proposed to measure the similarity between two clusters and indicates the potential probability of the two clusters being merged. The attraction function only considers the relative distribution of their nearest neighbors between two clusters in a vector space while other criteria, such as the well-known cluster-based cosine similarity function, measures the absolute distance between two clusters. This makes the attraction function effectively apply to the cases where different clusters may have very different distance variation. In each step, a kNN connection graph, consisting of several sets of kNN-connected clusters, is first constructed from the given clusters using a kNN algorithm and the concept of "kNN-connectedness". For each set of kNN-connected clusters, the attraction degree between any two clusters is calculated and several top connected cluster pairs will be merged. In this way, the iteration number can be largely reduced and the clustering process can be much speeded. Evaluation on a news document corpus shows that the kNN connection graph-based hierarchical document clustering algorithm can achieve better performance than the famous k-means clustering algorithm while reducing the iteration number sharply in comparison with normal hierarchical document clustering.

1 Introduction

With the dramatic increase in the amount of textual information available in the Web and digital archives, there has been growing interest in massive text information processing and management. Document clustering is a convenient way

* This research was supported by the High Technology Plan of Jiangsu Province, China under Grant No.2005020.

Y. Matsumoto et al. (Eds.): ICCPOL 2006, LNAI 4285, pp. 120–130, 2006.

and often the first step to help people organize and extract valuable information effectively and efficiently from various available information resources. As a key research topic in the machine learning research field, document clustering belongs to unsupervised learning. That is, document clustering automatically organize a set of documents into different clusters according to their similarity without an annotated training corpus.

In literature, there are many works in document clustering [Zamir et al 1998; Steinbach et al 2000] and can be classified into two categories: flat document clustering [Hartigan et al 1979; Kanungo et al 2002], which iteratively organizes a set of documents into flat-distributed clusters, and hierarchical document clustering [Willett 1988; Zhao et al 2002], which iteratively organizes a set of documents into hierarchically connected clusters through a tree structure. This paper will focus on hierarchical document clustering. In general, tree-structured clusters achieved from hierarchical document clustering can be flattened into flat-distributed clusters according to users' or applications' demand.

Hierarchical document clustering [Willett 1988; Zhao et al 2002] can be classified into two categories: the top-down approach via splitting and the bottom-up approach via merging.

- The top-down approach initializes a set of documents as a single big cluster. In each step, it chooses a most appropriate cluster using some criterion (e.g. information gain) and splits it into two clusters. The iterative process stops when some conditions meet.
- The bottom-up approach initializes each given document as a singleton cluster. In each step, it chooses two most similar clusters using a similarity function (e.g. the cosine similarity function) and merges them into a cluster. The iterative process stops when only one big cluster exists.

Moreover, hierarchical document clustering can be done via single-link, complete link and average link [Schutze 2005]:

- Single-link hierarchical clustering merges in each step the two clusters with the smallest minimum pair-wise distance.
- Complete-link hierarchical clustering merges in each step the two clusters with the smallest maximum pair-wise distance.
- Average-link clustering merges in each step the two clusters with the highest cohesion. It is a compromise between the sensitivity of complete-link clustering to outliers and the tendency of single-link clustering to form long chains that do not correspond to the intuitive notion of clusters as compact, spherical objects.

This paper will focus on bottom-up hierarchical document clustering. In comparison with the top-down approach, the bottom-up approach biases more on local information and less on global information. This can help find smaller but more similar clusters. However, this also makes it suffer from early merging errors since such early merging errors will be carried forward. Another disadvantage is that it is computationally expensive to compute the similarity between any two clusters, especially when there exist tens of thousands of documents. Finally, only two clusters are normally merged in each iterative process.

In order to resolve above problems in bottom-up hierarchical document clustering, much research has been done. Wei et al (2005) combined flat document clustering with hierarchical document clustering. They first applied k-means clustering to organize given documents into many small clusters and then used bottom-up hierarchical document clustering to iteratively merge them into a hierarchical tree structure. Zhen (2006) merged and split clusters in a dynamic way to resolve the early errors in top-down hierarchical document clustering, which is also applicable to bottom-up hierarchical document clustering. Wu et al (2004) proposed an efficient algorithm to largely reduce computational time and memory requirement. This is done by automatically detecting possible overlaps among different clusters in different stages.

This paper tackles above problems by proposing a novel bottom-up hierarchical document clustering algorithm. This is done via a concept of "kNN-connectedness", which measures the mutual connectedness of clusters in kNNs, and a kNN connection graph, which organizes given clusters into several sets of kNN-connected clusters. In such a graph, a connection between any two clusters only exists in the kNN-connected clusters of the same set. Moreover, a new kNN-based attraction function is proposed to measure the similarity between two clusters and indicates the potential probability of the two clusters being merged. The attraction function only considers the relative distribution of their nearest neighbors between two clusters in a vector space while other criteria, such as the well-known cluster-based cosine similarity function, measures the absolute distance between two clusters. This makes the attraction function effectively apply to the cases where different clusters may have very different distance variation. In each step, a kNN connection graph, consisting of several sets of kNN-connected clusters, is first constructed from the given clusters using a kNN algorithm and the concept of "kNN-connectedness". For each set of kNN-connected clusters, the attraction degree between any two clusters is calculated and several top connected cluster pairs will be merged. In this way, the iteration number can be largely reduced. As a result, the clustering process can be much speeded and better performance can be achieved through the new kNN-based attraction function.

The rest of this paper is organized as follows. Section 2 describes the kNN connection graph while the bottom-up hierarchical document clustering algorithm is presented in Section 3. Finally, we evaluate our algorithm in Section 4 and conclude this paper in Section 5.

2 kNN Connection Graph

The bottom-up hierarchical document clustering algorithm proposed in this paper is based on a kNN connection graph, which is constructed in each iterative process and consists of two parts: a set of notes, each of which represents a document cluster, and a set of connections, each of which represents the kNN-based attraction degree between the corresponding two clusters. A connection between two clusters only exists when they are "kNN-connected" via a kNN propagation sequence in the kNN connection graph and its attraction degree is computed using the kNNs of the two clusters.

In a kNN connection graph, two clusters c_i and c_j is defined as "directly kNN-connected" if $c_j \in kNN(c_i)$ and $c_i \in kNN(c_j)$, and defined as "kNN-connected" if there exist a sequence of clusters $c_{k_1}, c_{k_2} \cdots c_{k_n}$ and $(c_i, c_{k_1}), (c_i, c_{k_2}), \ldots,$ (c_{k_n}, c_j) are all "directly kNN-connected". Here, $kNN(c_i)$ indicates the k nearest neighbors (including c_i itself) for a given cluster c_i in the kNN algorithm where a cluster is represented by the average center of all the documents included in the cluster and the cosine similarity function is applied to measure the similarity between any two clusters. Obviously, the "kNN-connectedness" among clusters are symmetric and transitive. Using the concept of "kNN-connectedness", all the clusters in a given kNN connection graph can be organized into several sets of kNN-connected clusters.

Given above definitions, the kNN-based attraction degree between any two clusters c_i and c_j is computed by the following function, with the denominator calculated over the union set of their kNNs and the numerator calculated over the cross set of their kNNs:

$$kNNAttraction(c_i, c_j) = \frac{\sum\limits_{c' \in kNN(c_i) \cap kNN(c_j)} sim(c', c_i) \cdot sim(c', c_j)}{\sum\limits_{c' \in kNN(c_i) \cup kNN(c_j)} sim(c', c_i) \cdot sim(c', c_j)} \quad (1)$$

The above formula for the attraction function measures the shared clusters between the two sets of kNNs for the two clusters, weighted by their respective similarities with the two clusters. Similar to the kNN algorithm, a cluster is represented by the average center of all the documents included in the cluster and the cosine similarity function is applied to calculate the similarity $sim(c_i, c_j)$ in the above formula. Obviously, the attraction function-based hierarchical document clustering belongs to the average-link category. The intuition behind is that, the more shared between the two kNNs of two clusters, the more attractive of the two clusters and the more probable of their merging. Especially, they have the attraction degree of 1 if the two clusters have the same set of kNN, and 0 if they share no nearest neighbors. The most appealing characteristics of this new kNN-based attraction function is that it only considers the relative distribution of their nearest neighbors between two clusters in a vector space while other criteria, such as the well-known cluster-based cosine similarity function, measures the absolute distance between two clusters. This makes the attraction function apply to the cases where different clusters may have very different distance variation. Another advantage is that it not only considers the similarity between the given two clusters but also takes into account the effect of their respective kNNs.

3 kNN Connection Graph-Based Hierarchical Document Clustering

Figure 1 shows the overall algorithm of our KNN connection graph-based bottom-up hierarchical document clustering algorithm. Given a set of documents, it first treats each document as a singleton cluster, and then successively merges clusters until all documents have been merged into a single remaining cluster. Normally a hierarchical document clustering algorithm merges two clusters according to their similarity in each iterative process. One advantage of our algorithm is that it can merge more clusters in each iterative process. This is done by propagating kNN-connected clusters using the kNN connection graph.

```
Algorithm: fnKNNCG-HierarchicalClustering(D, mergedclusterCollections)
Input: a set of given documents D
Output: hierarchically distributed clusters mergedclusterCollections
BEGIN
  Generate a set of basic singleton clusters basicClusterCollection, each of which
    includes a document in the set of given documents D.
  Initialize mergedclusterCollections[ 0 ] = basicClusterCollection
  Initialize the hierarchical level hlevel = 0
  DO
  BEGIN
    Build a kNN connected graph kNN-CG
        given mergedclusterCollections[hlevel]
    Generate a set of merged clusters from a set of the basic clusters given the
      kNNCF by calling fnGenerateClustersFromBasicClusters ( kNN-CG,
          mergedclusterCollections[hlevel] , mergedclusterCollections[hlevel+1])
      mergedclusterCollections[hlevel]= mergedclusterCollections[hlevel+1]
      hlevel++
  END
  UNTIL |mergedclusterCollections[hlevel]|=1
  Return mergedclusterCollections
END Algorithm
```

Fig. 1. Algorithm: kNN connection graph-based bottom-up hierarchical document clustering

In each iterative process, given a set of clusters, a kNN connection graph is first generated as follows:

- Determine $kNN(c_i)$ for each cluster c_i in the set of given clusters. In this paper, the cluster c_i itself is also included in $kNN(c_i)$.

- Determine $kNNDC(c_i)$ the set of directly kNN-connected clusters for each cluster c_i by checking the direct kNN-connectedness of the cluster c_i with each of the remaining clusters.
- Determine $kNNC(c_i)$ the set of kNN-connected clusters for each cluster c_i by checking the kNN-connectedness of the cluster c_i with each of the remaining clusters through a directly kNN-connected cluster propagation path. In this way, all the clusters in the kNN-connection graph are grouped into several sets (e.g. M) of kNNCs.
- Compute the attraction degree between each pair of c_i and c_j in each kNNC as shown in Formula (1). By restricting a connection between two clusters in a kNNC, we can largely reduce the computational overload and thus speed up the clustering process.

Then, using the kNN connection graph constructed as above, the given clusters are merged into a new set of clusters. Figure 2 shows the algorithm for the merging process. Since $|kNNC|$ the number of clusters in a kNNC may be quite different for different kNNCs, for each kNNC, $|kNNC|/N$ top ranked pairs are retrieved according to their kNN-based attraction degrees without cluster overlapping[1] and the two clusters in each of the retrieved pairs are merged into a bigger cluster. Here, N controls the merging rate. Obviously, at most 2/N of clusters can be merged pairwisely in each iterative process. In this way, the number of the iterative processes can be largely reduced. The iteration stops until there is only one cluster left.

Similar with other bottom-up hierarchical document clustering algorithms, this algorithm also suffers from the computational problem since it needs to calculate the similarity between each pair of clusters in determining nearest neighbors in the kNN algorithm and the attraction function between two clusters in each set of kNN-connected clusters. Normally, there are tens of thousands of documents to be clustered. This makes our algorithm computationally demanding at early stages. In order to resolve this problem, we can combine it with flat document clustering as described in Wei et al (2005) by first applying a flat document clustering algorithm (e.g. k-means clustering) to organize given documents into a certain number of basic clusters, such as k. And if the value of k is n/m, where n is the number of the total documents, then the hierarchical document algorithm's complexity will decrease from $O(n^2 \log n)$ to $O(\frac{1}{m^2}n^2 \log n)$.

The major advantage of our algorithm is that the new kNN-based attraction function only considers the relative distribution to their nearest neighbors between two clusters in a vector space while other criteria, such as the cosine similarity, measures the absolute distance between two clusters. This makes our algorithm apply to the cases where different clusters may have very different distance variation. Moreove, the attraction function not only considers the similarity between the given two clusters but also takes into account the effect of their respective kNNs. Another advantage is that our algorithm can merge many similar clusters in each iterative process. This can largely reduce the iteration

[1] That is, if one cluster in a cluster pair has occurred in a higher ranked cluster pair, this pair will not be included in the top ranked pairs.

```
Algorithm: fnGenerateClustersFromBasicClusters( kNN-CG,
        basicClusterCollection, mergedClusterCollection )
Input:  A set of basic clusters basicClusterCollection
        A kNN connection graph kNN-CG given basicClusterCollection
Output: A set of merged clusters mergedClusterCollection
BEGIN
  Initialize the set of merged clusters mergedClusterCollection as empty
  FOR each set of kNN-connected clusters (kNNC) in the kNN-CG
  BEGIN
    Find |kNNC|/N top ranked cluster pairs without cluster overlapping
    Merge each of above top ranked cluster pairs into one cluster and put
        it into mergedClusterCollection
    Delete those clusters in the top ranked cluster pairs from
        basicClusterCollection
    Move the remaining clusters to mergedClusterCollection
  END FOR
  Return mergedClusterCollection
End Algorithm
```

Fig. 2. Algorithm: Generating a set of merged clusters from a set of the clusters in a given kNN connection graph

number and thus much speed up the overall clustering process. The third advantage is that the final clustering output does not depend on the processing sequence of given documents. This is due to that our algorithm is based on the kNN connection graph, which captures the kNN-connectedness among clusters. This makes our approach stable, which only depends on the set of given documents, k(the number of nearest neighbors in kNN) and the number N in controlling the number of top ranked clusters to be merged in each step. In this paper, they are set to 5 and 4 respectively.

4 Experimentation

The kNN connection graph-based hierarchical document clustering is evaluated, using precision, recall and F-measure, against the FudanUniv news corpus. This corpus contains 2816 documents and has been manually classified into 10 classes: education, sports, military, arts, politics, transportation, environment, computer, economics and medicine.

There exists one problem when assessing the agreement between the clustering output and a manually annotated corpus since there is no corresponding class label for each cluster in the clustering output. To resolve the problem, we construct one (a cluster in the clustering output) to one (a class in the annotated corpus) mapping by using a contingency table T, where each entry t_{ij} gives the number of instances that belong to both the i-th estimated cluster and j-th ground truth class. Moreover, to ensure that any two clusters do not share the same class label, we adopt a permutation procedure to find a one-to-one mapping from the ground truth classes TC to the estimated clustering output EC using the hierarchical cluster output without cluster overlapping. In this way, the one-to-one mapping can be performed, which can be formulated as the function $\hat{\Omega} = \arg\max_{\Omega} \sum_{j=1}^{|TC|} t_{\Omega(j)j}$, while $\Omega(j)$ is the index of estimated cluster associated with the j-th class.

Furthermore, we adopt several evaluation measurements as applied in [Hammouda et al 2004] for document clustering:

- Precision of a class, which measures the ratio of members of a cluster in the associated class:

$$P(j) = \Pr ecision(i,j) = \frac{N_{i,j}}{N_i} \tag{2}$$

  Where N_{ij} is the number of members of cluster i in the associated class j and N_i is the number of members of cluster i.
- Recall of a class, which measures the ratio of members of the associated class and a cluster:

$$R(j) = Recall(i,j) = \frac{N_{i,j}}{N_j} \tag{3}$$

  Where N_{ij} is the number of members of cluster i in the associated class j and N_j is the number of members of class$j$.
- F-measure of a class, which integrates the precision and recall:

$$F(j) = \frac{2 * P(j) * R(j)}{P(j) + R(j)} \tag{4}$$

- Overall Precison/Recall/F-measure, which, for the clustering result C, measures the weighted average of the Precison/Recall/F-measure for each class j:

$$P(C) = \frac{\sum_j |c_j| * P(j)}{\sum_j |c_j|} \tag{5}$$

$$R(C) = \frac{\sum_j |c_j| * R(j)}{\sum_j |c_j|} \tag{6}$$

$$F(C) = \frac{\sum_j |c_j| * F(j)}{\sum_j |c_j|} \tag{7}$$

Finally, all the documents are represented using the vector space model with each feature in a vector of a document representing a word in the document. Since there are no separators between Chinese words, the ICTCLAS Chinese word segmentation system [Zhang et al 2003] is applied to segment a Chinese sentence into a sequence of words. The weight of each word in a document is normalized using the standard tf.idf formula.

Table 1 shows the performance of our algorithm for each class on the FudanUniv news corpus using the kNN-based attraction function. It shows that our approach achieves the F-measure of 75.32. It also compares the attraction function with the general cluster-based (cosine) similarity function. It shows that the kNN-based attraction function performs better than the general cluster-based cosine similarity function.

Table 1. Comparison of our kNN connection graph-based hierarchical clustering using the kNN-based attraction function and the cluster-based similarity function

Category	#Doc	kNN-based Attraction Degree			Cluster-based Similarity		
		R(%)	P(%)	F1	R(%)	P(%)	F1
Education	220	75.00	95.93	84.18	83.64	95.34	89.10
Sports	450	95.56	93.68	94.61	96.89	96.67	96.78
Military	249	35.34	98.88	52.07	34.94	98.86	51.63
Arts	248	62.10	95.06	75.12	76.61	99.48	86.56
Politics	505	79.21	76.92	78.05	91.68	76.23	83.42
Transportation	214	69.16	100	81.77	67.29	98.63	80.00
Environment	201	33.33	95.71	49.45	36.32	96.05	52.71
Computer	200	74.00	98.01	84.33	55.00	100	70.97
Economics	325	63.69	83.13	72.12	40.92	83.12	54.85
Medicine	204	43.13	98.88	60.07	42.65	92.55	58.39
Overall	**2816**	**63.05**	**93.62**	**75.32**	**62.59**	**93.69**	**75.01**

Table 2 shows the effect of applying k-means clustering at the initial stage by first clustering given documents into 500 classes and then applying the proposed algorithm in this paper. It shows that that our algorithm achieves the F-measure of 72.75 using the kNN-based attraction function. It also shows that the kNN-based attraction function much outperforms the general similarity function by 2.93 in F-measure.

Comparing Table 1 and Table 2, we can find that, although initial k-means clustering can largely speedup the whole clustering process, it much reduces the performance. It also shows that the kNN-based attraction function is much more robust to the initial merging errors caused by k-means clustering. It is also interesting to note that, with initial k-means clustering, the kNN-based attraction function slightly increases the recall and significantly decrease the precision due to its aggressive merging strategy while the cluster-based similarity function much decrease both the recall and the precision. Finally, Table 3 compares our algorithm with the famous k-means clustering algorithm [Kunungo et al 2002]. For comparison, the cluster number in the k-means clustering algorithm

Table 2. Comparison of our kNN connection graph-based hierarchical clustering with initial k-means clustering into 500 classes using the kNN-based attraction function and the cluster-based similarity function

Category	#Doc	kNN-based Attraction Degree			Cluster-based Similarity		
		R(%)	P(%)	F1	R(%)	P(%)	F1
Education	220	76.82	91.35	83.46	79.09	90.16	84.26
Sports	450	89.33	91.16	90.24	96.22	94.34	95.27
Military	249	33.73	93.33	49.56	23.29	96.67	37.54
Arts	248	65.32	96.43	77.88	80.24	96.60	87.67
Politics	505	83.96	78.66	81.23	66.93	83.46	74.29
Transportation	214	67.76	97.32	79.89	40.19	96.63	56.77
Environment	201	38.31	88.51	53.47	36.32	89.02	51.59
Computer	200	81.00	59.78	68.79	66.00	97.78	78.81
Economics	325	65.85	63.69	64.75	42.15	76.97	54.47
Medicine	204	50.00	53.97	51.91	50.49	53.64	52.02
Overall	**2816**	**65.21**	**81.42**	**72.75**	**58.09**	**87.53**	**69.82**

Table 3. Performance of k-means clustering

Category	#Doc	R(%)	P(%)	F1
Education	220	90.91	93.90	92.38
Sports	450	81.78	98.66	89.43
Military	249	46.18	28.12	34.95
Arts	248	93.15	93.15	93.15
Politics	505	49.70	90.29	64.11
Transportation	214	80.37	69.64	74.62
Environment	201	41.79	29.17	34.36
Computer	200	80.50	54.95	65.31
Economics	325	57.54	85.39	68.75
Medicine	204	49.02	40.32	44.25
Overall	**2816**	**67.09**	**68.36**	**68.20**

is set to the true cluster number of the FudanUniv news corpus. It shows that k-means clustering achieves the F-measure of 68.20. Comparing Table 1 and Table 3, we can find that our kNN connection graph-based hierarchical clustering significantly outperforms k-means clustering by about 7 in F-measure.

5 Conclusion

This paper proposes a novel bottom-up hierarchical document clustering algorithm. Through a KNN connect graph, this algorithm can simultaneously merge several clusters through capturing the kNN-connectedness among clusters. In this way, it can largely reduce the iteration number and much speed up the clustering process. Moreover, a novel kNN-based attraction function is proposed to determine the merging of two clusters. It only considers the relative distribution to their nearest neighbors between two clusters in a vector space while other criteria, such as the cosine similarity, measures the absolute distance between two clusters. This makes our algorithm apply to the cases where different clusters may have very different distance variation. Moreover, the attraction function not only considers the similarity between the given two clusters but also takes into account the effect of their respective kNNs. Evaluation on a news document corpus shows that our algorithm performs better than the widely used k-means clustering algorithm. It also shows that the new kNN-based attraction function can better measure the similarity between two clusters and is more robust to the early merging errors than the general cluster-based similarity function.

In the future work, we will explore our algorithm and the kNN-based attraction function in more corpora with a large number of documents, e.g. the Reuters corpus.

References

1. Breiman L. Bagging Predictors. Machine Learning, 24, 123-140. 1996
2. Croft W.B. Clustering large files of codument using the single-link method. Journal of the American Society for Information Science. 1977
3. K. M. Hammouda, M. S. Kamel. Efficient Phrase-Based Document Indexing for Web Document Clustering, IEEE Transactions on Knowledge and Data Engineering, vol.16, no.10, pp:1279-1296, October 2004
4. Hartigan J.A. and Wong M.A. Algorithm AS 136: A K-means clustering algorithm. Applied Statistics. 1979
5. Kanungo T., Mount D.M., Netanyahu N.S., Piatko C.D. etc. An Efficient k-Means Clustering Algorithm: Analysis and Implementation. IEEE Transactions on Pattern Analysis and Machine Intelligence. 2002
6. J. W. Han and M. Kamber. Data Mining: Concepts and Techniques[M]. CA: Morgan Kaufmann, San Francisco, 2001
7. Wei J.H., He P.L. and Sun Y.H. Research on Text Hierarchical Clustering Algorithm based on K-Means [J]. Computer Applications, 2005, 25(10): 2323-2324
8. Zhen T. Research of Clustering Algorithm Based on Hierarchical and Partitioning Method [J]. Computer engineering and Applications, 2006, 8: 182-184
9. Wu F. and Li S.J. An Efficient Hierarchical Clustering Algorithm [J]. Computer Engineering, 2004, 30(9): 70-71
10. Schutze H. Single-link, complete link and average-link. http://www-csli.stanford.edu/ schuetze/completelink.html Accessed 20/05/2006
11. Steinbach M. Karipis G. A comparison of document clustering techniques. KDD workshop on text mining 2000
12. Willett P. Recent trends in hierarchical document clustering: a critical review. Information Processing and Management. 1988
13. Zamir O. and Etzioni O. Web document clustering: a feasibility demonstration. SIGIR'1998
14. Zhang H.P., Yu H.K. Xiong D.Y. and Liu Q. HHMM-based Chinese Lexical Analyzer ICTCLAS. Proceedings of the Second SIGHAN Workshop on Chinese Language Processing, pp.184-187. 2003
15. Zhao Y. and Karipis G. Evaluation of hierarchical document clustering for document datasets. CIKM'2002.

The Great Importance of Cross-Document Relationships for Multi-document Summarization

Xiaojun Wan, Jianwu Yang, and Jianguo Xiao

Institute of Computer Science and Technology,
Peking University, Beijing 100871, China
{wanxiaojun, yangjianwu, xiaojianguo}@icst.pku.edu.cn

Abstract. Graph-based methods have been developed for multi-document summarization in recent years and they make use of the relationships between sentences in a graph-based ranking algorithm to extract salient sentences. This paper proposes to differentiate the cross-document relationships and the within-document relationships between sentences for multi-document summarization. The two kinds of relationships between sentences are deemed to have unequal contributions in the graph-based ranking algorithm. We apply the graph-based ranking algorithm based on each kind of sentence relationships and explore their relative importance for multi-document summarization. Experimental results on DUC 2002 and DUC 2004 data demonstrate the great importance of the cross-document relationships between sentences for multi-document summarization. Even the system based only on the cross-document relation-ships can perform better than or at least as well as the systems based on both kinds of relationships between sentences.

1 Introduction

Automated multi-document summarization has drawn much attention in recent years. Multi-document summary is usually used to provide concise topic description about a cluster of documents and facilitate the users to browse the document cluster. For example, a number of news services, such as Google News (http:// news.google.com), NewsBlaster (http://www1.cs.columbia.edu/nlp/newsblaster/), have been developed to group news articles into news topics, and then produce a short summary for each news topic. The users can easily understand the topic they have interest in by taking a look at the short summary.

A particular challenge for multi-document summarization is that the document set might contain much information unrelated to the main topic. Hence we need effective summarization methods to analyze the information stored in different documents and extract the important information related to the main topic. In other words, a good summary is expected to preserve the globally important information contained in the documents as much as possible, and at the same time keep the information as novel as possible. In recent years, multi-document summarization has been widely explored in the natural language processing and information retrieval communities. A series of workshops and conferences on automatic text summarization (e.g. DUC[1]), special

[1] http://duc.nist.gov

Y. Matsumoto et al. (Eds.): ICCPOL 2006, LNAI 4285, pp. 131 – 138, 2006.

topic sessions in ACL, COLING, and SIGIR have advanced the technology and produced a couple of experimental online systems. Extraction-based summarization methods have been widely explored by [3, 5, 6, 8, 13, 14].

In recent years, graph-based methods [4, 10, 11] have been proposed for multi-document summarization based on sentence relationships. All these methods make use of the relationships between sentences and select sentences according to the "votes" or "recommendations" from their neighboring sentences, which is similar to PageRank [2] and HITS [7]. However, all the methods have not differentiated different kinds of relationships between sentences, i.e. cross-document relationships and within-document relationships. They all assume that the two kinds of sentence relationships are equally important, which is in fact inappropriate. In this study, we investigate the relative importance of the cross-document relationships and the within-document relationships between sentences in an extended affinity graph based approach. The approach extends previous work by treating each kind of sentence relationship as a separate "modality" and computing sentence information richness based on each "modality". Also, the approach applies a diversity penalty process to capture the novelty of a sentence. Experiments on DUC 2002 and DUC 2004 data are performed and we find that the cross-document relationships between sentences are very important for multi-document summarization. The system based only on the cross-document relationships can always perform better than or at least as well as the systems based on both the cross-document relationships and the within-document relationships between sentences.

The rest of this paper is organized as follows: The proposed affinity graph based summarization approaches are presented in Section 2. The experiments and results are given in Section 3. Lastly, we conclude our paper in Section 4.

2 The Affinity Graph Based Approach

The proposed approach is an extension of previous graph-based summarization methods [4, 10, 11] and consists of the following three steps: (1) different affinity graphs are built to reflect different kinds of relationships between the sentences in the document set respectively; (2) the information richness of the sentences is computed based on each affinity graph and the final information richness of the sentences is either one of the computed information richness or a linear combination of them; (3) based on the whole affinity graph and the information richness scores, a diversity penalty is imposed on the sentences and the affinity rank score of each sentence is obtained to reflect both information richness and information novelty of the sentence. The sentences with high affinity rank scores are chosen to produce the summary.

2.1 Affinity Graph Building

Given a sentence collection $S=\{s_i \mid 1\leq i\leq n\}$, the affinity weight $aff(s_i, s_j)$ between a sentence pair of s_i and s_j is calculated using the Cosine measure [1]. The weight associated with term t is calculated with the tf_t*isf_t formula, where tf_t is the frequency of term t in the corresponding sentence and isf_t is the inverse sentence frequency of term t, i.e. $1+\log(N/n_t)$, where N is the total number of sentences and n_t is the number of the sentences containing term t. If sentences are considered as nodes, the sentence collection can be modeled as an undirected graph by generating a link between two sentences if their affinity weight exceeds 0, i.e. an undirected link between s_i and s_j ($i \neq j$) with affinity weight $aff(s_i,s_j)$ is constructed if $aff(s_i,s_j)>0$; otherwise no link is constructed. Thus, we construct an undirected graph G reflecting the semantic relationship between sentences by their content similarity. The graph G contains all kinds of links between sentences and is called as the Whole Affinity Graph. We use an adjacency (affinity) matrix $M=(M_{i,j})_{n\times n}$ to describe the whole affinity graph G with each entry corresponding to the weight of a link in the graph, i.e. $M_{i,j} = aff(s_i,s_j)$ for $i \neq j$. Then M is normalized to $\tilde{M}$ to make the sum of each row equal to 1.

Similar to the above process, the other two affinity graphs G_{intra} and G_{inter} are also built: the within-document affinity graph G_{intra} is to include only within-document links between sentences (the entries of cross-document links are set to 0); the cross-document affinity graph G_{inter} is to include only cross-document links between sentences (the entries of within-document links are set to 0). Note that given a sentence pair of s_i and s_j, if s_i and s_j belong to different documents, the link between s_i and s_j is a cross-document link (relationship); otherwise, the link is a within-document link (relationship). The corresponding adjacency (affinity) matrices of G_{intra} and G_{inter} are denoted by M_{intra} and M_{inter} respectively. In fact, M_{intra} and M_{inter} can be extracted from M and we have $M=M_{intra}+M_{inter}$. Similar to equation (2), M_{inter} and M_{intra} are respectively normalized to $\tilde{M}_{intra}$ and $\tilde{M}_{inter}$ to make the sum of each row equal to 1.

2.2 Information Richness Computation

Based on the whole affinity graph G, the information richness score $InfoRich_{all}(s_i)$ for a sentence s_i can be deduced from those of all other sentences linked with it and it can be formulated in a recursive form as follows:

$$InfoRich_{all}(s_i) = d \cdot \sum_{all\ j\neq i} InfoRich_{all}(s_j) \cdot \tilde{M}_{j,i} + \frac{(1-d)}{n} \qquad (1)$$

where d is the damping factor set to 0.85. The convergence of the iteration algorithm is achieved when the difference between the information richness scores computed at two successive iterations for any sentences falls below a given threshold (0.0001 in this study).

Similarly, the information richness score for a sentence s_i can be deduced based on either the within-document affinity graph G_{intra} or the cross-document affinity graph G_{inter} and two information richness scores $InfoRich_{intra}(s_i)$ and $InfoRich_{inter}(s_i)$ can be obtained for the sentence s_i, respectively. The final information richness $InfoRich(s_i)$ of a sentence s_i can be either $InfoRich_{all}(s_i)$, $InfoRich_{intra}(s_i)$ or $InfoRich_{inter}(s_i)$, or the linear combination of $InfoRich_{intra}(s_i)$ and $InfoRich_{inter}(s_i)$ as follows:

$$InfoRich(s_i) = \lambda \cdot InfoRich_{intra}(s_i) + (1-\lambda) \cdot InfoRich_{inter}(s_i) \qquad (2)$$

where $\lambda \in [0,1]$ is a weighting parameter, specifying the relative contributions to the final information richness of sentences from the cross-document relationships and the within-document relationships between sentences. If $\lambda=0$, $InfoRich(s_i)$ is equal to $InfoRich_{inter}(s_i)$; if $\lambda=1$, $InfoRich(s_i)$ is equal to $InfoRich_{intra}(s_i)$; and if $\lambda=0.5$, the cross-document relationships and the within-document relationships are assumed to be equally important.

Note that all previous graph based summarization methods have $InfoRich(s_i)= InfoRich_{all}(s_i)$.

2.3 Diversity Penalty Imposition

Based on the whole affinity graph *G* and obtained final information richness scores, a greedy algorithm [15] is applied to impose diversity penalty and compute the final affinity rank scores of sentences as follows:

1. Initialize two sets $A=\emptyset$, $B=\{s_i \mid i=1,2,\ldots,n\}$, and each sentence's affinity rank score is initialized to its information richness score, i.e. $ARScore(s_i) = InfoRich(s_i)$, $i=1,2,\ldots n$.
2. Sort the sentences in *B* by their current affinity rank scores in descending order.
3. Suppose s_i is the highest ranked sentence, i.e. the first sentence in the ranked list. Move sentence s_i from *B* to *A*, and then a diversity penalty is imposed to the affinity rank score of each sentence linked with s_i as follows:
   For each sentence s_j in *B*, we have
   $$ARScore(s_j) = ARScore(s_j) - \omega \cdot \tilde{M}_{j,i} \cdot InfoRich(s_i) \qquad (3)$$
   where $\omega>0$ is the penalty degree factor. The larger ω is, the greater penalty is imposed to the affinity rank score. If $\omega=0$, no diversity penalty is imposed at all.
4. Go to step 2 and iterate until $B=\emptyset$ or the iteration count reaches a predefined maximum number.

After the affinity rank scores are obtained for all sentences, the sentences with highest affinity rank scores are chosen to produce the summary according to the summary length limit.

3 Experiments

3.1 Experimental Setup

Generic multi-document summarization has been one of the fundamental tasks in DUC 2001, DUC 2002 and DUC 2004 (i.e. task 2 in DUC 2001, task 2 in DUC 2002 and task 2 in DUC 2004), and we use DUC 2001 data as training set and DUC 2002 and DUC 2004 data as test sets in our experiments. In DUC 2002, 59 TREC document sets (D088 is excluded from the original 60 document sets by NIST) of approximately 10 documents each were provided and generic abstracts of each document set with lengths of approximately 100 words or less were required to be created. In DUC 2004, 50 TDT (Topic Detection and Tracking) document clusters were provided and a short summary with lengths of 665 bytes or less is required to be created. Note that the TDT topic would not be input to the system. In the process of affinity graph building, the stop words were removed and Porter's stemmer [12] was used for word stemming.

For evaluation, we use the ROUGE [9] evaluation toolkit[2], which is widely adopted by DUC for automatic summarization evaluation. It measures summary quality by

[2] We use ROUGEeval-1.4.2 downloaded from http://haydn.isi.edu/ROUGE/

counting overlapping units such as the n-gram, word sequences and word pairs between the candidate summary and the reference summary. ROUGE toolkit reports separate scores for 1, 2, 3 and 4-gram, and also for longest common subsequence co-occurrences. Among these different scores, unigram-based ROUGE score (ROUGE-1) has been shown to agree with human judgement most (Lin and Hovy, 2003). We show three ROUGE metrics in the experimental results: ROUGE-1 (unigram-based), ROUGE-2 (bigram-based), and ROUGE-W (based on weighted longest common subsequence, weight=1.2). In order to truncate summaries longer than length limit, we use the "-l" or "-b" option in ROUGE toolkit and we also use the "-m" option for word stemming.

3.2 Experimental Results

The affinity graph based systems are compared with top 3 performing systems and two baseline systems (i.e. the lead baseline and the coverage baseline) on task 2 of DUC 2002 and task 2 of DUC 2004 respectively. The top three systems are the systems with highest ROUGE scores, chosen from those performing systems in the tasks of DUC 2002 and DUC 2004 respectively. The lead baseline and coverage baseline are two baselines employed in the multi-document summarization tasks of DUC.

The following four affinity graph based systems are investigated: 1) **Inter-Link:** The system computes the information richness of a sentence based only on the cross-document relationships between sentences, i.e. $InfoRich(si)= InfoRich_{inter}(si)$; 2) **Intra-Link:** The system compute the information richness of a sentence based only on the within-document relationships between sentences, i.e. $InfoRich(s_i)= InfoRich_{intra}(s_i)$; 3) **Union-link:** The system computes the information richness $InfoRich_{inter}(s_i)$ and $InfoRich_{intra}(s_i)$ of a sentence s_i based on the cross-document relationships and the within-document relationships between sentences respectively, and then combines them to get the final information richness score. Typically, we let λ=0.5 to make the two kind of relationships equally important. i.e., $InfoRich(s_i)= 0.5*InfoRich_{intra}(s_i)+0.5* InfoRich_{inter}(s_i)$; 4) **Uniform Link:** The system computes the information richness of a sentence based on the whole affinity graph without differentiating the cross-document relationships and the within-document relationships, as in previous graph based summarization methods, i.e. $InfoRich(s_i)= InfoRich_{all}(s_i)$.

Table 1. System comparison on task 2 of DUC 2002

System	ROUGE-1	ROUGE-2	ROUGE-W
Inter-Link	**0.38668**	**0.08582**	**0.12629**
Union-Link	0.37904	0.07793	0.12293
Uniform Link	0.37621	0.08321	0.12230
Intra-Link	0.35382	0.06305	0.11427
S26	0.35151	0.07642	0.11448
S19	0.34504	0.07936	0.11332
S28	0.34355	0.07521	0.10956
Coverage	0.32894	0.07148	0.10847
Lead	0.28684	0.05283	0.09525

Table 2. System comparison on task 2 of DUC 2004

System	ROUGE-1	ROUGE-2	ROUGE-W
Inter-Link	**0.41338**	**0.09474**	**0.12603**
Uniform Link	0.41072	0.09188	0.12430
Union-Link	0.40510	0.08803	0.12309
S65	0.38232	0.09219	0.11528
Intra-Link	0.37609	0.07097	0.11426
S104	0.37436	0.08544	0.11305
S35	0.37427	0.08364	0.11561
Coverage	0.34882	0.07189	0.10622
Lead	0.32420	0.06409	0.09905

Tables 1 and 2 show the comparison results on DUC 2002 and DUC 2004, respectively. The factor ω is tuned on DUC 2001 and set to 8. We can see from the tables that all the three affinity graph based systems considering the cross-document relationships between sentences (i.e. "Inter-Link", "Union-Link" and "Uniform Link") much outperform the top performing systems and baseline systems. Among the four affinity graph based systems, the system based only on the cross-document relationships between sentences (i.e. "Inter-Link") performs best on both DUC 2002 and DUC 2004 tasks over all three metrics, while the system based only on the within-document relationships between sentences (i.e. "Intra-Link") performs worst. The above observations demonstrate the great importance of the cross-document relationships between sentences for multi-document summarization.

Figures 1 to 4 show the comparison results of the four affinity graph based systems under different values of the penalty degree factor ω. Here, the ROUGE-W comparison results are omitted due to page limit. Seen from the figures, the systems considering the cross-document relationships between sentences always outperform the system based only on the within-document relationships between sentences (i.e. "Intra-Link") under different values of ω. Among the three systems considering the cross-document relationships between sentences, the system based only on the cross-document relationships between sentences (i.e. "Inter-Link") performs better than or

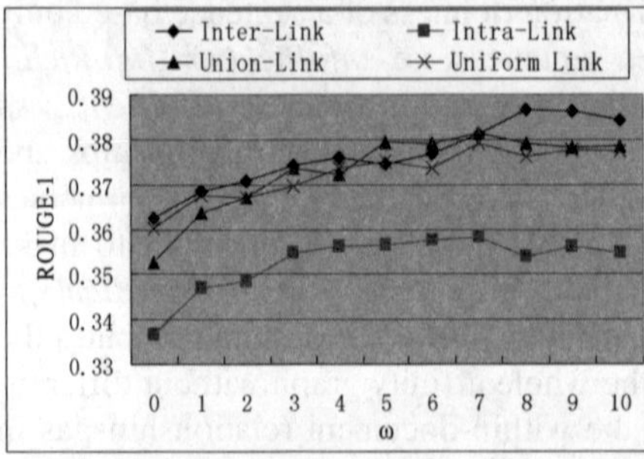

Fig. 1. ROUGE-1 comparison on task 2 of DUC2002

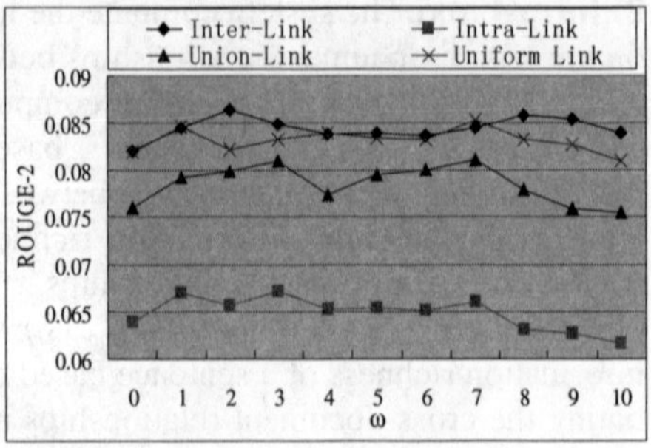

Fig. 2. ROUGE-2 comparison on task 2 of DUC2002

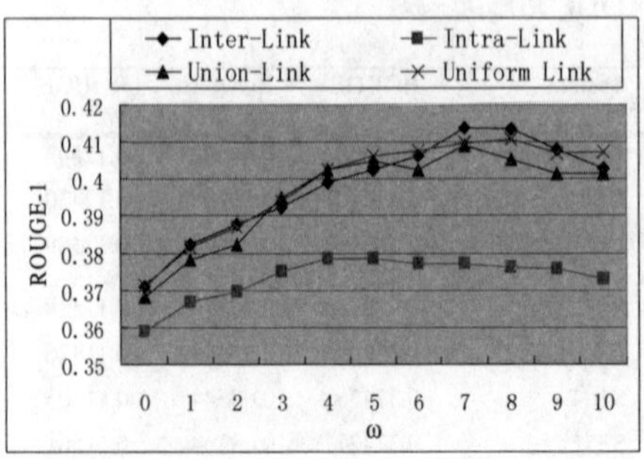

Fig. 3. ROUGE-1 comparison on task 2 of DUC2004

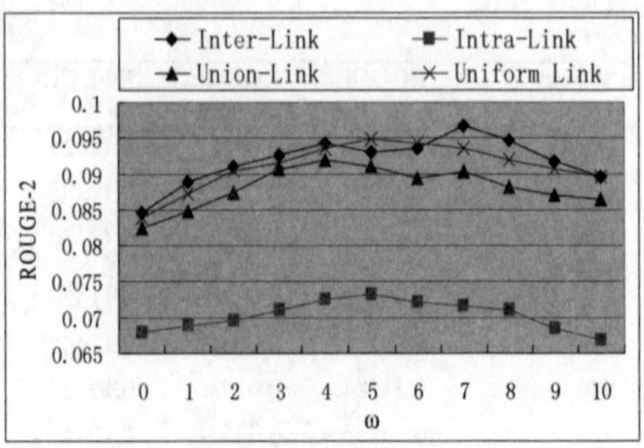

Fig. 4. ROUGE-2 comparison on task 2 of DUC2004

at least as well as the other two systems (i.e. "Union-Link" and "Uniform Link") under different values of ω. This further validates the great importance of the cross-document relationships betweens sentences for multi-document summarization.

In order to investigate how the relative contributions from the cross-document relationships and the within-document relationships between sentences influence the summarization performance, Figures 5-6 show the performances of the "Union-Link" system under different values of the weighting parameter λ. The ROUGE-1 and ROUGE-2 performances of the system without diversity penalty imposition (ω=0) and the system with diversity imposition (ω=8) on DUC 2002 and DUC 2004 are shown in the figures. Seen from Figures 5-6, it is clear that the performance values of the systems decrease with the increase of λ on both DUC 2002 and DUC 2004, which demonstrates that the less relative contributions are given to the cross-document relationships between sentences, the worse the system performance is. The cross-document relationships between sentences are much more important than the within-document relationships between sentences.

The experimental results demonstrate the great importance of the cross-document relationships between sentences for multi-document summarization, which can be explained by the essence of multi-document summarization. The aim of multi-document summarization is to extract important information from the whole document set, in other words, the information in the summary should be globally important on the whole document set. The information contained in a globally informative sentence will be also expressed in the sentences of other documents and the votes or recommendations of neighbors in other documents are more important than the votes or recommendations of neighbors in the same document.

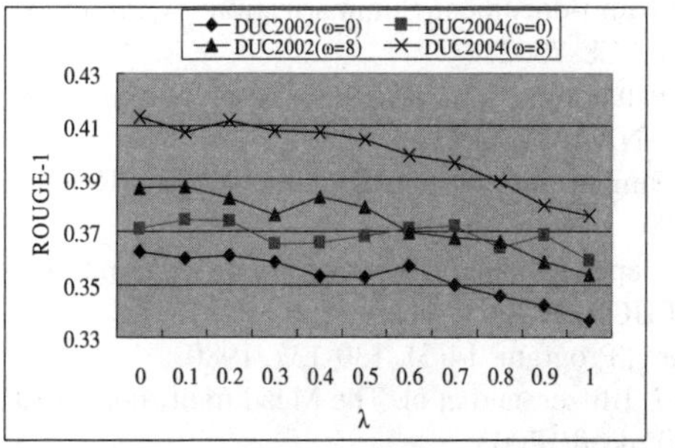

Fig. 5. ROUGE-1 performance of "Union-Link" system under different values of λ

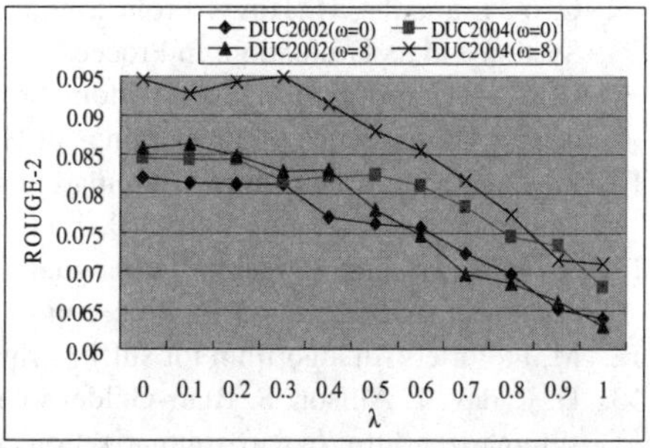

Fig. 6. ROUGE-2 performance of "Union-Link" system under different values of λ

4 Conclusion and Future Work

In this paper we differentiate the two kinds of relationships between sentences for affinity graph based multi-document summarization, i.e. the cross-document relationships and the within-document relationships. Experimental results demonstrate the great importance of the cross-document relationships between sentences. The system can achieve best performance even based only on the cross-document relationships between sentences. Though the experiments were performed on English data sets, it is deemed that similar results and conclusions will be obtained for other languages because the summarization methods explored in this study do not make any use of the characteristics of the specific language.

We will further investigate the importance of the cross-document relationships between sentences in the task of topic-focused multi-document summarization in future work.

References

1. R. Baeza-Yates and B. Ribeiro-Neto. Modern Information Retrival. ACM Press and Addison Wesley, 1999.
2. S. Brin and L. Page. The anatomy of a large-scale hypertextual Web search engine. Computer Networks and ISDN Systems, 30:1–7, 1984.
3. J. Carbonell and J. Goldstein. The use of MMR, diversity-based reranking for reordering documents and producing summaries. In Proceedings of SIGIR'1998.
4. G. Erkan and D. Radev. LexPageRank: prestige in multi-document text summarization. In Proceedings of EMNLP'04
5. S. Harabagiu and F. Lacatusu. Topic themes for multi-document summarization. In Proceedings of SIGIR'05, Salvador, Brazil, 202-209, 2005.
6. H. Hardy, N. Shimizu, T. Strzalkowski, L. Ting, G. B. Wise, and X. Zhang. Cross-document summarization by concept classification. In Proceedings of SIGIR'02, Tampere, Finland, 2002.
7. J. M. Kleinberg. Authoritative sources in a hyperlinked environment. Journal of the ACM, 46(5):604-632, 1999.
8. C.-Y. Lin and E.H. Hovy. From Single to Multi-document Summarization: A Prototype System and its Evaluation. In Proceedings of ACL-2002.
9. C.-Y. Lin and E.H. Hovy. Automatic Evaluation of Summaries Using N-gram Co-occurrence Statistics. In Proceedings of HLT-NAACL 2003.
10. I. Mani and E. Bloedorn. Summarizing Similarities and Differences Among Related Documents. Information Retrieval, 1(1), 2000.
11. R. Mihalcea and P. Tarau. A language independent algorithm for single and multiple document summarization. In Proceedings of IJCNLP'2005.
12. M. F. Porter. An algorithm for suffix stripping. Program, 14(3): 130-137, 1980.
13. D. Radev, T. Allison, S. Blair-Goldensohn, J. Blitzer, and et al. The Mead multi-document summarizer. http://www.summarization.com/mead/, 2003.
14. D. R. Radev, H. Y. Jing, M. Stys and D. Tam. Centroid-based summarization of multiple documents. Information Processing and Management, 40: 919-938, 2004.
15. B. Zhang, H. Li, Y. Liu, L. Ji, W. Xi, W. Fan, Z. Chen, and W.-Y. Ma. 2005. Improving web search results using affinity graph. In Proceedings of SIGIR'05.

The Effects of Computer Assisted Instruction to Train People with Reading Disabilities Recognizing Chinese Characters

Wan-Chih Sun[1], Tsung-Ren Yang[2], Chih-Chin Liang[3], Ping-Yu Hsu[3], and Yuh-Wei Kung[4]

[1] Resource Room, Taipei Bo-Ai Elementary School, No.20, Lane 95, Songren Rd., Sinyi Disrict, Taipei City, Taiwan 110, R.O.C.
lgcwow@gmail.com

[2] Department of Special Education, National Taipei University of Education, No.134, Sec. 2, Heping E. Rd., Da-an District, Taipei City, Taiwan 106, R.O.C.
try@tea.ntue.edu.tw

[3] Department of Management, National Central University, No. 300, Jung-da Rd., Jung-li City, Taoyuan, Taiwan 320, R.O.C.
{92441022, pyhsu}@mgt.ncu.edu.tw

[4] Department of Applied Foreign Languages, Nanya Institute of Technology, No. 414, Sec. 3, Chung-Shan E. Rd., Jung-li City, Taoyuan, Taiwan 320, R.O.C.
parrot@nanya.edu.tw

Abstract. Chinese stem-deriving instruction has been proved to effectively help people with reading disabilities recognize Chinese characters. With the applications and development of information technology, cybernetic Chinese stem-deriving instruction can help more people with reading disabilities learn Chinese characters and peruse articles more effectively. In this study, we develop computer-assisted instruction method for Chinese stem-deriving instruction and compare three teaching strategies. In this work, we recruit three elementary students with reading disabilities as participants, and evaluate the effectiveness of instructing with a proposed teaching strategy.

Keywords: Education, Reading Disabilities, Recognition of Chinese Characters, Interactive Learning Environment, Stem-Deriving Instruction.

1 Introduction

Common instructions used to recognize Chinese characters could be classified as two approaches: character-distributed instruction, and character-centralized instruction [1]. The character-distributed instruction means a student learns Chinese characters by reading the full text of an article. Whenever a student understands the content of the article, s/he recognizes the overall meaning of each sentence and then come to understand each character. A teacher makes students understand Chinese characters mainly

Y. Matsumoto et al. (Eds.): ICCPOL 2006, LNAI 4285, pp. 139 – 148, 2006.

through this instruction [1]. In this case, a student can be instructed to understand the form, articulation, and the meaning of Chinese words. On the other hand, character-centralized instruction deploys the method to classify Chinese characters with the similar visual-orthographic and the same phonetic through which a student can memorize characters and to learn the words. Through Chinese Stem-Deriving Instruction (CSDI), a student can easily associate an unfamiliar Chinese character by combining a familiar stem character (the equivalent with a "root" in English) with a radical or a component that is derived from the character-centralized instruction. However, to learn through CSDI, a student needs to recognize few basic Chinese characters [1].

The CSDI has also been proved very effective for students with reading disabilities to recognize Chinese characters [1]. Previous studies on CSDI have also showed that the character-centralized instruction is feasible for instructing student with poor short-term memory through the CSDI [2].

A previous study has proved the effectiveness of instructing students by the proposed computer-assisted application of the Computer-Assisted Chinese Stem-Deriving Instruction (C-CSDI) was as good as instructing students by the Chinese Stem-Derived Instruction taught by a teacher (T-CSDI) [3]. That is, the effectiveness of instructing through the C-CSDI and the T-CSDI is similar. In Addition, the costs for instruction can be reduced because students can be taught concurrently, and the instruction person-time can be reduced by adopting the computer-assisted instruction [4]. Although a teacher can control the emotion of a student and adjust the teaching strategy in time through the T-CSDI, one teacher would not be able to take care of many students with reading disabilities at the same time. Sometimes, students have difficulties in concentrating on learning. Without sufficient involvement of teachers, adopting the computer-assisted instruction alone would not be able to ensure the effectiveness of multimedia learning. In order to make sure students with reading disabilities learn with attention, a teacher must be properly involved in a computerized instructional application. However, the role of a teacher must be clarified when students are learning through a computerized instructional tool in the classroom [5]. In this study, we propose a teaching strategy: the Teacher Involved Computer-Assisted Chinese Stem-Deriving Instruction (TC-CSDI) to evaluate the effectiveness of instruction, discuss the human interaction design, and provide suggestions and teaching methods for further studies.

2 Background

2.1 People with Reading Disabilities and Chinese Stem-Deriving Instruction

Generally, the people with reading disabilities pay much time and attention on recognizing characters when reading an article [6]. They easily tend to fail to read and understand the article because they are busy in recognizing unfamiliar words and may not be able to understand the meaning and context of the article well [7]. The Chinese characters consist of different visual form and articulation. Word meanings are difficult to learn. Recently, the studies show the Chinese Stem-Deriving Instruction (CSDI), one of the character-centralized instructions, can be used to help people with reading disabilities recognize Chinese characters.

Fig. 1 shows the conceptual diagram of CSDI. One stem character combines different radical characters can help students understand unfamiliar characters. For example, "菁" is composed of "艹" and "青". The Chinese character "青" is a stem character and "艹" is a radical character. CSDI, combining a stem character with a radical or a component, can be used to drive students to recognize unfamiliar characters through familiar characters [7]. Students can easily comprehend relationships among visual form, articulation, and word meanings of Chinese characters, which can be derived from one stem character combined with radical characters.

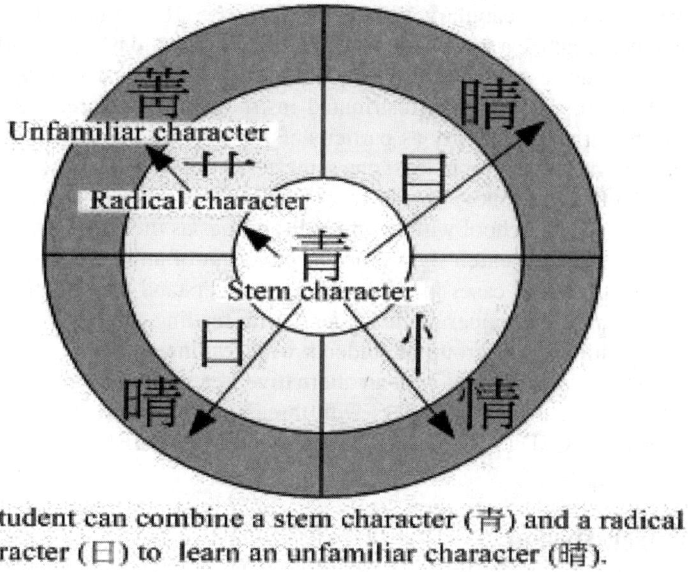

Fig. 1. The conceptual diagram of CSDI

2.2 Computer-Assisted Instruction

Computer-Assisted Instruction (CAI) is used as an instructional tool adopted to improve learning effectiveness, such as understanding form, articulation, and word meanings, in various areas. The instructional tool can provide immediate feedback, monitor individual learning performance, and help people study repeatedly. Moreover, learning motivation can be maintained by the CAI effectiveness, especially for elementary students [1]. CAI can be a means of instruction for teachers to apply an effective teaching strategy by using computers, and using time and other resources wisely [5].

The teacher-involved model is very important for building up a computerized application [8]. The CAI is designed for students' needs without teachers' comments. However, the software affects designers, trainers (or teachers) and users (or students) in the software-developing process) [8].

Students with reading disabilities are unable to pay full attention to CAI without teachers' reminders at all times. Therefore, to build up an application of C-CSDI, a teacher must participate in the whole process of developing software to ensure the application works to instruct students when the teacher is involved [3].

3 Research Method

In Taiwan, although students start learning Chinese characters at the first grade, CSDI cannot be used for instructing students at this grade level. The students must reach a basic level of word vocabulary before receiving CSDI, while first-grade students normally only recognize a few characters and therefore would not be qualified to be the participants in this research. In Taiwan, normal students usually recognize Chinese characters through the character-distributed instruction. Therefore, this study adopts students with learning disability as participants. In addition, a teacher of special education must teach students for at least one semester in order to diagnose the student with reading disabilities in Chinese characters [1]. Therefore, this work selects upper grade students from primary school with learning disabilities as the participants.

The single subject research shows the effectiveness of an instruction can be proved through three successful cases [4]. Therefore, to understand the effectiveness of three different types of CSDI, upper grade students with reading disabilities were chosen as participants. With a small group of students with reading disability, this study adopts single subject research method with an alternative treatments design in experiments to compare the instructing effectiveness with three teaching instruction types: T-CSDI, C-CSDI, and TC-CSDI. The results are presented as statistic analysis and visual analysis.

4 Research Design

This work adopts single subject research method and selects three participants from a pool of 12 students with reading disabilities in one school elementary school in Taipei, Taiwan [3]. These selected participants are qualified to be research subjects because of the following:

- their intelligence quotients are above 70,
- the percentile of their scores in the Grade Chinese Characters Reading Test is under 25 percentile, and
- their reading disabilities are not caused by sensory deficit, emotional disturbance, cultural disadvantages, and inappropriate instructions [9],[10].

Table 1 shows the background information of the three student subjects in this study. This study adopts three instruction approaches: T-CSDI, C-CSDI, and TC-CSDI, to instruct the participants to lean Chinese characters [1]. Each participant has been instructed ten times with T-CSDI, C-CSDI, and TC-CSDI, respectively. The order of three instructions is randomly selected in each instructing session by playing a dice :

---

[1] The detail design of the proposed cybernetic instructional tool has been described [3].

number one and number two indicate a student must be instructed through T-CSDI, number three and number four through C-CSDI, and others through TC-CSDI. A teacher teaches each participant one session a day.

- Using T-CSDI, a teacher follows all instructing procedures, including oral introduction, pre-test, instruction, review, character game, and test. The phase of character game is used for consolidating effects of learning.
- Using C-CSDI, instruction procedure includes pre-test, instruction, review, and character game are operated through a computer. Moreover, the teacher operates the phase of test merely.
- Using TC-CSDI, a teacher introduces the completely instructing process including software instruction of TC-CSDI during the phase of oral introduction. The phases of pre-test, instruction review and character game are operated. A teacher assists participants during instruction and guides participants on the phase of character game in the original design.

Table 1. Participants' profile

		Participant A	Participant B	Participant C
Age		8	8	8
Grade		3rd	3rd	3rd
Gender		Girl	Boy	Boy
The Wechsler Intelligence Scale for Children-III	Verbal	92	89	95
	Performance Scale	102	73	116
	Full score	96	79	105
The Grade Chinese Characters Reading Raw Test /Percentile Rank		49/24	22/2	36/8

First, a pilot test was carried to test the feasibility of the instruction approaches. We found that a teacher could not interrupt the thinking process of participants on the phase of computer instruction from the pilot study. The flow theory shows whenever a person is using a computer, the person might be in the flow experience and the learning effects are reduced if a teacher interrupts those using computer [11]. Therefore, this work suggested that a teacher focus on the first phase of oral introduction to attract participants and the phase of character game in order to reinforce the memory of instructed Chinese characters.

Fig. 2 shows the proposed design of the TC-CSDI. The dotted arrow "assist" means a teacher does not interrupt the teaching process through computer and only assists users when necessary. For example, helping poor readers focus on learning when they are distracted, providing more examples to illustrate the relationship clearly between stems and radicals, and answering students' questions when the teacher adopts TC-CSDI in this study. That is, a teacher helps a participant only when any participant has problem in TC-CSDI.

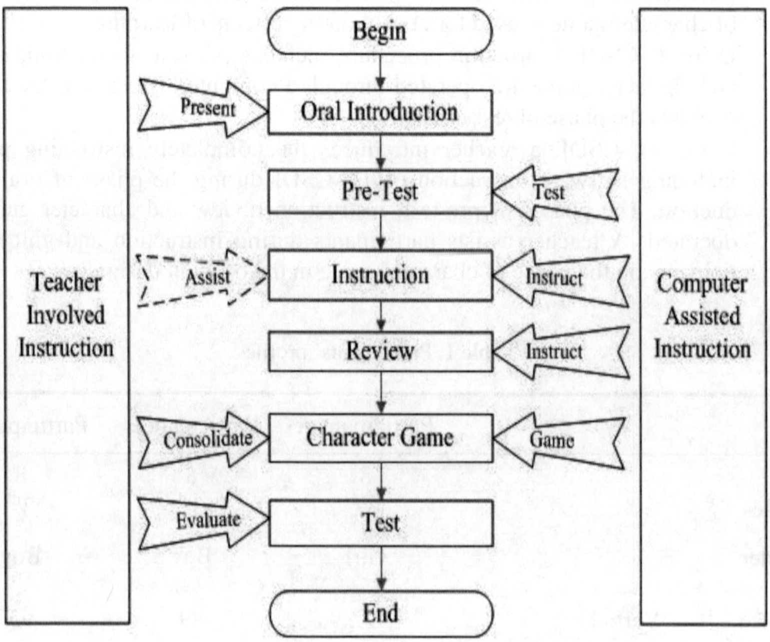

Fig. 2. The Design of the TC-CSDI

Table 2. The Selected Set of Chinese Characters

T-CSDI	C-CSDI	TC-CSDI
重：種、動、衝	生：星、姓、產	寸：射、村、守
丁：頂、訂、釘	交：校、咬、較	欠：吹、資、飲
至：屋、室、到	昜：湯、陽、場	且：祖、粗、助
同：洞、筒、銅	豆：凱、短、逗	亥：孩、該、刻
乍：作、昨、炸	取：趣、最、聚	殳：設、殺、毅
能：熊、態、罷	告：造、酷、浩	白：百、拍、怕
亥：孩、該、刻	皮；被、破、波	車：連、揮、陣
勺：約、釣、的	雚：歡、觀、勸	立：拉、位、啦
子：季、仔、孫	支：枝、翅、鼓	合：給、答、拿
丩：糾、收、叫	青：請、精、睛	古：苦、胡、故

The learning of recognizing characters is unreversed. That is, the effect of recognizing a character through instructing one time is the same as instructing many times. Each set of Chinese characters is only adopted to one instruction once, whenever a set is used to instruct one participant. No matter which instruction is selected, the participant must study another set of Chinese characters. This study selects 30 sets of character through a test of 36 characters used to test 40 students without learning disabilities [5]. Table 2 shows the selected sets of Chinese characters.

5 Results

The learning effectiveness of each teaching strategy, T-CSDI, C-CSDI, and TC-CSDI, of each participant is analyzed by applying statistic analysis and visual analysis. Table 3 presents the average result of ten-time instruction with T-CSDI, C-CSDI, and TC-CSDI.

The average results indicate that the effectiveness of adopting TC-CSDI is better than C-CSDI, which in turn is better than T-CSDI. Whereas, a previous study shows that the effectiveness of T-CSDI and C-CSDI is not significantly different [3]. The experimental results of previous study and this study are different.

Table 3 shows that participant B performs worst among three participants on intelligence quotients and percentile of the result of the graded Chinese Characters Reading Test. Additionally, participant C is the best one. The instructing result of participant A shows 5.5 points gain (about 0.3 times of standard deviation of C-CSDI) from adopting C-CSDI to adopting TC-CSDI. The result of participant B increases 9.38 points (0.34 times of standard deviation of C-CSDI) from adopting C-CSDI to adopting TC-CSDI. The result of participant C is increased 0.5 points (0.03 times of standard deviation of C-CSDI) from adopting C-CSDI to adopting TC-CSDI. Hence, the rising trend seems to imply that the TC-CSDI might be efficient if a participant is a poorer reader. That is, whenever a teacher assists the poorer reader, the improvement is greater than a better reader.

Table 3. The average results of participants

Participant	T-CSDI	C-CSDI	TC-CSDI	Standard Deviation Of C-CSDI
A	74.50	91.00	96.50	18.55
B	65.25	74.95	84.33	27.59
C	84.00	91.75	92.25	18.64
Average	74.58	85.90	91.03	

Fig. 3 shows the result of instructing participant A in each test. However, the visual analysis shows the results of adopting TC-CSDI are within top 20% and the results of adopting T-CSDI grows up stably. The variation of the results of C-CSDI is large. On average, the teacher helps participant A three times during the process of TC-CSDI. The fifth test through C-CSDI shows a different result from the ones of other two instructions. It should be caused by an emotional problem: participant A was distracted by computer games he just had played and the C-CSDI was the first adopted instruction at that time.

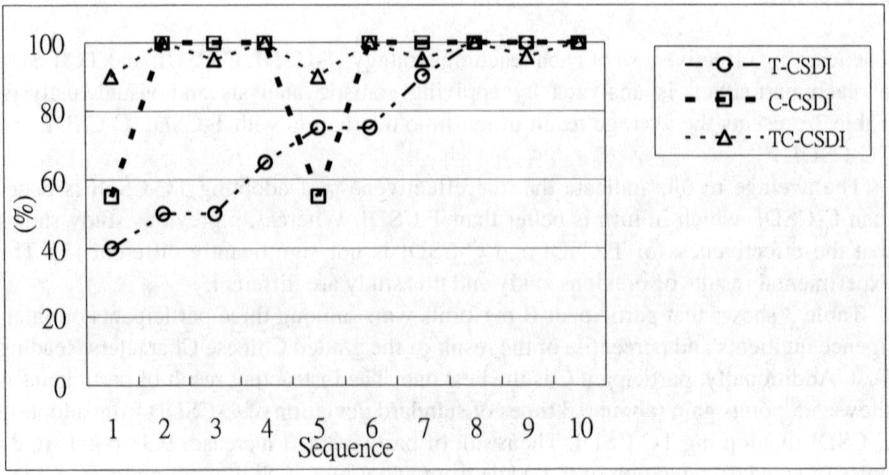

Fig. 3. Result of instructing participant A in each test

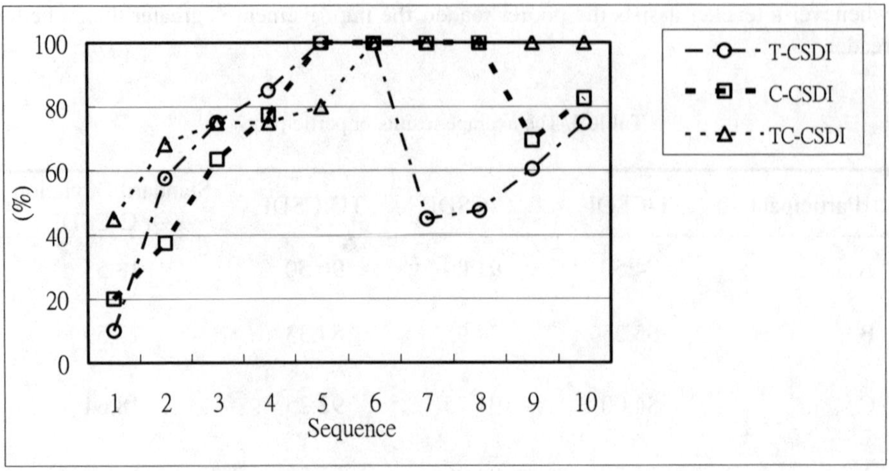

Fig. 4. Result of instructing participant B in each test

Fig. 4 shows the beginning results of participant B are not good, because participant B is not familiar with these instructions at the beginning from observation. However, participant B is familiar with the TC-CSDI in the fifth test, so the results of adopting TC-CSDI are shown 100% since the fifth test. The variation of the results of C-CSDI is large. The results of T-CSDI descend from 100% to almost 40% at the sixth test, because the participant is eager to be instructed by C-CSDI or TC-CSDI at that time. In average, the teacher helps participant B seven times during the process of TC-CSDI.

Fig. 5 shows the result of instructing participant C in each test. The results of T-CSDI, C-CSDI and TC-CSDI are similar. However, the results of TC-CSDI are within top 20%. In average, the teacher helps participant C one time during the process of TC-CSDI.

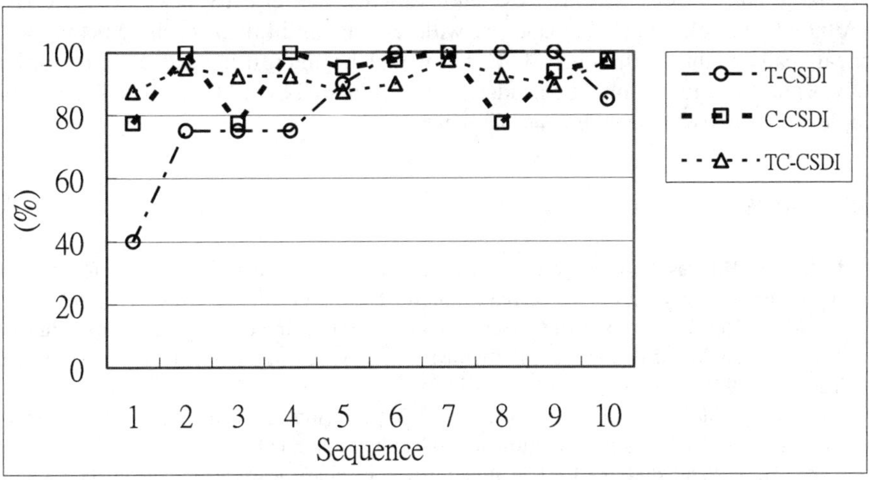

Fig. 5. Result of instructing participant C in each test

Therefore, for these participants, adopting TC-CSDI exhibited higher scores from CSDI than using T-CSDI and C-CSDI.

6 Conclusion and Recommendations

In the past, no proper computer system could assist a teacher to help people with reading disabilities recognize Chinese characters. In this study, we apply computer multimedia to help teachers instruct people with reading disabilities, and compare the difference in effectiveness of three instruction methods.

This study proposes a novel instruction, Teacher-Involved Computer-Assisted Chinese Stem-Deriving Instruction (TC-CSDI), to build up a multimedia application for people with reading disabilities for learning Chinese characters. Finally, this work found the effects of instructing through TC-CSDI for the participant B is better than other participants are. The improved effects of participant B is 0.34 times of standard deviation of C-CSDI from adopting C-CSDI to TC-CSDI that is greater than the improved results of participant A and participant C. Additionally, the IQ and initial reading ability of participant B is the lowest and participant B is easy to be distracted. He needs the teacher's assistance to concentrate on the instructional application. However, the result of adopting TC-CSDI to teach participant B is better than to teach participant A and participant C. We believe that the poorer reader needs teacher's guidance when using computer instruction software, and nevertheless, it needs to be studied in further researches. This work also suggests a proper role of a teacher within the process of instructing students with TC-CSDI.

This and previous works adopt single subject experiments. As the results of both works, to instruct a student through a computer only is not worse than through a teacher. However, we cannot draw a general conclusion that is statistically sound due to the small sample size that we have. Therefore, we recommend that following: 1) enlarge the sample size to test and validate the effectiveness of TC-CSDI. 2) Although the selected participants are with reading disabilitics, teaching people with typical reading ability through TC-CSDI would be interesting and warrant further study. 3) Further study would be needed to determine if TC-CSDI could be feasible for people whose native language is not Chinese.

References

1. Hall, T.E., Hughes, C.A., Filbert, M.: Computer assisted instruction in reading for students with learning disabilities：A research synthesis. Ed Treat Child, 23 (2000) 173-193.
2. Lu, M.C.: The Effectiveness of Chinese Stem-Deriving Instruction on Elementary Students with Severely Word-Recognition Difficulties. National Taiwan Normal University, Taipei, Taiwan (1995).
3. Sun, W. C., Yang, T. R., Liang, C. C., Hsu, P. Y.: Developing a computer-assisted program on Chinese stem-deriving instruction: a pilot study. Spec Ed Q, 90 (2004) 13-18.
4. Teng, H. Y.: The study of CAI in functional vocabulary learning for elementary school Students with moderate mental retardation. National Hualien Teachers College, Taipei, Taiwan (2002).
5. Huang, D.H.: The Study on Improving the Word-Recognition Learning of Elementary Students with Severely Word-Recognition Difficulties. National Taipei University of Education, Taipei, Taiwan (2003).
6. McArthur, G.M., Hogben, J.H., Edwards, V.T., Heath, S.M., Mengler, E.D.:On the "Specifics" of Specific Reading Disability and Specific Language Impairment. J Child Psychology and Psychiatry Allied Discipline, 41 (2000) 869-874.
7. Liu, C. Y.: A Integrated Analysis of the Effects of Word Recognition Teaching. J Secon Ed, 9 (2002) 121-152.
8. Hinostroza, J.E., Mellar, H.G.: Pedagogy embedded in educational software design: report of a case study. Comput Ed, 37 (2001) 27-40.
9. McCandliss, B.D., Noble, K.G.: The development of reading impairment: A cognitive neurosience model. Mental Retardation and Developmental Disabilities Research Reviews, 9 (2003) 196-205.
10. Wu, Y.C.: A Study of Family Surrounding Factors as Related to Children's Reading Abilities. Bulletin of Edu Psyc, 34 (2002) 1-19.
11. Novak, T.P., Hoffman, D.L., Duhachek, A.: The Influence of Goal-Directed and Experiential Activities on Online Flow Experiences. J Consum Psyc, 13 (2003) 3-16.

Discrimination-Based Feature Selection for Multinomial Naïve Bayes Text Classification

Jingbo Zhu, Huizhen Wang, and Xijuan Zhang

Natural Language Processing Laboratory
Institute of Computer Software and Theory, Northeastern University, Shenyang, P.R. China
zhujingbo@mail.neu.edu.cn

Abstract. In this paper we focus on the problem of class discrimination issues to improve performance of text classification, and study a discrimination-based feature selection technique in which the features are selected based on the criterion of enlarging separation among competing classes, referred to as discrimination capability. The proposed approach discards features with small discrimination capability measured by Gaussian divergence, so as to enhance the robustness and the discrimination power of the text classification system. To evaluation its performance, some comparison experiments of multinomial naïve Bayes classifier model are constructed on Newsgroup and Ruters21578 data collection. Experimental results show that on Newsgroup data set divergence measure outperforms MI measure, and has slight better performance than DF measure, and outperforms both measures on Ruters21578 data set. It shows that discrimination-based feature selection method has good contributions to enhance discrimination power of text classification model.

1 Introduction

With the popularity of the World Wide Web, there is an increasing need to provide some effective content-based text processing techniques for handling these huge amounts of unstructured online text data in the Internet. Text classification is a very useful technique as a part of text processing systems, such as text indexing for information retrieval or filtering. The task of text classification is to automatically assign one or more pre-defined categories to natural language texts based on their content. When provided with enough labeled training data, a variety of techniques for supervised learning algorithms have demonstrated remarkable performance for text classification[1][2][3][4][5], such as KNN, Rocchio, SVM, decision tree, Maximum Entropy, naïve Bayes models.

A common problem of text classification is high-dimensional sparse feature space of document representation which is formed using bag-of-words model. To solve the problem, there are two common effective ways, namely feature selection and feature extraction. The term feature selection refers to algorithms that output a subset of the input feature set. Feature extraction algorithms create new features based on transformations or combinations of the original feature set. This paper focuses on feature selection issues. In the procedure of text classification, instead of using all

Y. Matsumoto et al. (Eds.): ICCPOL 2006, LNAI 4285, pp. 149 – 156, 2006.

words in the vocabulary as features, some feature selection methods such as document frequency, Chi statistic, information gain, term strength and mutual information methods are used to select a subset that performs the best under the classification system[6]. It can reduce not only the cost of classification, but in some cases it can also provide a better classification accuracy due to finite sample size effects[7].

In this paper, we focus on the problem of class discrimination issues to improve performance of text classification. Because of difficulty of distinguishing confusable classes, to obtain correct rank among all competing classes, one way is to seek some effective techniques that enlarge separation between the correct class and other competing classes. To achieve this goal, we study a discrimination-based feature selection technique in which the features are selected based on the criterion of enlarging separation among competing classes. We study a discrimination-based feature selection method which discards features with small discrimination capability measured by Gaussian divergence between all possible class pairs in the training data so as to enhance the robustness and the discrimination power of the text classification system.

In section 2, the baseline system is introduced briefly. Class discrimination issues in designing a classifier are discussed in section 3. In section 4 we present the concept of Gaussian divergence. To implement feature selection procedure, a maxmin algorithm is described in section 5. Data collection and experimental results are shown in section 6. At last, we address conclusions and future work in section 7.

2 Baseline System

In recent years Naïve Bayes(NB) approaches have been applied for text classification, and found to perform well. There are two different models in common use, both of which make "naïve Bayes assumption", called multi-variate Bernoulli model and multinomial model. McCallum and Nigam's paper[5] reported that the multinomial NB model usually performs better than the multi-variate Bernoulli NB model. In this paper we use multinomial NB approach for text classification. We only describe multinomial NB model briefly here since full details have been presented in paper [5]. The basic idea in naïve Bayes approaches is to use the joint probabilities of words and categories to estimate the probabilities of categories when a document is given. Suppose that a document is represented as a vector $d_i=\{x_{i1},x_{i2},\ldots x_{i|V|}\}$, where x_{it} indicates how often the word w_t occurs in d_i. Given a document d_i for classification, so the most likely class c^* for a document d_i could be computed as

$$c^*(d_i) = \arg\max_j p(c_j)p(d_i \mid c_j) = \arg\max_j p(c_j)\prod_{t=1}^{|V|} p(w_t \mid c_j)^{n(w_t,d_i)} \quad (1)$$

Where $p(c_j)$ is the class prior probabilities, $n(w_t|d_i)$ is the frequency of word w_t in document d_i, $|V|$ is the size of vocabulary V, w_t is the tth word in the vocabulary, and $P(w_t|c_j)$ thus represents the probability that a randomly drawn word from a randomly drawn document in category c_j will be the word w_t. We can calculate Bayes-optimal estimates for these parameters from a set of labeled training data.

Document frequency thresholding(DF) and mutual information(MI) are two criterions commonly used to perform scoring functions for feature selection in text classification in some literatures. In text classification experiments, feature selection is done by using DF and MI measures, respectively, comparing with our divergence-based measure. Details about DF and MI measures are discussed in the yang's paper[6]. Here we only briefly introduce them as follows.

DF thresholding is a simple technique for vocabulary reduction with approximate linear computational complexity. The basic assumption considered is that rare terms are either non-informative for class prediction, or not influential in global performance. In DF-based feature selection procedure, a feature would be discarded if its document frequency was less than a predetermined threshold.

Suppose that let t be a term and c be a class, A be the number of times t and c co-occur, B be the number of times t occurs without c, C be the number of times c occurs without t, and N be the total number of documents. The mutual information criterion between t and c could be estimated using

$$\mathrm{I(t,c)} \approx \log \frac{A \times N}{(A+C) \times (A+B)} \tag{2}$$

In comparison experiments, to perform a global feature selection, we use $\mathrm{MI}_{\max}$ measure as follows:

$$I_{\max}(t) = \max_{i=1}^{m}\{I(t, c_i)\} \tag{3}$$

3 Discrimination Issues

In general, a text classification system uses direct rank ordering method to assign a correct class to input text in which all competing classes are ordered decreasingly. We always consider the top-1 candidate class as the correct assignment in single-label classification task. Intuitively, a good performance of a classifier could be achieved if the correct rank ordering can be obtained. Unfortunately, the metric for measuring the rank ordering of all competing classes is usually unknown. So to obtain the correct rank ordering, we are interested in finding a discrimination function that well preserves the correct rank ordering. We assume that each of the classes to be compared is characterized by a classification model such as naïve Bayes models, and the parameters of the classification model are estimated from a given training data[8].

The goal of text classification system is to achieve a better performance in the testing data. In other words, we hope to obtain the correct ranking order in the testing set. It means that we must deal with the possible mismatch of parameters of NB classifier between the training set and testing set. One way to achieve the goal is to enlarge the interclass distance, and reduce the intra-class variance so that a maximum separation between the correct class and other competing classes is obtained. By reserving a large tolerance region in the neighborhood of the decision boundary between two competing classes, we think that the discrimination power of the classifier would be increasing[8]. This goal can be achieved by using discrimination-based feature selection approaches which we focus on in this paper.

4 Gaussian Divergence

In this section, we present the concept of divergence which is a measure of dissimilarity between two classes, and could be used to determine feature ranking and to evaluate the effectiveness of class discrimination[9]. Suppose that let $p_i(X)$ and $p_j(X)$ be the probability density functions of class C_i and class C_j, respectively. For a sample X, the discriminating information for class C_i versus class C_j may be measured by the logarithm of the likelihood ratio, and the averaged discrimination information for class C_i is given by[8]

$$I_{ij}(X) = \int_X p_i(X) \cdot \ln(p_i(X) / p_j(X)) dX \tag{4}$$

To make the form symmetric to classes C_i and C_j, the total averaged information for discriminating class C_i from C_j, often referred to as divergence which is given by

$$D_{ij}(X) = I_{ij}(X) + I_{ji}(X) \tag{5}$$

If $p_i(X)$ and $p_j(X)$ are known exactly, $I_{ij}(X)$ has a close relationship with the recognition rate for discriminating classes C_i and C_j. The system with a larger $I_{ij}(X)$ often has a small error rate. If the probability densities $p_i(X)$ and $p_j(X)$ are multivariate Gaussian densities with mean vector $\mu_i(X)$ and $\mu_j(X)$, and covariance matrix $V_i(X)$ and $V_j(X)$, respectively, then divergence $D_{ij}(X)$, referred to as Gaussian divergence used in this paper, could be calculated by[9]

$$\begin{aligned} D_{ij}(X) &= \frac{1}{2} \cdot tr[(V_i(X) - V_j(X)) \cdot (V_j^{-1}(X) - V_i^{-1}(X))] \\ &+ \frac{1}{2} \cdot tr[(V_i^{-1}(X) - V_j^{-1}(X)) \cdot (\mu_i(X) - \mu_j(X)) \cdot (\mu_i(X) - \mu_j(X))'] \end{aligned} \tag{6}$$

where tr(A) is the trace of matrix A.

5 Feature Selection

Without any loss of generality, we define the problem of feature selection procedure as follows: Let Y be the original feature set with cardinality n. To find the best feature subset $X\,(\subseteq Y)$ of size d, we first define the feature selection criterion function for the subset X be denoted by $J(X)$. We consider a higher value of J to indicate a better feature subset. So the best subset $X^*\,(\subseteq Y)$ of size d could be formed by

$$\mathrm{J}(\mathrm{X}^*) = \max_{X \subseteq Y, |X| = d} J(X) \tag{7}$$

Now we see that an exhaustive searching procedure is required to produce the optimal feature subset[7]. Making such exhaustive searching is impractical for even moderate values of *n*. But in most cases, evaluating such criterion in (7) requires costly and completely computations for each possible subset, possibly turning feature subset selection into a combinatorial problem. In the feature selection for text classification, one common way to overcome this exhaustive searching problem is that a conditional independence assumption is introduced. In practice, this assumption is rarely met. But we could see that this hypothesis allows the selection of feature subset of any size without the need for an exhaustive search[10].

In the paper[11], Schneider proposed a feature selection score for text classification based on the KL-divergence between the distribution of words in training documents and their classes. To alleviate computation complexity of KL-divergence estimation, an approximation estimation method was adopted under two assumptions: the number of occurrences of a feature is the same in all documents that contain the feature, and all documents in the same class have the same length. However, in practice, both conditions are always not satisfied well.

In our approach, we focus on discrimination capability of features, and consider that goodness of a feature subset is measured though a criterion based on class separability same as the concept of discrimination capability. We think some features which will lead to a large divergence are more important ones, since they carry more discriminatory information. Thus, we may rank the importance of each feature according to its associated divergence. Any feature that makes a small contribution to the total divergence may be discarded in the procedure of feature selection.

To implement the feature selection, we adapt "maximin" algorithm which was originally used to recognition of E-set letters in speech recognition filed by Su[8]. The problem of original "maximin" algorithm to feature selection could be defined as follows: For a given feature set, it first computes the worst case divergence value among all class pairs, and then finds the best subset that maximizes the worst divergence among all possible subset. To solve the exhaustive searching problem in feature selection procedure, the class conditional independence assumption is introduced in our algorithm shown in table 1.

Table 1. Description of our maximin algorithm

```
MaximinFeatureSelection(features, classes, d)
Begin
   For each feature in features
      Evaluate mean vector and the covariance matrix for this feature of each class in classes;
      Using (6) to evaluate the Gaussian divergences of this feature between all possible class
pairs in classes;
      Find the smallest Gaussian divergence value of this feature as D_min;
   EndFor
   Sort all D_min in descending order, then the sequence of features is obtained.
   Return top-d features as results
End
```

6 Experimental Results

6.1 Comparison Experiment on Newsgroups

The first data set is the Newsgroups set which contains approximately 20,000 newsgroup documents. It is partitioned evenly across 20 different newsgroups. In the data preprocessing procedure, we discard the subject line and remove the words that occur only once in the data set or on a stoplist with no stemming, after this processing there are 62264 words left. In the experiment, we use five trials with 20% of the data held out for testing. Experimental results are reported as average classification accuracy across trials.

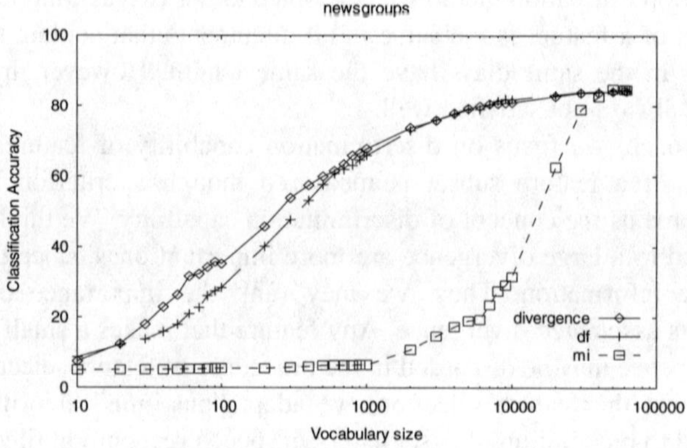

Fig. 1. A comparison of three feature selection methods for different vocabulary size on the Newsgroup data set

Figure 1 shows results on the Newsgroup data set, and shows for all feature selection methods NB classification model do best at the maximum vocabulary sizes. Our divergence-based method outperforms greatly MI measure, and has slight better performance than DF measure. As discussed in the paper[6], a weakness of MI measure is that the score is strongly influenced by the marginal probabilities of terms. Maybe rare terms will have a higher score than common terms. In yang's comparison experimental results[6], DF measure shows good performance for text classification task. However, in Newgroups data set, many of the categories belong to confusion classes, such as comp.* discussion group including five categories: *comp.graphics, comp.os.ms-windows. misc, comp.sys.ibm.pc.hardware, comp.sys. mac.hardware, and comp.windows.x.* Our preliminary Experiments showed that it is difficult to classify well these confusable classes in classification procedure. Our divergence-based method is a discrimination-based method which focuses on discrimination capability between all possible class pairs. In our approach, an important feature will be selected if it has higher discrimination capability measured by Gaussian divergence between all possible class pairs. We think these features with high discrimination capability have good contributions to improve the classification performance.

6.2 Comparison Experiment on Reuters-21578

The second comparison experimental results are based on the Reuters-21578 widely used in the literatures. The ModApte train/test split is used to divide Reuter-21578 into training and test documents. The data set contains 12902 newswire articles in 135 overlapping topic categories. In our comparison experiment, only ten topic categories including *acq, corn, crude, earn, grain, interest, money-fx, ship, trade,* and *wheat* are used to build binary classifiers. In the data preprocessing, words on a stopilist are removed, and we do not use stemming. The result vocabulary has 23444 words. Results on Reuter-21578 are shown as macro-averaged precision-recall breakeven points, a standard information retrieval measure for binary classification. The precision-recall breakeven point is the value at which precision and recall are equal.

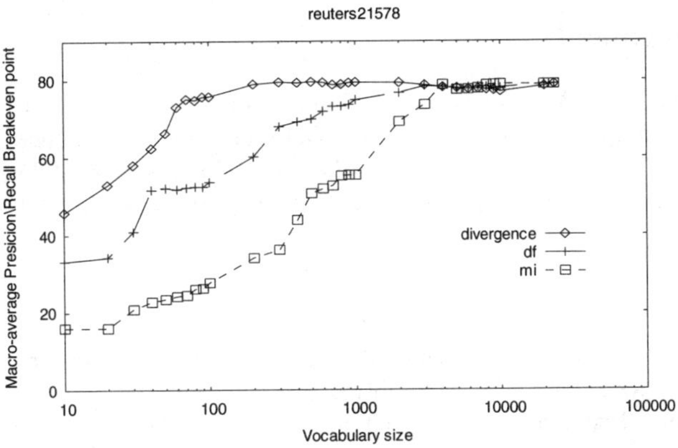

Fig. 2. A comparison of three feature selection methods for different vocabulary size on the Reuter-21578 data set

Figure 2 shows that our divergence measure outperforms DF and MI measures. Through data analysis, we find the phenomenon of confusion class in Reuter-21578 is very serious, because many stories are assigned to multiple categories. Using the ModApte split, for example, *corn, grain, wheat* categories evidently belong to confusion classes. It is very difficult to effectively distinguish these classes. As mentioned above, our divergence-based approach focuses on discrimination capability of a feature to improve discrimination power of text classification model. We think divergence measure has better discrimination capability for feature selection than DF and MI measures. From Figure 2 we could see a surprised phenomena that DF and MI measures do best at the maximum vocabulary sizes; however, divergence measure reaches the best performance when vocabulary size is around 500. It is claimed that Gaussian divergence method is able to discard less discriminative features to improve the classification performance. In our intuitive opinions, enlarging vocabulary size should improve quality because more discriminative features are available when vocabulary becomes bigger. But the fact is not true. We think the main reasons are that in Reuter categorization tasks, for several of the categories high accuracy can be obtained with only a handful of words, sometimes even the single word that is the title of the category[2][5]. Our results are consistent with such results.

7 Conclusion

In this paper we discuss the problem of class discrimination issues in designing text classifier, and study a discrimination-based feature selection technique in which a feature will be selected according to its discrimination capability. The higher discrimination capability is, the more important is. The proposed approach discards features with small discrimination capability measured by Gaussian divergence in the feature selection procedure, so as to enhance the robustness and the discrimination power of the text classification system. Experimental results show Discrimination-based feature selection method has good contributions to enhance discrimination power of text classification model than DF and MI measure. In the future, we will study further Gaussian Mixture Model(GMM) for text classification.

Acknowledgements

We thank Prof. Keh-Yih Su for some kind discussions related to this work. This research was supported in part by the National Natural Science Foundation of China(No.60473140), 985 project of Northeastern University(No.985-2-DB-C03) and by Program for New Century Excellent Talents in University(No.NCET-05-0287).

References

1. D. Lewis, R. Schapire, J. Callan, and R. Papka, Training Algorithms for Linear Text Classifiers, Proceedings of ACM SIGIR, pp.298-306, 1996.
2. T. Joachims, Text categorization with Support Vector Machines: Learning with many relevant features. In Machine Learning: ECML-98, Tenth European Conference on Machine Learning, pp. 137--142. 1998
3. D. Lewis, A Comparison of Two Learning Algorithms for Text Categorization, Symposium on Document Analysis and IR, 1994
4. K. Nigam, J. Lafferty, and A. McCallum, Using maximum entropy for text classification. In IJCAI-99 Workshop on Machine Learning for Information Filtering, p 61-67, 1999
5. McCallum and K. Nigam. A comparison of event models for naive bayes text classification. In AAAI-98 Workshop on Learning for Text Categorization, 1998.
6. Yiming Y. and J. O. Pedersen, A comparative study on feature selection in text categorization, in 14th international conference on machine learning, p412-420, 1997
7. Anil Jain and Douglas Zongker, Feature selection: evaluation, application, and small sample performance, IEEE transactions on pattern analysis and machine intelligence, 19(2), p153-158, 1997
8. K.Y. Su, C.H. Lee, Speech recognition using weighted HMM and subspace projection approach, IEEE transactions on speech and audio processing, 2(1), p69-79, 1994
9. Julius T.Tol, and Rafael C. Gonzalez, Pattern recognition Principles, Addison-Wesley publishing company, 1974
10. Macro Bressan and Jordi Vitria, On the selection and classification of independent features, IEEE transactions on pattern analysis and machine intelligence, 25(10), p1312-1317, 2003
11. Karl-Michael Schneider, A new feature selection score for multinomial naïve Bayes text classification based on KL-divergence, 42nd Annual meeting of the association for computational linguistics, 2004

A Comparative Study on Chinese Word Clustering*

Bo Wang and Houfeng Wang

Institute of Computational Linguistics, School of Electronic Engineering and Computer Science, Peking University, Beijing, 100871, China
{wangbo, wanghf}@pku.edu.cn

Abstract. This paper evaluates four unsupervised Chinese word clustering methods, respectively maximum mutual information (MMI), function word (FW), high frequent word (HFW), and word cluster (WC). Two evaluation measures, part-of-speech (POS) precision and semantic precision, are employed. Testing results show that MMI reaches the best performance: 79.09% on POS precision and 49.75% on semantic precision, while the other three exceed 51.09% and 29.78% respectively. When applying word clusters generated by the methods mentioned above to the alignment-based automatic Chinese syntactic induction, the performance is further improved.

Keywords: word clustering, syntactic parsing, alignment-based learning, unsupervised learning.

1 Introduction

The main problem in constructing language models is data sparseness, for many events with little probability don't happen in a sample set. Word clustering methods might solve such problem to some extent and facilitate syntactic parsing as well.

There has been an amount of previous work using local distributional information to cluster. Finch and Chater[3] utilized a set of features derived from the co-occurrence statistics of common words, for sufficiently frequent words, to produce satisfactory categories. Brown et al [6] made use of a mass of data and a well-founded information theoretic model to induce large numbers of plausible semantic and syntactic clusters.

In this paper, we mainly focus on evaluating performance of four methods employed to Chinese word clustering. The first one is maximum mutual information (MMI) based on the n-gram class model proposed by Brown[6], where we employed a bigram model. The second is called function word (FW), which uses distribution information of function words with respect to a given content word to characterize the content word's context distribution. The third is high frequent word (HFW), in which the most frequent 1000 words are selected to depict the two dimensional probability distribution of each word's local context. The last one is word cluster (WC), in which each word's context

* This work is supported by National Natural Science Foundation of China (No. 60473138, No.60675035) and National Social Science Foundation of China (No. 05BYY043).

Y. Matsumoto et al. (Eds.): ICCPOL 2006, LNAI 4285, pp. 157 – 164, 2006.

distribution is calculated by the two dimensional probability distribution over the two clusters before and after it.

The rest of the paper is organized as follows. Section 2 describes four word clustering methods in detail; section 3 gives the experimental evaluation of the four methods, and demonstrates the improvement of performance when applying word clusters to alignment-based learning method [11][12] of Chinese syntactic parsing; the last section comes to a conclusion.

2 Word Clustering Methods

Four clustering methods of Chinese words, MMI, FW, HFW and WC, are examined in this paper respectively.

2.1 MMI

Cross entropy or perplexity is one of the evaluation criteria for a language model, which measures the uncertainty of it. This paper adopts a partition function π, which maps a word w_i to its cluster c_i, to reduce the perplexity of a n-gram language model. Considering a bigram language model, the cross entropy of the training corpus $L = w_1 w_2 \cdots w_N$ can be calculated using (1).

$$H(L,\pi)=-\frac{1}{N}\log p(w_1 w_2 \cdots w_N)\approx-\frac{1}{N-1}\sum_{w_1 w_2} C(w_1 w_2)\log p(w_2|w_1). \tag{1}$$

Suppose that the probability of one word in the current cluster only depends on the cluster which the previous word belongs to, then (1) can be rewritten as (2).

$$H(L,\pi)=-\frac{1}{N-1}\sum_{w_1 w_2} C(w_1 w_2)\log\left(p(c_2|c_1)p(w_2|c_2)\right). \tag{2}$$

Further, (2) can be reduced as (3).

$$\begin{aligned} H(L,\pi) &\approx -\left[\sum_{w_1 w_2}\frac{C(w_1 w_2)}{N-1}\left[\log p(w_2|c_2)+\log p(c_2)\right]+\sum_{w_1 w_2}\frac{C(w_1 w_2)}{N-1}\left[\log p(c_2|c_1)-\log p(c_2)\right]\right] \\ &= -\left[\sum_{w_2}\frac{\sum_{w_1} C(w_1 w_2)}{N-1}\log p(w_2|c_2)p(c_2)+\sum_{c_1 c_2}\frac{C(c_1 c_2)}{N-1}\log\frac{p(c_2|c_1)}{p(c_2)}\right]=H(w)-I(c_1,c_2) \end{aligned} \tag{3}$$

(3) indicates that the cross entropy can be minimized by an appropriate partition function. As a result, the clusters which maximize the mutual information can be considered optimum clustering result. In practice, word clusters can be obtained by clustering. For a bigram language model, an exchange algorithm was proposed to implement it, which works as Fig1.

```
Repeat
  For every word w in corpus
    Get the current cluster cur to which w belongs
    For every cluster c in corpus
      Calculate the increase of mutual information if w is moved from cur to c
    Determine the target cluster tar to which if w is moved from cur the mutual
    information will increase fastest.
    Move w from cur to tar
Until the stopping condition is satisfied.
```

Fig. 1. Implementation for MMI

Suppose all words will be partitioned into G clusters, the most frequent G-1 words, as an initial condition, will be assigned to its own cluster respectively, and the remaining words are assigned to cluster G. The stopping criterion is either having executed a pre-specified number of iterations or no more words are moved.

2.2 FW

Function words play a very import part in parsing since they contain rich syntactic information. As a result a content word can be characterized by the distribution of all function words with respect to it. Some typical Chinese function words are listed in Table1.

Table 1. Some typical function words in our experiment

方位词(orientation word)：左 右 东 南 东部 西部 南部 北部 中部…
助词(auxiliary word)：的 了 等 地 着 …
介词(preposition)：在 对 为 与 从 自 自从 以 把 向 …
连词(conjunction)：关于 为了 按照 依照 因为 除了 …

The distribution of a given function word relative to a given content word is calculated according to the former occurrence frequency information in certain window of the latter. The window size is defined as the distance (measured in number of words) that either side of the content word we consider. Take a toy example on the following sentence *NBA 球星麦迪曾在中国呆过几天*. Suppose the window size is 4, the occurrence of function word *在* with respect to content word *中国* can be calculated as Table2. To sum up, the occurrence frequency of a given function word can be calculated by window. For the corpus of *People's Daily* in January of 1998, the occurrence frequency of *在* in relation to *中国* is illustrated as Fig2.

Each relative position in the window of a given content word can be treated as a separate dimension, and as a result, it's possible to represent the above graph as a vector by putting together the occurrence frequency in each dimension, hence, the distribution of function word *在* with respect to content word *中国* can be calculated as Fig3.

Table 2. Demonstrating how to calculate occurrence of 在 relative to 中国

Word	NBA	球星	麦迪	曾	在	中国	呆过	几	天
Position	1	2	3	4	5	6	7	8	9
Relative Position		-4	-3	-2	-1	0	1	2	3
Occurrence Freq		0	0	0	1	0	0	0	0

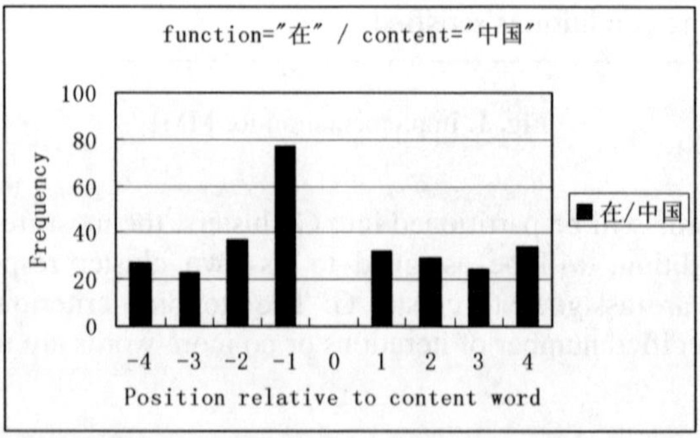

Fig. 2. Histogram shows how 在 occurs in the window of 中国, with window size 4

Relative Position	Frequency
-4	27
-3	23
-2	37
-1	77
0	0
1	32
2	29
3	24
4	34

27
23
37
77
0
32
29
24
34

Fig. 3. Demonstrating how the distribution is represented as a vector

The distribution profile for a given content word can be represented by a vector, which can be created by concatenating the distribution information of all the functions with respect to it. After the distribution information is calculated, the distribution vector for each content word should be normalized and a partition-based clustering method is applied to clustering words.

2.3 HFW

The context of a given word can be represented by its local context for simplification, which is an order pair of the previous word and the next word of it. Therefore the context distribution of a given word can be represented by a two dimensional probability distribution over a two dimensional stochastic variable <*PREVIOUS, NEXT*>.

The Kullback-Leibler distance can be used to describe the similarity of the distribution of two words' context, see (4).

$$D(p_1 \| p_2) = \sum_{i=1}^{V} p_1(i) \log \frac{p_1(i)}{p_2(i)} \cdot \tag{4}$$

Here, V denotes all the words that appear in local context of word w, p_1 denotes the probability distribution of some word. To acquire a symmetric distance measure, divergence is adopted as the measurement of the distance (5).

$$Div(p_1, p_2) = D(p_1 \| p_2) + D(p_2 \| p_1) \cdot \tag{5}$$

Suppose the previous word is independent of the next one, the distance of two words can be defined as (6).

$$Dist(w_1, w_2) = Div(p_1^{left}, p_2^{left}) + Div(p_1^{right}, p_2^{right}) \ . \tag{6}$$

2.4 WC

There are two drawbacks for HFW. First, the data is too sparse to estimate the context distribution adequately for any but the most frequent words. Second, the two positions of one word's local context are not dependent. Both problems can be solved by means of clusters: appropriate the context distribution of a given word as a probability distribution over ordered pairs of clusters multiplied by the conditional distributions of the words given in the clusters as formula (7).

$$p(\langle w_1, w_2 \rangle) = p(\langle c(w_1), c(w_2) \rangle)\, p(w_1 | c(w_1))\, p(w_2 | c(w_2)) \cdot \tag{7}$$

We write $c(w)$ for the cluster word w is in, or for brevity c_1 for $c(w_1)$, and so on. Note that under this model, when the conditional distribution is the same for p_1 and p_2, Kullback-Leibler distance can be simplified as (8).

$$KL(p_1 \| p_2) = \sum_{c_1, c_2} p_1(\langle c_1, c_2 \rangle) \log \frac{p_1(\langle c_1, c_2 \rangle)}{p_2(\langle c_1, c_2 \rangle)} \tag{8}$$

(8) means that the divergence between the distribution over all words is just that of the context distribution over the clusters. Based on the context distribution represented by word cluster, we utilized an iterative algorithm to obtain word clusters.

In practice we have a preliminary clustering no very rare words will be included, and some common words will also not be assigned owing to that they are ambiguous or they have idiosyncratic distributional properties. The more to be pointed out is, clustering quality may be depended on the selection of initial candidate.

3 Experimental Evaluation

3.1 Performance Measures

In our experiment, we proposed two evaluation criteria: part-of-speech (POS) precision and semantic precision.

The automatic approach to evaluate a given cluster based on POS works as follows: For each word, lookup in the corpus and find all the tags that it has been assigned to; determine the most common tag for the cluster; calculate the ratio of the number of words in the cluster that possess the most common tag over the total number of words in the cluster and return it as a percentage.

A Chinese thesaurus dictionary 同义词词林(TongYiCiCiLin) is employed to calculate semantic similarity, which is formed in tree structure, the words which have the same ancestor are considered similar in semantic meaning. The first-level nodes were selected as delegates of semantic meaning. The semantic precision is defined as the ratio of the number of the most common semantic delegate to the total number of words in the cluster.

3.2 Experimental Settings

We selected *Peoples' Daily* in January of 1998 as the training corpus, which was segmented and POS tagged in advance. The total number of words is 1093083. In FW, 233 different function words were contained, the window size was set to 2, i.e., the distance of two words were measured by Manhattan distance. The most 1000 frequent words were used to represent the context probability distribution for high frequent words method. In order to evaluate clustering results fairly, the same number of words and clusters was pre-specified. For each clustering method, 4096 words were clustered within 120 clusters.

3.3 Experimental Results

Table3 shows that POS precision for each method surpasses 50% and semantic precision is above 30%. For MMI, the semantic precision even reaches 50% and POS precision 79.1%, especially, it doesn't rely on any initial knowledge while the other three are dependent of some initial selection, in which inappropriate initial selection may lead to some wrong clustering result.

Table4 shows some clusters from the MMI's clustering result at random, which can precisely make human name, location name, and other proper nouns, time words, and

Table 3. Comparative results of various word clustering methods

Method	HFW(%)	WC(%)	FW(%)	MMI(%)
POS Precision	53.09	51.09	62.61	**79.09**
Semantic Precision	29.78	32.63	36.68	**49.75**

Table 4. Some clusters extracted at random by means of MMI

到 抵 赶赴 进驻 往 于 至 …	这部 这次 这个 这家 这项 ...
阿尔及利亚 阿根廷 奥地利 澳大利亚 ...	打破 带动 抵制 调动 夺取 遏制 ...
局长 秘书长 会长 社长 省长 首相 司长..	除夕 春节 虎年 佳节 年初 年底 农历 ..
规范 和谐 合格 合理 缓慢 混乱 激烈 ...	财产 成本 抵押 贷款 工资 金额利率 ...
10 月 11 月 12 月 1 月 2 月 3 月 4 月 ...	邹家华 朱镕基 叶利钦 希拉克 吴邦国 ...
省人大 省政府 书记处 铁道部 外交部...	膨胀 上升 上涨 通胀 下跌 下滑 增产 ...
球员 师生 投资者 消费者 志愿者 老百姓	石油 烟草 邮电 油气 医疗 医药 ...

title cluster respectively. In addition, many words in one cluster are similar in meaning. For instance, one cluster*{财产,成本,抵押,贷款,工资,金额,利率,利润,投资,关税,…}* means bund in finance.

3.4 Applying Word Clusters to Chinese Syntactic Parsing

We have applied alignment-based method [11][12], which induces constituent by means of aligning all the sentence pairs and considers the different parts as constituents, to induce Chinese syntactic structures and got satisfactory results. The alignment of two sentences can be defined as a list of pairs of similar words in them. Take sentence *A*= *北京是中国的首都* and sentence *B*=*朝鲜是中国的邻居* for example, the alignment is a list *(<是, 是>,<中国, 中国>,<的, 的>)*. Nevertheless,*北京* and *朝鲜*, both standing for location is not aligned. Resorting to the syntactic categories of word clusters, the alignments of all sentence pairs could be adjusted and the performance of syntactic parsing may be improved.

Table5 gives the experimental result of alignment-based Chinese syntactic parsing and that of introducing word cluster into alignment. FW, MMI, HFW, and WC indicate that using corresponding clustering result to adjust the alignment while NC means no cluster is introduced. The measure metrics Precision, Recall and FScore of NC are smaller than the others. Therefore, evidently, quality of syntactic parsing can be improved by means of word clusters, one reason for no drastic improvement in performance is that the training corpus of Chinese sentence bank and People's Daily are not tagged by the same standard, and another is that the word clusters in our Chinese sentence corpus are very sparse and the Chinese sentence corpus is not trained to partition words.

Table 5. Comparison result of Chinese syntactic parsing when word cluster is introduced

Method	FW(%)	MMI(%)	HFW(%)	WC(%)	NC(%)
Precision	28.18	28.19	**28.39**	28.01	27.58
Recall	46.54	**46.70**	46.62	46.62	46.38
FScore	35.11	35.15	**35.29**	35.00	35.00

4 Conclusion

Word clustering is a means of constructing small but effective language models, which in some cases could improve N-gram language models and syntactic induction of sentences. In this paper we concentrated on unsupervised learning of Chinese word clustering, evaluated four different methods and found that MMI and FB could identify interesting word clusters with the same syntactic function and somewhat similar semantic meaning, the POS precision is 79.09% and the semantic precision 49.75%. When word clusters were applied to alignment-based learning of Chinese syntactic parsing, the performance is further improved. In future, we will experiment on more initial selections to improve the performance of FB and WC, combine the clustering results and syntactic parsing of Chinese and extract translation template by means of bilingual word clusters.

References

1. Andrew Roberts .Automatic Acquisition of Word Classification using Distributional Analysis of Content Words with Respect to Function Words , November 17 2002.
2. Clark A. Unsupervised induction of stochastic context-free grammars using distributional clustering. In: Proc. of CoNLL 2001, July 2001, Toulouse, France.105-112.
3. Finch, S., & Chater, N.(1992a). Bootstrapping syntactic categories. In Proceedings of the 14th Annual Meeting of the Cognitive Science Society, pp. 820–825.
4. F. Pereira and N. Tibshy and L. Lillian .Distributional Clustering of English Words ,CL, 1993.
5. Klein D. The Unsupervised learning of natural language structure. PHD thesis, Stanford University. 2005.
6. Peter F. Brown, Vincent J. Della Pietra, Peter V. deSouza, Jenifier C. Lai, and Robert L. Mercer. Class-based n-gram models of natural language. In Proceedings of the IBM Natural Language ITL, pages 283-298, Paris, France, March.
7. Rile Hu, Chengqing Zong and Bo Xu.Semi-automatic Acquisition of Translation Templates from Monolingual Unannotated Corpora. pages 163-173, IEEE 2003.
8. Sch¨utze, H. (1997). Ambiguity Resolution in Language Learning. CSLI Publications
9. Siwen Yu. The Grammatical Knowledge-base of Contemporary Chinese-A Complete Specification, publishing company of Tsinghua University, 2003.
10. Sven Martin , Jorg Liermann , and Hermann Ney. Algorithms For Bigram And Trigram Word Clustering, October 05 1995.
11. Van Zaanen, M. and Adriaans, P. Alignment-Based Learning versus EMILE: A comparison. In Proc. of the Belgian-Dutch Conference on Artificial Intelligence (BNAIC); Amsterdam, the Netherlands, 2001, 315–322.
12. Van Zaanen, M. ABL: Alignment-based learning. In Proceedings of the 18th International Conference on Computational Linguistics (COLING 18),2000, 961–967.
13. Ye-Yi Wang, John Lafferty, and Alex Waibel. 1996. Word clustering with parallel spoken language corpora. In Proceedings of the 4th International Conference on Spoken Language Procesing (ICSLP'96), pages 2364--2367.

Populating FrameNet with Chinese Verbs Mapping Bilingual Ontological WordNet with FrameNet

Ian C. Chow[1] and Jonathan J. Webster[2]

[1] Department of Chinese, Translation and Linguistics
City university of Hong Kong, Tat Chee Avenue, Kowloon, Hong Kong
ianchow@cityu.edu.hk
[2] The Halliday Centre for Intelligent Applications of Language Studies
City University of Hong Kong, Tat Chee Avenue, Kowloon, Hong Kong
ctjjw@cityu.edu.hk

Abstract. This paper describes the construction of a linguistic knowledge base using Frame Semantics, instantiated with Chinese Verbs imported from the Chinese-English Bilingual Ontological WordNet (BOW). The goal is to use this knowledge base to assist with semantic role labeling. This is accomplished through the mapping of FrameNet and WordNet and a novel verb selection restriction using both the WordNet inter-relations and the concept classification in the Suggested Upper Merged Ontology (SUMO). The FrameNet WordNet mapping provides a channel for Chinese verbs to interface with Frame Semantics. By taking the mappings between verbs and frames as learning data, we attempt to identify subsuming SUMO concepts for each frame and further identify more Chinese verbs on the basis of synset inter-relations in WordNet.

Keywords: Frame Semantics, WordNet, Bilingual Ontological WordNet, Suggested Upper Merged Ontology.

1 Introduction

The goal of semantic role labeling is to identify the semantic relations between words in a text, providing a semantic interpretation of what roles are denoted by the words with respect to the types of event evoked by the main verb. FrameNet, WordNet are different lexical resources representing different levels of linguistic information: FrameNet covers a wide range of event types represented in frames with indication of Frame Element (FE), i.e. the Semantic Roles for the event, Phrase Type (PT) and their Grammatical Function (GF). The lexical coverage of FrameNet is sense-specific; however, the amount is relatively small compared to many lexicographical resources such as WordNet. WordNet organizes words into synonym sets with relation links between them. We aim at combining the two resources together in order to provide a comprehensive knowledge base to serve as the basis for identifying the range of possible semantic relations between words.

Y. Matsumoto et al. (Eds.): ICCPOL 2006, LNAI 4285, pp. 165 – 172, 2006.

2 Knowledge Base for SRL

The major problem encountered in semantic role labeling is the identification of the appropriate range of possible of semantic roles. Sentence level knowledge is based on the FrameNet frame. In order to draw the appropriate frame, we would need a lexical-semantic verb classification based on the event type evoked by the verb. Word level knowledge is from WordNet.

2.1 FrameNet

FrameNet[2] is based on the theory of Frame Semantics. A frame represents a scenario in terms of the interaction of its participants, and these participants play certain roles. Verbs, nouns can be used to identify a frame and the annotated sentences in each frame show the possible semantic roles for a given target word. We rely on the association of verbs and frames in FrameNet to generate a SRL assisted knowledge base. Currently, there are around 3000 verbs attached to 320 different frames.

2.2 Chinese-English Bilingual Ontological WordNet

The Bilingual Ontological WordNet (BOW) was constructed by the Academia of Sinica[3]. The English Chinese translation equivalence database is used in bootstrapping a Chinese WordNet with English WordNet. It is produced by the WordNet team at Academia Sinica base on WordNet 1.6. The BOW also links with the SUMO ontology. There are more than 12610 verb synsets in WN 1.6 and each synset represents a lexical concept shared by the synonyms included.

2.3 FrameNet WordNet Mappings

Shi & Mihalcea's [7] FrameNet WordNet Verb Mapping (FnWnVerbMap) has mapped the verb covered by FrameNet with WordNet. Around 3,600 verb senses from WordNet have been mapped to over 300 frames. Their work has suggested the addition of Chinese verbs to the verb coverage in FrameNet. The frame mapped English synsets are taken to locate the equivalent Chinese verbs in our integration work.

2.4 SUMO

The Suggested Upper Merged Ontology [12] is created at Teknowledge Corporation. The ontology, which contains 1000 terms and 4000 assertions, provides definitions for general purpose terms and acts as the basis for constructing more specific domain ontologies. Niles & Pease [9] mapped WordNet synsets and SUMO concepts. We attempt to identify the SUMO conceptualization of a FrameNet frame by a statistical distribution method with the aim of assigning more synsets to frames, thereby expanding the verb coverage.

3 Integration of the Resources

Since BOW is base on WordNet 1.6, the equivalent synsets in WordNet 2.0 must first be established in order to combine with the FrameNet WordNet Mapping (FnWnVerbMap) which is based on WordNet 2.0.

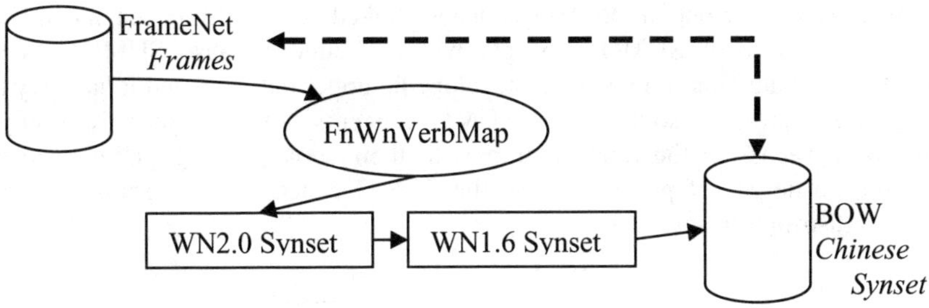

Fig. 1. The Mapping framework of FrameNet and BOW

The FnWnVerbMap is based on the verb-senses included in the lexical coverage of FrameNet and VerbNet. Owing to the relatively and more extensive wider coverage of WordNet, many synsets are not mapped with any frame. In order to expand the verb coverage of frame, we suggest an automatic assignment of synsets to their corresponding frame via WordNet synset relations and SUMO conceptualization.

3.1 Expanding FrameNet Coverage with Synsets

Based on the above mapping, each FrameNet frame corresponds to some synset(s). With these frame-mapped synsets, we may activate more synsets which are related to in a non-transitive manner. The major verb synset relations in WordNet [6] include hypernym, troponym(hyponym), entailment and antonym, cause and etc. These related synsets are semantically related to the frame bridged by the mapping. Others, however, may be inappropriate to the frame depending on the scenario evoked by the verb.

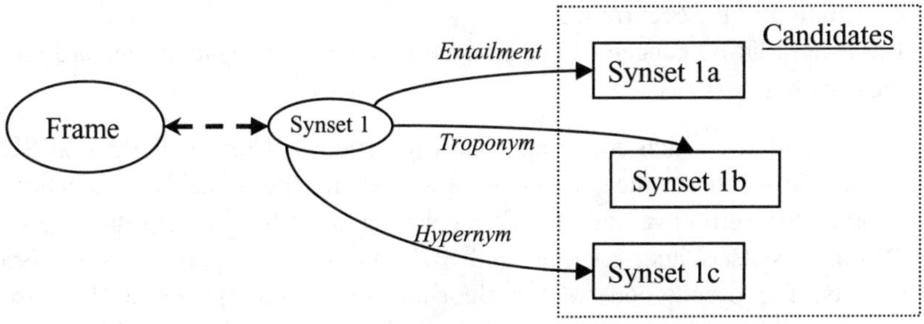

Fig. 2. Searching for potential synsets as candidates for frame mapping

In order to appropriately identify suitable synsets, we subsume SUMO concepts to FrameNet frames and take this information as selection criteria among the candidates in order to filter out incorrect synsets.

3.2 Populating Frames with Synsets by SUMO

WordNet synsets have been mapped with SUMO concepts. Upon the FnWnVerbMap, we can retrieve a group of SUMO concepts linked to the frame via the mapped synsets. A variety of SUMO concepts will be found because SUMO is not a linguistic-motivated ontology but an encyclopedic upper ontology and it has very rich concept taxonomy. As some of the SUMO concepts have a higher frequency of occurrence, we assign the relevant frames to their corresponding SUMO concepts based on the degree of prominence of the SUMO concepts as determined using a statistical distribution approach.

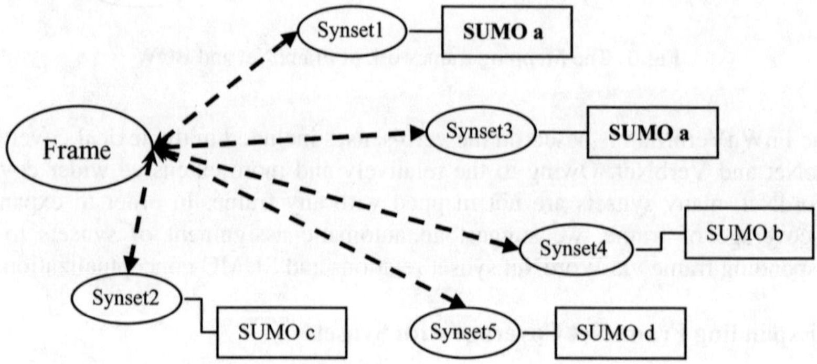

Fig. 3. Retrieving SUMO concepts subsumed by frame via WordNet

The SUMO conceptualization of a frame is classified into 3 types using a statistical distribution approach.

1. Core SUMO concepts with a positive standard score greater or equal to 1 in the distribution of the occurrence.
2. Peripheral SUMO concepts with a positive standard score between 0 and 1 in the distribution of the occurrence.
3. Irrelevant SUMO concepts have an occurrence with a negative standard score in the distribution.

Synset candidates which correspond with the frame's Core or Peripheral SUMO concepts are designates as the corresponding synsets for the frame. In other words, the expansion of the verb coverage of a FrameNet frames relies on semantic links from both WordNet synsets inter-relations and SUMO concept mapping between Frames and Synsets. The pseudo-code of the algorithm used to assign FrameNet semantic frames to WordNet verb synsets is described in Figure 4 and illustrated in Figure 5.

```
For each frame F
    For each synset mapped with F = (FnSYN)
    array frm-SUMO [ ] = SUMO(FnSYN)

    Core SUMO = SUMO with standard score >=1 in frm-SUMO []
    Peripheral SUMO = SUMO with standard score >0 & <1 in frm-SUMO []
    Irrelevant SUMO = SUMO with standard score <0 in frm-SUMO []
    Candidates (CandSYN) = hypernym(FnSYN) and troponym(FnSYN)
                            and entailment(FnSYN)
    if     SUMO(CandSYN) = Core SUMO or SUMO(CandSYN) = Peripheral SUMO
    then   assign Frame F to CandSYN
```

Fig. 4. Algorithm for mapping more WordNet synsets to FramNet frames

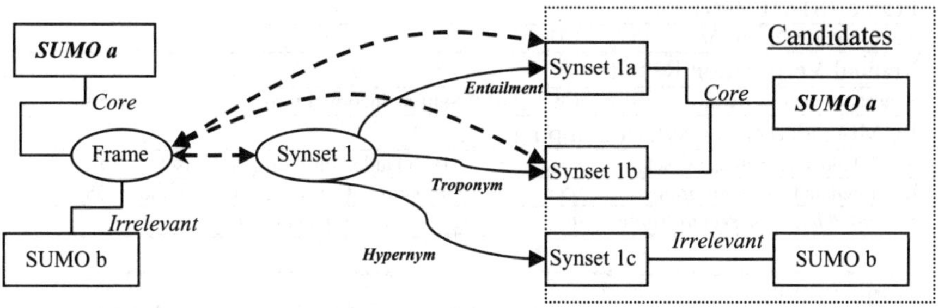

Fig. 5. Populating FrameNet frames with WordNet synsets according to the consistency of SUMO conceptualization

3.3 Evaluation on Recall and Precision

We present three test case frames for evaluation of the accuracy of synset selection. The three frames are:

STATEMENT, core FE, the Semantic Labels, includes Speaker, Message and Medium.
GIVING, core FE includes Donor, Recipient and Theme
ATTACK, core FE includes Assailant and Victim

The mapping result is shown in Table 1. We manually verified the synset in the 3 test cases as the golden standard. The result of recruiting appropriate synsets based on SUMO-aided frame identification is satisfying. In the STATEMENT frame, no incorrect synsets was recruited (Precision: 100%); in the ATTACK frame, there was 1 error out of the 61 candidates (Precision: 98.36%), and in the GIVING frame, 8 incorrect synsets were recruited out of the 91 candidate synsets (Precision: 91.21%).

The methodology has a very high accuracy on filtering out incorrect synsets among the candidates. In STATEMENT frame, all 58 incorrect synsets were filtered out; In the GIVING frame, 43 out of 51 incorrect synsets was filtered (84.31%) and the ATTACK frame filtered out 26 of the 27 incorrect synsets (96.29%).

By employing the algorithm, we have successfully expanded the verb coverage of FrameNet frames with WordNet synsets. Taking the STATEMENT frame by way of illustration, we proceed to implement a SUMO learning process returning 95 new synsets for the frame. It is noted that more than 1 lexeme may be included by each synsets which means it yields more than 95 new Chinese verbs for the verb coverage of the frame.

However, the recall percentage of these 3 test cases is not very high. In the STATEMENT frame, only 95 correct synsets out of the 139 correct synsets were recruited (68.34%), 31 out of 40 correct synsets were recruited (77.5%) in the GIVING frame and 13 out of 34 (38.24%) in the ATTACK frame. In the other words, there is room for improvement in assigning more synsets to FrameNet frames.

Table 1. Mapping result of frames STATEMENT, GIVING, ATTACK

Frame : STATEMENT			
No. of Synset Candidates	197		
Manual Verification Result			
Synsets correct to frame	139	Synsets incorrect to frame	58
SUMO aided frame synset mapping			
SUMO aided recruited Synset	95	SUMO aided un-recruited Synset	102
Recruited and correct to frame	95	Un-recruited and incorrect to frame	58
Recruited but incorrect to frame	***0***	*Un-recruited but correct to frame*	*44*

Frame : GIVING			
No. of Synset Candidates	91		
Manual Verification Result			
Correct Synset to frame	40	Incorrect Synset to frame	51
SUMO aided frame synset mapping			
SUMO aided recruited Synset	39	SUMO aided un-recruited Synset	52
Recruited and correct to frame	31	Un-recruited and incorrect to frame	43
Recruited but incorrect to frame	***8***	*Un-recruited but correct to frame*	*9*

Frame : ATTACK			
No. of Synset Candidates	61		
Manual Verification Result			
Synsets appropriate to frame	34	Synsets inappropriate to frame	27
SUMO aided frame synset mapping			
SUMO aided recruited Synset	14	SUMO aided un-recruited Synset	47
Recruited and correct to frame	13	Un-recruited and incorrect to frame	26
Recruited but incorrect to frame	***1***	*Un-recruited but correct to frame*	*21*

The high precision shows the positive potential of using SUMO concept as selection criteria for mapping WordNet synsets with FrameNet frames. The relatively unsatisfying recall rate implies certain short-comings in the methodology. SUMO is a conceptual upper ontology and its concepts denote world meaning; FrameNet frames represent a sentence-level scenario. Mapping ontological concept (SUMO) with the semantic interpretation of text (Frame Semantics) mediated by a statistical distribution of the grammatical realization of the text (Word Meaning -Verb) may be insufficient. The suggested mapping algorithm is a novel mapping approach in ontology engineering and here is proved requires much improvement although the precision rate is satisfying. Looking into the test case with the worst recall rate, the ATTACK frame, it is noted that the 21 un-recruited but correct synsets falls into four groups of SUMO concepts: *Impacting (17), Motion(1), Regulatory Process(1) & Shooting (2).* It is noted that the SUMO concept *Impacting* has a meaning similar to the ATTACK scenario but was not determined as the Frame core or peripheral SUMO. Similar situation occurred in the other two test cases. The major SUMO concepts in the un-recalled data of STATEMENT frame are: *ContentDevelopment & Declaring & Speaking*; and for GIVING frame are: *Putting & Lending*. These SUMO concepts do have a meaning similar to the scenario construed by the frame.

Hence, it shows that mapping between concepts from different knowledge base should not only rely on the degree of occurrence of its text realization but also to searching among all concepts of the mapping source.

4 Conclusion

The project of assigning BOW Chinese synsets to FrameNet is currently under development. The high accuracy of the Frame–Synsets Mapping yields a useful knowledge base for identifying the range of possible semantic relations between words for a given target verb. We aim at mapping all WordNet synsets to their corresponding FrameNet frame, providing a comprehensive knowledge base for semantic role labeling. The result of the knowledge base includes FrameNet frames knowledge, English verb lexical data from WN 2.0 and Chinese verb lexical data from BOW. Inspired by similar work [11] of drawing Chinese lexical data to FrameNet, representative Chinese example sentence should be included in the knowledge base making it a useful tool for dealing semantic parsing.

References

1. Bateman, John A. 1990 "Upper modeling organizing knowledge for natural language processing." *In 5th International Workshop on Natural Language Generation.* June, 1990, Pittsburg. PA.
2. Charles J. Fillmore, Christopher R. Johnson, and Miriam R.L. Petruck. 2003 "Background to FrameNet" *International Journal of Lexicography* 2003 16: 235-250
3. Chu-Ren Huang, Ru-Yng Chang, Shiang-Bin Lee. 2004 "Sinica BOW (Bilingual Ontological Wordnet): Integration of Bilingual WordNet and SUMO". *4th International Conference on Language Resources and Evaluation (LREC2004).* Lisbon. Portugal. 26-28 May, 2004."
4. Chow, Ian C. 2005. "Automating the Import of Lexical Data into a Relational Network" *The 32nd Linguistics Association of United States and Canada Forum.* LACUS 32nd Forum, Hanover, USA. Aug 2005
5. Chow, Ian C. & Wong, Tak Ming. 2006 "Axiomatizing Relational Network for Knowledge Engineering - Exploring WordNet and FrameNet". *The 2006 IEEE International Conference on Information Reuse and Integration.* Hawaii, USA. Sep 2006

6. Fellbaum, Christiane. 1998 WordNet An Electronic Lexical Database MIT Press. Cambridge.
7. Lei Shi and Rada Mihalcea. 2005. "Putting Pieces Together: Combining FrameNet, VerbNet and WordNet for Robust Semantic Parsing", *Cicling 2005, Mexico*
8. Levin, Beth. 1993. English Verb Class and Alternations: A Preliminary Investigation. Chicago: University of Chicago Press.
9. Niles, I., and Pease, A. 2003. "Linking Lexicons and Ontologies: Mapping WordNet to the Suggested Upper Merged Ontology", Proceedings of the IEEE International Conference on Information and Knowledge Engineering, pp 412-416.
10. Oltramari, A. 2006. "LexiPass methodology: a conceptual path from frames to senses and back" To apper *in Proceedings of LREC 2006* (Fifth international conference on Language Resources and Evaluation).
11. Pascale Fung and Benfeng Chen. 2004. "BiFrameNet: Bilingual Frame Semantics Resource Construction by Cross-lingual Induction". *In Proceedings of the 20th International Conference on Computational Linguistics (COLING 2004)*, Geneva, Switzerland: August, 2004.
12. SUO, (2003), The IEEE Standard Upper Ontology web site, http://suo.ieee.org
13. Webster, Jonathan J & Chow, Ian C. 2005. "Mapping WordNet to a Relational Network" *In Proceedings of IEEE Natural Language Processing and Knowledge Engineering 2005 (IEEE NLP-KE'05)*, WuHan. Oct 2005

Collecting Novel Technical Terms from the Web by Estimating Domain Specificity of a Term

Takehito Utsuro[1], Mitsuhiro Kida[2], Masatsugu Tonoike[3], and Satoshi Sato[4]

[1] Graduate School of Systems and Information Engineering, University of Tsukuba, 1-1-1, Tennodai, Tsukuba, 305-8573, Japan
[2] Nintendo Co., Ltd., 11-1, Hokotate-cho, Kamitoba, Minami-ku, Kyoto-shi, 601-8116 Japan
[3] Graduate School of Informatics, Kyoto University, Yoshida-Honmachi, Sakyo-ku, Kyoto 606-8501, Japan
[4] Graduate School of Engineering, Nagoya University, Furo-cho, Chikusa-ku, Nagoya 464-8603, Japan

Abstract. This paper proposes a method of domain specificity estimation of technical terms using the Web. In the proposed method, it is assumed that, for a certain technical domain, a list of known technical terms of the domain is given. Technical documents of the domain are collected through the Web search engine, which are then used for generating a vector space model for the domain. The domain specificity of a target term is estimated according to the distribution of the domain of the sample pages of the target term. We apply this technique of estimating domain specificity of a term to the task of discovering novel technical terms that are not included in any of existing lexicons of technical terms of the domain. Out of randomly selected 1,000 candidates of technical terms per a domain, we discovered about 100 $\sim$ 200 novel technical terms.

1 Introduction

Lexicons of technical terms are one of the most important language resources both for human use and for computational research areas such as information retrieval and natural language processing. Among various research issues regarding technical terms, full-/semi-automatic compilation of technical term lexicon is one of the central issues. In various research fields, novel technologies are invented every year, and related research areas around such novel technologies keep growing. Along with such invention of technologies, novel technical terms are created year by year. Considering such a situation, it requires a huge cost for manually compiling lexicons of technical terms for hundreds of thousands of technical domains. Therefore, it is inevitable to invent a technique of full-/semi-automatic compilation of technical term lexicons for various technical domains.

The whole task of compiling a technical term lexicon can be roughly decomposed into two sub-processes: (1) collecting candidates of technical terms of a technical domain, and, (2) judging whether each candidate is actually a technical term of the target technical domain. The technique of the first sub-process is closely related to research on automatic term recognition, and has been relatively well studied so far (e.g., [5]). On the other hand, the technique of the second

Y. Matsumoto et al. (Eds.): ICCPOL 2006, LNAI 4285, pp. 173–180, 2006.

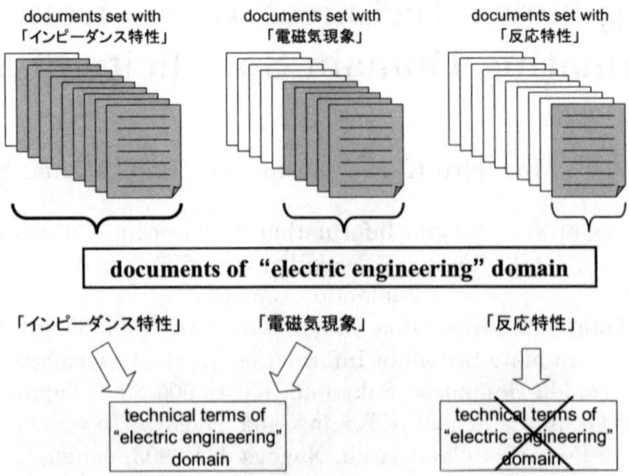

Fig. 1. Degree of Specificity of a Term based on the Domain of the Documents (Example terms: *impedance characteristic*, *electromagnetism*, and *response characteristic*)

sub-process has not been studied well so far. Exceptional cases are works such as [1,2], where their techniques are mainly based on the tendency of technical terms appearing in technical documents of limited domains rather than in documents of daily use such as newspaper and magazine articles. Although the underlying idea of those previous works is very interesting, those works are quite limited in that they require existence of certain amount of technical domain corpus. It is not practical for manually collecting technical domain corpus for hundreds of thousands of technical domains. Therefore, as for the second sub-process here, it is very important to invent a technique for automatically classifying the domain of a technical term.

Based on this observation, among several key issues regarding the second sub-process above, this paper mainly focuses on the issue of estimating the domain specificity of a term. In this paper, supposing that a target technical term and a technical domain are given, we propose a technique of automatically estimating the specificity of the target term with respect to the target domain. Here, the domain specificity of the term is judged among the following three levels: i) the term mostly appears in the target domain, ii) the term generally appears in the target domain as well as in other domains, iii) the term generally does not appear in the target domain.

The key idea of the proposed technique is as follows. In the proposed technique, we assume that sample technical terms of the target domain are available. Using such sample terms with search engine queries, we first collect a corpus of the target domain from the Web. In a similar way, we also collect sample pages that include the target term from the Web. Then, the similarities of the contents of the documents are measured between the corpus of the target domain and each of the sample pages that include the target term. Finally, the domain specificity of the target term is estimated according to the distribution of the domain of those sample pages.

Figure 1 illustrates rough idea of this technique. Among the three example (Japanese) terms, the first term (*impedance characteristic*) mostly appears in the documents of the *"electric engineering"* domain on the Web. In the case of the second term (*electromagnetism*), about half of sample pages collected from the Web can be regarded as in the *"electric engineering"* domain, while the rest are not. On the other hand, in the case of the last term (*response characteristic*), only a few of the sample pages can be regarded as in the *"electric engineering"* domain. In our technique, such difference of the distribution can be easily identified, and the domain specificities of those three terms are estimated.

As experimental evaluation, we first evaluate the proposed technique of estimating domain specificity of a term using manually constructed development and evaluation term sets, where we achieved mostly 90% precision/recall (details are presented in [6]). Furthermore, in this paper, we present the result of applying this technique of estimating domain specificity of a term to the task of discovering novel technical terms that are not included in any of existing lexicons of technical terms of the domain. Candidates of technical terms are first collected from the Web corpus of the target domain. Then, about 70~80 % of those candidates are excluded by roughly judging the domain of their constituent words. Finally, out of randomly selected 1,000 candidates of technical terms per a domain, we discovered about 100 ~ 200 novel technical terms that are not included in any of existing lexicons of the domain, where we achieved about 75% precision and 80% recall.

2 Domain Specificity Estimation of Technical Terms Using Documents Collected from the Web

In this section, we first describe the proposed technique of estimating domain specificity of a term using the Web.

2.1 Outline

Here, we estimate the domain specificity of a term t with respect to a domain C, supposing that the term t and the domain C are given. Generally speaking, the coarsest-grained classification of domain specificity of a term is binary classification, namely, the class of terms that are used in a certain technical domain, vs. the class of terms that are *not* used in a certain technical domain. In this paper, we further classify the degree $g(t, C)$ of the domain specificity into the following three levels:

$$g(t,C)=\begin{cases} + \text{ (t mostly appears in the documents of the domain C.)} \\ \pm \text{ (t generally appears in the documents of the domain C as well as} \\ \quad \text{in those of the domains other than C.)} \\ - \text{ (t generally does not appear in the documents of the domain C.)} \end{cases}$$

(When we simply classify domain specificity of a term into two classes with the coarsest-grained binary classification above, we regard those with domain

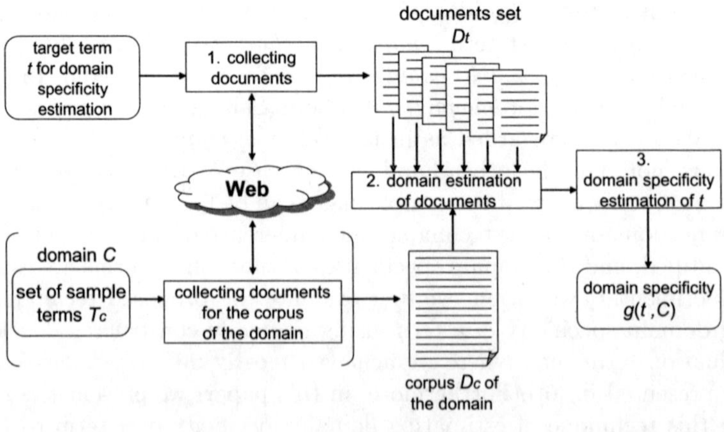

Fig. 2. Domain Specificity Estimation of Terms based on Web documents

specificity '+' or '±' as those that are used in the domain, and those with domain specificity '−' as those that are *not* used in the domain.)

The input and output of the process of domain specificity estimation of a term t with respect to the domain C are given below:

input	target term t for domain specificity estimation, set T_C of sample terms of the domain C
output	domain specificity $g(t, C)$ of t with respect to C

The process of domain specificity estimation of a term is illustrated in Figure 2, where the whole process can be decomposed into two sub-processes: (a) that of constructing the corpus D_C of the domain C, and (b) that of estimating the specificity of a term t with respect to the domain C. In the process of domain specificity estimation, the domain of documents including the target term t is estimated, and the domain specificity of t is judged according to the distribution of the domains of the documents including t. The details of those two sub-processes are described in the followings.

2.2 Constructing the Corpus of the Domain

When constructing the corpus D_C of the domain C using the set T_C of sample terms of the domain C, first, for each term s in the set T_C, we collect into a set D_s the top 100 pages obtained from search engine queries that include the term s.[1] The search engine queries here are designed so that documents that describe the technical term s are ranked high. When constructing a corpus of the Japanese language, the search engine "goo"[2] is used. The specific queries that are used in

[1] Related techniques for automatically constructing the corpus of the domain using the sample terms of the domain include those presented in [4,3]. We are planning to evaluate the performance of those related techniques and compare them with the one employed in this paper.

[2] http://www.goo.ne.jp/

this search engine are phrases with topic-marking postpositional particles such as "s-toha," "s-toiu," "s-wa," and an adnominal phrase "s-no," and "s."

Then, union of the sets D_s for each s is constructed and denoted as $D(T_C)$:

$$D(T_C) = \bigcup_{s \in T_C} D_s$$

Finally, in order to exclude noise texts from the set $D(T_C)$, the documents in the set $D(T_C)$ are ranked according to the number of sample terms (of the set T_C) that are included in each document. Through a preliminary experiment, we decided here that it is enough to keep top 500 documents, and regard them as the corpus D_C of the domain C.[3]

2.3 Domain Specificity Estimation of Technical Terms

Given the corpus D_C of the domain C, domain specificity of a term t with respect to a domain C is estimated through the following three steps:

Step 1. Collecting documents that include the term t from the Web, and constructing the set D_t of those documents.

Step 2. For each document in the set D_t, estimating its domain by measuring similarity against the corpus D_C of the domain C. Then, given a certain lower bound L of document similarity, from D_t, extracting documents with large enough similarity values into a set $D_t(C, L)$.

Step 3. Estimating the domain specificity $g(t, C)$ of t using the document set $D_t(C, L)$ constructed in the step 2.

Details of those three steps are given below:

Collecting Web Documents Including the Target Term. For each target term t, documents that include t are collected from the Web. According to a procedure that is similar to that of constructing the corpus of the domain C described in section 2.2, the top 100 pages obtained with search engine queries are collected into a set D_t.

Domain Estimation of Documents. For each document in the set D_t, its domain is estimated by measuring similarity against the corpus D_C of the domain C. Then, given a certain lower bound L of document similarity, documents with large enough similarity values are extracted from D_t into the set $D_t(C, L)$ [6].

Domain Specificity Estimation of a Term. The domain specificity of the term t with respect to the domain C is estimated using the document sets D_t

[3] In our evaluation, about 80$\sim$90 % of the documents of D_C are actually those of the domain C. Even with D_C having all of its documents as of the domain C, we achieved almost the same performance of domain specificity estimation of a term.

and $D_t(C, L)$. Here, this is done by simply calculating the following ratio r_L of the numbers of the documents within the two sets:

$$r_L = \frac{|D_t(C, L)|}{|D_t|}$$

Then, by introducing the two thresholds $a(\pm)$ and $a(+)$ for the ratio r_L, the specificity $g(t, C)$ of t is estimated with the following three levels:

$$g(t, C) = \begin{cases} + \ (a(+) \leq r_L) \\ \pm \ (a(\pm) \leq r_L < a(+)) \\ - \ (r_L < a(\pm)) \end{cases}$$

In experimental evaluation of section 4, as in the case of the lower bound L of the document similarity, the two thresholds $a(\pm)$ and $a(+)$ are also determined using the development term set mentioned above.

3 Collecting Novel Technical Terms of a Domain from the Web

This section illustrates how to apply the technique of domain specificity estimation of technical terms to the task of discovering novel technical terms that are not included in any of existing lexicons of technical terms of the domain. First, as shown in Figure 3, from the corpus D_C of the domain C, candidates of technical terms are collected. In the case of the Japanese language, as candidates of novel technical terms, we collect compound nouns with frequency counts five or more,

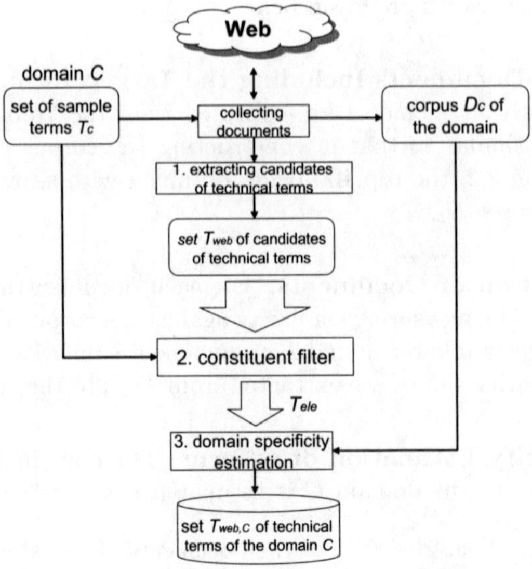

Fig. 3. Collecting Novel Technical Terms of a Domain from the Web

consisting of more than one noun. Here, we collect compound nouns which are not included in any of existing lexicons of technical terms of the domain. Then, after excluding terms which do not share constituent nouns against the sample terms of the given set T_C, the domain specificity of the remaining terms are automatically estimated. Finally, we regard terms with domain specificity '+' or '±' as those that are used in the domain, and collect them into the set $T_{web,C}$.

4 Experimental Evaluation

We evaluate the proposed method with five sample domains, namely, *"electric engineering"*, *"optics"*, *"aerospace engineering"*, *"nucleonics"*, and *"astronomy"*. For each domain C of those five domains, the set T_C of sample (Japanese) terms is constructed by randomly selecting 100 terms[4] from an existing (Japanese) lexicon of technical terms for human use. We evaluate the results of discovering novel technical terms that are not included in any of existing lexicons of technical terms of the domain. First, Table 1 compares the numbers of candidates of novel technical terms collected from the Web, with those after excluding terms which do not share constituent nouns against the sample terms of the given set T_C. As shown in the table, about 70∼80 % of the candidates are excluded, while the rate of technical terms within the remaining candidates increased. This result clearly shows the effectiveness of the constituent noun filtering technique in reducing the computational time of discovering fixed number of novel technical terms. Then, per a domain, we randomly select 1,000 of those remaining candidates, and estimate their domain specificity by the proposed method. After manually judging the domain specificity of those 1,000 terms, we measure the precision/recall of the proposed method as in Table 2, where we achieved about 75% precision and 80% recall. Here, however, as candidates of technical terms, we simply collect compound nouns, where sometimes their term unit is not correct since the technical term candidate could be with a certain prefix or suffix. Considering this fact, Table 2 also gives the term unit correct rate for those with domain specificity '+' or '±'. Finally, taking this term unit correct rate into account, we can

Table 1. Changes in Number of Technical Term Candidates with Constituent Filter

	before filtering		after filtering	
	# of candidates	# of tech. terms (estimated) (%)	# of candidates	# of tech. terms (estimated) (%)
electric engineering	24,460	1,272 (5.2)	6,623	848 (12.8)
optics	29,090	1,047 (3.6)	6,985	866 (12.4)
aerospace engineering	41,279	660 (1.6)	6,364	458 (7.2)
nucleonics	40,439	890 (2.2)	10,834	650 (6.0)
astronomy	29,240	1,170 (4.0)	5,491	659 (12.0)

[4] Through a preliminary experiment, we conclude that it is not necessary to start with the set T_C of sample terms which has more than 100 sample terms. The number of minimum requirement for the size of T_C varies according to domains.

Table 2. Precision/recall of Collecting Novel Technical Terms

(a) with threshold $a(\pm)$

	precision	recall	term unit correct rate
electric engineering	0.754(399/529)	0.828(399/482)	0.393(157/399)
optics	0.766(454/593)	0.875(454/519)	0.368(167/454)
aerospace engineering	0.797(408/512)	0.739(408/552)	0.402(164/408)
nucleonics	0.685(470/686)	0.953(470/493)	0.377(177/470)
astronomy	0.747(480/643)	0.945(480/508)	0.475(228/480)

(b) with threshold $a(+)$

	precision	recall	term unit correct rate
electric engineering	0.697(168/241)	0.853(168/197)	0.494(83/168)
optics	0.743(234/315)	0.932(234/251)	0.453(106/234)
aerospace engineering	0.666(277/416)	0.936(277/296)	0.502(139/277)
nucleonics	0.580(362/624)	0.981(362/369)	0.406(147/362)
astronomy	0.763(350/459)	0.888(350/394)	0.520(182/350)

conclude that, out of the 1,000 candidates, we discovered about 100 $\sim$ 200 novel technical terms that are not included in any of existing lexicons of the domain. This result clearly supports the effectiveness of the proposed technique for the purpose of full-/semi-automatic compilation of technical term lexicons.

5 Concluding Remarks

This paper proposed a method of domain specificity estimation of technical terms using the Web. We then applied this technique of estimating domain specificity of a term to the task of discovering novel technical terms that are not included in any of existing lexicons of technical terms of the domain.

References

1. T. M. Chung. A corpus comparison approach for terminology extraction. *Terminology*, 9(2):221–246, 2004.
2. P. Drouin. Term extraction using non-technical corpora as a point of leverage. *Terminology*, 9(1):99–117, 2003.
3. C.-C. Huang, K.-M. Lin, and L.-F. Chien. Automatic training corpora acquisition through Web mining. In *Proceedings of IEEE/WIC/ACM International Conference on Web Intelligence*, pages 193–199, 2005.
4. B. Liu, X. Li, W. S. Lee, and P. S. Yu. Text classification by labeling words. In *Proceedings of the 19th AAAI*, pages 425–430, 2004.
5. H. Nakagawa and T. Mori. Automatic term recognition based on statistics of compound nouns and their components. *Terminology*, 9(2):201–219, 2003.
6. T. Utsuro, M. Kida, M. Tonoike, and S. Sato. Towards automatic domain classification of technical terms: Estimating domain specificity of a term using the Web. In *AIRS 2006*, LNCS: Vol. 4182, pages 633–641. Springer, 2006.

Building Document Graphs for Multiple News Articles Summarization: An Event-Based Approach

Wei Xu[1], Chunfa Yuan[1], Wenjie Li[2], Mingli Wu[2], and Kam-Fai Wong[3]

[1] Department of Computer Science and Technology
Tsinghua University, China
vivian00@mails.tsinghua.edu.cn, cfyuan@mail.tsinghua.edu.cn
[2] Department of Computing,
The Hong Kong Polytechnic University, Hong Kong
{cswjli, csmlwu}@comp.polyu.edu.hk
[3] Department of System Engineering,
The Chinese University of Hong Kong, Hong Kong
kfwong@se.cuhk.edu.hk

Abstract. Since most of news articles report several events and these events are referred in many related documents, we propose an event-based approach to visualize documents as graph on different conceptual granularities. With graph-based ranking algorithm, we illustrate the application of document graph to multi-document summarization. Experiments on DUC data indicate that our approach is competitive with state-of-the-art summarization techniques. This graphical representation which does not require training corpora can be potentially adapted to other languages.

1 Introduction

The main issue of extractive summarization is how to judge the important concept that should be described in the summary. Existing Graph-based ranking algorithms are used to simulating the functioning of human intelligence and are proved to be efficient to identify the salient elements from graph. A graphic representation of documents provides a natural way to model textual units and the relationships that interconnect them on different levels of abstraction. According to the fact that most of news articles report several events and these events are referred in many other documents that are related to the topic, it is better to build event-centric graphs by choosing textual units as event elements (including actions and the entities that participate in the events), events or sentences containing events. In addition, graph solves the problem of reduplicate information by assessing weights of links between nodes.

In this paper, we propose to extract event information and derive intra-event relations between event elements in news articles without deep natural language processing techniques. A weighted document graph is then built to represent the cohesive structure of text, specially emphasizing on events. We evaluate the capability of graph representations on multiple news articles summarization with PageRank [1] ranking algorithms. To focus on the efficiency and potential of event-centric document graphs, we do not consider the other features known to be helpful when creating summaries. We close with the discussion of future work.

Y. Matsumoto et al. (Eds.): ICCPOL 2006, LNAI 4285, pp. 181 – 188, 2006.

2 Related Work

Graph is a relational structure capable of representing the meaning and construction of cohesive text with associative or semantic information, corresponding naturally to human memory. Text visualization has been used to represent the underlying mathematical structure of a text or a group of texts [8]. At the same time, graph-based ranking algorithms has been successfully used in hyperlink analysis [1] and social networks [2], and recently turned into application on natural language processing. These algorithms decide on the importance of a node within a graph through link structure, rather than relying only on local node-specific information.

Extractive summarization emphasizes on how to determine salient pieces from original documents and therefore benefits much from graph-based ranking algorithm. To rank entire sentences for sentence extraction, most of previous works add a node to the graph for each sentence in the text. Different measurements are used to determine how to represent sentence and how to define connections between sentences. The similarity between two sentences according to their term vectors is used to generate links and define link strength in [4]. Similarly, [3] weighed links by the content overlap of two sentences normalized by the length of each sentence. Yoshioka and Haraguchi [6] went one step further taking events into consideration. Two sentences are linked when they share similar events, which are mostly judged by the similarity of words and consistency of date. However, choosing sentences as nodes within graph limits the representation ability of information in documents and the flexibility for further applications. In [5], the importance of the verbs and nouns constructing events is evaluated with PageRank as individual nodes aligned by their dependence relations. Unfortunately, dependency analysis requires syntax processing techniques.

Event-based summarization has been investigated in recent research. As introduced above, [5] and [6] both extracted events information by dependency structure of sentences and then formed a graph for summarization. In contrast, Filatova and Hatzivassiloglou suggested extracting atomic events to capture information about name entities and the relationships between these name entities, avoiding deep structure analysis of sentences [7]. They evaluated sentences only by times of appearance of pairs of name entities and atomic event connectors. The proposed approach claimed to out-perform conventional tf*idf approach on summarization and demonstrated that defining events based on named entities is feasible. However, their event definition is too strict to capture adequate information from texts.

Our work differs from these previous studies in two key respects. First, we propose a novel approach to extract semi-structured events with shallow natural language processing. Second, we build event-centric document graphs to make conceptual information visible and rank textual units for summarization on different granularities.

3 Event-Based Document Graph

3.1 Extraction of Event

Events described in texts link major elements of events (people, companies, locations, times etc.) through actions. In this paper, we use the definition of event proposed in

[8]. Events are anchored on major elements representing as named entities and high frequently occurring nouns, kind of named entities that can not be marked by general named entity taggers. A verb or an action noun is deemed as an event term only when it appears at least once between two named entities. Event terms roughly relate to the actions of events. Thus, we extract events based on named entities and co-occurrence of event elements without syntactic analysis.

Events are extracted from documents by using following steps:

1. Mark texts with named entities and POS tags.
2. Add a frequent noun into the set of named entities (NE) when its appearance times are above a certain threshold.
3. Detect pairs of named entities in every sentence and extract verbs and action nouns as event terms (ET), ignoring stopwords.
4. Scan documents again to extract events as event terms with adjacent named entities. These events take the form as triple $(et_x \mid ne_i, ne_j)$, if the event terms between a pair of named entities; or as couple $(et_y \mid ne_k)$, if the event terms is neighboring with only one named entity in a sentence.

Original:
The <Organization>Justice Department</Organization> and the 20 states <VB>suing</VB> <Organization>Microsoft</Organization> believe that the tape will <VB>strengthen</VB> their <HN>case</HN> because it shows <Person>Gates</Person> saying he was not <VB>involved</VB> in plans to take what the <HN>government</HN> alleges were illegal steps to <VB>stifle</VB> <AN>competition</AN> in the Internet <HN>software</HN> <HN>market</HN>.

Events:		
	1.	{sue | Justice Department, Microsoft}
	2.	{strengthen | Microsoft, case}
	3.	{involve | Gates, government}
	4.5.	{stifle, compete | government, software}

Fig. 1. Example of Event Extraction from a sentence

This approach complements the advantages of statistical techniques and captures semantic information as well. Figure 1 shows an original sentence of news article and five extracted events. The event "sue" represents the structure of Subject-Verb-Object (SVO), whereas the other four events only carry partial relationship of SVO, and "software" is not as proper as "the Internet software market". However, graph-based ranking algorithm calculates the weights of nodes and roughly gets rid of unimportant event elements and extra elements added by mistake.

3.2 Building Document Graph

To form the document graph, we take these events by choosing event elements (event terms and named entities) as nodes. The edges between event elements are established by co-occurrence in a same event. A piece of a graph built by our system for cluster d30026 (DUC 2004) is shown in Figure 2.

The document graph is weighted but undirected. Different from previous work on intra-event relevance [7] [9], the relationship between event elements is measured not only by counting how many times they co-occur in events, but also by taking linguistic structure of sentence into consideration. We observe in real texts that two named entities can be far apart in a long sentence and more than one event terms emerge between them (e.g. "stifle" and "compete" event in Figure 1; event terms in joined rectangles in Figure 2). These adjacent event terms which are associated with same pair of named entities are mostly because of complicate sentence structure, such as subordinate clause. The strength of link between action and named entity within an event is indicated as $L_{event}(et_x, ne_i) = L_{event}(ne_i, et_x) = 1/n$, when n is the number of adjacent event terms between the same named entity (pair). The weight of connection within graph is calculated as $\mathrm{R}(et_x, ne_i) = \mathrm{R}(ne_i, et_x) = \sum L_{event}(ne_i, et_x)$ .Figure 3 enlarges a part of document graph in Figure 2 to show the weight of each edge.

S1: The Justice Department and the 20 states suing Microsoft believe that the tape will strengthen their case because it shows Gates saying he was not involved in plans to take what the government alleges were illegal steps to stifle competition in the Internet software market.
S2: It showed a few brief clips of a point in the deposition when Gates was asked about a meeting on June 21, 1995, at which, the government alleges, Microsoft offered to divide the browser market with Netscape and to make an investment in the company, which is its chief rival in that market.
S3: In the taped deposition, Gates says he recalled being asked by one of his subordinates whether he thought it made sense to invest in Netscape.
S4: But in an e-mail on May 31, 1995, Gates urged an alliance with Netscape.
S5: The contradiction between Gates' deposition and his e-mail, though, does not of itself speak to the issue of whether Microsoft made an illegal offer to Netscape.

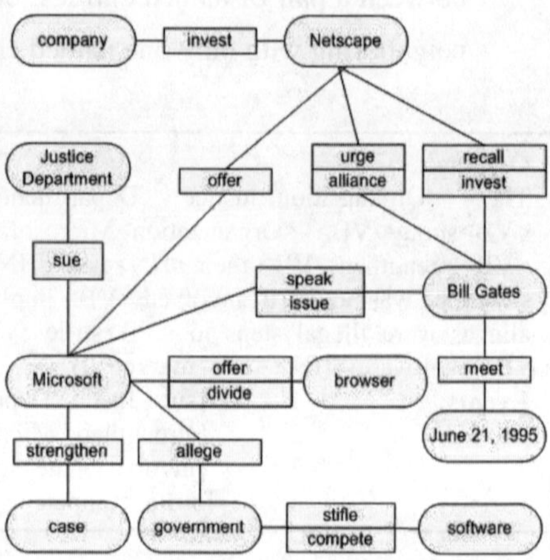

Fig. 2. Document Graph Fragment, on event element level

Since these events are commonly related with one another semantically, temporally, spatially, causally or conditionally, especially when the documents are under the same or related topic, we can derive intra-event relevance between two event terms or two named entities from document graph.

$$\mathrm{R}(et_x, et_y) = [\sum_{ne_i \in \mathrm{NE}(et_x) \cap \mathrm{NE}(et_y)} \mathrm{R}(et_x, ne_i) \cdot \sum_{ne_i} \mathrm{R}(ne_i, et_y)]^{1/2} \tag{E1}$$

$$\mathrm{R}(ne_i, ne_j) = [\sum_{et_x \in \mathrm{ET}(ne_i) \cap \mathrm{ET}(ne_j)} \mathrm{R}(ne_i, et_x) \cdot \sum_{et_x} \mathrm{R}(et_x, ne_j)]^{1/2} \tag{E2}$$

Where $\mathrm{NE}(et_x)$ is the set of named entities et_x associates; $\mathrm{ET}(ne_i)$ is the set of event terms ne_i associate.

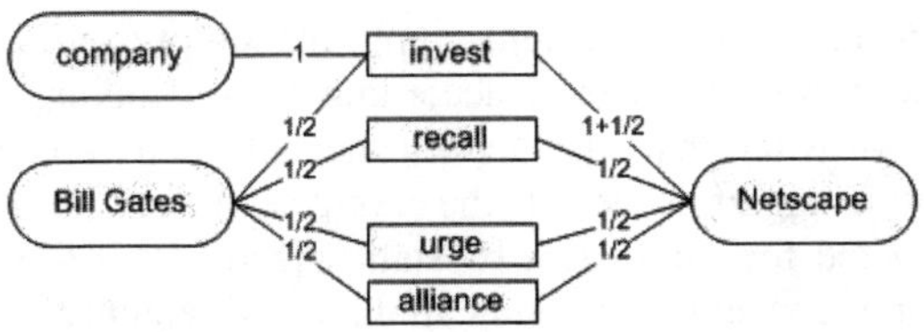

Fig. 3. Weight of link between event terms and named entities

For the convenience to observe organization of document and to investigate certain event or specific sentence with associated contextual information in the future, we design to form document graph on event and sentence level. To determine the strength of events, we have two choices. One is to use a simple cosine similarity based on a measure of event elements overlap and the other is to use the cross strength of relation between event elements. In this paper, we consider only events and neglect other words, thus the second approach is better to make use of event relevancy. As shown in Figure 4 and Figure 5, relations of events are measured by sum all the weights of connections between event elements and similarly, relations of sentence by weights of connections between events.

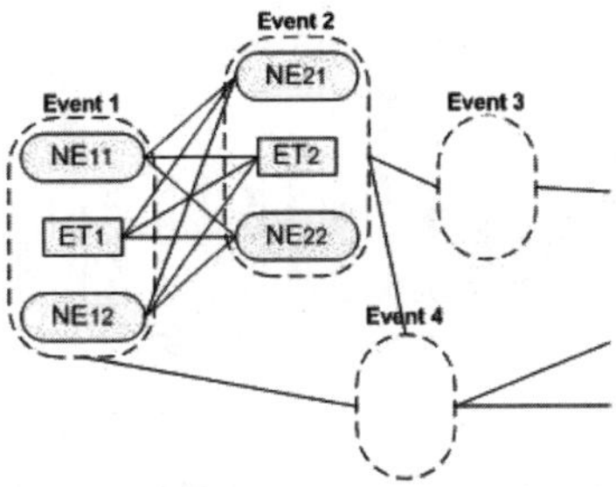

Fig. 4. Sketch Map of Document Graph, on event level

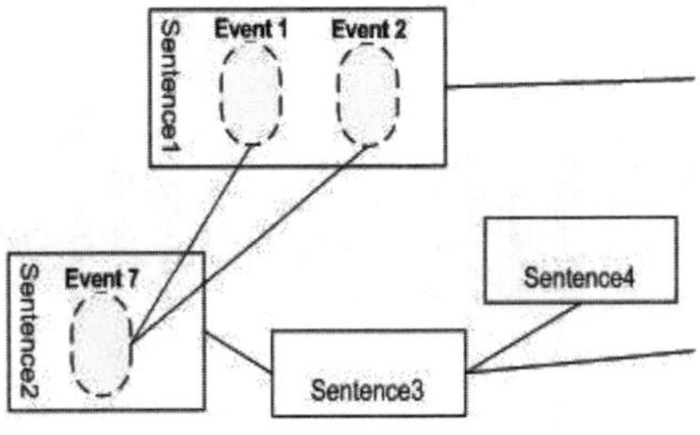

Fig. 5. Sketch Map of Document Graph, on sentence level

3.3 Node Scoring with PageRank for Summarization

To score the significance of nodes in a document graph, our system uses the PageRank algorithm [1]. The thrust of PageRank is that when a node links to more other nodes or links to another "important" node, it becomes more "important". A ranking process starts by assigning arbitrary values to each node in graph and followed by several iterations until convergence.

The formula for calculating Pagerank of a certain node n is given as follows:

$$\mathrm{PR}(node_n) = (1-d) + d \sum_{node_i \in L} \frac{\mathrm{PR}(node_n)}{\mathrm{R}(node_i, node_n)} \qquad \text{(E3)}$$

where L is the set of nodes linking into node n
d is a dampening factor, set to 0.85 experimentally

For different granularity of document graph, the significances of event elements, events and sentences are then scored according to the linking structure and edge weights respectively. After that, the significance of each sentence is obtained by simply summing the significance of the event elements or events it contains. Sentences are extracted for summaries by static greedy algorithm [7], if and only if they cover the most of concepts, removing all duplicate sentences.

With ranking algorithm for graph, process of extractive summarization can be fully unsupervised without training on corpora. Moreover, we can further realize information fusion, sentence compression and sentence generation in the future.

4 Experiments and Discussions

We test our event-based graphical approach by the task of multi-document summarization in DUC 2001(task 2) and DUC 2004(task 2). The documents are pre-processed with GATE to recognize named entities, verbs and nouns.

In order to evaluate the quality of the generated summaries, we use the automatic summary evaluation metric, ROUGE [10]. This metric is found to be highly correlated with human judgments.

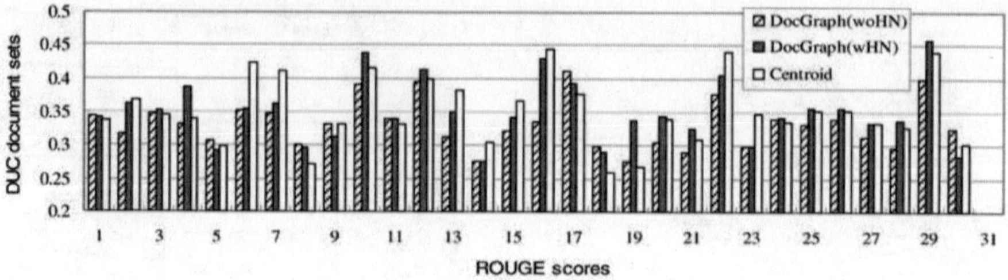

Fig. 6. ROUGE scores, Document Graph (with and without high frequency noun) vs. Centroid

In our first experiment our approach is evaluated on 200-words summaries of DUC 2001. We determine the salient concept by document graph on event element level. We compare the ROUGE scores of adding frequent nouns or not to the set of named entities to our system. A baseline is also included as Centroid-based summarization, which is a widely used and very challenging baseline in the text summarization community [11]. ROUGE scores are reported for each document set rather than average score because ROUGE scores depend on each particular document set (Figure 6). Finally, for 18 sets (60%) out of the 30 document sets, the summary created according to document graph with frequent nouns receives higher ROUGE score than Centroid-based approach. By taking high frequent nouns into the consideration, great improvement is achieved in 20 sets (66.7%) and 5% increase of ROUGE score is gained on average. The advantage of graph-based approach over Centroid is that it indicates redundant information by link weight and prevents improper high idf scores from rare words that are unrelated to the topic.

Next, we compare two methods to measure the strength of relationship between event elements, one is proposed in previous work by times of co-occurrence in events, the other is new in this paper splitting the weight in same named entity pair. As shown in Table 1, a slight improvement is achieved by the new approach. Besides we evaluate this adjustment on different strategies on deriving event relevance by graph-based ranking algorithm in [9], and prove that improvement is slight but constant.

As discussed before, document graph can be constructed by choosing different kinds of nodes. Table 2 shows the result by ranking text units for summarization on different granularity. The advantage of representing with separated actions and entity nodes over simply combining them into event or sentence node is to provide a convenient and effective way for analyzing the relevance between conceptual information. At the same time, the graph on event or sentence level helps people to observe and investigate documents more conveniently.

Table 1. ROUGE scores using different methods to weigh relations in graph

	DUC 2001		DUC 2004	
	co-occurrence times	split weight in same pair	co-occurrence times	split weight in same pair
ROUGE-1	0.35212	0.35250	0.32718	0.33255
ROUGE-2	0.07107	0.07179	0.07027	0.07357
ROUGE-W	0.13603	0.12901	0.12691	0.12949

Table 2. ROUGE scores according to document graph on different level (DUC 2001)

granularity	event elements	event	sentence
ROUGE-1	0.35212	0.33348	0.33957
ROUGE-2	0.07107	0.05886	0.06609
ROUGE-W	0.13603	0.12120	0.12387

5 Conclusion

In this paper, we propose a new approach to present documents by event-based graph and illustrate the application to text summarization. The extraction of event is considered to include basic concepts in news articles as actions and named entities. Document graph makes use of the associations of event elements based on co-occurrence to avoid complex natural language processing techniques. Graph-based ranking algorithm is put forward to determine salience of text units for extractive summarization. The experiment results indicate that this mixed approach of statistics and linguistics is competitive with up-to-date techniques on multiple news articles summarization.

The graph constructed in this way allow further complex processing, such as improving the coherence of summaries by relations and compressing the original

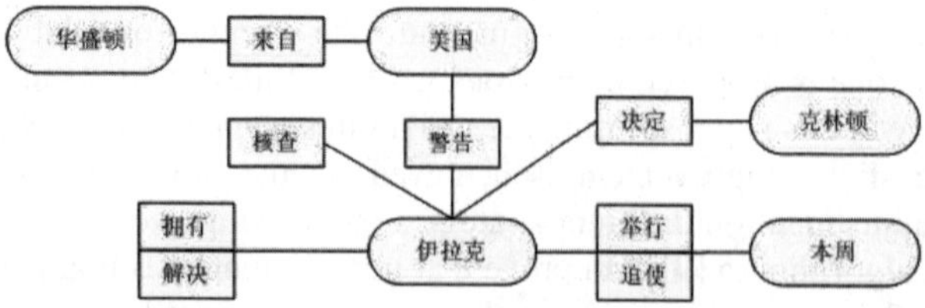

Fig. 7. Document Graph Fragment on Chinese Text

sentences by cutting inessential fragments in the graph. Another advantage of the graph-based document representation and ranking algorithms is that they exclusively rely on the text itself and do not require any training corpora. As a result, our approach can be adapted to other languages. In fact, we have recently attempted to apply the similar method to the texts in Chinese and shown a potential success in summarization (Figure 7).

Acknowledgments. The work presented in this paper is supported partially by National Natural Science Foundation of China (reference number: NSFC 60573186), partially by Research Grants Council on Hong Kong (reference number CERG PolyU5181/03E) and partially by the CUHK strategic grant (# 4410001).

References

1. Page, L., Brin, S.: The Anatomy of a Large-scale Hypertextual Web Search Engine. Computer Networks and ISDN Systems 30 (1998) 107-117
2. Dom, B., Eiron, I., Cozzi, A., Shang, Y.: Graph-based ranking algorithms for e-mail expertise analysis. In Proceedings of the 8th ACM SIGMOD workshop on Research Issues in Data Mining and Knowledge Discovery (2003) 42-48
3. Mihalcea, R.: Graph-based Ranking Algorithms for Sentence Extraction, Applied to Text Summarization. In Proceedings of the 42nd Annual Meeting of the Association for Computational Linguistics (2004) 170-173
4. Erkan, G., Radev D.R.: 2004. LexRank: Graph-based lexical as Salience in Text Summarization. Journal of Artificial Intelligence Research 22 (2004) 457-479
5. Vanderwende, L., Banko, M., Menezes, A.: Event-Centric Summary Generation. In Proceedings of the Document Understanding Conference Workshop (2004)
6. Yoshioka, M., Haraguchi, M.: Multiple News Articles Summarization based on Event Reference Information. In Working Notes of the 4th NTCIR Workshop (2004)
7. Filatova, E., Hatzivassiloglou, V.: Event-based Extractive Summarization. In Proceedings of ACL Workshop on Summarization (2004) 104-111
8. Bradley, J., Rockwell, G.: What Scientific Visualization Teaches Us about Text Analysis. In ALLC/ACH Conference (1994)
9. Li, W., Xu, W., Wu, M., Yuan, C., Lu, Q.: Extractive Summarization using Inter- and Intra- Event Relevance. In Proceedings of COLING-ACL (2006)
10. Lin, C., Hovy, E.: Automatic Evaluation of Summaries Using N-gram Co-occurrence Statistics. In Proceeding of HLT-NAACL (2003) 71-78
11. Radev, D.R., Jing, H., Stys, M., Tam D.: Centroid-based Summarization of Multiple Documents. Information Processing and Management. 40 (2004) 919-938

A Probabilistic Feature Based Maximum Entropy Model for Chinese Named Entity Recognition

Suxiang Zhang[1,2], Xiaojie Wang[1], Juan Wen[1], Ying Qin[1], and Yixin Zhong[1]

[1] School of Information Engineering of Beijing University of Posts and Telecommunications, 100876 Beijing, China
[2] Department of Electronic and Communication Engineering of North China Electric Power University, 071003 Baoding, China
zsuxiang@163.com

Abstract. This paper proposes a probabilistic feature based Maximum Entropy (ME) model for Chinese named entity recognition. Where, probabilistic feature functions are used instead of binary feature functions, it is one of the several differences between this model and the most of the previous ME based model. We also explore several new features in our model, which includes confidence functions, position of features etc. Like those in some previous works, we use sub-models to model Chinese Person Names, Foreign Names, location name and organization name respectively, but we bring some new techniques in these sub-models. Experimental results show our ME model combining above new elements brings significant improvements.

Keywords: Maximum entropy, Named entity recognition, probabilistic feature and evaluation.

1 Introduction

Named Entity Recognition (NER) is one of important task in Natural Language Processing. It plays an important role in Information Retrieval and Extraction, Question Answer, Machine Translation and so on. Lots of works have been done on NER. Generally, the approaches to NER can be classified into two sets, rule based approaches and machine learning based approaches. [Grishman, et al. 1995] [Krupka, et al. 1998] used rules to identify Named Entity (NE). Since rules were induced manually, such kind of approach is time-consuming and expensive. As the large NE tagged corpora are becoming available, machine learning based approach have been received more and more attentions.

[Sekine, et al.1998] used Decision Tree for NER. [Bikel, et al. 1997] modeled NER task using Hidden Markov Model(HMM). Maximum Entropy Model[Borthwick, et al. 1999][Mikheev, et al.1998] have been also proposed to solve the problem of NER. For Chinese, Role-tagging method was proposed in [ZHANG Hua-ping, LIU Qun, 2004]. [Lv Ya-juan, Zhao Tie-jun, 2001] used dynamic programming.

NE is diverse, person name, location name, organization name and so on. It has been shown that different models have different strength on different NE types.

Y. Matsumoto et al. (Eds.): ICCPOL 2006, LNAI 4285, pp. 189 – 196, 2006.

Combination of several sub-models to cope with different kind of NE respectively is an effective way to improve the performance of NER. For example, [Chen, et al. 1998] used statistical-based model on person name recognition, while rule based models are used on Location and Organization recognition.

No matter which model is used, almost all NER models recognize NE by mining the intrinsic features inside the name and the contextual features around the name [Borthwick, 1999] [Youzheng Wu, Jun Zhao, et al 2005]. How to make use of different kinds of features effectively is an important factor in improving the performance of NER.

This paper proposes a probabilistic feature based maximum entropy approach to NER. Where, probabilistic feature functions are used instead of binary feature functions, it is one of the several differences between this model and the most of the previous ME based model. We also explore several new features in our model, which includes confidence functions, position of features etc. Like those in some previous works, we use sub-models to model Chinese Person Names, Foreign Names, location name and organization name respectively, but we bring some new techniques in these sub-models. Experimental results show our ME model combining above new elements brings significant improvements to NER task. We achieved best F-measure in MSRA NER task of SIGHAN06 contest

2 Probabilistic Feature Based Maximum Entropy Model

The recognition of time and numbers is comparatively simple and can be implemented via finite state automata. This paper focuses on the recognition of person name, location name and organization name.

Our work is under the framework of Maximum Entropy (ME), but differs from most of the previous approaches in several ways. We first give a brief introduction of standard ME model.

We consider a random process which produces an output y, a member of a finite set Y. The process may be influenced by some contextual information x when it generate y, x is a member of a finite set X. Our task is to construct a stochastic model that accurately represents the behavior of the random process. Such a model is a method of estimating the conditional probability of y when contextual x is given. Then according to ME principle, the optimal parametric form of a model $p(y|x)$ is given by (1).

$$p_\lambda(y|x) = \frac{1}{Z_\lambda(x)} \exp(-\sum_i \lambda_i f_i(x, y)) \tag{1}$$

Where $f_i(x, y)$ is a feature function, $Z_\lambda(x)$ is a normalizing constant determined by the requirement that $\sum_y p_\lambda(y|x) = 1$ for all x

$$Z_\lambda(x) = \sum_y \exp(-\sum_i \lambda_i f_i(x, y)) \tag{2}$$

In standard ME, a feature functions is a binary function, for example, if we use CPN denotes the Chinese person name, SN denotes Surname, a typical feature is:

$$f_i(x,y)=\begin{cases}1 & if \quad y=CPN \quad and \quad x=SN\\ 0 & \quad otherwise\end{cases} \tag{3}$$

But in Chinese, firstly, most of words used as surname are also used as normal words. The probabilities are different for them to be used as surname. Furthermore, a surname is not always followed by a given name, both cases are not binary. To model these phenomena, we give probability values to features, instead of binary values.

For example, a feature function can be set value as follows:

$$f_i(x,y)=\begin{cases}0.985 \ \text{if} \ y=\text{CPN and } x=\text{郭}\\ 0 \ \ otherwise\end{cases} \tag{4}$$

Or

$$f_i(x,y)=\begin{cases}0.01805 \ \text{if} \ y=\text{CPN} \ and \ \text{x}=\text{于}\\ 0 \ \ otherwise\end{cases} \tag{5}$$

Probability values are estimated statistically from corpus. In this way, we distinguish different SN by using different probability values feature. This gives the model capability of exploring finer-grain difference in instances. For example, in (4) and (5), both 郭 and 于 are Chinese surname, but they have different probabilities as surnames. That is to say, we can get more information from instances.

Chinese characters used for translating foreign personal name are different from those in Chinese personal name. We built the foreign name model by collecting suffixes, prefixes, frequently-used characters, estimate their probabilities used in foreign personal name. These probabilities also used in model as probability features.

We also design a confidence function for a candidate person name sequence $W = C_1C_2...C_n$ to help model to estimate the probability of W as a person name. C_i is a character. Let f_{1F} is the conditional probability $p(w_{first} \mid person)$ of the C_1 which can be the first character of a person name, f_{iM} is the conditional probability $p(w_{iM} \mid person)$ of the C_i which can be the middle character of a person name, f_{nE} is the conditional probability $p(w_{last} \mid person)$ of the C_n which can be the last character of a person name. So the confidence function is

$$K(w, PERSON) = f_{1F} + \sum_{2<=i<=n-1} f_{iM} + f_{nE} \tag{6}$$

This function is included in ME frame as a feature.

3 Person Name Recognition and Location Name Recognition

Candidate person name collection is the first step of NER. Since the ambiguity of Chinese word segmentation always exists. We propose some patterns for model different kind of segmentation ambiguity. Some labels are used to express specific roles of Chinese characters in person names.

We have seven patterns to collect candidate Chinese person name; first two patterns are non-ambiguity, while the others model some possible ambiguity in Chinese person name brought by word segmenters.

(1) We use BCD pattern to model a Chinese person name which is composed of three Chinese characters. For example: "江*(B)* 泽*(C)* 民*(D)*". Where B is the surname of the Chinese person name. C is the first given name. D is the last given name.

(2) We use BD pattern to model a Chinese person name which is composed of two Chinese characters. For example "李*(B)* 鹏*(D)*".

(3) BCH pattern is used to model an ambiguous case where the last given name and its next Chinese character can be a Chinese word. For example: "江*(B)* 泽*(C)* 民主*(H)* 席". 江泽民 is a person name, but民主 is a Chinese word, a segmenter may combines 民 with 主 as a word before NER.

(4) UCD pattern models an ambiguous case where the surname and its previous character are combined into a word. For example:"武汉 市 长孙*(U)* 大*(C)* 亮*(D)*". Where 孙大亮 is a name, while长孙 is a Chinese word.

(5) In BE pattern, the first given name and the last given name can be a word. For example: "记者 余*(B)* 清楚*(E)* 报道". Here, 余清楚 is a person name, while清楚 is a Chinese word.

(6) In UD pattern, the family name and the first given name is a word. For example: "记者 白天（*U*）亮*(D)* 报道". Where白天亮 is a person name, while白天 is a normal word.

(7) Sometimes, a Chinese personal name may be composed of only two given names, for example "小 力*(C)* 达 *(D)*", where力达 is a person name, and we use CD to denote this pattern.

We made a statistical data from the People's Daily (2000) to achieve word or character lists which belong to the class "B, C, D, U, H, E" respectively.

We collect candidate Chinese person names based on the feature patterns. Moreover, contextual words of the candidate person name are also used for model.

For Chinese person name recognition, a problem here is how we can know whether a person name is composed of two or three Chinese characters. We used another technology to help boundary detection. We use co-occurrence count of a candidate person name and its next character (or word) to decide the length of the Chinese personal name. For example:

"李超为宁波拿下了一分"

In this sentence, we collected a candidate Chinese person name "李超为", then we should make a decision whether the last character "为" belongs to this personal name or not. To do it, we just count following two co-occurrences and compare them. Here, we have

$$number(NR\ \text{宁波}) < number(NR\ \text{为}) \qquad (7)$$

So, "为" is not included in the personal name, "李超" is a correct choice.

We implement Chinese person name recognition and foreign person name separately. These two kinds of names are very different in Chinese, but sometimes, we should pay special attention to distinguishing them. Some foreign person names include Chinese surnames in it, which are important cues for our model to recognize Chinese person names. In this case, a piece of the foreign person name may often be recognized as a Chinese person name.

In following example, the tagger *nr* means a person name.

"俄罗斯 政府 驻 车臣 副 代表 兹韦列夫/nr 和 格罗兹尼 市长 助理 哈布谢耶夫/nr 五月 三十日 晚 在 格罗兹尼 市郊 被 炸 身亡， 市长 莫赫恰耶夫/nr 身 负 重伤。", "韦", "谢", "赫" and "莫" are Chinese surname, when we recognized the Chinese name firstly, we will choose "韦列夫", "谢耶夫" and "莫赫恰" as three candidate Chinese person names, but兹韦列夫, 哈布谢耶夫and莫赫恰耶夫are three foreign names, so, we design a flexible method to identify foreign personal name in this case.

For a ambiguous Chinese character like "谢" or "马" which can be both in a Chinese name and a foreign name, our model will search forward and backward in context to find some other Chinese characters which belong to Chinese person name or foreign person names. According to the results of collection, the model will choose Chinese person model or foreign person model to identify the candidate person name. Experiments show that our method is promising, the recall and precision have been improved.

Location names often end with the some specific words like "省/province". Location name recognition is similar to person name in our model. The difference between them is different search direction when collecting candidate entity. The ME method and the confidence function are also used for location name recognition. The confidence function for location name is as equation (8).

$$K(w,location) = f_{1F} + \sum_{2<=i<=n-1} f_{iM} + f_{nE} \tag{8}$$

Feature of person name and location name are selected as follows: CPN denotes the Chinese person name; FPN denotes the foreign person name; LN denotes the location name;

Type	Feature list
CPN	Context information w_{i-1}, w_{i+1}
	Surname
	The first given name
	The last given name
	Context semantic
	The probability of a surname
	The probability of a given name
	Confidence function
FPN	Context information w_{i-1}, w_{i+1}
	The probability of character
	Confidence function
LN	Context information $w_{i-2}, w_{i-1}, w_{i+1}, w_{i+2}$
	The probability of character
	Confidence function
	Keyword list such as "省, 市"

4 Organization Recognition

Organization name recognition is very different from other kinds of entities. An organization is often composed of several characters or words. According to our statistical data. The maximum length of an organization is 7. At first, we collect candidate organization; secondly, the candidate phrase can be inputted into a sub-model to decide whether it is an organization.

Based on People's Daily (2000), we compute the probability $p(w_i \mid ORG)$ of each word of an organization, and the probability $p(w_{keyword} \mid ORG)$ of each keyword for an organization.

In a string W=W_1...$W_{keyword}$, Where $W_i (1 < i <= 6)$ is the word that belongs to an organization name, $W_{keyword}$ is a keyword of organization like "公司/company".

The probability of a candidate as an organization name is defined as in (9).

$$prog(W) = \frac{1}{n} \times \sqrt[n]{p(w_{keyword})\prod_{i=1}^{n-1} p(w_i)} \tag{9}$$

We compute the probability of a candidate organization. If *prog(W)*> α (a given constant), we then tag BIO information for it using ME model. BIO is usually used in chunk analysis. We here employ it on organization name recognition. Where B denotes the begin word (or character) of an organization name, I is a middle word (or character), L is the last word (or character) and O is not in an organization name. We only used the positive instances to train the model for organization name recognition. Output probability is used to assign tags.

Feature selection is a very important factor in ME model, we used some ordered features to recognize B, I, L and O as follows.

P_0	the current word
P_{-1}	the previous word
P_{+1}	the next word
$P_{0\text{-}pos}$	the POS of the current word
$P_{-1\text{-}tag}$	the label of the previous word
$P_{-1\text{-}pos}$	the POS of the previous word
$P_{+1\text{-}pos}$	the POS of the next word
	Feature order

5 Experiments and Results

We implement several experiments to test our model.

First, we want to see if our probability feature function based ME model can outperform normal binary feature function based ME model. We use 7M corpus of one-month People's Daily (January, 1998).

The 5-cross validation experimental result of Chinese person name is shown in Table1. The label of R, P and F show the recall, precision and F-measure respectively.

Table 1. Result between binary value and probability value

	Binary value (R/P/F)	Probability value (R/P/F)
1	0.80102/0.96435/0.87517	0.9194/0.98087/0.94916
2	0.79699/0.96358/0.8724	0.9134/0.97556/0.94346
3	0.80478/0.96865/0.8791	0.9228/0.98367/0.95225
4	0.80973/0.9665/0.8812	0.92132/0.98149/0.95041
5	0.809/0.95988/0.87802	0.92355/0.9743/0.94824

From the Table1, We can find the performance of probability feature is better than that in binary case, the F-measure improved 7.2%.

Therefore, we use the probability feature function in following experiment result.

We then used the first five month's text of 2000 People's Daily as our training data. Data of June is used as test data. The experimental results are shown in Table2.

Table 2. Person and Location recognition result

	R	P	F
Person	90.35%	93.75%	92.02%
Location	87.92%	90.25%	89.07%

Thirdly, we want to see if the probability threshold is helpful to BIO tagging. We implement a pair of contrastive experiments. One is with probability threshold filtering (PTF) before BIO tagging, and another is without. The experimental results are shown in Table.3

Table 3. Organization recognition result

ORG	R	P	F
BIO	83.75%	67.26%	74.60%
BIO+PTF	86.81%	89.68%	88.22%

Where, we set the threshold as 0.02. We find the F-measure improves more than 18% by using probability threshold filtering. This means, probability threshold filtering is very effective

We took part in the SIGHAN (2006) entity recognition open track contest for Microsoft Research Asia Research (MSAR) corpus, and achieved the highest F-measure.

Table 4. The official SIGHAN evaluation result for entity recognition in the open track

Type	R	P	F
Person	95.39%	96.71%	96.04%
Location	87.77%	93.06%	90.34%
Organization	87.68%	84.20%	85.90%

6 Conclusions

In this paper, we propose a probability feature based ME model for Chinese named entity recognition. Where, probabilistic feature functions are used instead of binary feature functions. We also explore several new features in our model, which includes confidence functions, position of features, probability threshold etc. We use several sub-models to model Chinese Person Names, Foreign Names, location name and organization name respectively. Experimental results show our ME model combining above new elements brings significant improvements.

Acknowledgment

The research work was supported by the China National Project 863 (the 863 Program) under the grant No.2001AA114210 and MOE funded project (MZ115-022): "Tools for Chinese and Minority Language Processing".

References

1. Ralph Grishman and Beth Sundheim: Design of the MUC-6 evaluation. In: 6th Message Understanding Conference, Columbia, MD. (1995)
2. Krupka, G. R. and Hausman, K. IsoQuest: Inc.: Description of the NetOwl TM Extractor System as Used for MUC-7. In Proceedings of the MUC-7, (1998)
3. Sekine S., Grishman R. and Shinou H. A decision tree method for finding and classifying names in Japanese texts. In: Proceedings of the Sixth Workshop on Very Large Corpora, Canada, (1998)
4. D.M. Bikel, Scott Miller, Richard Schwartz, Ralph Weischedel: Nymble: a High-Performance Learning Name-finder. In: Fifth Conference on Applied Natural Language Processing, (published by ACL). (1997) 194-201
5. Borthwick .A. A Maximum Entropy Approach to Named Entity Recognition. PhD Dissertation. (1999).
6. Mikheev A., Grover C. and Moens M: Description of the LTG System Used for MUC-7. In: Proceedings of 7th Message Understanding Conference (MUC-7), (1998)
7. Hua-ping ZHANG and Qun LIU. Automatic Recognition of Chinese Personal Name Based on Role Tagging. CHINESE JOURNAL OF COMPUTERS (2004), Vol (27), 85-91
8. YaJuan Lv, Tie-jun Zhao, Mu-yun Yang, Yu Hao, Li Sheng, Leveled unknown Chinese Words resolution by dynamic programming. Journal Information Processing, (2001), 15(1): 28-33
9. A. L. Berger, S. A. Della Pietra, and V. J. Della Pietra. A maximum entropy approach to natural language processing. Computational Linguistics, (1996) 22(1), 39-71
10. Youzheng Wu, Jun Zhao, Bo Xu and Hao Yu. Chinese Named Entity Recognition Based on Multiple Features. Proceedings of Human Language Technology Conference and Conference on Empirical Methods in Natural Language Processing (HLT/EMNLP), (2005) 427–434, Vancouver, October 2005.
11. H.H. Chen, et al: Description of the NTU System Used for MET2. In: Proceedings of the Seventh Message Understanding Conference. (1998)

Correcting Bound Document Images Based on Automatic and Robust Curved Text Lines Estimation

Yichao Ma, Chunheng Wang, and Ruwei Dai

Laboratory of Complex System and Intelligent Science
Institute of Automation, Chinese Academy of Science
Zhongguancun East Rd, No.95, Haidian Dist, Beijing, 100080, P.R. China

Abstract. Geometric distortion often occurs when taking images of bound documents. This phenomenon greatly impairs recognition accuracy. In this paper, a new one-image based method is proposed to correct geometric distortion in bound document images. According to this method, the document image is binarized first. Next, curved text-line features are extracted. Thirdly, locally optimized text curves are detected using a graph model. Finally, the technique of texture warping is applied to correct the image. Experimental results show that images restored by our proposed method can achieve good perception and recognition results.

1 Introduction

When imaging bound documents, such as books, originally straight text lines often appear as curves in the images. This geometric distortion impairs recognition accuracy greatly, because current OCR systems oftern require text lines be straight. Therefore, it is a necessary process for character recognition system in both scanner and camera images to make text lines straight.

In the literature, several methods have been proposed to deal with this kind of distortion. However, there has not been a good way to automatically and robustly rectify bound document images so far. [1] [2] devised special camera equipments to acquire document images. The depth of an image can be measured using these tools and 3D shape model be reconstructed. Although these methods are useful in some given applications, the requirements of camera calibration and some special tools do limit their use in daily life. [3][4] proposed effective shape from shading method to handle scanned document images. However, it is hard to deal with lighting information for camera images. Some approaches are based on one document image [5,6,7,8,9,10]. In these papers, curved text lines are usually fitted to represent the distortion in the image. And then the image is restored based on some predefined model. This one-image approach is more applicable, since no special devices, camera parameters or extra images are needed. In this kind of method, it is the crucial step to locate the curved text lines automatically and accurately.

In [5], one pixel width text line features are extracted first. Then searching is done from right to left based on the predefined step and angle. Next, two typical directrixes are chosen to model the bound document as a general cylinder surface. Finally, the rectification is performed in the whole image based on this model. In [6], first, four corner

Y. Matsumoto et al. (Eds.): ICCPOL 2006, LNAI 4285, pp. 197–204, 2006.

points of the text region need to be specified manually. Then a local adaptive cumulative projection is applied to detect possible starting points and starting orientations. Similar method is used to trace curved text lines. After text lines are detected, a source and target mesh are constructed correspondingly. Lastly, the images are corrected by a two-pass image warping algorithm. Zhang [7] also presented a method based on regression of curved text lines. This method is especially for scanned images. Shade area and clean area are detected first. Then projection and connected components analysis are used to detect curved text lines in each region respectively. After that, text lines are fitted by regression method. However, these existing methods either need human interaction or are not robust in complicate camera captured document images.

In this paper, we propose a new one-image based bound document image correction method. Our key idea is to introduce an automatic and robust correction. In order to do the correction precisely, an graph model is proposed to locate the text lines. The whole algorithm mainly includes the following steps. Firstly, thc document image is binarized using BST method to compensate nonuniform lighting or shade. Next, feature points are extracted from the minified image and locally optimized text curves are detected using a graph model. Finally, the distorted image is restored by using the technique of texture warping.

This paper is organized as follows. In section 2, the detail of our correction method is presented. Section 3 analyze the experiment results. And conclusion is presented in section 4.

2 New Document Image Dewarping Method

2.1 Text Line Feature Extraction and Preprocessing

A version of Seeger's Background Surface Thresholding method(BST) [11] is implemented. Text areas are labeled first. Then, continuous background intensities are estimated and thresholding is performed at the background plus some offset. This binarization method can handle light variation in the document image, and produce a less noisy image. After that, we propose to exploit the same observation as [5]: at a certain distance, text lines can be seen as line segments. Our method is based on this idea that working as low resolution makes it possible to extract text line as segments. So the binarized image is down sampled. Some geometric rules are then used to filter out non-text blocks. Afterwards, text lines' middle position in each column are recorded to form the feature image of text lines.

In addition, the obtained bound document images often include some parts of the neighboring page, as the example image shown in Fig.1. Some documents also have multiple columns. To deal with these, a simple projection profile based method is applied to analyze document layout in the feature image. Attached neighboring parts are screened out, and different columns are separated before being sent to the following process. A curved document image and its corresponding feature image after layout analysis are shown in Fig.1.

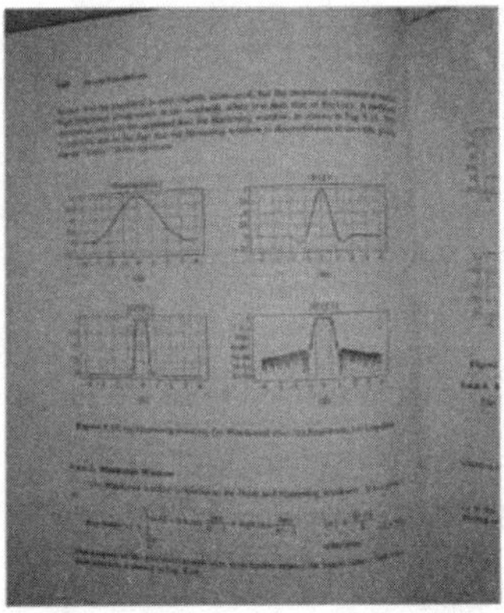

(a) A bound document image example

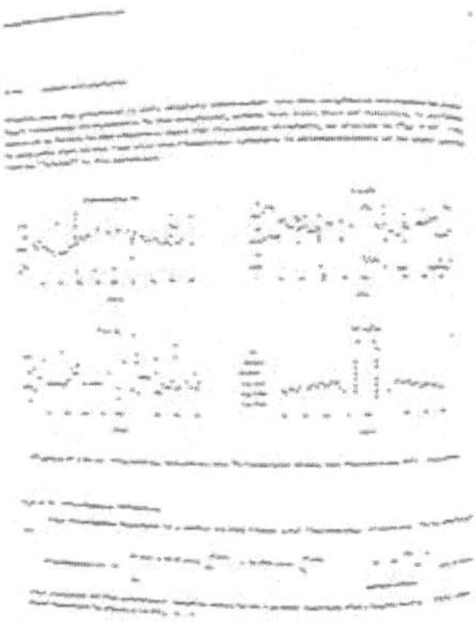

(b) The feature image after layout analysis

Fig. 1. An example of bound document image and its corresponding feature image

2.2 Graph Model of Text Line Features

Graph based method is widely used in the line or curve detection problems [12]. We can see from the feature image in Fig.1 that text lines are represented as one pixel width line segments, and segments on the same text line do not overlap with each other. In view of these, an oriented graph $G = \{V, E\}$ is adopted to represent the feature image. Here V is the set of nodes represent the segments' vertexes, $E \subseteq V \times V$ is the set of directed edges that represent line segments. Define the orientation of all the edges $e \in E$ is from left to right, i.e. for two vertexes on an edge, v_l is the vertex on the left, v_r is the vertex on the right, then $(v_l - v_r)$ is an edge, $(v_r - v_l)$ is not. Given the edges, we need to cluster them into different groups, and edges in each group represent a curved text line. Some definitions are necessary for further processing.

Definition 1. *Two edges e_1 and e_2 are connectible only if there is a path between all the four vertexes on these two edges.*

Fig.2 gives some examples to illustrate Definition 1. In (a), if adding an edge between v_2 and v_3, there is a path (v_1-v_2-v_3-v_4) through the four vertexes. While, in (b), there is no possible path through the four vertexes(The reason is the direction of an edge is from left to right). So the two edges in (a) can be connected, those in (b) cannot be connected.

If two edges could be connected, their connection coefficient is defined as $c = \Delta d + \mu(\theta_1 + \theta_2)$ where Δd is the Euclidean distance between two vertexes to be connected, θ_1, θ_2 are the deviation angles between two directed edges, as shown in Fig.3, μ is a jagged penalty parameter, it controls the preference between distance and smoothness, a bigger μ means smoothing connection is preferable.

Definition 2. *Two connectible edges e_i and e_j can be merged only if*

$$c_{ij} = \min\{c_{i.}\} \ and \ \min\{c_{j.}\} \ and \ c_{ij} < T$$

(a) These two edges are connectible. There is a path through the four vertexes(v_1-v_2-v_3-v_4).

(b) These two edges are not connectible. We cannot find a path through the four vertexes.

Fig. 2. The connectiblity of edges. (The direction of an edge is defined from left to right.)

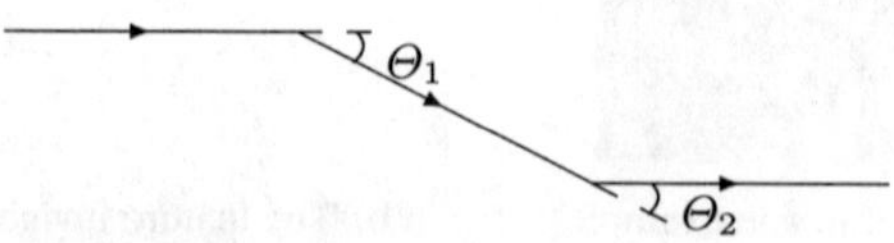

Fig. 3. Definition of deviation angles

where c_{ij} *is the connection coefficient between edge* e_i *and* e_j*, and* $c_{i\cdot}, c_{j\cdot}$ *are the connection coefficients of edge* e_i, e_j *to other edges respectively,* T *is a preset threshold to prevent over-connection.*

If two edges could be merged, it means they can be clustered into one group. The whole procedure of edge clustering is as follows.

Algorithm of clustering:

```
compute the connection coefficient set of the edges, S = {connection coefficients}
c_ij = min{S}
while ( c_ij is less than T )
    cluster e_i and e_j into one group
    merge e_i and e_j, update S
    c_ij = min{S}
end
output the clustering result
```

2.3 Computation Time Reduction

For a document image consists of many text lines, the number of line segments in the feature image is very large, i.e. $|E|$ is large. It requires long computation time to calculate the connection coefficient set. While, a curved document image has the property that: vertically, edges on the same text line only span across a narrow strip. Thus, computing the connection coefficient of two edges far away in the vertical direction is unnecessary. Moreover, separating the image into horizontal strips only influences text lines near the strip's boundary. In practical application, we separate the image into k horizontal strips. First, clustering is done in each strip. Then edges near the strip's boundary are collected to cluster with the previous merging result. Suppose the number of edges in an image is n, if processing is done in the whole image, then the computation complexity is $O(n(n+1)/2)$. When the image is evenly separated to k strips, i.e. the number of edges in each strip is n/k. Then the computation complexity is $O(n^2/k+n)$. Commonly, the computation time is reduced several times.

2.4 Curve Optimization and Image Restoration

After the above processes, line segments are clustered into different groups, i.e. feature points are clustered into different groups. Points in each group represent one curved text line. For simplicity, fourth degree polynomial $f_i(x) = a_i x^4 + b_i x^3 + c_i x^2 + d_i x + e_i$ is used to fit the least square regression curve. Choosing this model is to smooth the small fluctuations and simplify later processes. There are still two kinds of false detections must be handled. One is the short curves formed by points in figures or equations, they do not represent straight lines in the original image. A screening strategy is adopted to delete the points group which is shorter than $1/3$ of the image width. The other case is the occasional crossings between text lines. The local consistency between curvature of text lines is utilized to screen them. In a local region, the average slope in each curve column is computed, $\overline{\alpha} = \frac{\sum_{j \in local\ region} \alpha_j}{n_l}$, n_l is the number of lines in local region. Then the curve slope deviates from $\overline{\alpha}$ larger than a preset threshold θ is deleted.

A set of points are sampled on each curve according to the length of the curve. Suppose $P_s(x_s, y_s)$ is a sample point on $curve_i$ in the source image, its corresponding point in the destination image is $P_d(x_d, y_d)$. Then the position of P_d is defined as:

$$\begin{cases} x_d = \int_{start_i.x}^{x_s} \sqrt{1 + f_i'^2(x)}dx \\ y_d = \frac{1}{n_i} \sum\limits_{(x_i, y_i) \in curve_i} y_i \end{cases} \qquad (1)$$

where f_i is the coefficients of $curve_i$, as computed above, f_i' is its derivative, n_i is the number of points on $curve_i$, $start_i.x$ is the x position of the start point of $curve_i$ and (x_i, y_i) is a point on $curve_i$. After the corresponding points in the source image and the destination image are located, the technique of texture warping are used to restore the document image.

2.5 Experiment Result and Analysis

Experiments were made on 100 bound book document images taken from ordinary consumer digital cameras. The resolution of images ranges from 2 mega pixels to 4 mega pixels. Fig.4 shows some of the restoration results. Same parameters $\mu = 10, T = 20, n = 10$ are used for all the images.

OCR performance before and after rectification is compared based on ABBYY FineReader 8.0 for English documents and our OCR software for Chinese documents. Images are binarized using BST method first and the valid content region after layout analysis are sent into OCR software. The word accuracy measure is used as the evaluation metric: $accuracy = \frac{n-e}{n}$, where n is the number of characters in the document, and e is the number of falsely recognized characters. Here, only single character recognition result is considered. The comparison of recognition accuracy before and after

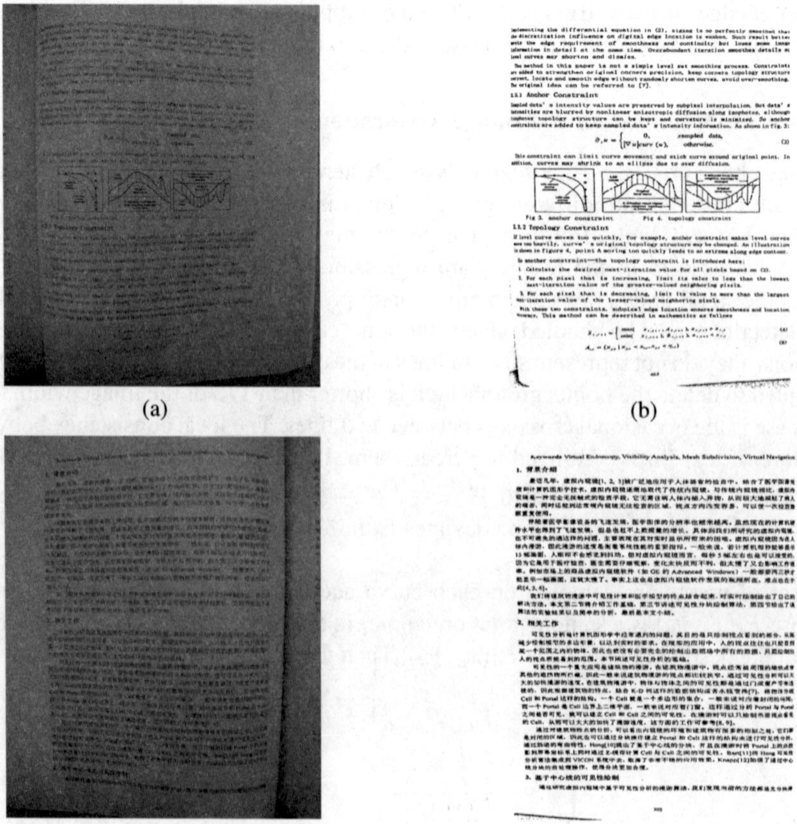

(a) (b) (c) (d)

Fig. 4. Some restoration results: (a), (c) are the original images, and (b), (d) are the corresponding results

rectification is shown in Table 1. As the layout analysis in current recognition software for Chinese document can not handle curved text lines well, the accuracy rate before rectification is very low. Therefore, we did not count it in the comparison. From Table 1 we could see that after rectification, recognition rate is improved 11 percent for English documents.

Table 1. Recognition accuracy rate before and after rectification

Characters type	Original(%)	after rectification(%)
English document	87.3	98.3
Chinese document	-	97.1

We also compared our rectification results with the similar function in OCR software ABBYY FineReader 8.0. Two metrics are used to evaluate the performance. First, we use the percent of correctly rectified lines(PCRL) as a metric. As for geometric distorted document images, the main problem for OCR systems is that curved text lines make the layout analysis and line separation processes hard. Preprocessing methods in most OCR systems depend on straight text lines. Even if some OCR systems that could handle curved lines, the layout reconstruction result is not satisfying. Texts on different lines are often falsely placed on the same line or texts on the same line are placed as superscripts or subscripts. So making curved text lines straight and horizontal is an important task for rectification algorithms. We propose the PCRL metric using the layout rebuilding function in OCR software, the number of lines in the document and the lines correctly rebuilding by OCR software are both counted.

$$PCRL = \frac{correctly\ rebuilt\ lines}{number\ of\ lines}$$

The comparison between our method and ABBYY's is shown in Table 2. In the experiment, formula lines are not counted in.

Table 2. Comparison between ABBYY FineReader 8.0 and our method

Metrics	ABBYY's method(%)	our method(%)
PCRL	75.6	97.8
Recognition accuracy	94.4	98.3

The ABBYY Fine Reader 8.0 is efficient for camera images, its preprocessing method does not depend on straight text lines. So its single character recognition accuracy is still high in the condition of curved text lines. Both methods are not sensitive to the image resolution, so the images are analyzed together. For single column document, our method outperforms ABBYY's method both in the objective metric and the perceptual results. For the two column document, ABBYY's method is better, it can perform rectification in the whole image scope.

2.6 Conclusion and Future Research

In this paper, we have proposed a correction method for bound document image. It follows the one image approach. That is only one camera or scanned image is needed. It takes the assumption that text lines are originally horizontal. In this method, the curved text-line features are first extracted from the binarized image, and curves of separate text lines are detected automatically using a graph model approach. Next, the technique of texture warping is utilized to restore the image. Experiments on OCR accuracy before and after restoration and comparison with ABBYY's method show the validity of the method.

In the future research, we intend to examine some global models to deal with multi-column documents, as well as short text lines at the top or bottom of a document image.

References

1. Brown, M.S., Seales, W.: Image restoration of arbitrarily warped documents. IEEE Transcations on Pattern Analysis and Machine Intelligence **26**(2) (2004) 1295–1306
2. Yamashita, A., Kawarago, A., Kaneko, T., Miura, K.T.: Shape reconstruction and image restoration for non-flat surfaces of documents with a stereo vision system. In: Proceedings of the Seventeenth International Conference on Pattern Recognition. Volume 1., IEEE (2004) 482–485
3. Wada, T., Ukida, H., Matsuyama, T.: Shape from shading with interreflections under a proximal light source: Distortion-free copying of an unfolded book. International Journal of Computer Vision **24**(2) (1997) 125–135
4. Tan, C., Zhang, L., Zhang, Z., Xia, T.: Restoring warped document images through 3d shape modeling. IEEE Transcations on Pattern Analysis and Machine Intelligence **28**(2) (2006) 195–208
5. Cao, H., Ding, X., Liu, C.: Rectifying the bound document image captured by the camera: a model based approach. In: Proceedings of the Seventh International Conference on Document Analysis and Recognition. Volume 1., IEEE (2003) 71–75
6. Wu, C., Agam, G.: Document image de-warping for text/graphics recognition. In: Proceedings of Joint IAPR 2002 and SPR 2002 Windsor. (2002) 348–357
7. Zhang, Z., Tan, C.l.: Correcting document image warping based on regression of curved text lines. In: Proceedings of the Seventh International Conference on Document Analysis and Recognition. (2003) 589–593
8. Zhang, L., Tan, C.l.: Warped image restoration with applications to digital libraries. In: Proceedings of the Eighth International Conference on Document Analysis and Recognition. (2005) 192–196
9. Ulges, A., Lampert, C.H., Breuel, T.M.: Document image dewarping using robust estimation of curled text lines. In: Proceedings of the Eighth International Conference on Document Analysis and Recognition. (2005) 1001–1005
10. Ezaki, H., Uchida, S., Asano, A., Sakoe, H.: Dewarping of document image by global optimization. In: Proceedings of the Eighth International Conference on Document Analysis and Recognition. (2005) 302–306
11. Seeger, M., Dance, C.: Binarising camera images for ocr. In: Proceedings of the Sixth International Conference on Document Analysis and Recognition. (2001) 54–58
12. Chen, M., Cheng, Z., Liu, Y.: A robust algorithm of principal curve detection. In: Proceedings of the Seventeenth International Conference on Pattern Recognition. (2004)

Cluster-Based Patent Retrieval Using International Patent Classification System

Jungi Kim[1], In-Su Kang[2], and Jong-Hyeok Lee[1]

[1] Division of Electrical and Computer Engineering
Pohang University of Science and Technology (POSTECH)
Advanced Information Technology Research Center (AITrc)
{yangpa, jhlee}@postech.ac.kr
[2] Information System Research Laboratory
Korea Institute of Science and Technology Information (KISTI)
dbaisk@kisti.re.kr

Abstract. A patent collection provides a great test-bed for cluster-based information retrieval. International Patent Classification (IPC) system provides a hierarchical taxonomy with 5 levels of specificity. We regard IPC codes of patent applications as cluster information, manually assigned by patent officers according to their subjects. Such manual cluster provides advantages over automatically built clusters using document term similarities. There are previous researches that successfully apply cluster-based retrieval models using language modeling. We develop cluster-based language models that employ advantages of having manually clustered documents.

Keywords: cluster-based retrieval, patent retrieval, invalidity search, international patent classification.

1 Introduction

Patent applications have different characteristics from other document types such as newspapers or web documents. Patents are generally very long and verbose, and their sizes are much more variable (Iwayama et al., 2003). Patent applications are well-structured and size of the collection is enormous; there are about 5 million U.S. patents, or 100-200 gigabytes of text, which are made up of hundred fields of textual or non-textual information (Larkey, 1998). One can take advantage of its structure for better retrieval or use it as a realistic-sized test collection. Also, patents provide a great test bed for exploring new ideas for manual clusters. Patent applications have one or more manually assigned International Patent Classification (IPC). Potential usefulness of using clusters for information retrieval has long been suggested and explored with no conclusive findings of its benefits (Liu and Croft, 2004). It is only recent that some promising results of cluster-based retrieval using statistical language modeling are reported (Kurland and Lee, 2004; Liu and Croft, 2004).

Kang et al. (2006) is the first to use IPC system as manual clusters for searching patent documents. They define two roles of cluster model: smoothing-oriented and topic-oriented. Their cluster model based on statistical language modeling is used either to smooth document language model or as an independent topic model which is

Y. Matsumoto et al. (Eds.): ICCPOL 2006, LNAI 4285, pp. 205 – 212, 2006.

interpolated with document model for final scores of retrieved documents. They report smoothing document model with cluster model does increase the retrieval performance marginally, and more gain in performance is obtained by interpolating document and cluster model.

Our work extends Kang et al. (2006)'s topic-oriented model with consideration of the characteristics of manually clustered documents. We show and discuss the cluster characteristics obtained by statistically analyzing the corpus. Then, we propose new models that incorporate the information of cluster size and similarity of manually built clusters into a cluster-based retrieval model. Finally, we present and discuss our experimental results and our conclusion.

2 International Patent Classification

2.1 International Patent Classification System

International Patent Classification System (IPC) is a 5-level hierarchical taxonomy for sorting patent applications administered by World Intellectual Property Organization (WIPO). The 5 levels of IPC are: section, class, subclass, main group, and subgroup. A patent application is hand-assigned to one or more appropriate IPC codes by human examiners. There are many overlapping categories in IPC taxonomy, and it is possible that an application has many IPC codes of different subgroups, groups, subclasses, classes, or sections. We consider IPC at each level as a cluster of documents, although for our experiments we only use IPC Code at level 5.

2.2 Statistics of IPC Clusters

To better view the characteristics of manual cluster of the patent applications and to compare with that of automatically built clusters, statistics of IPC cluster size and cluster memberships of patent documents are collected from the NTCIR-4 patent document collection.

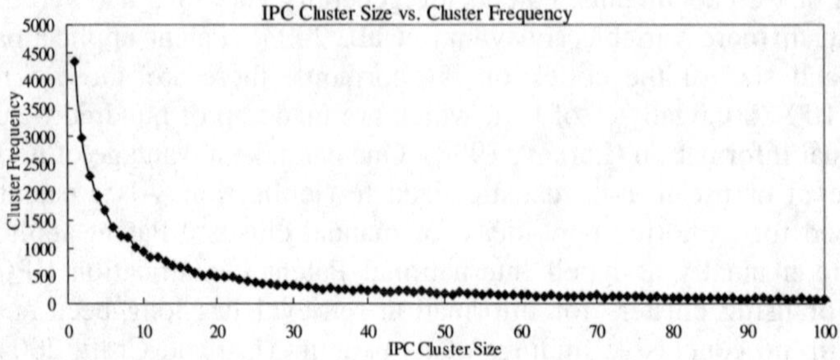

Fig. 1. IPC Cluster Size vs. Cluster Frequency at IPC level 5

The size and the document memberships of IPC clusters are very different from that of automatic clusters over which we generally have controlled sizes or a fixed number of clusters a document can belong.

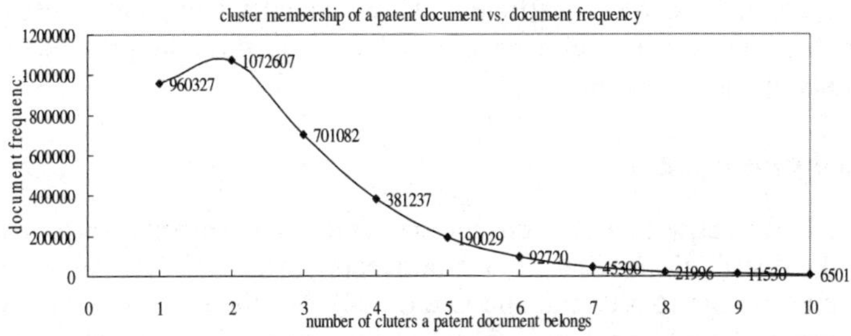

Fig. 2. Cluster Membership of Document vs. Document Frequency at IPC level 5

Figure 1 shows frequencies of different IPC cluster sizes at IPC level 5. Cluster size varies from 1 to 85355 documents, and the average number of documents per cluster is about 172 with the standard deviation of 859. Statistics and distribution of cluster sizes at different IPC level of a different patent collection are also available in the literature written by Fall et al. (2003).

Figure 2 shows how many IPC codes a document belongs and frequencies of such documents. A patent document has at least 1 IPC code and at most 91 IPC codes. On average, a patent document belongs to 2.6 IPC clusters at level 5, and the standard deviation is about 3.

Distributions of cluster size and cluster membership of documents are similar to that of Zipf's: few occur very often, while others occur rarely.

3 Cluster Characteristics

3.1 Automatic Clustering

Many researchers have tried different clustering methods to create automatic clusters of documents, among which are agglomerative algorithms that belong in a hierarchical clustering (van Rijsbergen and Croft, 1975), partitional clustering such as k-means (Liu and Croft, 2004), or different methods that uses common classification algorithms such as k-Nearest Neighbors algorithms (Kurland and Lee, 2004).

There has not been any research on the characteristics of automatically built clusters. However, from the algorithmic and the probabilistic-point of view, we can expect the properties for some of the methods.

For k-Means algorithm, if we assume a document has an equal chance to be assigned to any of the clusters, size of a cluster should follow a binomial distribution. Each document has probability of 1/k and there are n independent trials where n is the number of documents, hence the binomial distribution B (n, p=1/k) describes the distribution of cluster sizes and gives the expected mean of n * 1/k, or n/k, with the variance of n * 1/k * (n–1)/k, or $(n^2\text{-}n)/k^2$.

k-Nearest Neighbors algorithms has a fixed size of k + 1 for a cluster. If a cluster is built for every document, there are n numbers of overlapping clusters.

Similarity measure plays an important role in the quality of automatically built clusters. Automatic clustering uses measures of inter-document similarity such as

Dice's coefficient, Jaccard's coefficient, Cosine coefficient, and Kullback-Liebler (KL) divergence. However, such measures solely depend on the presence of terms of the documents in consideration.

3.2 Manual Clustering

In section 2.2, we show that IPC cluster size and cluster membership of documents have Zipf-like distributions. Cluster size noticeably affects the cluster-based retrieval performance according to Kurland and Lee (2004). For their experiments, while small sizes of clusters (4 or 5) do better than baseline cluster-less model, cluster size 40 gives superior retrieval performance.

We argue that cluster size is also important in manual cluster-based retrievals and propose methods to normalize cluster size. One way is to make an artificial cluster with a pre-determined size; if a cluster is too small, expand the cluster by merging with neighbor clusters, in case of IPC clusters, merging into a higher level, and if the cluster is too big, in result of sizing up or in its original state, use automatic clustering techniques to separate the target number of most relevant documents. Doing so can take advantages of manual clustering and automatic clustering: an accurate relevance measure and the control over cluster sizes. Another simple but crude method is to use only clusters of wanted sizes and ignore the rest.

Unlike automatically built clusters, IPC clusters are clustered based on topic, not terms. Each document is hand-assigned to its appropriate clusters based on its subject. Although we may not directly know how the inter-document similarity is measured, we can infer the similarities of clusters using cluster overlap size: the number of documents clusters share. Clusters of more similar topics should have more documents in common. For normalization purposes on cluster size, we use Dice's coefficient for similarity score.

4 Cluster-Based Language Models

4.1 Cluster-Less Language Model

We set cluster-less language model as our baseline to compare cluster-less model with cluster-based model. We take the performance of Kang et al.'s Jelinek-Mercer Smoothed unigram maximum likelihood language model:

$$P_{LM}(Q|D)=\prod_{q\in Q}[(1-\lambda)P_{ml}(q|D)+\lambda P_{ml}(q|Coll)]^{freq(q)} \quad (1)$$

where Q and D, coll are query, document, and collection model respectively, λ is a smoothing parameter for Jelinek-Mercer smoothing, and q is a query term. P_{ml} indicates the probability induced from maximum likelihood and freq(q) is the number of times a term q appears in D. Best retrieval performance, 21.93 in MAP, is obtained when λ =0.2.

4.2 Interpolation Model

Kang et al. (2006) defines two cluster-based models: smoothing-oriented and topic-oriented. Each model uses the cluster model generated by language modeling for different purposes. Smoothing-oriented model uses cluster model for smoothing document language model before smoothing with collection model, while topic-oriented model interpolates document model and cluster model after smoothing each model with collection model.

Our cluster-based models are extended version of the topic-oriented model which performed better than smoothing-oriented model.

Topic-oriented model, which we will refer to as interpolation model is:

$$P_{IM}(Q \mid D) = \prod_{q \in Q} \left[(1-\beta) P_{LM}(q \mid D) + \beta \frac{\sum_{C \in cluster(D)} P_{LM}(q \mid C)}{|cluster(D)|} \right]^{freq(q)} \quad (2)$$

and

$$P_{LM}(q \mid D) = (1-\lambda_1) P_{ml}(q \mid D) + \lambda_1 P_{ml}(q \mid Coll), \quad (3)$$

$$P_{LM}(q \mid C) = (1-\lambda_2) P_{ml}(q \mid C) + \lambda_2 P_{ml}(q \mid Coll), \quad (4)$$

where β is an interpolation parameter, C is a cluster, and cluster(D) is a set of clusters that document D belongs.

For our experiments, both λ1 and λ2 are set to 0.2, and β to 0.1 which give the best performance in Kang et al.'s work for the interpolation model. Following models extends the interpolation model and have the same parameter settings.

4.3 Cluster Size-Limit Model

As Kurland and Lee (2004) have done, we try to prevent too many irrelevant documents from being added by restraining on cluster size. As described in 3.2, the best way to achieve this is to control the degree of relevance, one of which being the number of documents in a cluster. It is possible to normalize the sizes of manual clusters as described in 3.2, however, since implementing and carrying out such method is complicated and takes a long time, we simply add a size-limit parameter so that any clusters having size larger than the parameter are ignored entirely. Relevant documents in large clusters may not be benefited. Nonetheless, for experimental purposes, it should be sufficient enough to show how cluster size affects the retrieval performance.

4.4 Cluster Expansion Model

We infer the cluster similarity information from the corpus as described in section 3.2. We expand initially retrieved clusters by adding similar clusters based on their topics. The score of the added cluster is calculated averaging the scores of clusters that expand it. Since initial retrieval scores about 40,000 IPC clusters out of 42239 clusters, for expansion, we limit the number of clusters we expand. There are two parameters: the number of top clusters from which clusters are expanded, and the number of clusters to expand from each top cluster.

Experimental Setup. For our test collection, we use NTCIR-4 patent which contains 1,707,185 Japanese patent applications. The test collection has 101 search topics which are patent applications rejected by Japanese Patent Office, of which 32 main topics are used for our experiments.

Among various sections a patent application has, claim, date of filing, and detailed description are of our interest. Claim sections and dates of filing of rejected patent applications are used as queries. Claim, date of filing, and detailed description sections are extracted to represent documents.

Relevant judgment set are prior arts that invalidates the topic patent application. Hence, only the documents that are filed before the topic application can appear in relevant document set.

For index and query terms, character bigrams of Japanese, numbers, and English words are used.

5 Experimental Results

5.1 Size-Limit Model

As table 1 shows, the cluster size plays an important role in cluster-based retrieval. Retrieval performance using only clusters with small sizes stayed around that of cluster-less baseline. With the increasing cluster sizes, however, MAP fluctuates quite a bit, indicating, as the number of retrieved relevant documents shows, that some informative clusters are added at some size-limit value, but irrelevant documents are brought in with increased size-limit.

5.2 Cluster Expansion Model

We expected Cluster Expansion Model to perform well. However, it performed even worse than Cluster-less Model. The poor performance can be explained in several

Table 1. Performance of Size-Limit Model and reported in MAP and number of relevant documents retrieved

Cluster Size-Limit	MAP	Rel. Ret.
10	0.2187	117
50	0.212	117
100	0.2171	117
300	0.2296	120
500	0.2137	121
700	0.2135	123
1000	0.2325	123
1500	0.2278	123
2000	0.2283	122
2500	0.2276	122
3000	0.2264	119
3500	0.2264	119
4000	0.2265	120
4500	0.2266	120
5000	0.2266	120

Table 2. Performance of Cluster Expansion Model and reported in MAP and number of relevant documents retrieved at different number of top IPC clusters and number of expanded clusters

Cluster Size-Limit	Num. Top Clusters	Num. Expanded Clusters	MAP	Rel. Ret.
1000	100	1	0.2161	115
1000	100	5	0.2154	119
1000	100	10	0.209	114
1000	1000	1	0.2126	121
1000	1000	5	**0.2273**	118
1000	1000	10	0.2195	117
1000	10000	1	0.2243	117
1000	10000	5	0.2143	119
1000	10000	10	0.2091	118
1000	∞	1	0.2183	114
1000	∞	5	0.2119	121
1000	∞	10	0.2127	119
∞	100	1	0.2137	116
∞	100	5	0.2148	116
∞	100	10	0.2126	118
∞	1000	1	0.2098	116
∞	1000	5	0.2177	117
∞	1000	10	**0.2181**	117
∞	10000	1	0.2148	120
∞	10000	5	0.2052	116
∞	10000	10	0.2056	116
∞	∞	1	0.2138	116
∞	∞	5	0.2037	117
∞	∞	10	0.2038	116

ways. First, parameters are too coarse. Although the number of top clusters and the number of expanded clusters seemed reasonable from the authors' point of view, the range defined by experimenter may not cover the optimum parameters. Also, the ratio of document model and cluster model of the Interpolation Model was fixed throughout the experiments, but reducing the number of clusters decreased the portion of cluster model in the final score of the Interpolation Model. At different number of clusters, One needs to search the optimal ratio exhaustively.

The effect of changing the number of clustered used and the number of expanded cluster, however, is well demonstrated.

6 Conclusions

We have proposed new models for cluster-based patent retrieval using International Patent Classification system. We first showed manual clusters are statistically different from automatically built clusters. As pointed out by other literatures, cluster size plays an important role in cluster-based retrieval, and we were able to show that it applies to manual cluster as well. With such knowledge, we proposed models more suitable and beneficial for manual cluster-based retrieval and show justifiable results.

Currently we are working on normalizing clusters for a target size and investigating the optimal size for cluster-based retrieval. Also, we are devising a more appropriate and general model for retrieving manually clustered documents. Cluster Expansion Model can also apply to automatically clustered documents and we plan carry out such experimentation, soon.

Acknowledgments. This work was supported by the Korea Science and Engineering Foundation (KOSEF) through the Advanced Information Technology Research Center (AITrc), and also partially by the BK 21 Project in 2006.

References

W. Bruce Croft.1980. *A model of cluster searching based on classification*. Information Systems, 5, 189-195.

Abdelmoula El-Hamdouchi, and Peter Willett. 1989. *Comparison of hierarchic agglomerative clustering methods for document retrieval*. The Computer Journal, 323, 220-227.

C. J. Fall, A. Torcsvari, K. Benzineb, and G. Karetka. 2003. *Automated Categorization in the Internation Patent Classification*. SIGIR Forum 37(1): 10-25.

Atsushi Fujii, Makoto Iwayama, and Noriko Kando. 2004. *Overview of patent retrieval task at NTCIR-4*. Working Notes for the Fourth NTCIR Workshop Meeting pp. 225-232.

Djoerd Hiemstra. 2001. Using language models for information retrieval. PhD Thesis, University of Twente.

Makoto Iwayama, Atsushi Fujii, Noriko Kando, and Yuzo Marukawa. 2003. *An empirical study on retrieval models for different document genres: patents and newspaper articles*. In Proceedings of the 26th Annual International ACM SIGIR Conference on Research and Development in Information Retrieval pp. 251-258.

In-Su Kang, Seung-Hoon Na, Jungi Kim, and Jong-Hyeok Lee. 2006. *Cluster-based Patent Retrieval*. Information Processing and Management.

Oren Kurland and Lillian Lee. 2004. Corpus Structure, Language Models, and Ad Hoc Information Retrieval. Proceedings of the 27th Annual International ACM SIGIR Conference on Research and Development in Information Retrieval.

Leah S. Larkey. 1998. Some *Issues in the Automatic Classification of U.S. patents*. In Working Notes of the Workshop on Learning for Text Categorization, 15th National Conference on Artificial Intelligence (AAAi-98).

Leah S. Larkey. 1999. *A patent search and classification system*. In Proceedings of the fourth ACM Conference on Digital Libraries pp. 179-187.

Xiaoyong Liu, and W. Bruce Croft. 2004. *Cluster-based retrieval using language models*. In Proceedings of the 27th Annual International ACM SIGIR Conference on Research and Development in Information Retrieval, pp. 186-193.

Jay M. Ponte and W. Bruce Croft. 1998. *A language modeling approach to information retrieval*. Proceedings of the 21st Annual International ACM SIGIR Conference on Research and Development in Information Retrieval, pp. 275-281.

C.J. van Rijsbergen, 1979. *Information Retrieval*. Butterworth-Heinemann, Newton, MA.

Peter Willett. 1988. *Recent trends in hierarchic document clustering: a critical review*. Information Processing and Management, 24(5):577-597.

Chengxiang Zhai and John Lafferty. 2001. *A study of smoothing methods for language models applied to Ad Hoc information retrieval*. In Proceedings of the 24th Annual International ACM SIGIR Conference on Research and Development in Information Retrieval, pp. 334-342.

Word Error Correction of Continuous Speech Recognition Using WEB Documents for Spoken Document Indexing

Hiromitsu Nishizaki and Yoshihiro Sekiguchi

Interdisciplinary Graduate School of Medicine and Engineering,
University of Yamanashi, Kofu, Yamanashi 400-8511, Japan

Abstract. This paper describes an error correction method of continuous speech recognition using WEB documents for spoken documents indexing. We performed an experiment of error correction for news speech automatically transcribed, where we focused on especially proper nouns. Two LVCSR systems were used to detect correctly and incorrectly recognized words. Keywords for the Internet search engine were selected among the correctly transcribed words, then correct candidates for the mis-recognized words were obtained in retrieved documents. A Dynamic Programming (DP) technique with a confusion matrix was utilized to compare the candidates with the mis-recognized words. In results of experiment of error correction, recognition rate of proper nouns achieved improvement of about 10% by using WEB documents.

1 Introduction

Recently, environments have been prepared in which a large number of audio and multimedia archives such as video tapes, digital libraries and so on can be easily used. Especially, a rapidly increasing number of spoken documents, such as broadcast radio and television programs, are archived, many of which can be accessed through the Internet. The needs for retrieving such speech information have been more growing, by the day, while it is definitely true that an effective retrieval technique is lacking at present. Moreover, and the development of technology for retrieving such speech information is becoming more and more important.

In the TREC Spoken Document Retrieval track[1], a number of studies were presented on the subject of English and Mandarin broadcast news documents. A standard approach to spoken document retrieval (SDR) is to automatically transcribe spoken documents into word sequences, which can be directly matched against queries.

Robinson et al. [2] have proposed a retrieval method that is robust for word recognition errors in case of transcribing spoken documents. This method uses parallel text-formed documents whose contents are similar to contents of the spoken documents. Hauptmann et al. [3] and Jourlin et al. [4] have investigated SDR performance using various indexes. Each index has various recognition performances when spoken documents were transcribed. In this approach, however,

Y. Matsumoto et al. (Eds.): ICCPOL 2006, LNAI 4285, pp. 213–221, 2006.

a serious problem arises when both the queries and the documents include out-of-vocabulary (OOV) keywords , where matching against OOV keywords always fails, because the OOV keyword can not be transcribed as a word. Many previous studies handled an OOV problem in spoken document retrieval. In most of them, spoken documents were not transcribed into word sequences, but into phoneme / syllable sequences using phoneme / syllable recognizers[5,6,7,8]. Wechsler et al.[5] used a phoneme recognizer to transcribe spoken documents. K. Ng et al.[6] worked on the use of sub-word unit representation based on phonemes in spoken document retrieval. H. M. Wang[7] also investigated syllable-based indexing for retrieval of spoken documents in Mandarin Chinese using a syllable lattice. C. Ng et al. [8] reported, however, that better retrieval performance in terms of average precision was obtained using word-based indexing rather than using phoneme/syllable based indexing. Furthermore, word based retrieval must be faster than that based on phoneme/syllable.

For spoken document retrieval, a mis-recognition problem is also fatal as far as word based indexing is used. It is well known that a voting scheme such as ROVER (Recognizer Output Voting Error Reduction) for combining multiple speech recognizers' outputs can achieve word error reduction[9]. We [10] have proposed a spoken document retrieval method which is robust for mis-recognition and OOV words. In [10], we prepared two types of indexes, the one was a word-based index, the other was a syllable-based index. Our final goal in this study is to make a refined word-based index of spoken documents. So, this paper describes word error correction of outputs of large vocabulary continuous speech recognition (LVCSR) system by using WEB documents for spoken document indexing. Especially, we focus on a correction of mis-recognized proper nouns that are likely to be keywords for retrieving documents. Considering a case of automatically indexing news speeches from TV and radio, we can find the text-formed documents, whose contents are similar to the news speeches, on the Internet. So, the proposed method may be very effective. In the result of experiment of error correction, our technique significantly improved the recognition rate of proper nouns on a TV news dictation.

2 A Correction Using WEB Documents

2.1 Overview of the Process

Figure 1 shows an overview of the error correction process we proposed. In this framework, a document contains some sentences is processed as a processing unit instead of a sentence unit.

First, word or word sequence candidates to be corrected are detected by two LVCSR systems. At the same time, transcribed words or word sequences that are probably correct are also detected. Those words are used to retrieve WEB documents as a query. Next, after retrieving WEB documents using the words, correct word candidates which may be substituted for mis-recognized words that are extracted from the WEB documents. Then, each incorrect word candidate is matched against a correct word candidate using a syllable-based DP matching

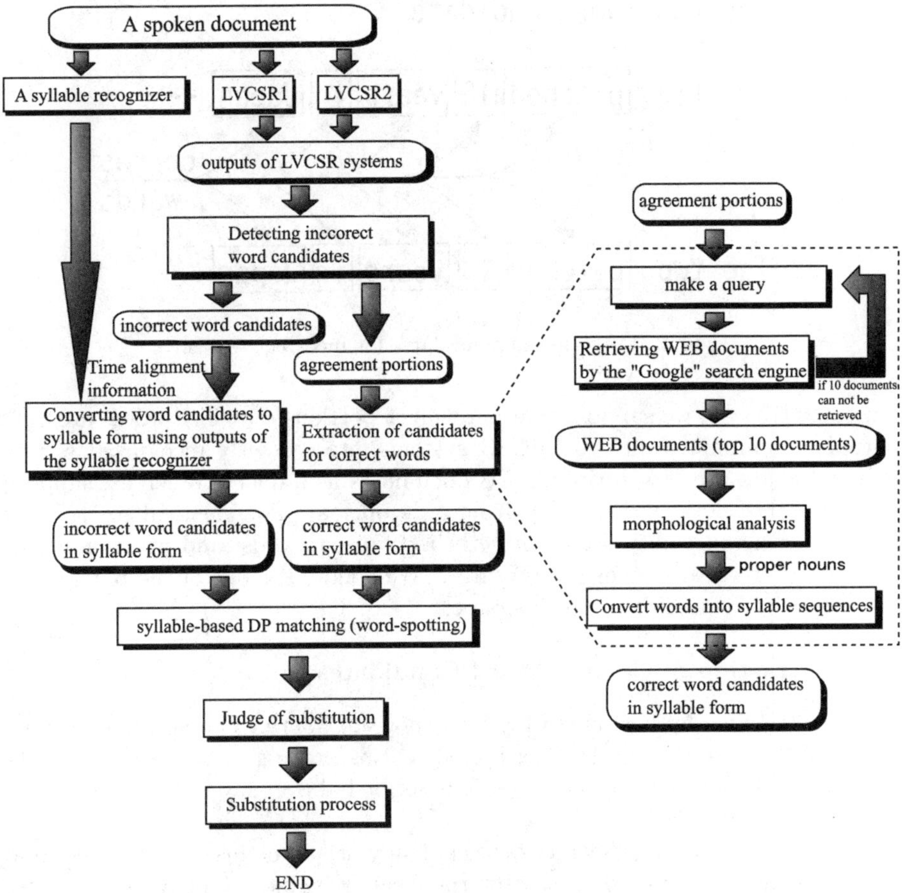

Fig. 1. An overview of the error correction process

technique. If a incorrect word candidate is identical to a correct word candidate, the incorrect word candidate is replaced by the correct one.

2.2 Detection of Incorrect Word Candidates

To detect incorrect words or word sequences that are mis-recognized, we uses two LVCSR systems, the one named "SPOJUS" which has been developed in Toyohashi University of Technology[11], as well as the one named "Julius" which is provided by IPA Japanese dictation free software project [12]. Both decoders are composed of two decoding passes, where the first pass uses the word bigram, and the second pass uses the word trigram. Our previous study reported that the agreement between the outputs with different decoders can achieve quite reliable confidence on broadcast news speech and reading speech of newspaper

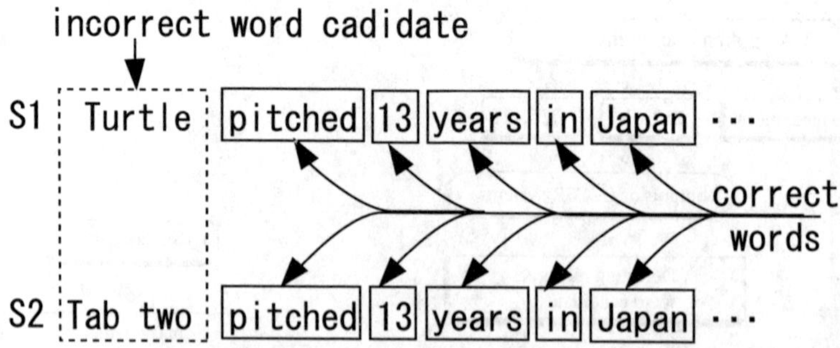

Fig. 2. Detection of candidates for incorrect words

sentences [13]. So, we suppose that agreement portions, in which words are outputted in common from two LVCSR systems, are correctly recognized words, then those are used to retrieve WEB documents as a query. Whereas, suppose that the other potions, i.e. words from a decoder are not identical to the ones from the other decoder, are incorrectly recognized words, and are candidates should be corrected as shown in Figure 2. We denote the one of the portions as "incorrect word candidate" in this paper.

2.3 Extraction of Correct Word Candidates

The flow among a broken line in Figure 1 shows extracting candidates of correct words. We denote the word set extracted by this process as "correct word candidates". The incorrect word candidate described above may be replaced by the correct word candidate.

Correct word candidates which consist of only proper nouns are extracted from WEB documents retrieved by the Internet search engine "Google". A query for "Google" consists of words which are proper nouns and outputted in common from two LVCSR systems when a spoken document is transcribed. The query is automatically composed. If 10 documents can not be retrieved from "Google", a word randomly selected is removed from the query. Then, the new query is used again in the search engine. Those words included in the query have high confidence because about 95% of proper nouns which are transcribed in common by two LVCSR systems are correct.

Finally, the correct word candidates are converted into syllable sequences.

2.4 Substitution Process

In the substitution process, first, incorrect word candidates are matched against correct word candidates using a syllable-based DP matching to investigate adequacy for the substitution. In other words, a word-spotting is performed to check that whether a correct word candidate is in a portion of incorrect word sequences or not. The substitution process is made story by story.

Figure 3 shows an overview of the syllable based DP matching. Let us denote a syllable sequence, which is detected as a incorrect word candidates, as $X = \{x_1, x_2, \cdots, x_I\}$ (I is the number of syllable in X, $x_i \in X$) and syllable sequence of a correct word candidate as $Y = \{y_1, y_2, \cdots, y_J\}$ (J is the number of syllable in Y, $y_j \in Y$). As shown in Figure 3, we apply the syllable based DP matching on the syllable sequence X against the syllable sequence Y. In this DP matching, the likelihood $G(i,j)$ of arriving at the point (i,j) in a DP lattice matrix is maximized:

$$G(i,j) = \begin{cases} 1.0 \quad (i \geq 0, j = 0) \\ G(0, j-1) \cdot P(y_j, \phi) \quad (i = 0, j > 0) \\ \max \begin{cases} G(i-1, j) \cdot P(\phi, x_i) \\ G(i-1, j-1) \cdot P(y_j, x_i) \\ G(i, j-1) \cdot P(y_j, \phi) \\ \qquad\qquad (i > 0, j > 0) \end{cases} \end{cases}$$

where $P(r,h)$ is the probability of mis-recognizing syllable r as h. $P(r,\phi)$ means the probability of insertion error of syllable r, and $P(\phi, h)$ is the probability of deletion error of syllable h. Syllable sequences are obtained using a syllable recognizer for a development data described below in section 3.1. By comparing those sequences with the references, the syllable recognition error confusion matrix is calculated as bellow:

$$P(r,h) = \frac{C(r,h)}{\sum_{k \in h} C(r,k)}.$$

where $C(r,h)$ is the number to misrecognize syllable r as h. When the DP score $G(i,J)$ is larger than a given threshold, we judge that the incorrect word candidate is identical to the correct word candidate. The threshold is decided based on an experiment in our previous work[10].

3 Experiment of Error Correction

3.1 Data Set

The speech data used in this work is Japanese NHK (Japan Broadcasting Corp.) broadcast TV news from June 1st to July 14th, 1996. The data is divided into 2 portions. The one (June 1st) is an evaluation set for error correction, the other (from June 2nd to 14th July) is an development set for calculating the confusion probability $P(r,h)$ described in section 2.4. The evaluation speech data is partitioned into separate news stories[1], and includes 18 stories (175 sentences, 435 proper nouns) in total, half of which are utterances with noise such as background music. It is very difficult to transcribe certain portions of those spoken documents, because utterances with dialogue speech and field reports are also included in this data.

[1] A story is regarded as a document.

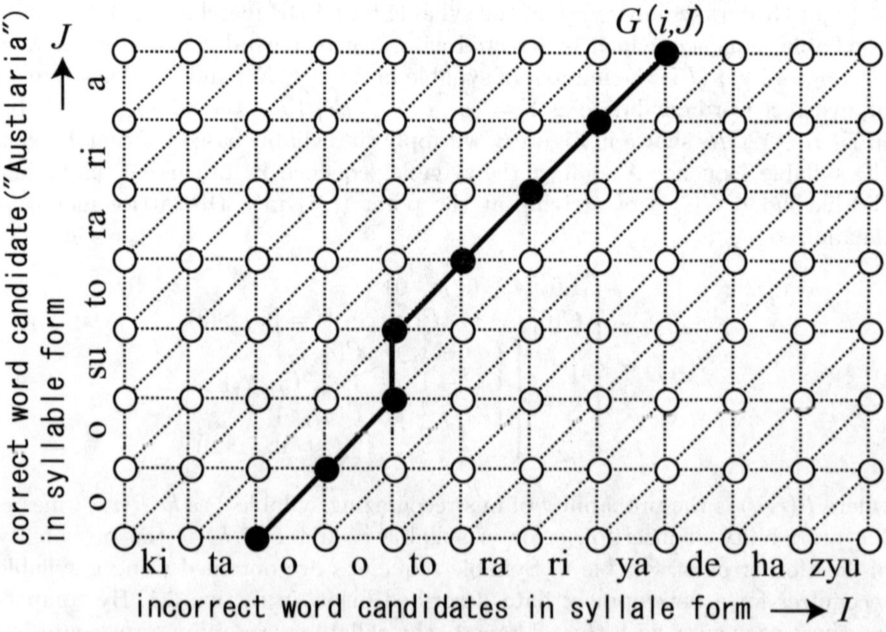

Fig. 3. A syllable-based DP matching

3.2 Recognition Performance of Each Decoder

We uses two LVCSR systems described in Section 2.2. As the language models for both recognizers, the word bigram and trigram language models for 20K vocabulary size (coverage of 96.7%) are trained using 5 years Japanese NHK broadcast news scripts (1992∼1996, approximately 120,000 sentences). The acoustic models are based on a Gaussian mixture HMM. We used esyllable-based HMMs. Speaker-independent acoustic models were trained by using read speech (about 20,000 sentences uttered by 180 male speakers; JNAS[15]). The feature parameters consist of 12 dimensional mel frequency cepstrum coefficients (MFCC), delta 12 dimensions, delta delta 12 dimensions, delta powers, and delta delta powers [2]

The sampling frequency is 16 kHz, and the frame is shifted by 10 ms at every frame.

Table 1 shows word-based recognition rates of each decoder for the target spoken documents. "Proper nouns" denotes the correct rates of proper nouns only.

3.3 Experiment on Error Detection

We evaluate a performance of detecting incorrect word candidates by using two LVCSR systems.

[2] Delta delta MFCC and delta delta powers are only used in SPOJUS.

Table 1. Word-based recognition rates[%]

LVCSR	Corr.	Acc.	Proper nouns
Julius	68.9	63.6	76.6
SPOJUS	71.8	56.1	70.8

Table 2. Performance of Error detection

word corr. rate of PNs in common	94.5%
num. of incorrect word candidates	755
num. of PNs to be corrected in incorr. candidates	130
num. of all PNs to be corrected	162 words
num. of incorrectly recognized PNs	146 words
num. of incorrectly recognized PNs in common	16 words

Table 2 shows the performance of error correction. The recognition rate of proper nouns which are recognized by two LVCSR systems in common is 94.5% (273 / 289 = 0.945%). The number of incorrect word candidates which include proper nouns required error correction is 130 among all automatically detected incorrect word candidates, that is 755[3]. As there is the case of including a few mis-recognized proper nouns in a incorrect word candidate, the total number of automatically detected proper nouns should be corrected is 146, whereas the total number of manually detected ones is 162 words. Our proposed error detection approach achieves detection rate of 90.1% (146 / 162 = 0.901) of proper nouns incorrectly recognized.

We investigated whether WEB documents including the correct word candidates are truly retrieved or not, when a query used to retrieve the WEB documents is made from the common words from two LVCSR systems. In the error correction experiment described below, the query for the search engine is automatically made. However, manually selecting keywords from the outputs of LVCSR systems as a query are used for WEB retrieval in this section. This is to simply inspect how WEB documents including correct word candidates are retrieved by the transcribed words. The result claims that the search engine can retrieve WEB documents which include 139 correct proper nouns corresponding to the incorrect word candidates that are not correctly transcribed. In short, there is a fair possibility of correcting the mis-recognized proper nouns of 95.2% (139 / 146 = 0.952) using the WEB documents.

3.4 Experiment on Error Correction

The Error correction performance is evaluated throughout the series of the process. Table 3 shows the result of error correction experiment. The number

[3] In this paper, the substitution process is performed on all incorrect word candidates. Naturally, it is necessary to detect whether proper nouns are in a incorrect word candidates or not.

Table 3. Performance of Error detection

num. of incorrectly recognized PNs	146 words
num. of correct PNs included in WEB documents	76 words
num. of PNs correctly replaced	29 words

of correct proper nouns which are included in the WEB documents retrieved by the "Google" search engine, where a query is automatically composed, is 76 words compared with 146 proper nouns that should be corrected. Proper nouns of 52.1% (75 / 146 = 0.521) can be automatically retrieved from the WEB documents. In the case of using the queries manually composed, proper nouns of 95.2% can be covered descrived above on section 3.3. This shows that a new retrieval method of WEB documents are required such as how to make a query, using new search engine and so on.

The number of proper nouns, which are correctly replaced as incorrect word candidates using the syllable-based DP matching, is 29 words. The rate that incorrect word candidates are replaced with correct proper nouns is 19.9% (29 / 146 = 0.199). In 29 words, 12 words can not be correctly transcribed by the both the LVCSR systems. In other words, 17 words of the remainder are correctly recognized by either LVCSR system.

Considering the case that words, which are outputted from both the LVCSR systems, are registered into a word-based index, and it is not necessary to agree between those words from both the LVCSR systems, 75 words in 146 proper nouns can be corrected. As the all proper nouns from both LVCSR systems are entered into a index, there is a risk of degrading retrieval performance of spoken documents. However, our previous works [14] clearly claimed that redundant words in a index did not fatally injure the retrieval performance.

Finally, the proposed error correction technique achieves improvement on the recognition rate of the proper nouns in each LVCSR system to 80.0% from 76.6% (Julius) and 70.8% (SPOJUS), respectively.

4 Conclusions

This paper proposes the error correction technique of outputs of the LVCSR systems for spoken documents indexing. The technique consists of three conponents, the first process is to detect portions of incorrect word or word sequences by using the agreement of outputs between two LVCSR systems. The next is to retrieve WEB documents which are related to the incorrect words, then, extract correct words which should be replaced as the incorrect words from the documents. The final process is to substitute the incorrect words with the correct words by the syllable-based word-spotting technique. In the result of error correction experiment, the proposed technique significantly improves the recognition rate of the proper nouns in each LVCSR system.

In the future work, we are goint to perform the retrieval experiment of spoken documents using a index which contains corrected words by this techniques. In addition, we would like to develop more refined method that make a query for retrieving WEB documents.

References

1. J.Garofolo, C.G.P.Auzanne, and E.Voorhees. "The TREC SDR Track: A Success Story", In Proc. of the 8th Text Retrieval Conference, pages 107–129, 2000.
2. T.Robinson, D. Abberley, D. Kirby, and Steve Renals. "Recognition, indexing and retrieval of British broadcast news with the THISL system", In Proc. of EuroSpeech'99, pp. 1267-1270, 1999.
3. A.G.Hauptmann and H.D. Wactlar."Indexing and search of multimodal information", In Proc. of ICASSP'97, pp.195-198, 1997.
4. Pierre Jourlin, Sue E. Johnson, Karen Sparck Jones, and Philip C. Woodland. "Spoken document representations for probabilistic retrieval", Speech Communication, Vol.32, No. 1-2, pp. 21-36, 2000.
5. M. Wechsler, E. Munteanu, and P. Schauble. New Techniques for Open-vocabulary Spoken Document Retrieval. *In Proceedings of the SIGIR'98*, pages 20–27, 1998.
6. K. Ng and V. W. Zue. Subword-based Approaches for Spoken Document Retrieval. *Speech Communication*, 32(3):157–186, 2000.
7. H. min Wang. Experiments in Syllable-based Retrieval of Broadcast News Speech in Mandarin Chinese. *Speech Communication*, 32(1-2):49–60, 2000.
8. C. Ng, R. Wilkinson, and J. Zobel. "Experiments in Spoken Document Retrieval using Phoneme N-grams",Speech Communication, Vol.32, No.1-2, pp.61–77, 2000.
9. J.G.Fiscus. A Post-processing System to Yield Reduced Word Error Rates: Recognizer Output Voting Error Reduction (ROVER). *In Proceedings of the IEEE Workshop on Automatic Speech Recognition and Understanding*, pages 347–354, 1997.
10. Hiromitsu Nishizaki and Seiichi Nakagawa. "Japanese Spoken Document Retrieval Considering OOV Keywords Using LVCSR System with OOV Detection Processing", Proc. of Human Language Technology Conference 2002, pp. 144–151, 2002.3
11. A.Kai, Y.Hirose, and S.Nakagawa, "Dealing with out-of-vocabulary words and speech disfluencies in an n-gram based speech unde rstanding system," *ICSLP'98*, 1998, pp. 2427–2430.
12. T.Kawahara, T.Kobayashi, K.Takeda, N.Minematsu, K.Itoh, M.Yamamoto, A.Yamamoto, T.Utsuro, and K.Shikano, "Sharable software repository for japanese large vocabulary continuous speech recognition," *ICSLP'98*, 1998, pp. 763–766.
13. Takehito Utsuro, Tetsuji Harada, Hiromitsu Nishizaki and Seiichi Nakagawa, "A Confidence Measure Based on Agreement among Multiple LVCSR Models – Correlation between Pair of Acoustic Models and Confidence –", Proc. of ICSLP2002, pp. 701–704, 2002.9
14. Hiromitsu Nishizaki and Seiichi Nakagawa, "A System for Retrieving Broadcast News Speech Documents Using Voice Input Keywords and Similarity between Words"Proc. of ICSLP2000,Vol.3, Vol.3, pp. 1073-1076, 2000.10
15. K.Itoh, M.Yamamoto, K.Takeda, T.Takezawa, T.Matsuoka, T.Kobayashi, K.Shikano, and S.Itahashi, "JNAS: Japanese speech corpus for large vocabulary continuous speech recognition research," Journal of the Acoustical Society of Japan (E), vol.20, no.3, pp.199–206, 1999.3

Extracting English-Korean Transliteration Pairs from Web Corpora

Jong-Hoon Oh and Hitoshi Isahara

Computational Linguistics Group
National Institute of Information and Communications Technology (NICT)
3-5 Hikaridai, Seika-cho, Soraku-gun, Kyoto 619-0289 Japan
{rovellia, isahara}@nict.go.jp

Abstract. Transliteration pair acquisition has received significant attention as a technique for constructing up-to-date transliteration lexicons, and for supporting machine translation and cross-language information retrieval. Previous studies on transliteration pair acquisition focused on only the phonetic similarity model but seldom considered the real-usage of transliterations in texts. Moreover, previous web-based validation models considered only one-way validation (validation from the viewpoint of a source term) rather than joint validation between a source term and a target term. To address these problems, we propose a novel transliteration pair acquisition model that extracts transliteration pairs from the Web and validates the pairs by combining the phonetic similarity and joint web-validation models. Experiments demonstrated that our transliteration pair acquisition model was effective.

1 Introduction

Transliteration, "phonetic translation" or "translation by sound", is used to translate proper names and technical terms especially from Latin languages to non-Latin languages, such as from English to Korean, Japanese, or Chinese. For example, the English word *data* is transliterated into Korean as 'de-i-teo'[1]. Transliterations are one of the main sources of the Out-of-Vocabulary problem [1]. *Transliteration pair acquisition* extracts transliteration pairs from bilingual corpora to construct up-to-date transliteration lexicons, and to support machine translation and cross-language information retrieval [1]. Most of the previous work on *transliteration pair acquisition* [2,3,4,5,6,7,8] has focused on only the phonetic similarity model for validating transliteration pairs. They extracted transliteration pair candidates from bilingual corpora and then validated them by using their own phonetic similarity model. However, the transliteration pair validation depending on only the phonetic similarity model has limitations, because it does not consider the real-usage of transliterations in texts.

[1] In this paper, target language transliterations are represented in their Romanized form with single quotation marks and hyphens between syllables.

Y. Matsumoto et al. (Eds.): ICCPOL 2006, LNAI 4285, pp. 222–233, 2006.

As the Web becomes one of the main knowledge sources for natural language processing, many researchers have tried to extract and validate translation lexicons using the Web [9, 10, 11, 12, 13]. They extracted translation lexicons from anchor texts (hyperlink texts) [11, 12] and Web search results [10, 13], and validated the extracted translation lexicons by using Web frequency (the number of Web pages retrieved by Web search engines). However, they only validated whether a target term (t) was likely to be a counterpart of a source term (s) — this can be regarded as one-way validation from s to t — rather than both whether t is likely to be a counterpart of s and whether s is likely to be a counterpart of t — bi-directional validation or joint validation between s and t.

To address these problems, we propose a novel transliteration pair acquisition model that extracts transliteration pairs from the Web (Web search results) and validates the pairs by combining phonetic similarity and joint Web-validation models.

This paper is organized as follows. In section 2, we discuss related work. In sections 3 and 4, we show extraction of transliteration pair candidates and their validation. In section 5, we discuss our experiments. We then conclude in section 6.

2 Related Work

There have been several works on transliteration pair acquisition [2,3,4,5,6,7,8]. The previous work extracted transliteration pairs from bilingual texts such as bilingual newspapers and used their own phonetic similarity model to validate transliteration pairs. The phonetic similarity model falls into two categories — 1) monolingual comparison approach and 2) bilingual comparison approach — according to the way of calculating the phonetic similarity. For given two words in different languages, the monolingual comparison approach [2, 4, 5, 7, 8] transformed a word in one language into phonetically equivalent one in the other language using machine transliteration. Then it calculated string similarity between two words written in the same language — one is original word and the other is generated by machine transliteration. The bilingual comparison approach directly calculated phonetic similarity between two words in different languages [3, 6, 8]. Generally, the basic framework of a machine transliteration model in the monolingual comparison approach and a phonetic similarity model in the bilingual comparison approach is similar to each other — the two models can be trained by the same transliteration lexicons. Therefore, the two approaches do not show significant difference in performance if the two models in the two approaches are trained by the same transliteration lexicons [8]. However, previous work on transliteration pair acquisition has some drawbacks. First, they extracted transliteration pairs from the limited size of bilingual corpora. This makes it difficult to extract up-to-date transliteration pairs, which we run across routinely in Web pages. Second, they just rely on phonetic similarity to validate transliteration pairs. This results in producing wrong transliteration pairs, where a target term is not used in target language texts.

There have been several Web-based translation validation models [10, 11, 12, 13]. Qu & Grefenstette [10] used Web frequency (the number of Web pages) to validate Japanese romanji-kanji conversion. They retrieved Web pages by using both romanji and kanji (or both source term and target term) as a query for Web search engines. Then they used the number of the retrieved Web pages to filter out wrong Japanese kanji. Lu *et al.* [11, 12] and Wang *et al.* [13] extracted translation lexicons from the Web and validated the translation lexicons by using chi-square (χ^2) test in Eq. (1). For source term s and target term t, they transformed the conventional chi-square (χ^2) test into the similarity measure described in Eq. (1) [11, 12, 13].

$$\chi^2(s,t) = \frac{N \times (a \times d - b \times c)^2}{(a+b) \times (a+c) \times (b+d) \times (c+d)} \tag{1}$$

Each parameter in Eq. (1) was represented by the number of Web pages retrieved by a Web search engine as follows.

- a: the number of Web pages containing both s and t;
- b: the number of Web pages containing s but not t;
- c: the number of Web pages containing t but not s
- d: the number of Web pages containing neither s nor t
- $N = a + b + c + d$

Because Web search engines can usually accept Boolean queries, the four parameters (a, b, c, and d) in Eq. (1) are obtained by retrieving Web pages with the following four Boolean queries, "$s \bigcap t$", "$s \bigcap \neg t$", "$\neg s \bigcap t$", and "$\neg s \bigcap \neg t$", each of which corresponds to a query for estimating a, b, c and d, respectively. The Web-based translation validation models [10, 11, 12, 13], however, just focused on validating whether target term (t) is most likely to be a counterpart of source term (s). However, we need to validate whether t is likely to be the counterpart of s and vice versa to improve the validation ability — bi-directional validation or joint validation between s and t.

3 Candidate Extraction: Transliteration Boundary Detection

We used Web search results as a language resource for extracting transliteration pairs. We used a given English term as a query for a Web search engine and then extracted transliterations corresponding to the English term from the Web search result. Web search results output by Web search engines usually contain a series of snippets[2] of Web documents retrieved by the Web search engines. Korean Web pages are usually composed of rich texts in a mixture of Korean (main language) and English (auxiliary language), where English is used for auxiliary description or translation of Korean terms in Korean texts. For example, the

[2] "Snippet" refers to the title and summary of each Web document in the Web search results.

Web search result for the English term *synapse* are shown in Fig. 1. The Web search result contains the underlined English term *synapse* and the underlined Korean transliteration 'si-naep-seu', which is the counterpart of *synapse*. The transliteration and its corresponding English term in the Web search result indicate that Web search results as a source of transliteration pairs can be very useful.

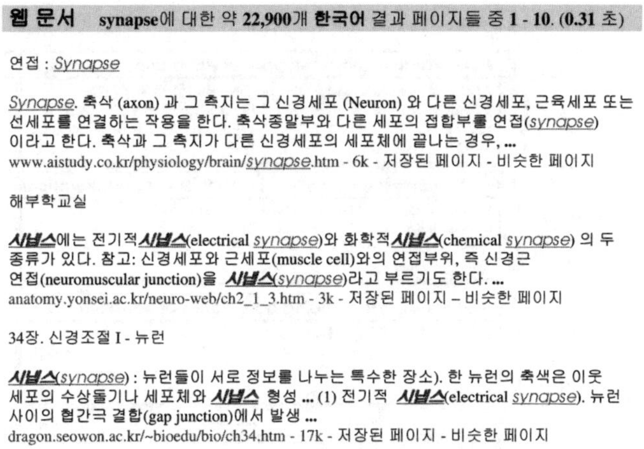

Fig. 1. Web search result for *synapse*

Our candidate extraction is based on a transliteration boundary detection algorithm. The algorithm recognizes the beginning and ending boundaries of transliterations corresponding to given English terms. It makes use of phonetic similarity between the beginnings and endings of an English term and those of a Korean term. First, a phonetic similarity model trained by a transliteration lexicon generates candidates of the beginning and ending boundaries of transliterations corresponding to an English term. Note that the candidates are a set of Korean syllables, which phonetically correspond to the beginnings and endings of an English term. Let $e = eg_1, ..., eg_n$ be an English term composed of n English letters, KS be a set of Korean syllables, and eg_0 and eg_{n+1} be dummy letters representing the start and end of words, respectively. Then, candidates of the beginning and ending boundaries are generated with Eq. (2).

$$
\begin{aligned}
Pr(KS|eg_0, \cdots, eg_3) > \theta \\
Pr(KS|eg_{n-2}, \cdots, eg_{n+1}) > \theta
\end{aligned}
\tag{2}
$$

Second, the algorithm looks for the transliteration boundaries in texts, which are snippets in Web search results. Once the algorithm recognizes the boundaries, it finally extractstransliteration candidates with the recognized boundaries.

We applied two constraints, a length constraint and a language constraint, to avoid redundant transliteration candidates. The length constraint means that a transliteration candidate should satisfy the condition, $m < 2 \times n$ and $n < 2 \times m$, where n is the length of an English term (the number of English letters) and m is the length of a Korean transliteration candidate (the number of graphemes in a transliteration candidate). The language constraint means that all characters in recognized transliteration candidates should be Korean (excluding English letters, numbers, and other special symbols).

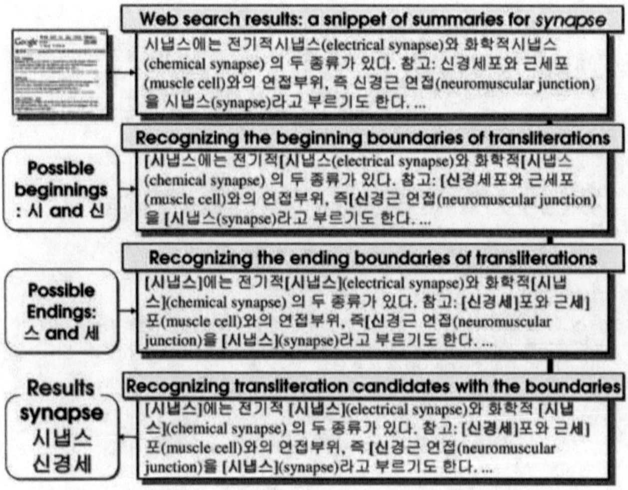

Fig. 2. "Candidate Extraction" through transliteration boundary detection

"Candidate Extraction" for the English term *synapse* is shown in Fig. 2. Our transliteration boundary detection algorithm generates candidates for the beginning boundaries (*Possible beginnings*: 'si' and 'sin') and those for ending boundaries (*Possible endings*: 'seu' and 'se') with Eq. (2). Then, the algorithm finds the candidates represented by the square brackets, where the open bracket represents the beginning boundary and the close bracket represents the ending boundary. Finally, the algorithm recognizes two transliteration candidates, 'si-naep-seu' and 'sin-gyeong-se' (in the square brackets) satisfying the length and language constraints. The transliteration pair candidates are (*synapse*, 'si-naep-seu') and (*synapse*, 'sin-gyeong-se').

4 Candidate Validation: Phonetic Similarity and Joint Web Validation Models

Once candidates for transliteration pairs are recognized, they are validated by the phonetic similarity (S_{PSM}) and joint Web-validation (S_{Joint}) models. Our validation model based on the two models is represented in Eq. (3). In this section, we describe the two models in more detail.

$$S_{TV}(e,k) = S_{PSM}(e,k) \times S_{Joint}(e,k). \tag{3}$$

4.1 Phonetic Similarity Model: S_{PSM}

Let $e = eg_1, \cdots, eg_m$ be an English term composed of m English letters (or graphemes) and $k = kg_1, \cdots, kg_l$ be a Korean term composed of l Korean graphemes. The phonetic similarity model phonetically compares e with k with the assumption that k will be phonetically similar to e if k originated from e or k is the transliteration of e. Our phonetic similarity model directly calculates the phonetic similarity between English and Korean terms in candidates for transliteration pairs, as in [3, 6]. The phonetic similarity model can be represented with the conditional probability $Pr(k|e)$, as shown in Eq. (4). $Pr(k|e)$ is estimated with kc_i, which is a chunk of Korean graphemes corresponding to eg_i. So $Pr(kg_1^l|eg_1^m)$ can be rewritten as $Pr(kc_1^m|eg_1^m)$. Then, $Pr(kc_1^m|eg_1^m)$ can be simplified into a series of products of $Pr(kc_i|eg_{i-2}^{i+2})$ with the assumption that kc_i depends on $eg_{i-2}, \cdots, eg_{i+2} = eg_{i-2}^{i+2}$.

$$S_{PSM}(e,k) = \sqrt[|e|]{Pr(k|e)} \tag{4}$$

$$Pr(k|e) = Pr(kg_1^l|eg_1^m) \approx Pr(kc_1^m|eg_1^m) \approx \prod Pr(kc_i|eg_{i-2}^{i+2}). \tag{5}$$

4.2 Joint Web Validation Model: S_{Joint}

Let e be an English term, K be a set of Korean transliteration candidates extracted from Web search results for the query e, k_i be the ith Korean transliteration candidate in K, E_i be a set of English candidates extracted from Web search results for the query k_i, and e_{ij} be the jth English candidate in E_i. The assumption underlined in the joint Web validation model is that e will be the most relevant counterpart of k_i and vice versa if they are the correct transliteration pair. With this assumption, we applied forward and backward validation to e and k_i, as shown in Fig. 3. Forward validation determines "how likely k_i among elements of K is to be a counterpart of e" — it validates candidates of transliteration pairs from the viewpoint of e. Backward validation determines "how likely e among elements of E_i is to be a counterpart of k_i" — it validates candidates of transliteration pairs from the viewpoint k_i. The possible counterparts of k_i, E_i, are extracted from the Web search results for the query k_i in a similar way described in Section 3. All English terms in the Web search results for the query k_i are extracted and then E_i is constructed with the extracted English terms, which satisfy three conditions — the length constraint, the language constraint (excluding Korean graphemes, numbers, and other special symbols), and Eq. (2). The joint Web validation model can be represented with Eq. (6), where $S_{BPS}(e,k_i)$ and $S_{BPS}(k_i,e)$ represent forward and backward validation, respectively.

$$S_{Joint}(e,k_i) = S_{BPS}(e,k_i) \times S_{BPS}(k_i,e) \tag{6}$$

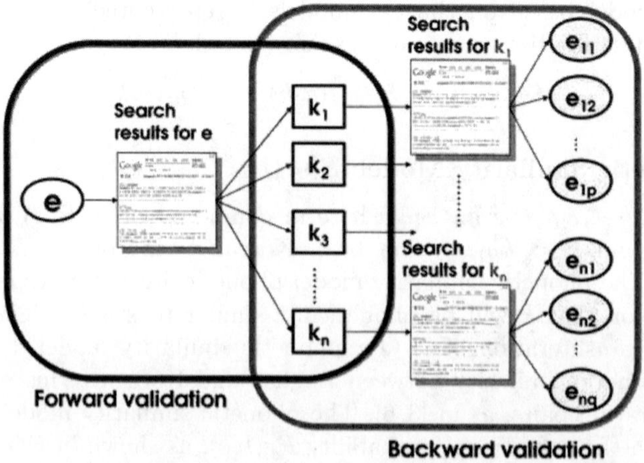

Fig. 3. Illustration of the joint Web validation model

We observed that e and k tend to be close together in texts of Web pages if they are counterparts of each other, such as 'si-naep-seu' (*synapse*) in Fig. 1. To retrieve such Web pages, we used "BILINGUAL PHRASAL SEARCH (BPS)", where phrases composed of e and k are used as queries for a Web search engine. There have been several Web-based translation or transliteration validation models, which are based on Web frequency derived from "BILINGUAL KEYWORD SEARCH (BKS)" [10, 11, 12, 13]. Let e and k be a English term and a Korean term, and '[e k]' or '[k e]' and 'e AND k' or 'k AND e' be queries for BPS and BKS, respectively. Then, the difference between BPS and BKS can be represented as Fig. 4. '[e k]' or '[k e]' retrieves Web pages where '[e k]' or '[k e]' exists as phrases; while 'e AND k' retrieves Web pages if e and k simply exist in the same document. Note that BKS frequently retrieves Web pages, where e and k have little co-relation, because BKS does not consider distance between e and k. Web frequency based on such Web pages makes it difficult to correctly validate transliteration pairs. However, BPS can address the problems by applying the constraint that e and k should be a phrase in the retrieved Web pages. Therefore, the number of Web pages retrieved by BPS is more reliable for validating transliteration pairs. For these reason, BPS is more suitable for our transliteration pair validation.

Let $W_{BPS}(e,k)$ be the sum of the Web frequencies retrieved by '[e k]' and '[k e]'. Then S_{BPS} can be represented as Eq. (7), where $S_{BPS}(e,k_i)$ and $S_{BPS}(k_i,e)$ represent forward validation and backward validation, respectively. Note that $S_{BPS}(k_i,e) = 0$ if E_i does not contain e.

$$S_{BPS}(e,k_i) = \frac{W_{BPS}(e,k_i)}{\sum_{k_j \in K} W_{BPS}(e,k_j)} \tag{7}$$

$$S_{BPS}(k_i,e) = \frac{W_{BPS}(e,k_i)}{\sum_{e_{il} \in E_i} W_{BPS}(e_{il},k_i)}.$$

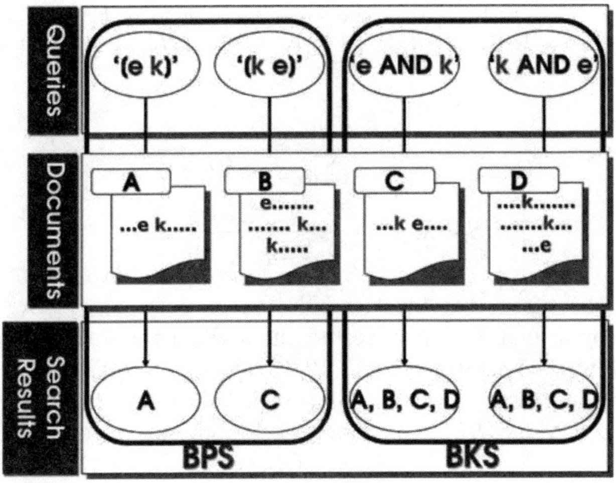

Fig. 4. Difference between BPS and BKS

5 Experiments

5.1 Experimental Setup

Our experiments were done on English-to-Korean transliteration pair acquisition. The test set [14] consisted of 7,172 English-Korean transliteration pairs — the number of training data was about 6,000 and the number of blind test data was about 1,000. The test set covered proper names, technical terms, and general terms. We trained Eqs. (2) and (4) with the training data. In the evaluation, we used k-fold cross-validation ($k = 7$). The test set was divided into k subsets. Each one was used for testing, while the remainder was used for training. Then, the average performance across all the k trials was computed. We experimentally set θ in Eq. (2) to 0.001 and the number of Web pages output by a Web search engine (used for "Candidate extraction") to 100 through cross-validation.

We experimented with the contribution of each validation model (S_{BKS}, S_{χ^2}, S_{BPS}, S_{PSM}, S_{Joint}, and S_{TV}) along with the effects of each Web search method (BPS and BKS) on transliteration pair validation. In this experiment, we focused on comparing three Web-based validation models (S_{BKS}, S_{χ^2}, and S_{BPS}) with each other. The two existing Web-based validation models, S_{BKS} [10] and S_{χ^2} [11, 12, 13], were compared to our S_{BPS} to show the effectiveness of BPS and S_{BPS}. S_{BKS} in Eq. (8) is similar to S_{BPS} in Eq. (7); but S_{BKS} uses W_{BKS} (Web frequency derived from BKS).

$$S_{BKS}(e, k_i) = \frac{W_{BKS}(e, k_i)}{\sum_{k_j \in K} W_{BKS}(e, k_j)} \tag{8}$$

$$S_{BKS}(k_i, e) = \frac{W_{BKS}(e, k_i)}{\sum_{e_{il} \in E_i} W_{BKS}(e_{il}, k_i)}$$

We modeled $\chi^2(e,k)$ in Eq. (9) in the same manner as in Eq. (1) [11,12,13], but we normalized $\chi^2(e,k)$ as S_{χ^2}. Like in Eq. (7), $S_{BKS}(k_i,e)=0$ and $S_{\chi^2}(k_i,e)=0$ if E_i does not contain e.

$$S_{\chi^2}(e,k_i)=\frac{\chi^2(e,k_i)}{\sum_{k_j\in K}\chi^2(e,k_j)} \tag{9}$$

$$S_{\chi^2}(k_i,e)=\frac{\chi^2(e,k_i)}{\sum_{e_{il}\in E_i}\chi^2(e_{il},k_i)}$$

We compare the three Web-based validation models (S_{BKS}, S_{χ^2}, and S_{BPS}) under three conditions. First, we tested them when the forward validation was used alone as in previous work [10,11,12,13] (S_{BKS}, S_{χ^2}, and S_{BPS} in Table 1). Second, we tested them with our joint Web validation model ($S_{Joint}(S_{BKS})$, $S_{Joint}(S_{\chi^2})$, and $S_{Joint}(S_{BPS})$ in Table 1). Finally, we tested them by applying S_{TV} ($S_{TV}(S_{BKS})$, $S_{TV}(S_{\chi^2})$, and $S_{TV}(S_{BPS})$ in Table 1). We evaluated the performance with pair accuracy within the Top-1, Top-2, Top-3 and Top-5 ranks. The pair accuracy is the proportion of extracted correct transliteration pairs to transliteration pairs in the blind test data.

5.2 Results

The performance of the validation models is shown in Table 1. First, S_{BPS} outperformed both S_{BKS} (about 53% in Top-1) and S_{χ^2} (about 157% in Top-1)[3]. The noise caused by BKS made it difficult for S_{BKS} and S_{χ^2} to correctly validate transliteration pairs; while S_{BPS} effectively filtered out wrong transliteration pairs by using reliable Web pages derived from BPS. S_{χ^2} and S_{BKS} usually made errors when transliteration candidate k_i appears much more frequently than the correct transliteration in the Web; k_i had more chances to appear with e, although k_i is not a counterpart of e. Second, the joint Web-validation model significantly boosted the performance regardless of S_{BKS}, S_{χ^2}, or S_{BPS} compared to the case when we used the forward validation alone[4]. More specifically, the performance of S_{BKS} and S_{χ^2} was much more improved than that of S_{BPS} by the joint Web-validation model. The main reason of the performance improvement was that we could not find English term e in the Web search results retrieved by Korean transliteration candidate k_i, when transliteration pair (e,k_i) was a wrong pair.

Many errors of $S_{Joint}(S_{\chi^2})$ and $S_{Joint}(S_{BPS})$ in the Top-1 were caused by substrings of the correct transliteration or strings composed of the correct transliteration and Korean postposition. For example, transliteration candidates for the English term *synapse* were ranked by each model as follows. Note that the underlined Korean term 'si-naep-seu' is the correct transliteration and 'si-naep-seu-e' ('e' is a postposition) is incorrect.

[3] A one-tail paired t-test showed that the results of S_{BPS} were always significantly better than those of S_{χ^2} and S_{BKS} (level of significance = 0.001).

[4] A one-tail paired t-test showed that the results of joint Web validation were always significantly better than those of forward validation (level of significance = 0.001).

Table 1. Performance depending on transliteration pair validation models

Models		Top-1 (%)	Top-2 (%)	Top-3 (%)	Top-5 (%)
S_{PSM}		70.34	83.37	87.01	89.51
S_{χ^2} [11,12,13]		28.53	45.57	56.50	69.63
S_{BKS} [10]		48.31	65.44	73.70	81.51
S_{BPS}		73.80	84.12	86.52	88.23
S_{Joint}	S_{χ^2}	54.98	76.55	84.47	88.73
	S_{BKS}	73.41	85.53	88.13	89.33
	S_{BPS}	78.28	86.14	88.15	89.33
S_{TV}	S_{χ^2}	62.42	81.05	86.81	89.47
	S_{BKS}	76.45	87.06	88.93	89.70
	S_{BPS}	**80.98**	**87.99**	**89.25**	**89.75**

– S_{χ^2}: 'si-naep-seu-e' (3rd), **'si-naep-seu'** (15th)
– S_{BKS}: 'si-naep-seu-e' (1st), **'si-naep-seu'** (2nd)
– S_{BPS}: **'si-naep-seu'** (1st), 'si-naep-seu-e' (15th)
– $S_{Joint}(S_{\chi^2})$: 'si-naep-seu-e' (1st), **'si-naep-seu'** (3rd)
– $S_{Joint}(S_{BKS})$: 'si-naep-seu-e' (1st), **'si-naep-seu'** (2nd)
– $S_{Joint}(S_{BPS})$: **'si-naep-seu'** (1st), 'si-naep-seu-e' (3rd)

Here, S_{χ^2} produced S_{χ^2}(*synapse*, 'si-naep-seu') $= 0.007 <$ S_{χ^2}(*synapse*, 'si-naep-seu-e') $= 0.1190$ by forward validation and S_{χ^2}('si-naep-seu', *synapse*) $= 0.008$ $<$ S_{χ^2}('si-naep-seu-e', *synapse*) $= 0.3330$ by backward validation. Similar to S_{χ^2}, S_{BKS}(*synapse*, 'si-naep-seu-e') and S_{BKS}('si-naep-seu-e', *synapse*) were highest in both forward and backward validation. Finally, S_{TV} showed the best performance[5]. Actually, both S_{PSM} and S_{Joint} by themselves performed well but a more powerful transliteration pair validation could be had by combining them. S_{PSM} effectively filtered out wrong Korean transliterations, which were phonetically dissimilar to English terms and could not be discarded by S_{Joint}.

5.3 Error Analysis

We analyzed errors caused by absence of correct transliteration pairs in the extracted transliteration pair candidates — the errors occupied about 10% of test data (724 among 7,172). We defined two error types — **retrieval error** and **extraction error** – causing the errors. The retrieval error occurred when there are not the correct Korean transliterations in the Web pages retrieved by a Web search engine (e.g. *ceiling*, *chemoreceptor*, and *gibbsite*) — it occupied about 6% of test data (402 among 7,172). However, we find that most of the retrieval error can be addressed by increasing *snippet size* (the number of Web pages

[5] A one-tail paired t-test showed that the results of S_{TV} were always significantly better than those of joint Web validation and forward validation (level of significance = 0.001).

retrieved by a Web search engine)[6]. The extraction error occurred when the "Candidate extraction" step did not recognize a correct Korean transliteration even though the correct one existed in the Web search result — it occupied about 4% of test data (322 among 7,172). θ in Eq. (2) mainly caused the error — as θ increases, the number of the extraction error also increases[7]. We can decrease the extraction error if we assign θ to lower values.

6 Conclusion

We proposed a novel approach to transliteration pair acquisition from the Web. Our system retrieved Web documents with English terms as queries for a Web search engine and then extracted candidates of transliteration pairs in the Web search results. Then our system validated the candidates by the combining phonetic similarity and joint Web-validation models. Experimental results can be summarized as follows. First, both the phonetic similarity and joint Web validation models are very useful for transliteration pair validation. By combining phonetic similarity and the joint Web validation model, our proposed method achieved higher performance (about 81% pair accuracy) than each individual model. Second, our joint Web-validation model more effectively validates transliteration pairs than previous Web-validation models [10,11,12,13], which rely on only the forward validation. Finally, BPS (BILINGUAL PHRASAL SEARCH) can give more reliable results than BKS (BILINGUAL KEYWORD SEARCH) [10,11,12,13].

Because experiments demonstrated that our joint-validation model based on BPS is very effective in transliteration pair validation, we expect that it is also useful for validating translation pairs. In future work, we plan to extend our method to translation pair acquisition and to transliteration pair acquisition between other language pairs, such as English and Japanese.

References

1. Fujii, A., Tetsuya, I.: Japanese/English cross-language information retrieval: Exploration of query translation and transliteration. Computers and the Humanities **35**(4) (2001) 389–420
2. Kang, B.J., Choi, K.S.: Two approaches for the resolution of word mismatch problem caused by English words and foreign words in Korean information retrieval. IJCPOL **14**(2) (2001)

[6] To investigate the effects of the number of retrieved Web pages on performance, we evaluated performance by changing the snippet size from 10 to 100 and fixing θ at 0.001. We get pair accuracy in Top-5 — 55.0% (*snippet size* = 10), 70.9% (*snippet size* = 20), 86.1% (*snippet size* = 60), and 90.1% (*snippet size* = 100).

[7] To investigate the effects of θ on performance, we evaluated performance by changing θ from 0.5 to 0.001 and fixing the snippet size at 100. We get pair accuracy in Top-5 — 60.6% ($\theta = 0.5$), 78.7% ($\theta = 0.1$), 86.1% ($\theta = 0.01$), and 90.1% ($\theta = 0.001$).

3. Brill, E., Kacmarcik, G., Brockett, C.: Automatically harvesting Katakana-English term pairs from search engine query logs. In: Proc. of NLPRS 2001. (2001) 393–399
4. Tsujii, K.: Automatic extraction of translational Japanese-Katakana and English word pairs from bilingual corpora. IJCPOL **15**(3) (2002) 261–279
5. Lee, C.J., Chang, J.S.: Acquisition of English-Chinese transliterated word pairs from parallel-aligned texts using a statistical machine transliteration model. In: Proc. of the HLT-NAACL 2003 Workshop on Building and using parallel texts. (2003) 96–103
6. Bilac, S., Tanaka, H.: Extracting transliteration pairs from comparable corpora. In: Proc. of Symposium on Large-Scale Knowledge Resources (LKR2005). (2005) 203–206
7. Oh, J.H., Choi, K.S.: Recognizing transliteration equivalents for enriching domain-specific thesauri. In: Proc. of the 3rd International WordNet Conference (GWC-06). (2006) 231–237
8. Oh, J.H., Choi, K.S., Isahara, H.: A hybrid model for extracting transliteration equivalents from parallel corpora. In: Proc. of the 9th International Conference on TEXT, SPEECH and DIALOGUE (TSD 2006). (2006)
9. Resnik, P., Smith, N.A.: The web as a parallel corpus. Computational Linguistics **29**(3) (2003) 349–380
10. Qu, Y., Grefenstette, G.: Finding ideographic representations of Japanese names written in Latin script via language identification and corpus validation. In: Proc. of ACL. (2004) 183–190
11. Lu, W.H., Chien, L.F., Lee, H.J.: Translation of web queries using anchor text mining. ACM Transactions on Asian Language Information Processing **1**(2) (2002) 159–172
12. Lu, W.H., Chien, L.F., Lee, H.J.: Anchor text mining for translation of web queries: A transitive translation approach. ACM Transactions on Information Systems **22**(2) (2004) 242–269
13. Wang, J.H., Teng, J.W., Lu, W.H., Chien, L.F.: Exploiting the web as the multilingual corpus for unknown query translation. Journal of the American Society for Information Science and Technology **57**(5) (2006) 660–670
14. Nam, Y.S.: Foreign dictionary. Sung An Dang (1997)

From Phoneme to Morpheme: Another Verification Using a Corpus

Kumiko Tanaka-Ishii and Zhihui Jin

Graduate School of Information Science and Technology
University of Tokyo
{kumiko, jin}@i.u-tokyo.ac.jp

Abstract. We scientifically test Harris's hypothesis that morpheme/word boundaries can be detected from changes in the complexity of phoneme sequences. We re-formulate his hypothesis from a more information theoretic viewpoint and use a corpus to test whether the hypothesis holds. We found that his hypothesis holds for morphemes, with an F-score of about 80%, in both English and Chinese. However, we obtained contrary results for English and Chinese with regard to word boundaries; this reflects a difference in the nature of the two languages.

1 Introduction

Zellig S. Harris, the most influential teacher of Chomsky, wrote "From Phoneme to Morpheme" in 1955 [1]. Harris studied language from a computational viewpoint, and the spirit of his approach was taken up by Chomsky and is seen in Chomsky's well known linguistic theories [2]. In his paper, Harris suggests that morpheme boundaries can be detected by observing the *successor counts* of phoneme sequences. A successor of an utterance of a given length n is the phoneme that follows the utterance. A successor count is the number of different successors, and it is obtained by going through many utterances that start with the given utterance. For example, given a short sentence "He is clever" /hiyzklevər/, utterances coming after /h/ (such as "hot coffee", "How are you") are collected and the successor counts are measured. Then utterances coming after /hi/ (such as "hit it", "he is good") are collected and successor counts are measured. This is repeated for /hiy/, /hiyz/, /hiyzk/, and so on. Harris describes what is observed when this successor count shift is monitored:

> When this count is made for each n of the utterance, it is found to rise and fall a number of times. If we segment the test utterance after each peak, we will find that the cuts accord very well with the word boundaries and quite well with the morpheme boundaries of that utterance.

The idea behind Harris's hypothesis is illustrated in Figure 1. (This figure is further explained later in this paper).

When Harris was doing this research, computers were not used as personal tools, nor were there huge bodies of electronic data available. Therefore, he tested

Y. Matsumoto et al. (Eds.): ICCPOL 2006, LNAI 4285, pp. 234–244, 2006.

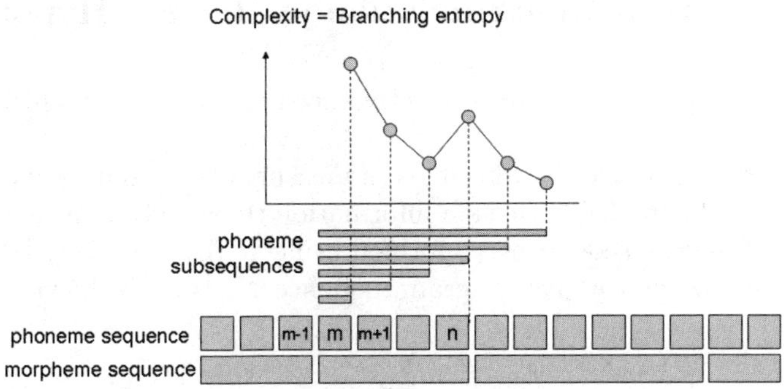

Fig. 1. Segmentation Example

his hypothesis by asking people to give as many utterances as possible that began with a certain utterance fragment. This method of asking people directly is the most accurate possible, but limits the scale of any experiment; Harris does not state how many people or how many phrases he considered when evaluating his hypothesis. We now have easy access to relatively powerful computers and large corpora to use as our tool and database, though, so we can test Harris's hypothesis using these means. In this paper, we first mathematically reformulate Harris's hypothesis from an information theoretic viewpoint and then discuss a large-scale evaluation of the reformulated hypothesis.

Harris's hypothesis is interesting because it fills the gap of "double articulation" — first described by Martinet [3] — meaning that language is segmented into two different units: phonemes without meaning and morphemes with meaning. Precisely, Harris's hypothesis can generally suggest that this gap is filled, but can be regarded more generally: how a *meaningful* unit is generated given a sequence. Language can be considered to have multiple layers — phoneme, morpheme, word, collocation and so on — with each layer, except that of a phoneme, formed of a sequence of meaningful units. Harris's hypothesis suggests that each layer is formed as chunks of a smaller layer. For example, if given a morphomeme sequence, we should be able to scan through the successor counts of morpheme sequences and find that the peaks are correlated with word boundaries. Indeed, studies in NLP have found that Chinese, which is written as ideogram (morpheme) sequences, can be segmented into words based on an hypothesis similar to that of Harris with a precision as high as about 90% [4] [5]. Another example is that for a given word sequence, scanning through the successor counts of the word sequence should reveal peaks correlated with collocation boundaries. This has been tested for collocation extraction [6], and has been applied in an application tool [7]. Thus, Harris's hypothesis can be generalized as a law governing language, and for segmenting out a meaningful unit at various levels. To date, however, the hypothesis has not been tested at this most basic level of phoneme to morpheme or phoneme to word using any of the corpora now widely available. Therefore, we have chosen to scientifically verify the extent to which the Harris hypothesis holds.

2 Mathematical Re-formulation of Harris's Hypothesis

The formulation presented here is exactly the same as the one which appeared in [8].

The successor count can be considered a measure of the complexity of the successor. This can be modeled through information theory. Given a set of elements χ and a set of n-gram sequences χ_n formed using χ, the *branching entropy* of an element occurring after a given n-gram sequence $X_n = x_n$ is defined as

$$H(X|X_n = x_n) = -\sum_{x \in \chi} P(x|x_n) \log P(x|x_n), \tag{1}$$

where $P(x) = P(X = x)$, $P(x|x_n) = P(X = x|X_n = x_n)$, and $P(X = x)$ indicates the probability of x occurring. The last term in this formula, $-\log P(x|x_n)$, indicates the information of a token of x coming after x_n, and thus the branching after x_n. $H(X|X_n = x_n)$, the local entropy value for a given x_n, indicates the average information of branching for a *specific* n-gram sequence x_n. For simplicity, we denote $H(X|X_n = x_n)$ as $h(x_n)$ for the rest of this paper.

Harris's hypothesis says that for any given $x_1, \ldots, x_i, x_{i+1}$, where x_i forms the prefix of x_{i+1}, $h(x_i)$ repeatedly falls and rises for $i = 1, \ldots,$ and the peak points are the boundaries of an element larger than the elements of χ. These falls can be explained in relation to another universal nature of language, $H(X|X_n)$, defined as

$$H(X|X_n) = -\sum_{x_n \in \chi_n} P(x_n) \sum_{x \in \chi} P(x|x_n) \log P(x|x_n). \tag{2}$$

This $H(X|X_n)$ is the average of $h(x_n)$; i.e., the average uncertainty of a successor for any subsequence of length n. For language data, it is known that the larger n becomes, the smaller $H(X|X_n)$ will be. For instance, in the case of the word "natural", it is easier to guess which character comes next given x_6="natura" than x_1="n". This fact means it is more effective and more natural to consider all *increasing* points rather than just peak points. Therefore, the hypothesis to be tested in this paper is

> If the complexity of successive tokens increases, the location of the increase is at a border. (X)

That is, we consider places at n where

$$h(x_n) > h(x_{n-1}). \tag{3}$$

This hypothesis (X) differs from Harris's hypothesis in two ways:

- Harris uses the successor variety count, but we use branching entropy.
- Harris uses the maximum points, but we use increasing points.

3 Experimental Setting

3.1 Procedure

To verify (X), we need test data and training data. As we want to apply (X) to two cases, "from phoneme to morpheme" and "from phoneme to word", a gold standard —data indicating the supposedly true segmentation boundaries— for word boundaries and one for morpheme boundaries are required for the test data.

The test data was first segmented at punctuations. These text segments are called *fragments* in this paper. Both the test data and the training data were then transcribed into phoneme sequences. Further details of the input data are explained in §4.1 and §5.1.

From this transcribed training data, we measured the branching entropy for all phoneme subsequences in the test data. As a subsequence becomes longer, the entropy approaches zero. Therefore, we set a maximum length value *maxlen* and considered only subsequences shorter than *maxlen*. For all of these subsequences, we obtained branching entropies from the training data.

Each fragment was then processed as follows to obtain the boundary, where $x_{m,n}$ indicates a subsequence of a given sequence x from offset m until n.

1. Set $m = 0$, $n = m + 1$.
2. Calculate $h(x_{m,n})$.
3. Compare the result with $h(x_{m,n-1})$. If $h(x_{m,n}) - h(x_{m,n-1}) > \textit{threshold}$, output n as the boundary.
4. If $n > m + \textit{maxlen}$, then $m = m + 1$, $n = m + 1$. Otherwise, set $n = n + 1$ and go to 2.

The *threshold* is an arbitrary threshold that we varied during our evaluation. This procedure is illustrated in Figure 1 for the case where m is fixed to a certain offset and n is successively increased in step 4 of the procedure. This algorithm is the simplest algorithm to test (X), where we just scan through all phoneme subsequences of length less than *maxlen*.

So far, we have considered only regular (forward) order processing: the branching entropy is calculated for *successive* elements of x_n. We can also consider the reverse order, which involqves calculating h for the *previous* element of x_n. In the case of the previous element, the question is whether the head of x_n forms the *beginning* of a context boundary. The above algorithm can then be applied in reverse. Even though language linearity suggests asymmetry between regular and reverse orders, we also apply the thus defined process in the reverse (backward) order, as reverse order processing was also suggested by Harris [1].

3.2 Precision and Recall

The output was evaluated in comparison with the gold standard. We calculated the precision and recall to test both the "from phoneme to word" and "from

Table 1. Phoneme transcription of some English examples

Example 1:	he HH IY1	is IH1 Z	clever K L EH1 V ER0	
Example 2:	I AY1	like L AY1 K	this DH IH1 S	book B UH1 K
Example 3:	I AY1	got G AA1 T	it IH1 T	

phoneme to morpheme" cases. Precision and recall were defined as

$$\text{Precision} = \frac{N_{correct}}{N_{test}} \tag{4}$$

$$\text{Recall} = \frac{N_{correct}}{N_{true}} \tag{5}$$

$$\text{F-score} = \frac{2 * Precision * Recall}{Precision + Recall}, \quad where \tag{6}$$

$N_{correct}$ is the number of correct boundaries in the result,
N_{test} is the number of boundaries in the test result, and,
N_{true} is the number of boundaries in the gold standard.

For example, if the boundary in the gold standard is "abc | def | ghi | jk", with a | indicating a boundary, there are three boundaries. Suppose that we obtained a result of "ab | cd | ef | ghi | jk", and two of the four reported boundaries are correct. The precision is then 50% ($=\frac{2}{4}$) and the recall is 67% ($=\frac{2}{3}$).

Note that as word boundaries are all included among the morpheme boundaries, the precision is always better for morpheme results than for word results because the numerator is larger for morphemes.

4 Verification in English

4.1 Data

We used 100 MB of data from the Wall Street Journal corpus as training data and 1 MB as test data. English text was transformed into phoneme sequences by using the CMU Pronouncing Dictionary [9]. Some examples of text and its transcription by CMU pronunciation description are shown in Table 1. A phoneme is described as a sequence of capital letters possibly followed by a number. There are 39 phonemes used in the CMU dictionary, which contains various words and their variants. If a word in a fragment (see §3.1) was not included in the CMU dictionary, we eliminated the whole fragment. Such eliminated fragments amounted to 23.4% of the test data, so the actual amount of test data was 766 Kbytes. Note that after the English was transformed into phoneme sequences, NO SPACES to indicate the word boundaries remained. Since punctuation was eliminated when we used fragments, the input test data uniformly consisted of phoneme sequences. The constant *maxlen* was set to 10.

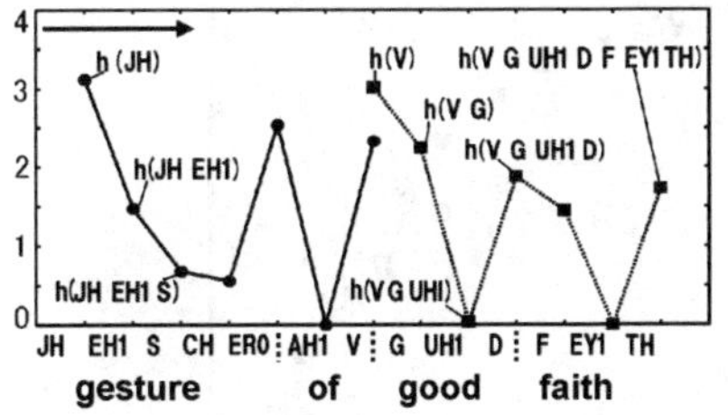

Fig. 2. Forward Example

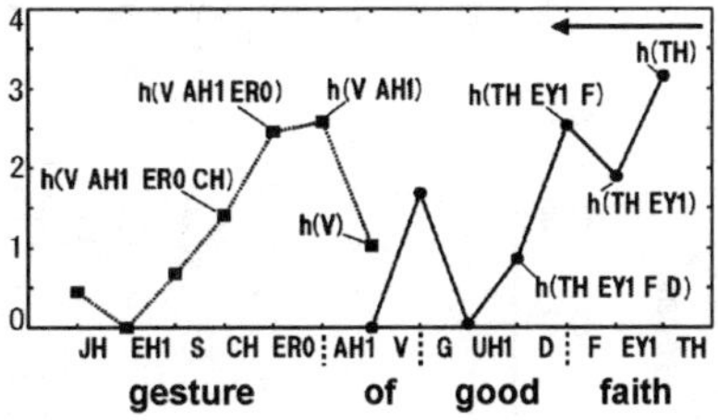

Fig. 3. Backward Example

Word boundaries in the original English text were considered the gold standard. However, it was difficult to obtain a complete set of morphemes in English and the only publicly available means to do so was the PC-KIMMO analysis tool [10]. However, this software works only for English text. As the CMU dictionary does not provide phoneme-to-text alignment, it would be difficult to automatically create a gold standard for 766 K of phoneme sequence data. Therefore, we went through the following procedure to obtain a very small gold standard for morphemes.

1. Randomly obtain 50 fragments containing only words registered in PC-KIMMO.
2. Process each fragment with PC-KIMMO and get the morpheme boundaries.
3. By looking at this result from PC-KIMMO, manually put morpheme boundaries into phoneme sequences transcribed from the 50 fragments.

4.2 Small Examples

Before going through our full-scale experiment, here we show that hypothesis (X) holds for a small example.

Figure 2 and Figure 3 each show an actual graph of the entropy shift for the input phrase "JH EH1 S CH ER0 AH1 V G UH1 D F EY1 TH" (*gesture of good faith*). The former shows the entropy shift for the forward case, and the latter shows the entropy shift for the backward case. Note that for the backward case, the branching entropy was calculated for phonemes *before* the x_n.

In Figure 2, there are two lines, one for the branching entropy after the substrings starting from JH. The leftmost line plots h(JH), h(JH EH1) ... h(JH EH1 S CH ER0 AH1 V). There are two increasing points, indicating that the phrase was segmented between ER0 and AH1 and after V. The second line plots h(V) ... h(V G UH1 D F EY1 TH). The increasing locations are between D and F, and after TH.

Figure 3 shows the entropy shift for the backward case. There are two lines, indicating the branching entropy before the substring ending with suffix TH. The rightmost line plots h(TH), h(TH EY1) ... h(TH EY1 F D UH1 G V) running from back to front. We can see increasing points (as seen from back to front) between F and D and between G and V. As before, the leftmost line starts from

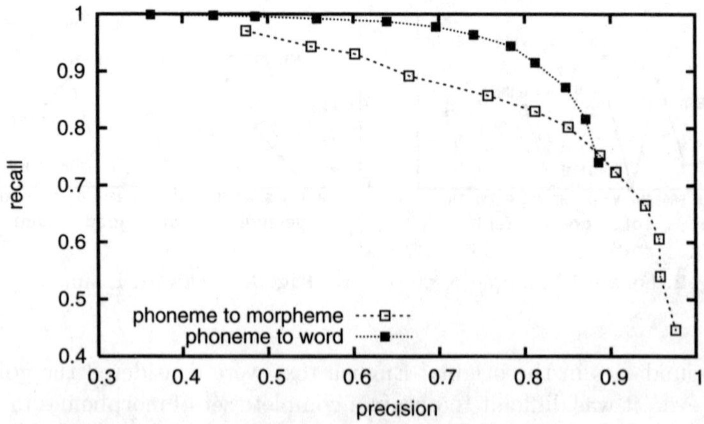

Fig. 4. Precision and Recall in English

V and runs from back to front, indicating boundaries between AH1 and ER0, and just before JH.

If we consider all the increasing points in the two forward and backward lines and take the union set of them, we obtain the correct segmentation as follows with | indicating the boundaries:

gesture | of | good | faith
JH EH1 S CH ER0 | AH1 V | G UH1 D | F EY1 TH

which is the 100% correct word segmentation in terms of both recall and precision. For this example, no morpheme boundaries were detected.

In fact, as there are 13 phonemes in this input, there should be 13 lines starting from each phoneme for all substrings. For readability, however, we only show two lines each for the forward and backward cases and set the maximum length of a line to 7, in experiments the length should be 10 because we took 10-grams out of the learning data. If we consider all the increasing points in all 13 lines and take the union set, then we again obtain 100% precision and recall. We find it amazing that all 13 lines indicate only correct word boundaries.

4.3 Larger-Scale Performance

Precision and recall were plotted by changing the threshold from 0.0 to 2.4, with an interval of 0.2 (Figure 4). Two lines are shown: one for the "from phoneme to word" case, and the other for the "from phoneme to morpheme" case.

For the word result, we obtained an F-score of 86.1% (precision=81.2%, recall=91.5%) at a threshold value of 1.6. Note how high this is: the input is a plain phoneme sequence without any spaces, yet the word boundaries were detected at 85%. At the highest threshold of 2.4, the precision was 89.2% with recall of 70%.

The 18.8% (=100-81.2) of boundaries which were not word boundaries included morpheme boundaries. The F-score for morpheme boundaries at a threshold of 1.6 was 80.4% (precision=90.5%, recall=72.3%). Thus, almost half were correct in-word morphemes (90.5-81.2=9.3% which is the half of 18.8%). As mentioned in §3.2, for a given threshold, the precision of the morpheme result is always higher than that for the word result; therefore, this performance decrease was due to the recall decrease.

This was partly due to limitations of the vocabulary in the corpus. For example, the word "absorb" contains two morphemes "ab" and "sorb", but when the corpus only has "absorb" as the word with the prefix "ab", the boundary after "ab" is missed. However, we consider this phenomenon to also be due to the nature of English in that the English language puts less emphasis on the morpheme as a unit compared with other languages such as Chinese, where morphemes appear more explicitly in the language system. Therefore, we next tested (X) on the Chinese language.

5 Verification in Chinese

5.1 Data

The training data for Chinese was 200 MB of unsegmented text taken from the Contemporary Chinese Corpus of the Center of Chinese Linguistics at Peking University. The test data was the 7.8 Mbyte manually segmented People's Daily corpus of Peking University [11]. Chinese is a suitable test language for testing (X) because the phonetic standard pinyin forms a good approximation of a phoneme sequence. Thus, Chinese text is converted into pinyin by using NJstar [12], a Chinese word processor software application, and the tonal number is eliminated. Pinyin can then be automatically separated into what corresponds to phonemes. For example, the Chinese pinyin "zi ran yu yan" (natural language) is decomposed into phonemes as "z i r a n y u y a n". Some special cases in Chinese pinyin had to be carefully processed during this decomposition. For example, the pinyin "u" following "j", "q", or "x" is actually pronounced as "v", so we decomposed "ju", "qu", and "xu" into "j v", "q v", and "x v", respectively. In the case of "yi", "yin", "ying", and "wu", the "y" and "w" are not pronounced, so "y" and "w" were eliminated. The constant *maxlen* was set to 12.

As the test data was manually segmented into words, they were considered the gold standard for words. As for the morphemes, ideogram boundaries were considered the gold standard.

5.2 Results

Our results are shown in Figure 5. The precision and recall were plotted by changing the threshold of the boundary detection condition from 0.0 to 2.4, with an interval of 0.2. Again, we show two lines, one corresponding to the word results and the other to the morpheme results.

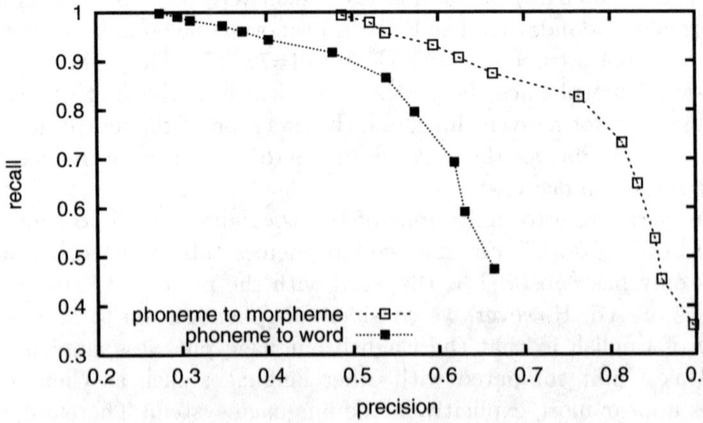

Fig. 5. Precision and Recall in Chinese

For morphemes, we obtained an F-score of 79.4% (precision=76.7%, recall=82.4%) at a threshold value of 1.2, which was similar to the English morpheme case. Therefore, from phoneme to morpheme, (X) seems to hold with an 80% F-score for both English and Chinese.

For words, however, we obtained an F-score of 66.9% (precision=54.7%, recall=86.1%) at a threshold value of 1.6. Comparing the two graphs, the word result can be obtained by shifting the morpheme result towards the left. This means the precision decreased, while the recall was similar for each threshold. At a threshold of 1.6, 42.8% (=100 - 55.3) of the places indicated were not word boundaries and 20% (precision is 83.5, 83.5-54.7 = 19.8) of these were correct morpheme boundaries. This is a drastic difference from the English case, which warrants discussion.

6 Discussion

We account for the above difference from English as follows. From phoneme to morpheme, the performance was the same for Chinese and English: recall and precision both being about 80%. However, there was a drastic difference for the word boundaries. For English, precision and recall was more than 80%, but precision decreased to 57% for Chinese, with half of the remaining 43% containing morpheme boundaries.

This, in fact, shows an important role morphemes play in Chinese. Chinese language is formed with the unit being ideograms and the pinyin for each ideogram forms a syllable. In this language, morphemes are far more explicit than in English. Every word is formed as a true combination of ideograms. Therefore, "from phoneme to morpheme" holds only for "morpheme" and word boundaries are formed at a higher level of "from morpheme to word". In our previous work where we applied (X) to the case "from morpheme to word" [4], the F-score was

83%, with precision of 88% and recall of 79%. In contrast, English had a higher result for the "from phoneme to word" case. This suggests that morphemes are less explicit as units, so word units should be formed directly "from phoneme".

How correct was Harris? Harris stated that *the cuts accord very well with the word boundaries and quite well with the morpheme boundaries of that utterance.* For morphemes, we obtained F-scores of about 80% for both English and Chinese, indicating that the morpheme boundaries were detected *quite well* and that (X) is valid in this case. With regard to words, it depended on the language. For English, the F-score was about 85%, with recall being above 90%, which indicates the word boundaries were detected *very well.* For Chinese, though, (X) does not hold and words seem to be formed "from morphemes". Therefore, Harris was probably unaware that his hypothesis not only applies for double articulation, but also applies more generally: that the hypothesis (X) might be a law which can be used to segment out a larger meaningful unit from a smaller unit chain.

7 Conclusion

We scientifically verified the validity of Harris's "from phoneme to morpheme" hypothesis using a large-scale corpus. Harris's hypothesis is that morpheme/word boundaries can be detected from changes in the complexity of phoneme sequences. After re-formulating Harris's hypothesis from an information theoretic viewpoint, we did a large-scale experiment on sequences in English and Chinese. Harris stated that morpheme boundaries could be detected "quite well" under his hypothesis, and we confirmed this with an F-score of about 80%. However, Harris also stated that word boundaries could be detected "very well"; although we confirmed this for English, with an F-score of 85%, we could not confirm it for Chinese. This result suggests that Chinese words are constructed from morphemes rather than from phonemes.

References

1. Harris, S.: From phoneme to morpheme. Language (1955) 190–222
2. Imai, K.: Dictionary of Chomsky. Taishukan (1986) in Japanese.
3. Martinet, A.: Elements de linguistique generale. Colin (1960)
4. Jin, Z., Tanaka-Ishii, K.: Unsupervised segmentation of chinese text by use of braching entropy. In: COLLING/ACL. (2006)
5. Huang, H., Powers, D.: Chinese word segmentation based on contexual entropy. In: Pacific Asian Conference on Language, Information and Computation. (2003)
6. Frantzi, T., Ananiadou, S.: Extracting nested collocations. 16th COLING (1996) 41–46
7. Tanaka-Ishii, K., Nakagawa, H.: A multilingual usage consultation tool based on internet searching -More than a search engine, less than QA-. In: WWW Conference. (2005) 363–371
8. Tanaka-Ishii, K.: Entropy as an indicator of context boundaries —an experiment using a web search engine —. In: IJCNLP. (2005) 93–105

9. Carnegie Mellon University: CMU pronouncing dictionary version 0.6 (2006) visited 2006, http://www.speech.cs.cmu.edu/cgi-bin/cmudict.
10. SIL:Pc-kimmoversion2,amorphologialparser(1995)http://www.sil.org/pckimmo/.
11. ICL: People's daily corpus,Beijing university (1999) http://www.icl.pku.edu.cn/icl_res/.
12. NJStar Software Corp: Njstar, chinese word processing software. (2006) http://www.njstar.com.

Chinese Abbreviation Identification Using Abbreviation-Template Features and Context Information*

Xu Sun and Houfeng Wang

Department of Computer Science and Technology
School of Electronic Engineering and Computer Science
Peking University, Beijing, 100871, China
sunxu@pku.edu.cn, wanghf@pku.edu.cn

Abstract. Chinese abbreviations are frequently used without being defined, which has brought much difficulty into NLP. In this study, the definition-independent abbreviation identification problem is proposed and resolved as a classification task in which abbreviation candidates are classified as either 'abbreviation' or 'non-abbreviation' according to the posterior probability. To meet our aim of identifying new abbreviations from existing ones, our solution is to add generalization capability to the abbreviation lexicon by replacing words with word classes and therefore create abbreviation-templates. By utilizing abbreviation-template features as well as context information, a SVM model is employed as the classifier. The evaluation on a raw Chinese corpus obtains an encouraging performance. Our experiments further demonstrate the improvement after integrating with morphological analysis, substring analysis and person name identification.

1 Introduction and Background

As a special form of unknown words, Chinese abbreviations are frequently used without being defined, which has brought much difficulty into natural language processing (NLP), especially for agglutinative languages such as Chinese, in that the problem is exacerbated by the lack of word boundaries. How to identify abbreviations[1] becomes a common problem within Chinese word segmentation, Chinese co-reference resolution, Chinese named-entity (NE) recognition, etc. Take for instance the NE recognition in which a large part of target NEs are abbreviated within the source texts, it is of necessity to retrieve NEs from those abbreviations. As a precondition for this task, however, there lies the above-mentioned more basic problem: How to retrieve abbreviations within agglutinative Chinese texts?

---

* Supported by National Social Science Foundation of China (No. 05BYY043) and National Natural Science Foundation of China (No. 60473138, No. 60675035).

[1] In this paper, the term *abbreviation* will always stand for *Chinese abbreviation* if there is no specific indication.

Y. Matsumoto et al. (Eds.): ICCPOL 2006, LNAI 4285, pp. 245 – 255, 2006.

To a large extent, the success of identifying English abbreviations[2] goes to two aspects: First, most of the English abbreviations contain uppercase letters (e.g., 'DNA'). Second, lots of English abbreviations are marked by parentheses, namely the pattern 'abbreviation (definition)' or 'definition (abbreviation)'. Taghva (1999), Yeates (1999), and Byrd (2001) utilized the uppercase information for their abbreviation acquisition, while Schwartz (2003), Chang (2002), and Zahariev (2004) suggested the parentheses. Unfortunately, for Chinese texts, there is no 'uppercase', and very few abbreviations are explicitly marked by parentheses. Additionally, in Chinese this issue is exacerbated by the ambiguity of word boundaries. Thereby, it is of extraordinary difficulty to extend the abbreviation identification techniques from English to Chinese.

The literature on Chinese abbreviation identification is relatively small. Sproat (2002) and Sun (2002) introduced heuristics for this study. Such heuristics, however, can easily break. Of the more recent research in the area, important work is that of Chang (2004), who presented a hidden Markov model (HMM) based approach for abbreviation identification. In the experiment to guess the abbreviations from given definitions, the accuracy rate is 72%. Yet we can see that the definition information is still of necessity in the task.

Hence, in this study, our motivation is to investigate a definition-independent approach so that the abbreviations can still be identified in texts where the availability of definitions is not guaranteed. Instead of relying on abbreviation-definition mapping, which is a typical technique being vastly used in previous definition-required studies, we add generalization capability to the abbreviation lexicon by replacing words with word classes and therefore create abbreviation-templates to meet our aim of identifying new abbreviations from existing ones.

As has been mentioned, automatic abbreviation identification is a key component in systems that handle the various extended tasks, such as automatic abbreviation expansion, co-reference resolution, named-entity recognition or automatic query expansion. The extended tasks, however, are not the main focus of this paper.

The rest of this paper is organized as follows. In next section, the system architecture is described. In section 3 and section 4, abbreviation disambiguation techniques and improvement solutions are respectively presented. The remainder of this paper is experiment results and conclusion.

2 System Overview

In our study, word segmentation and abbreviation identification are integrated into a unified framework in order to automatically extract abbreviations from raw text. As described in Gao (2003), we define Chinese words in this paper as one of the following four types as well: (1) entries in a lexicon (lexicon words below), (2) morphologically derived words, (3) factoids, and (4) named entities, because these four types of words have different functionalities in Chinese language processing, and are processed in different ways in our system.

---

[2] Here, *English abbreviation* is a general denotation, containing *acronym*.

In this paper, we made the following assumptions about abbreviations:

(1) The abbreviation length should be between two and five characters. The single-character abbreviations will not be considered in our system, because the number of single-character abbreviations (e.g., '法' for '法国') is much less than multi-character abbreviations and that it is possible to be enumerated, so that the their identification can be resolved by simply using a single-abbreviation list. On the other hand, the abbreviation containing more than five characters is very rare (less than 0.1%).

(2) An abbreviation contains no lexicon words (otherwise it will be segmented during word segmentation). Though actually there are abbreviations containing lexicon words, our experiment performed on an abbreviation set containing 5,121 entries shows that this kind of abbreviation is few (about 4%).

In our system, the abbreviation identification process contains three steps: (1) word segmentation and text normalization, (2) abbreviation candidate search, and (3) abbreviation disambiguation. A lexicon containing around 100K entries is employed for word segmentation, and during text normalization, the factoids are normalized and replaced with tags (e.g., 十二点三十分 '12:30' will be replaced by 'F_TIME'.). In our study, word segmentation is employed for two reasons: First, based on abbreviation assumption (2), abbreviation contains no lexicon words so that abbreviation candidates will be collected from unknown character sequences, and word segmentation is required for the retrieve of unknown sequences. Second, word segmentation is required for providing context information during abbreviation disambiguation.

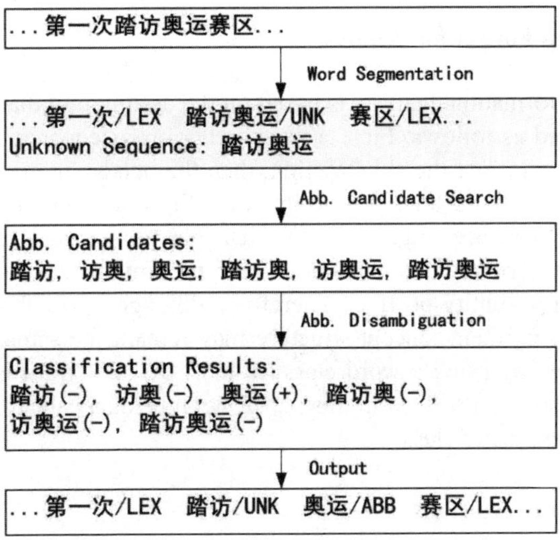

Fig. 1. Illustration of the overall system architecture

After unknown sequences are generated during word segmentation, searching abbreviation candidates is relatively simple. Based on abbreviation assumption (1), for each substring within an unknown sequence, if it measures up to the length constraints, it will be collected as an abbreviation candidate.

An abbreviation candidate may also relate to (be a sub-sequence or super-sequence of) a named-entity, a misspelled word, etc, thus abbreviation disambiguation is crucial in our system. Since a range of weak evidence must be combined in order to make a judgment, statistical techniques are ideal for this environment. The details of abbreviation disambiguation will be presented in next section.

The system overview is illustrated in Fig. 1. As we can see, '踏访奥运' is marked as an unknown sequence during word segmentation because it contains no lexicon word. Totally 6 abbreviation candidates are collected according to abbreviation assumption (1), thereafter, candidates are classified by the abbreviation disambiguation model, and eventually only '奥运' (abbreviated from奥林匹克运动会'Olympic Games') is classified as 'abbreviation'.

3 Abbreviation Disambiguation

We use the SVM (support vector machines) model to disambiguate abbreviation candidates. The key to the classification is to select discriminative features that effectively capture the distinction between 'abbreviation' and 'non-abbreviation'. Based on our own investigation of the abbreviations, two groups of features are employed: abbreviation formation analysis (the conceptual formation of abbreviations are modeled by using a class-based language model) and context information analysis.

3.1 Abbreviation Formation Analysis

The abbreviation formation analysis is based on the assumption that a Chinese abbreviation is generated as follows: First, a person chooses a sequence of concepts to set up the concept structure of the abbreviation; then the person attempts to express each concept by choosing characters. We assume that different abbreviations may share a common concept structure. E.g., 上影厂 'Shanghai-Film-Studio' and 北工大 'Beijing-Technology-University'[3] is generated from the same concept structure 'location + category/industry + entity-postfix'. Therefore, although new abbreviations are constantly being created, their concept structure may remain the same. In our system, a concept is modeled by using a word class (in most cases, a word is represented as a character in the abbreviations, so in this paper it can be also called 'abbreviated-word class' or simply 'character class').

Word Class of Abbreviations

An efficient method for word clustering has been introduced in Och (1999) for machine translation, and its main idea is adopted for determining word classes in our study. We use a statistical language model to estimate the probability $P(w_1^N)$ of the

[3] The formal name is 'Beijing University of Technology'.

word sequence $w_1^N = w_1 ... w_N$ of an abbreviation. A simple approximation of $P(w_1^N)$ is to model it as a product of bigram probabilities $P(w_1^N) = \prod_{i=1}^{N} P(w_i \mid w_{i-1})$. Rewriting the probability using classes we arrive at the following probability model:

$$P(w_1^N \mid C) := \prod_{i=1}^{N} P(C(w_i) \mid C(w_{i-1})) \cdot P(w_i \mid C(w_i)) \qquad (1)$$

where the function C maps words w to their class $C(w)$. In this model, we have two types of probabilities: the transition probability $P(C|C')$ for class C given its predecessor class C', and the membership probability $P(w|C)$ for word w given class C. To determine the optimal word classes C for a given abbreviation corpus (a manually collected abbreviation set containing 5,121 entries is used as the training data), we perform a maximum-likelihood estimation:

$$C_{opt} = \arg\max_{C} P(w_1^N \mid C) \qquad (2)$$

During the implementation, an efficient optimization algorithm for word clustering is the exchange algorithm (Martin et al., 1998). It is necessary to fix the number of classes in C in advance as the optimum is reached if every word is a class of its own. Considering the size of our training data, the default number of classes is set as 30 in our system.

Here two resulting word classes are selected for illustration. The first one is '班, 办, 部, 场, 厂, 池, 处, 局, 具, 圈, 势, 室, 署, 厅, 系, 校, 源, 院, 站'. As we can see, most of the entries are 'entity-postfix', which is frequently used as the last character in the formation of abbreviations. '埃, 北, 东, 低, 南, 西, 朝, 成, 川, 滇, 韩, 京, 兰, 黎, 柳, 闽, 欧, 葡, 深, 沈, 蜀, 湘, 亚, 燕, 粤, 云, 浙, 中' gives another example of our resulting word classes. As illustrated, most of these entries are 'location name'.

Abbreviation Formation Features

Three features are used for abbreviation formation analysis:

Abbreviation formation score: For an abbreviation candidate $w_1^N = w_1 ... w_N$, its abbreviation formation score will be estimated by using:

$$P(w_1^N \mid C_{abb}) := \prod_{i=1}^{N} P(C_{abb}(w_i) \mid C_{abb}(w_{i-1}), C_{abb}(w_{i-2})) \cdot P(w_i \mid C_{abb}(w_i)) \qquad (3)$$

Where C_{abb} is the partition of word classes trained from the above-mentioned abbreviation corpus. Trigram probabilities are used to estimate the transition probability (while only bigram probabilities are used for determining word classes in Eq. (1), and the difference comes from different compromise between efficiency and exactness). The trigram probabilities can be calculated by training on the same abbreviation set, and to further deal with the data sparseness, we use a standard backing off schema (Katz, 1987).

Word penalty: This feature counts the length in words of the target abbreviation candidate to balance the abbreviation formation score. Without this feature, the final abbreviation produced tends to be too short, because shorter items tend to get higher probability based on the language model.

Numeric information: Although factoids have been normalized during preprocessing, there are still many abbreviation candidates containing numeric characters. Most of these candidates are noise, while some of them are not (e.g., 二战 'the Second World War'). This feature records the number of numeric characters inside a candidate as well as their character position: 'BEGINNING', 'MIDDLE', or 'END'.

3.2 Context Information Analysis

In our study, two features of context information are adopted to improve abbreviation identification.

Contextual words: A large part of abbreviations are from named entities, and their contextual words have its own traits. This feature is used to record the left/right word and their corresponding length (number of characters) surrounding the abbreviation candidate. Note that the contextual words are ambiguous when both sides of this abbreviation candidate are unknown sequence. In such a case, this feature will not be chosen.

Frequency feature: This feature counts the occurrence of abbreviation candidate in the local document. It is used to discriminate abbreviation from random noise, in that in many cases such noise will occur only once on the document level.

4 Improvement

The abbreviation disambiguation model introduced in section 3 is selected as our baseline model, in which the following deficiencies emerged during experimental evaluation (the experimental result will be shown in section 5): First, morphologically derived abbreviations are neglected. Second, experiments showed that some substrings of the abbreviations were mistakenly classified as 'abbreviation'. Third, we find lots of noise coming from named-entities, especially person names (PN). Then, we will provide solutions to the three problems respectively.

4.1 Morphological Analysis

As described in Gao (2003), the morphologically derived words are generated using 4 morphological patterns: (1) affixation: 朋友们 (friend - plural) 'friends'; (2) head particle(i.e. expressions that are verb+comp): 走 'walk' + 出去 'out' ->走出去 'walk out'; (3) reduplication: 高兴 'happy' -> 高高兴兴 'happily' ; and (4) merging: 上班 'on duty' + 下班 'off duty' ->上下班 'on-off duty'.

Due to the reason that Chinese morphological rules are not as 'general' as their English counterparts, it is difficult to simply extend the well-known techniques from English (i.e., finite-state morphology) to Chinese. We use two different methods to solve those four morphological patterns: For morphological pattern of affixation, head particle and reduplication, we simply use the solution of extended lexicalization suggested in Gao (2003). On the other hand, morphological pattern of merging is a special form of abbreviating and is called 'morph-abbreviation'. The identification of 'morph-abbreviation' is integrated into our unified abbreviation identification model, and a new feature is employed:

Morph-abbreviation identification: For the target candidate with the consecutive character string of ABC, we first extend it to ACBC. This feature then returns 'TRUE' if both AC and BC are proved being lexicon words and 'FALSE' otherwise. In the case of returning 'TRUE', it is of large probability that ABC is a morph-abbreviation merged from two lexicon words, namely AC and BC.

4.2 Substring Analysis

Examination shows that some substrings of the abbreviations are incorrectly classified as 'abbreviation'. Those errors come from the traits of the conceptual formation of abbreviations. For instance, the substring '影厂' of the abbreviation '北影厂' is sharing the same conceptual structure with another factual abbreviation '师大': 'category/industry + entity-postfix'. As a result, it is possible that the substring '影厂' will get a high score during abbreviation formation analysis. In order to address this issue, a 'super-sequence diversity feature' is developed based on the 'diversity' difference between a factual abbreviation and its substrings.

First, using an illustration we briefly define 'left-minimum super-sequence' (LMS) and 'right-minimum super-sequence' (RMS): In the consecutive sequence of ABCD, ABC is the LMS for BC and BCD is the RMS for BC. For the overall occurrences of an abbreviation in the local text, their LMSs and RMSs tend to be inconsistent. Yet for the overall occurrences of a substring, either their LMSs or RMSs tend to keep the same form, and vice versa, in that the substring itself is not an independent term for denotation. Therefore, it is possible to develop a rule combining LMSs and RMSs for discriminating real abbreviations from their substrings:

Super-sequence diversity feature: Formally, this feature is scored by using *Div(A)*, and larger value of *Div(A)* would represent a larger degree of this special form of diversity:

$$Div(A) := \frac{type(LMS\_A) + type(RMS\_A)}{count(A)} \tag{4}$$

where the function *type(x)* represents the total kinds of inconsistent forms for *x*, and *count(x)* returns the overall occurrence-number of *x* on the document level.

4.3 Named-Entity Identification

We find that lots of noise comes from named entities, especially Chinese person names (CN) and transliterated foreign names (FN). The following heuristics are employed in our system to identify CNs and FNs inside the unknown sequences.

Chinese person names: As described in Gao (2003), Chinese PN consists of a family name F and a given name G, and is of the pattern F+G. Both F and G are of one or two characters long. We only consider PN candidates that begin with an F stored in the family name list (which contains 297 entries in our system). High frequency used G character were also stored in a given name list.

Transliterated foreign names: As described in Sproat (1996), FNs are usually transliterated using Chinese character strings whose sequential pronunciation mimics the source language pronunciation of the name. Since FNs can be of any length and their original pronunciation is effectively unlimited, the recognition of such names is tricky. Fortunately, there are only a few hundred Chinese characters that are particularly common in transliterations. Therefore, an FN candidate would be generated if it contains only characters stored in a transliterated name character list (containing 472 entries).

It should be emphasized that there is no feature dimension increase during integrating named-entity identification: it is employed only for candidate pruning.

5 Evaluation

Our experiment data comes from the People's Daily corpus (http://icl.pku.edu.cn). The selected data contains 20,063 sentences from 4,769 documents, which are divided into two sets: one from 3,146 documents, for training; and the other from 1,623 documents, for testing. The number of abbreviation tokens in the testing corpus is 4,941. The abbreviation tokens within the corpus have been manually annotated. The original corpus is already segmented, and in order to get unsegmented raw corpus we have completely erased the segmentation marks. In practice, a lexicon containing around 100K entries is used to segment the raw corpus. This lexicon contains pure words and there is no additional tag to indicate extra information. To keep our training and testing outcomes justifiable, all abbreviations from experimental corpus are removed from the lexicon.

The SVM model is employed for abbreviation disambiguation. The commonly used SVM model is a machine learning paradigm based on statistical theory. The SVM model calculates separating hyperplanes that maximize the margin between two sets of data points. While the basic training algorithm can only construct linear separators, kernel functions can be used to calculate scalar products in higher dimensional spaces.

To evaluate the performance of our system, we use F-measure. Based on the precision P and the recall R, the F-measure is defined as follows:

$$F = \frac{2 * P * R}{P + R} \quad (5)$$

5.1 Comparison of Kernel Functions

During tuning the SVM model[4], we select a linear function as the kernel function according to the experimental statistics, a part of which is shown in Table 1. It is

[4] We use the software SVM ^{light} (T. Joachims, 1999).

interesting to note that the linear kernel outperforms the Gaussian RBF kernel as well as the polynomial kernel, with the final F-measure of 73.7%. The reason might be the 'over fit' problem within the training of the RBF and polynomial kernels. Moreover, it should be emphasized that the linear kernel is efficient both in learning and classifying.

In order to deal with data sparseness problem, we discard those features occurring only once in the training data.

Table 1. Experimental results upon different SVM kernel functions[5]

Kernel Functions	P (%)	R (%)	F1 (%)	T-secs	C-secs
Linear Function	82.7	66.5	73.7	887	1
Radial Basis Function	83.2	65.3	73.2	16,550	517
Polynomial Function	82.4	66.1	73.4	14,863	473

5.2 Improvement Evaluation

We conducted incrementally the following four experiments:

(1) The SVM approach using features of abbreviation formation analysis together with context information analysis, which is selected as our baseline performance;
(2) Integrating the feature of morphological analysis (MA) into (1);
(3) Integrating the feature of substring analysis (SA) with (2);
(4) Integrating NE identification (NEI) with (3).

Both MA and SA will bring new features into our SVM classifier, while NEI is used only for candidate pruning. The details of incremental evaluation are shown in Table 2. As can be noticed, our baseline model reaches the F-measure of 64.0 %.

The integration of MA led to a slight better resulting F-measure. Primarily, the improvement results from the identification of morph-abbreviations. Unfortunately, we find that MA can sometimes make inaccurate identifications, which may undermine the improvement.

The system performance is also enhanced by the integration of the SA. However, this improvement is not as significant as our anticipation, the reason is that unfortunately some abbreviations occur only once throughout its context so that they can not be well discriminated from their substrings by the Super-sequence diversity feature employed by SA.

In our baseline model, the set of abbreviation candidates is large, and merely about 1/19 of them are real abbreviations. Thereby, the candidate pruning performed by NEI is crucial. In experiments, we found that by integrating the NEI, we not only achieved more efficient training and testing (in testing the abbreviation candidates are trimmed from 92,015 to 85,571 items), but also obtained significant higher F-measure. The

[5] *T-time* denotes *Training-time (s)*, and *C-time* denotes *Classifying-time (s)*. Experiments are performed on a 1.6G HZ CPU.

abbreviation recall rate increases from 55.5% to 66.5%. Its significant improvement is achieved through reducing the noise influence from NEs, in that the People's Daily Corpus is a news corpus, which tends to use a large number of Chinese person names and transliterated foreign names.

Table 2. The results of improvement evaluation

methods	Total Abbs	P (%)	R (%)	F1 (%)
Baseline	4,941	72.5	57.3	64.0
Baseline + MA		81.0	54.5	65.2
Baseline + MA + SA		81.4	55.5	66.0
Baseline + MA + SA + NEI		82.7	66.5	73.7

6 Conclusion and Future Work

In this paper, we proposed a supervised learning approach for automatic abbreviation identification in raw Chinese text. The definition-independent abbreviation identification problem is regarded as a classification problem in which an abbreviation candidates is classified into either 'abbreviation' or 'non-abbreviation' based on its posterior probability, and is integrated as part of a unified word segmentation model. The posterior probability is estimated by using abbreviation formation information and context information. It reaches an encouraging performance according to the experimental result upon the People's Daily Corpus.

Moreover, additional experiments further demonstrate the improvement after integrating with named entity identification (NEI), morphological analysis (MA) and substring analysis (SA). Morphological analysis helps the identification of morph-abbreviations, especially leads to a higher precision rate. The improvement of integrating SA is not as significant as our anticipation, because some abbreviations occur only once throughout its context so that the 'super-sequence diversity feature' becomes undiscriminating. On the other hand, significant improvement is obtained by integrating NEI.

In our future work, we will focus on fine-tuning our abbreviation formation model to further enhance its performance. Especially, we would like to investigate a more effective word clustering algorithm which enables a joint learning from both positive examples and negative examples, so that we can improve the quality of word classes and therefore generate more discriminating abbreviation-templates.

Acknowledgments

We would like to thank Galen Andrew for helpful suggestions on implementing word clustering algorithms and Sujian Li for helpful comments on earlier versions of this paper.

References

1. J.Chang, H.Schütze and R.Altman, Creating an online dictionary of abbreviations from MEDLINE, Journal of American Medical Information Association, 2002, 9(6), pp. 612-620.
2. Jianfeng Gao, Mu Li, and Changning Huang. Improved Source-channel Models for Chinese Word Segmentation. In Proceedings of the 41st Annual Meeting of Association for Computational Linguistics (ACL). July 8-10, 2003. Sapporo, Japan. pp. 272-279.
3. Jin-Shin Chang and Yu-Tso Lai. A Preliminary Study on Probabilistic Models for Chinese Abbreviations. In Proceedings of the Third SIGHAN Workshop on Chinese Language Learning, ACL, 2004, Barcelona, Spain, pp. 9-16.
4. Jian Sun, Jianfeng Gao, Lei Zhang, Ming Zhou and Chang-Ning Huang. Chinese Named Entity Identification Using Class-based Language Model. In Proc. of the 19th International Conference on Computational Linguistics, Taipei, 2002, pp. 967-973.
5. Katz, S.M. 1987. Estimation of Probabilities from Sparse Data for the Language Model Component of a Speech Recognizer. IEEE ASSP 35(3):400-401.
6. Och, Franz Josef. 1999. An efficient method for determining bilingual word classes. In EACL-99: Ninth Conference of the European Chapter of the Association for Computational Linguistics, pp. 71–76.
7. Richard Sproat, Chilin Shih, William Gale and Nancy Chang. 1996. A Stochastic Finite-State Word-Segmentation Algorithm for Chinese. Computational Linguistics. 22(3): 377-404.
8. Richard Sproat and Chilin Shih. 2002. Corpus-Based Methods in Chinese Morphology and Phonology. In: COLING-2002.
9. Schwartz, A. and Hearst, M. 2003. A simple algorithm for identifying abbreviation definitions in biomedical texts, Pacific Symposium on Biocomputing (PSB 2003), Kauai, Hawaii.
10. S.Martin, J.Liermann and H.Ney. 1998. Algorithms for Bigram and Trigram Word Clustering. Speech Communication, 24(1): 19-37, 1998.
11. Taghva, K. and Gilbreth, J. (1999), Recognizing acronyms and their definitions, International journal on Document Analysis and Recognition, pp. 191-198.
12. T. Joachims, Making large-Scale SVM Learning Practical. In: B. Schkopf and C. Burges and A. Smola (ed.), Advances in Kernel Methods - Support Vector Learning, MIT Press, 1999.
13. Yeates, S. (1999), Automatic extraction of acronyms from text. In Third New Zealand Computer Science Research Students' Conference, pp. 117-124.

Word Frequency Approximation for Chinese Using Raw, MM-Segmented and Manually Segmented Corpora*

Wei Qiao and Maosong Sun

National Lab. of Intelligent Technology & Systems,
Department of Computer Sci. & Tech.,
Tsinghua University, Beijing 100084, China
qiaow04@mails.tsinghua.edu.cn, sms@mail.tsinghua.edu.cn

Abstract. Word frequencies play important roles in many NLP-related applications. Word frequency estimation for Chinese remains a big challenge due to the characteristics of Chinese. An underlying fact is that a perfect word-segmented Chinese corpus never exists, and currently we only have raw corpora, which can be of arbitrarily large size, automatically word-segmented corpora derived from raw corpora, and a number of manually word-segmented corpora, with relatively smaller size, which are developed under various word segmentation standards by different researchers. In this paper we propose a new scheme to do word frequency approximation by combining the factors above. Experiments indicate that in most cases this scheme can benefit the word frequency estimation, though in other cases its performance is still not very satisfactory.

Keywords: word frequency estimation, raw corpus, automatically word-segmented corpus, manually word-segmented corpus.

1 Introduction

Word frequencies play important roles in many NLP-related applications, for example, TF in information retrieval. The estimation of word frequencies is easy for English whereas difficult for Chinese, because unlike English, there isn't spacing to explicitly delimit words in Chinese text. We therefore can not obtain word frequencies by simply counting word token occurrences in raw corpora.

Generally speaking, we need a 'perfect' (or, correct) manually word-segmented Chinese corpus to estimate word frequencies [5]. However, we face two fundamental difficulties. The first is that there exists serious inconsistency within/among manually segmented corpora, even when the same segmentation standard is

* The research is supported by the National Natural Science Foundation of China under grant number 60573187 and 60321002, and the Tsinghua-ALVIS Project co-sponsored by the National Natural Science Foundation of China under grant number 60520130299 and EU FP6.

Y. Matsumoto et al. (Eds.): ICCPOL 2006, LNAI 4285, pp. 256–267, 2006.

adopted for annotation. Due to the characteristics of Chinese word-formation [2,1], it is very tough to construct a 'fully' correct manually segmented corpus, although the definition of 'word' [14,13,11] seems very clear from the linguistic perspective. For example, a constituent, '猪肉', we can either consider it as a compounding word, *pork*, or consider it as a phrase consisting of two single-character words '猪'(pig) and '肉'(meat). Thus, the word frequency of '猪肉' could be pretty high if it is treated in the corpus in the former way, and could also be zero if treated in the latter way. The second difficulty is, according to Zipf's law, in order to obtain a statistically reliable word frequency estimation, even for a medium-sized Chinese wordlist, a balanced corpus with several hundred million characters, rather than several million characters, is required. But constructing such a huge manually segmented corpus is almost impossible, – it is both labor-intensive and time-consuming.

Since a 'perfect' manually segmented corpus is not feasible, (although the 'imperfect' manually segmented corpus is obviously useful for word frequency estimation), we have to in addition consider the possibility of making use of the following three types of corpora for the task here:

The first type is *'perfect' automatically segmented corpus*: Use a 'perfect' word segmenter to segment the corpus automatically, leading to a 'perfect' automatically segmented corpus. Then word frequencies can be easily estimated based on the corpus. Clearly, it would be ideal if a very powerful word segmenter is available [7]. Unfortunately, the state-of-the-art Chinese word segmenters are not satisfactory in performance. In the First International Chinese Word Segmentation Bakeoff in 2003 [8] organized by SIGHAN, the highest F-scores for word segmentation in the open test on four small-scale corpora were 95.9%, 95.6%, 90.4% and 91.2%, respectively. In the Second SIGHAN International Chinese Word Segmentation Bakeoff [3], the situation remains unchanged in nature, despite the minor increase in performance of word segmentation. A side-effect of such systems is that they try to solve segmentation ambiguities and recognize unknown words in context, producing a lot of unexpected inconsistencies in segmentation, which are obviously not favored by the task here.

The second type is *MM-segmented corpus*: Use 'Maximal matching'(MM), the most basic method for Chinese word segmentation, to segment the corpus automatically, then obtain the approximated word frequencies from the resulting corpus. [7] first used MM to handle large-scale texts. According to the direction of sentence scanning, MM can be further sub-categorized as forward MM (FMM) and backward MM (BMM). Experiments in [4] showed that MM is both effective and efficient (fast and easy to implement). [10] distinguished four cases in which FMM and BMM were both considered, and it provides a very strong evidence for supporting MM-based schemes to be reasonable estimations of word frequencies. Another advantage of MM-based schemes is their high consistency in word segmentation. The weak point of MM is that segmentation errors inevitably exist and when out-of-vocabulary words exist, the performance of MM will drop severely.

The third type is *raw corpus*: Use the frequency of a string of characters as an approximation (notice that we use the term 'approximation' here) of the word frequency of a constituent [9], which can be derived directly from any raw corpus. Obviously, its value is always larger than the value of word frequency for any word given a corpus. This scheme may over-estimate word frequencies seriously for some words (in particular for mono-syllabic words), but it has two good properties: the first one is that it is free from any kind of word segmentation errors; the second one is that this kind of corpus can be easily obtained and the size can be arbitrarily large.

According to the analysis above, for the task of word frequency estimation, a 'perfect' word-segmented corpus is ideal but, it doesn't exist, either manually or automatically - what we have are a variety of imperfect ones as well as raw corpora. Each type of corpora has its own advantages and drawbacks, so neither of them alone can fit the task of word frequency estimation. We have to consider a trade-off strategy which tries to utilize all the imperfect word-segmented corpora available so far, ranging from manually segmented corpora, MM-segmented corpora to raw corpora, and combine them to do sort of word frequency approximation, instead of word frequency estimation.

The remainder of this paper is organized as follows: Section 2 introduces the data set we used throughout the paper; Section 3 proposes the construct process of our trade-off scheme; Section 4 presents experiments to show the performance of the proposed scheme. And Section 5 concludes our work.

2 Data Set

In this section we introduce the corpora we used in our experiments throughout the paper.

First, two manually word-segmented corpora: The first one is the HUAYU corpus consisting of 1,763,762 characters, constructed by Tsinghua University and Beijing Language and Culture University. The second one is the BEIDA corpus consisting of 15,839,323 characters, constructed by Peking University. So the manually word-segmented corpora have totally 17,603,085 characters.

Second, the golden-standard corpus: We use a manually word-segmented corpus constructed by the National Institute of Applied Linguistics, denoted YUWEI, which contains 25,000,309 words with 51,311,659 characters. As the YUWEI corpus is sort of a noted authority and relatively large in size, we take it as golden-standard for our tests. An original wordlist is obtained from this corpus and the corresponding word frequencies can be obtained. We delete the words with frequency less than 4 from the original wordlist to form our final wordlist, which is denoted YWL and contains 99,660 entries.

Third, a raw corpus: We use a very large raw corpus, denoted RC, which contains 447,079,112 characters. Taking YWL as the wordlist, we obtain the frequency of a string of characters for each word from RC.

Fourth, MM-segmented corpora: In terms of YWL, we segment the raw corpus RC with FMM-segmenter and BMM-segmenter separately, resulting in two MM-segmented corpora.We denote them RC_FMM and RC_BMM respectively.

Thus in total, we have two moderate size manually-segmented corpora (HUAYU and BEIDA), one very large raw corpus (RC), two MM-segmented corpora (RC_FMM and RC_BMM), and a golden-standard corpus (YUWEI).

3 The Approximation Scheme

In this section we propose our trade-off scheme. In order to properly combine the five corpora which are of different size and different types, the combining process is organized in three steps. Firstly we combine the raw corpus and the two MM-segmented corpora. Secondly we combine the two manually segmented corpora. At last we combine the above two results and obtain the final approximation scheme. In the following, we introduce this step by step.

3.1 Combining Raw and MM-Segmented Corpora

From each of the three corpora: the raw corpus and the two MM-segmented corpora, we can obtain word frequency for each word $w_i (i = 1, 2, \ldots, 99660)$, respectively. We use the following symbols to clarify further descriptions:

$f_{FMM}(w_i)$: Word frequency of w_i obtained from RC_FMM.
$f_{BMM}(w_i)$: Word frequency of w_i obtained from RC_BMM.
$f_{RAW}(w_i)$: Frequency of a string of characters for w_i obtained from RC.

The work of [12] indicates that in the framework of MM, the average of $f_{FMM}(w_i)$ and $f_{BMM}(w_i)$ gives the best approximation of word frequencies for 1 to 4 character words. $f_{BMM}(w_i)$ is the best for 5 characters words, and $f_{RAW}(w_i)$ the best for words with word length 6 or above. We simply follow this conclusion here.

Using $F_{RFB}(w_i)$ to represent the result of word frequency approximation by jointly considering RC, RC_FMM and RC_BMM, we have:

For words with 1-4 characters:

$$F_{RFB}(w_i) = \frac{1}{2}\left[f_{FMM}(w_i) + f_{BMM}(w_i)\right] \quad (1)$$

For words with 5 characters:

$$F_{RFB}(w_i) = f_{BMM}(w_i) \quad (2)$$

For words with 6 or more than 6 characters:

$$F_{RFB}(w_i) = f_{RAW}(w_i) \quad (3)$$

This word frequency approximation scheme is called RFB.

3.2 Combining Manually Segmented Corpora

Having two manually segmented corpora HUAYU and BEIDA, we can obtain the word frequency for each word w_i in YWL. We denote the word frequency of (w_i) derived from these two corpora $f_{HUA}(w_i)$ and $f_{BEI}(w_i)$ respectively. We simply take the sum of the two values as the result of word frequency approximation in terms of these two manually segmented corpora, denoted $F_{HB}(w_i)$:

$$F_{HB}(w_i) = f_{HUA}(w_i) + f_{BEI}(w_i) \tag{4}$$

This word frequency approximation scheme is called HB.

3.3 Combining $F_{RFB}(w_i)$ and $F_{HB}(w_i)$

With two parts of combining results $F_{RFB}(w_i)$ and $F_{HB}(w_i)$, we come up with two problems:

The first one is, these two results come from the corpora with different sizes which are extremely unbalanced: one (HUAYU+BEIDA) is 17,603,085 characters while the other (RC) is 447,079,112 characters. So we can not directly combine the results from these two corpora. We thus need to introduce a parameter α to balance the corpus size.

It is naïve that we just take the ratio value of the two corpora size as the value of α. Later on, we will adjust α through experiments to receive the most appropriate value. At this stage, we just take the size ratio as the α value, so α=25.4.

We use C_0 to denote the total number of characters of the manually segmented corpora (HUAYU+BEIDA), and let C_1 denote the total number of characters of raw corpus (RC).

We expect to integrate the manually segmented corpora (HUAYU+BEIDA) and the raw corpus (RC) into a 'new' corpus of size $2C_0$. Thus the size of RC will be reduced to C_1/α . Accordingly the word frequency $F_{RFB}(w_i)$ should be changed to $F'_{RFB}(w_i)$:

$$F'_{RFB}(w_i) = F_{RFB}(w_i)/\alpha \tag{5}$$

In order to keep the whole corpus size to be $2C_0$ after integration, the final manually segmented corpus size, denoted C'_0, should be:

$$C'_0 = 2C_0 - C_1/\alpha \tag{6}$$

Thus the word frequency $F_{HB}(w_i)$ will in turn become $F'_{HB}(w_i)$:

$$F'_{HB}(w_i) = F_{HB}(w_i) \times \frac{2C_0 - C_1/\alpha}{C_0} \tag{7}$$

The second problem concerns an observation in Chinese, i.e., the smaller the word length is, the less reliable the approximation obtained from the raw corpus, and thus the larger the weight of the approximation result from $F_{HB}(w_i)$ should be. Here, we use a factor β as the weighting parameter. Experimentally, we set β as follows:

$$\beta = \begin{cases} 7 & \text{for one-character words} \\ 6 & \text{for two-character words} \\ 3 & \text{for three-character words} \\ 0 & \text{otherwise} \end{cases}$$

Taking the above two problems into consideration, and based on Equation 5 and Equation 7, the final word frequency approximation in terms of RC, RC_FMM and RC_BMM can be represented as:

$$F''_{RFB}(w_i) = F'_{RFB}(w_i) \times \frac{1}{1+\beta} = F_{RFB}(w_i) \times \frac{1}{\alpha(1+\beta)} \tag{8}$$

Correspondingly the final word frequency estimated by HUAYU and BEIDA should be:

$$F''_{HB}(w_i) = F_{HB}(w_i) \times \frac{2C_0 - \frac{C_1}{\alpha(1+\beta)}}{C_0} \tag{9}$$

Thus we get our final trade-off strategy, denoted $F_{RFB+HB}(w_i)$:

$$\begin{aligned} F_{RFB+HB}(w_i) &= F''_{HB}(w_i) + F''_{RFB}(w_i) \\ &= F_{HB}(w_i) \times \frac{2C_0 - \frac{C_1}{\alpha(1+\beta)}}{C_0} + F_{RFB}(w_i) \times \frac{1}{\alpha(1+\beta)} \\ &= F_{HB}(w_i) \times (1 + \frac{\beta}{1+\beta}) + F_{RFB}(w_i) \times \frac{1}{\alpha(1+\beta)} \end{aligned} \tag{10}$$

Note that for 4 or more than 4 characters words, β=0, thus in these cases Equation 10 reduces to Equation 11:

$$F_{RFB+HB}(w_i) = F_{HB}(w_i) + F_{RFB}(w_i) \times \frac{1}{\alpha} \tag{11}$$

This word frequency approximation scheme is called RFB+HB.

4 Experiments

In order to evaluate the performance of our trade-off scheme (RFB+HB), we conducted experiments from different perspectives. We compare this scheme with the other two schemes: the first one is the scheme using raw corpus and MM-segmented corpora (RFB), the second one is the scheme using only manually-segmented corpora (HB). Following experiments focus on these three schemes.

4.1 Perspective 1: The Spearman Coefficient of Rank Correlation

In terms of word frequencies derived from YUWEI, we can obtain a rank sequence for the 99,660 entries of YWL, denoted R_{YW}, which is in descending order of word frequencies. Similarly, we can also obtain a rank sequence for all these entries in terms of each of $F_{HB}(w_i)$,$F_{RFB}(w_i)$ and $F_{RFB+HB}(w_i)$, denoted R_{HB}, R_{RFB} and R_{RFB+HB} respectively. Every word w_i in YWL has its own rank numbers in R_{HB}, R_{RFB} and R_{RFB+HB}. We assign these rank numbers to w_i, with R_{YW} as a fixed index, resulting in three new rank sequences, denoted $R'_{HB}(w_i)$, $R'_{RFB}(w_i)$ and $R'_{RFB+HB}(w_i)$, accordingly.

Then we calculate the closeness between R_{YW} and each of $R'_{HB}(w_i)$, $R'_{RFB}(w_i)$ and $R'_{RFB+HB}(w_i)$, with R_{YW} as the standard rank sequence. We use the Spearman coefficient of rank correlation (SCRC) to measure the closeness between a pair of rank sequences over YWL, as given by:

$$SCRC \equiv 1 - 6 \sum_{i=1}^{99660} \frac{d_i^2}{N(N^2-1)} ,$$

where d_i is the difference between two rank numbers of w_i with respect to R_{YW} and R', N is the length of YWL, and R' is R'_{HB}, R'_{RFB} or R'_{RFB+HB}. Table 1 shows the values of SCRC(R_{YW}, R'), under $\alpha = 25.4$.

Table 1. SCRC values over YWL, under α=25.4

	(R_{YW}, R'_{HB})	(R_{YW}, R'_{RFB})	(R_{YW}, R'_{RFB+HB})
SCRC	0.675	0.704	0.732

The SCRC value of the proposed scheme is the biggest among the three, indicating that the rank sequence R'_{RFB+HB} is the closest to R_{YW} compared to the rank sequences R'_{HB} and R'_{RFB}.

We also conduct an experiment to determine the most adequate value for α regarding SCRC(R_{YW}, R'_{RFB+HB}).

Fig.1 shows that $\alpha = 6.5$ receives the highest SCRC value. So in the later experiments, we fix $\alpha = 6.5$.

To further observe the performance of the proposed scheme, we continue to carry out some experiments on subsets of YWL. Table 2 and Table 3 give the SCRC values for the top part of YWL with word frequencies ≥ 10 and ≥ 200 respectively.

In these two cases, the proposed scheme also outperforms the other two schemes.

The improvements of the proposed scheme compared to the other schemes over YWL under word frequencies ≥ 4, ≥ 10, and ≥ 200, are summarized in Table 4.

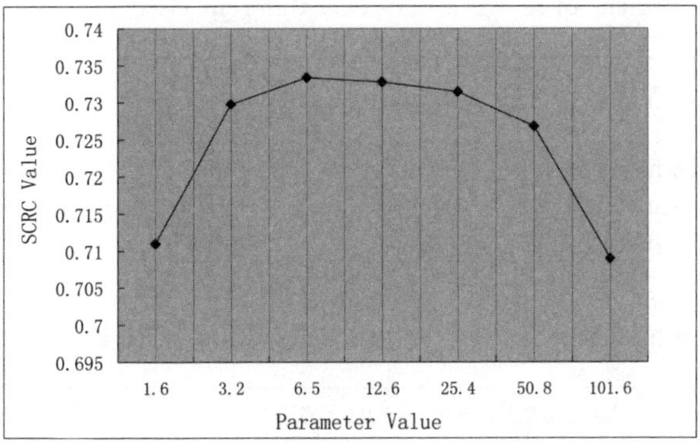

Fig. 1. SCRC value curve with respect to different α' values

Table 2. SCRC values over YWL for word frequency ≥ 10, under $\alpha = 6.5$

	(R_{YW}, R'_{HB})	(R_{YW}, R'_{RFB})	(R_{YW}, R'_{RFB+HB})
SCRC(word frequency≥10)	0.663	0.682	0.736

4.2 Perspective 2: Rank Sequence Deviation

Now we look at the performance of the proposed scheme in more detail, particularly its relationship with word length. We therefore define the rank sequence deviation $\sigma(R_{YW}, R')$ with respect to two rank sequences R_{YW} and R', $\sigma_{R'}$ for short, as $\sum_i |R'(w_i) - R_{YW}(w_i)|$(i over a subset of YWL), then calculate σ_{HB}, σ_{RFB} and σ_{HB+RFB}. The values of $(\sigma_{HB+RFB} - \sigma_{HB})/\sigma_{HB}$ and $(\sigma_{HB+RFB} - \sigma_{RFB})/\sigma_{RFB}$ present the varying rate of the σ value using our scheme compared to the other two schemes respectively, as listed in Table 5.

From Table 5 we can see, the proposed scheme receives the best results for 1 to 3 character words in YWL but for 4+ character words, it turns to be worse.

In order to further investigate the performance of our scheme, we divide the YWL words into three parts, i.e., high, medium and low frequency words. Fig.2 shows the coverage rate of top N frequent words to YUWEI.

Based on the coverage rate curve shown in Fig.2, we get the point HM to divide high and medium frequency words and the point ML to divide medium and low frequency words. Then we have:

High frequency words: Top 8,076 frequent words (1 $\sim$ HM), with word frequency > 281; Medium frequency words: the words from $8,077^{\text{th}}$ to $60,224^{\text{th}}$(HM

Table 3. SCRC values over YWL for word frequency ≥ 200, under $\alpha = 6.5$

	(R_{YW}, R'_{HB})	(R_{YW}, R'_{RFB})	(R_{YW}, R'_{RFB+HB})
SCRC(word frequency≥200)	0.680	0.708	0.771

Table 4. Improvement of SCRC values over different parts of YWL, under $\alpha = 6.5$

The part of YWL	No. of words	SCRC: $R'_{RFB+HB} - R'_{HB}$	SCRC: $R'_{RFB+HB} - R'_{RFB}$
Words with frequency≥4	99,660	0.057	0.028
Words with frequency≥10	68,100	0.073	0.054
Words with frequency≥200	10,528	0.091	0.063

Table 5. The comparison of rank sequence deviations with respect to word length

Word length	$\frac{\sigma_{HB+RFB} - \sigma_{HB}}{\sigma_{HB}}$	$\frac{\sigma_{HB+RFB} - \sigma_{RFB}}{\sigma_{RFB}}$	Is HB+RFB the best among three schemes?
1	-22.7%	-16.2%	√
2	-19.0%	-13.4%	√
3	-13.7%	-7.3%	√
4+	15.2%	17.5%	×

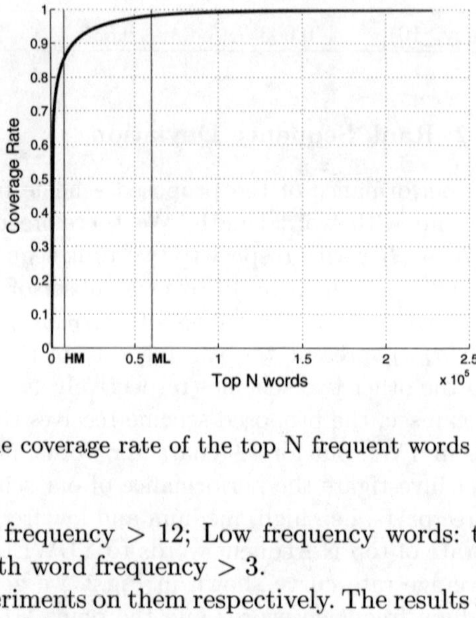

Fig. 2. The coverage rate of the top N frequent words to YUWEI

$\sim$ ML) with word frequency > 12; Low frequency words: the remained words (ML $\sim$ 99,660), with word frequency > 3.

Then we do experiments on them respectively. The results are given in Table 6, Table 7 and Table 8. We can see that in most cases, our scheme received the best results. But for the low frequency words, especially one character words and 4+ character words, the results turn to be worse.

4.3 Perspective 3: The Coverage Rate

We select the top 50,000 frequent words from R_{HB}, R_{RFB} and R_{HB+RFB}, then calculate the coverage rates of them over YUWEI. Table 9 gives the results.

In Table 9 we can see that the coverage rate of the proposed scheme increases 3.0% and 1.9% compared to HB and RFB respectively.

Table 6. The comparison of rank sequence deviations for high frequency words

	1 character words	2 character words	3 character words	4+ character words
$\frac{\sigma_{HB+RFB}-\sigma_{HB}}{\sigma_{HB}}$	-44.5%	-38.0%	-68.9%	-88.1%
$\frac{\sigma_{HB+RFB}-\sigma_{RFB}}{\sigma_{RFB}}$	-35.9%	-31.7%	-59.0%	-81.2%
Is HB+RFB the best among three schemes?	√	√	√	√

Table 7. The comparison of rank sequence deviations for medium frequency words

	1 character words	2 character words	3 character words	4+ character words
$\frac{\sigma_{HB+RFB}-\sigma_{HB}}{\sigma_{HB}}$	-33.0%	-14.5%	-7.5%	-13.4%
$\frac{\sigma_{HB+RFB}-\sigma_{RFB}}{\sigma_{RFB}}$	-18.1%	-7.3%	-9.1%	-10.0%
Is HB+RFB the best among three schemes?	√	√	√	√

Table 8. The comparison of rank sequence deviations for low frequency words

	1 character words	2 character words	3 character words	4+ character words
$\frac{\sigma_{HB+RFB}-\sigma_{HB}}{\sigma_{HB}}$	27.5%	-24.0%	-17.6%	49.1%
$\frac{\sigma_{HB+RFB}-\sigma_{RFB}}{\sigma_{RFB}}$	-3.1%	-10.9%	-6.2%	22.9%
Is HB+RFB the best among three schemes?	×	√	√	×

Table 9. Top 50,000 words coverage rate on YUWEI

Scheme	HB	RFB	$HB+RFB$
Coverage rate	94.1%	95.2%	97.1%

4.4 Sample Analysis

Now we choose R'_{HB} and R'_{HB+RFB} to make further comparison. Comparing against R'_{HB}, there are totally 57,024 words in R'_{HB+RFB} whose ranks are better adjusted (i.e., these ranks are closer to the standard sequence $R_{YW}(w_i)$ than their ranks in R'_{HB}), which we call positive samples; 42,619 words whose ranks are worse adjusted (i.e. these ranks are farther apart from the standard sequence $R_{YW}(w_i)$ than their ranks in R'_{HB}), which we call negative samples; 17 words have the same ranks in R'_{HB} and R'_{RFB+HB}. Table 10 and Table 11 show the

Table 10. The distribution of positive samples at different word frequency levels

Word frequency region	Total words	# of being better adjusted	Proportion
High frequency words	8,076	5,383	66.7%
Medium frequency words	52,148	28,399	54.5%
Low frequency words	39,436	23,242	58.9%

Table 11. The distribution of negative samples at different word frequency levels

Word frequency region	Total words	# of being worse adjusted	Proportion
High frequency words	8,076	2,679	33.2%
Medium frequency words	52,148	23,746	45.5%
Low frequency words	39,436	16,194	41.1%

distribution of positive samples and negative samples over different frequency regions (high, medium, and low), respectively.

Here we give some positive examples which are reasonably adjusted, such as '生物技术(biologic technology)', '知识经济(knowledge economy)', '信息高速公路(information thruway)' and '温室效应(greenhouse effect)'. These words have high frequency nowadays. When using our scheme, the ranks of this kind of words are properly adjusted ahead. We also give some negative examples, such as '周总理(Premier Zhou)', '中央红军(central red army)' and '西单商场(Xidan Market)'. These words are in rare use today, but our scheme made wrong decisions by adjusting them to higher rank positions, due to the fact that these words were frequently used historically, as reflected in RC, a very large raw corpus covering the linguistic phenomena of that time span more intensive than HUAYU, BEIDA, as well as YUWEI.

5 Conclusion and Future Work

In this paper we propose a trade-off scheme which jointly uses the raw corpus, MM-segmented corpora and manually segmented corpora to make approximation for word frequencies in Chinese. The experiments indicate that this new scheme can benefit the word frequency estimation, though in some cases it seems not very satisfactory, as indicated by '×' in Table 8. Besides, the experiments presented here are also very preliminary mainly due to the limited resources available. How to obtain a more accurate word frequency estimation for Chinese is still a big challenge.

References

1. Chen G.L.: On Chinese Morphology. In: Xuelin Publisher, Shanghai, (1994)
2. Dai X.L.: Chinese Morphology and its Interface with the Syntax. In: Ph.D Dissertation, Ohio State University, USA, (1992)

3. Emerson T.: The Second International Chinese Word Segmentation Bakeoff. In: Proceedings of the Third SIHAN Workshop on Chinese Language Processing. Jeju, Korea, (2005)
4. Liang N.Y.: CDWS: A Word Segmentation System for Written Chinese Texts. Journal of Chinese Information Processing. Vol. 1, No. 2, 44-52, (1987)
5. Liu E.S.: Frequency Dictionary of Chinese Words. Mouton and Co N.V. Publishers, (1973)
6. Liu K.Y.: Study on the Evaluation Technique for Word Segmentation of Contemporary Chinese. Applied Linguistics (Beijing). No. 1, 101-106,(1997)
7. Liu Y., Liang N.Y.: Counting Word Frequencies of Contemporary Chinese - An Engineering of Chinese Processing. Journal of Chinese Information Processing. Vol. 0, No. 1, 17-25, (1986)
8. Sproat R., Emerson T.: The First International Chinese Word Segmentation Bakeoff. Proceedings of the Second SIHAN Workshop on Chinese Language Processing. Sapporo, Japan, 133-143, (2003)
9. Sun M.S., Shen D.Y., T'sou B.K.Y.: Chinese Word Segmentation without Using Lexicon and Hand-crafted Training Data. Proceedings of 36th ACL and 17th COLING, 1265-1271, Montreal, Canada, (1998)
10. Sun M.S., T'sou B.K.Y.: Ambiguity Resolution in Chinese Word Segmentation. Proceedings of the 10th Pacific Asia Conference on Language, Information and Computation. Hong Kong, 121-126, (1995)
11. Sun M.S., Wang H. J. et al.: Wordlist of Contemporary Chinese for Information Processing. Applied Linguistics (Beijing), No. 4, (2001), 84-89
12. Sun M.S., Zhang Z.C., Benjamin KYT'sou., Lu Huaming.: Word Frequency Approximation for Chinese without Using Manually Annotated Corpus. In: Proceeding of 7th International Conference, CICLing 2006, Mexico, 105-116, (2006)
13. Tang T.C.: Chinese Morphology and Syntax. Vol. 3. Taiwan Student Publisher, Taipei, (1992)
14. Zhu D.X.: Lectures on Grammar. The Commercial Press, Beijing, (1982)

Identification of Maximal-Length Noun Phrases Based on Expanded Chunks and Classified Punctuations* in Chinese

Xue-Mei Bai[1], Jin-Ji Li[1], Dong-Il Kim[2], and Jong-Hyeok Lee[1]

[1] Department of Computer Science and Engineering, Electrical and Computer Engineering Division and Advanced Information Technology Research Center (AITrc), Pohang University of Science and Technology (POSTECH), San 31 Hyoja Dong, Pohang, 790-784, R. Korea
{xuemei, ljj, jhlee}@postech.ac.kr
[2] Language Engineering Institute, Department of Computer, Electron and Telecommunication Engineering, Yanbian University of Science and Technology (YUST), Yanji, Jilin, 133-000, P.R. China
dongil@ybust.edu.cn

Abstract. In general, there are two types of noun phrases (NP): Base Noun Phrase (BNP), and Maximal-Length Noun Phrase (MNP). MNP identification can largely reduce the complexity of full parsing, help analyze the general structure of complex sentences, and provide important clues for detecting main predicates in Chinese sentences. In this paper, we propose a 2-phase hybrid approach for MNP identification which adopts salient features such as expanded chunks and classified punctuations to improve performance. Experimental result shows a high quality performance of 89.66% in F_1-measure.

Keywords: Maximal-Length Noun Phrase (MNP), Expanded Chunk, Classified Punctuation.

1 Introduction

Text chunking is defined as dividing text into syntactically related non-overlapping groups of words which is often used as a pre-processing of full parsing problem [1]. Furthermore, NP identification is a challenging subtask in NLP because NPs are the basic components for conveying the meaning of sentences and appropriate translation units in machine translation. MNP refers to NPs which are not included in any other NPs. Specifically, MNP identification can largely reduce the complexity of full parsing problem, help analyze general structure of complex sentences, and provide important clues for detecting main predicates in Chinese sentences.

Usually, various constituents are included in MNP and there exists long dependency problem in detecting MNP, hence identification of MNP is more difficult than that of BNP. As an isolating language, Chinese has relatively few linguistic features for MNP identification. Hence, methods for selecting the features will be an important factor.

* The detailed explanation of Expanded Chunks and Classified Punctuations will be shown in Section 3.

Y. Matsumoto et al. (Eds.): ICCPOL 2006, LNAI 4285, pp. 268 – 276, 2006.

We propose an efficient 2-phase hybrid approach. In the first phase, detect the expanded chunks with heuristic rules and in the second phase, identify MNP with machine learning (ML) methods. Here, two salient features such as expanded chunks and classified punctuations are developed to improve performance.

The rest of the paper is organized as follows. Section 2 will briefly introduce the related works of MNP detection. Section 3 explains why we need expanded chunks and classified punctuations in our system. System architecture is described and experiment results are analyzed in Section 4 and 5, respectively. Finally, conclusion and future work will be given in Section 6.

2 Related Work

There are a few papers on the topic related to MNP identification [2, 3, 4, 5, 6, 7, and, 9]. After taking all the papers into consideration, we can conclude that MNP identification task is categorized into three types of method, such as rule-based, statistics-based and ML-based approach. Among them, the ML-based systems have relatively high performance and have become main stream [7 and 9]. Usually, features, such as POS and semantic codes are employed. Also the 2-phase ML method proves to be effective, which first detects BPs and then detects MNPs in [7].

The key points of related work are that base phrase (BP) chunking is very useful and ML-based approach is the main stream in MNP identification. However, a method for efficiently applying the BP information and deciding on what kind of features should be selected still remain as critical issues.

3 Why Expanded Chunks and Classified Punctuations?

As noted before, expanded chunks and classified punctuations are developed as effective features for detecting MNPs. The reasons are explained as follows.

3.1 Expanded Chunks

We put forward the definition of expanded chunks such as word groups which can be processed as one unit in MNP identification. Expanded chunks in Chinese include BPs, co-occurrence patterns, words surrounded by quotation marks or brackets, and words listed by slight pause marks. In other words, the expanded chunks are linguistic units which are definitely processed as one unit when identifying MNPs.

The usage of correctly identified expanded chunks will improve the efficiency of MNP identification, but there exists a risk when inappropriate expanded chunks are used. To effectively avoid the error propagation problem, we suggest a rule-based system to identify expanded chunks which guarantee high precision. The precision and recall of our rule-based system is 97% and 65%, respectively.

3.1.1 Base Phrases

Compared to English and other languages, definition of BP in Chinese is relatively ill-defined, although there are many researchers working on it. In this paper, we followed the BP definition of CoNLL-2000 shared task[1]. From 11 types of English BPs, PRT and UCP do not occur in Chinese sentences and there exists LP[2] which is a specific base phrase in Chinese. Hence, we defined 10 types of BPs in Chinese. Among these BPs, SBAR, PP, CONJP, INTJ, LST, and LP cannot provide useful clues in MNP identification, hence we will not take these BP information into consideration.

To maximize the usefulness of BP information, NP is sub-classified into common NP, proper NP, temporal NP, and quantifier NP and VP is sub-classified into common VP, copula VP, 'you(有)' VP, and adjective VP.

3.1.2 Co-occurrence Patterns[3]

Ex 1: [加工/NN 贸易/NN] [(在/P 广东/NR 外经济/NN 发展/NN 中/LC) 的/DEC 地位/NN][4]

In Chinese, there is a few co-occurrence patterns which are surrounded by a pair of words such as '在' and '中'. In [16], it is called Frame Phrase. Generally, a co-occurrence pattern could involve in an MNP or the inside of co-occurrence pattern could be considered as an MNP.

3.1.3 Words Surrounded by Quotation Marks or Brackets

Ex 2: (上海/NR 浦东/NR) 不/AD 是/VC 简单/VA 的/DEV 采取/VV [(“/PU 干/VV一/CD 段/M 时间/NN，/PU 等/P 积累/VV 了/AS 经验/NN 以后/LC 再/AD 制定/VV 法规/NN 条例/NN “/PU) 的/DEC 做法/NN]。 /PU

From the example, we can see that by pre-chunking the words surrounded by quotation marks, wider context information can be referenced by substituting the patterns by an MNP tag. Here, '做法' can reference the word '采取' which is far from the head word '做法' because the expanded chunks are handled as one unit.

3.1.4 Words Listed by Slight Pause Marks

Ex 3: (上海/NR 浦东/NR) (近年/NT) 来/LC 颁布/VV 实行/VV 了/AS [涉及/VV (经济/NN、 /PU 建设/NN 、/PU 规划/NN 、/PU 科技/NN 、/PU 文教/NN) 等/ETC 领域/NN 的/DEC 七十一/CD 件/M 法规性/NN 文件/NN] 。/PU

In this example, we pre-chunk the words listed by slight pause marks which play as a linguistic unit when detecting MNP.

---

[1] There are 11 types of BPs in English such as: NP, VP, PP, ADVP, SBAR, ADJP, PRT, CONJP, INTJ, LST, and UCP. The detailed description is in [8].

[2] LP is Location Phrase in Chinese.

[3] Chinese examples are extracted from Penn Chinese Treebank 4.0.

[4] Words surrounded by '[' and ']' is an MNP, and words surrounded by '(' and ')' belong to an expanded chunk. The POS tag information is described in [15].

3.2 Classified Punctuation

In linguistics, punctuations have different functions, but usually they are tagged in one POS tag such as ‘PU’ in Penn Chinese Treebank. Through corpus analysis, we found an interesting and useful distribution of Chinese punctuation. The result is shown in Table 1.

Table 1. Top 10 punctuation frequencies from the training corpus

Punctuation	Total Freq.	Inside MNP	Outside MNP	Rate of Outside MNP
Comma (,)	12, 695	404	12,291	96.82%
Period (。)	4,698	9	4,689	99.81%
Slight-pause mark (、)	2,725	2,306	419	15.38%
Brackets (「」)	1,666	1,217	449	26.95%
Question Mark (?)	308	12	296	96.10%
Semicolon (;)	302	10	292	96.69%
Colon (:)	191	9	182	95.29%
Quotation Mark (“ “)	163	144	19	11.66%
Exclamation Mark (!)	131	0	131	100%
Brackets(())	114	86	28	24.56%

Here, inside and outside the MNP indicate frequencies when punctuation is located in an MNP or out of an MNP, respectively. The rate of outside MNP shows a distinct tendency that punctuations could provide a meaningful clue in MNP identification. According to the functions of Chinese punctuation [14] and usages in Penn Chinese Treebank, we propose to classify Chinese punctuations into five groups as described in Table 2.

Table 2. Classified Chinese punctuation

Class	Punctuation
Group 1	Slight-pause Mark (、)
Group 2	Comma (,)
Group 3	Period (。), Question Mark (?), Exclamation Mark (!), Semicolon (;), Colon (:), ...
Group 4	Quotation Marks (“ ” ， ‘ ’ ， 《》， <>， 「」), Brackets ({}, ()), ...
Group 5	Hyphen (-), dash (--), apostrophe (‘), Slash Mark (/), dot (.), ...

4 System Architecture

As noted before, our system is a 2-phase hybrid approach. In the first phase, expanded chunks are identified by heuristic rules that can guarantee high precision. The second phase carries out ML-based MNP identification based on various linguistic features such as POS, expanded chunk, and classified punctuation.

To construct the MNP identification model, we employ IOBES tags which were introduced by [11]. Therefore, MNP identification can be transformed into a classification problem with IOBES tags. The detailed description is as follows.

Table 3. Types of IOBES tag in MNP identification

MNP Tag	Description
MNP_B	Current token is the start of an MNP consisting of more than one token.
MNP_E	Current token is the end of an MNP consisting of more than one token.
MNP_I	Current token is a middle of an MNP consisting of more than two tokens.
MNP_S	Current token is an MNP consisting of only one token.
MNP_O	Current token is outside of any MNP.

5 Experiment and Discussion

We used the Penn Chinese Treebank 4.0 as a training corpus. We selected 5,000 sentences in which the average length of sentences is about 32 words and contains 28,000 MNPs with the average length of 3 to 4 words. We employ a ML approach of Support Vector Machines (SVM) [13] and all of the experiments performed 10-fold cross validation. The performance is measured in precision, recall, and F_1-measure.

The experiments will focus on how effectively the 2-phase hybrid approach works and how useful the proposed linguistic features are. Also, optimal window size is discovered through experiments. Here, the window size is based on tokens which are segmented in Chinese word segmentation. The baseline system uses only POS as a feature without the classified punctuation information, which means that all of the punctuations are marked as 'PU' in the training corpus.

5.1 Using BPs and Expanded Chunks

Here, we will focus on whether the 2-phase hybrid approach is efficient or not and how useful the expanded chunks are. Table 4 shows the result of when pre-chunked BPs and expanded chunks are adopted as pre-processing in the first phase. The experiment result displays that expanded chunks are more useful than BPs. However, the improvement rate is not as high as we had expected. The reason lies in that our rule-based system has 97% of precision and 65% of recall. Low recall hinders more improvement in MNP identification.

Table 4. System performance when using BPs and expanded chunks

Window size	Baseline (F_1-Measure)	Baseline + BPs (F_1-Measure)	Baseline + Expanded Chunks (F_1-Measure)
3	84.02%	84.64%	86.01%
5	85.36%	86.02%	87.93%
7	86.65%	87.22%	88.36%
9	87.01%	87.52% (+0.51%)	88.42% (+1.41%)
11	87.01%	87.48%	88.36%

5.2 Using Classified Punctuations

As shown in Table 5, classified punctuations are proved to be powerful features which improve the performance by 2.57% when the window size is 9.

Table 5. System performance when using classified punctuation

Window size	Baseline (F_1-Measure)	Baseline + Classified Punctuations (F_1-Measure)
3	84.02%	85.90%
5	85.36%	88.33%
7	86.65%	89.14%
9	87.01%	89.58% (+ 2.57%)
11	87.01%	89.58%

5.3 Using Both Expanded Chunks and Classified Punctuations

Performance of each experiment converges when window size is 9. The following experiments are conducted with a fixed window size of 9. When expanded chunks and classified punctuations are both selected as features, performance is improved by 2.65% than the baseline model. Furthermore, the IOBES tag result is displayed in Table 7.

From Table 7, we can see that classification performance of MNP_B and MNP_S tag is relatively low than other tags. A more effective method is required to improve the identification of such tags.

Table 6. Performance of proposed approach (Window size = 9)

Window size	F_1-Measure
Baseline	87.01%
+ Classified Punctuations	89.58%
+ Expanded Chunks	89.66% (+2.65%)

Table 7. Performance of each IOBES tag of MNP (Window size = 9)

MNP Tag	Precision	Recall	F_1-Measure
MNP_B	84.48%	81.22%	82.99%
MNP_E	90.72%	93.52%	92.10%
MNP_S	86.35%	88.33%	87.33%
MNP_I	93.41%	87.66%	90.44%
MNP_O	93.19%	97.77%	95.43%

5.4 Error Analysis

To investigate which factors cause a main error, we report tag error distributions of IOBES that is classified by POS information in Table 8 and discover that the noun has the highest error rate. The result can be explained from two aspects. The first aspect is that noun could be tagged as one of the candidate tags such as MNP_B, MNP_E, MNP_I, and MNP_S, which further complicate the classification problem. The second is that Chinese is a topic-prominent and isolating language, which means that a topic or a subject could consist of MNPs without any inflections or functional words between them. Also, verb, adverb, and preposition appear less frequently in MNPs, but when they are included in MNPs, it is difficult to correctly detect them as components of MNPs. Here are two examples.

Table 8. Error distributions classified by POS information

POS	Freq.	Rate
Noun	5479	37.53%
Verb	2897	19.85%
Adverb	1263	8.65%
Preposition	993	6.80%
Others	3966	27.17%

Ex 4: [中国/NR 边境/NN 开放/NN 城市/NN] [经济/NN 发展/NN 成就/NN] 显著/VA 。/PU

Ex 5: [最/AD 常/AD 挂/VV 在/P 嘴/NN 上/LC 的/DEG 一/CD 句/M 话/NN] 是/VC……

In example 4, '中国边境开放城市' is a topic and '经济发展成就' is a subject and both of them are MNPs. However, the boundary between them cannot be detected by the features which we use currently. The fifth example contains adverbs such as '最' and '常' and these words do not appear frequently in an MNP. Our system cannot detect '最' and '常' as members of MNP.

Hence, these problems should reference more linguistic knowledge such as valency information and so on, and they remain as future works.

5.5 Comparative Experiments

To more fairly compare the performance of our proposed method, the comparative result is shown as Table 9. Our proposed system has the best performance.

Table 9. Comparative result with other systems

Methods	Precision	Recall	F_1-Measure
Yin [7]	77.1%	75.5%	76.3%
Zhou [2]	85.4%	82.3%	83.82%
Proposed Method	89.70%	89.70%	89.66%

We conduct a comparative experiment with the same method and the corpus used in [7] which shows a large improvement with our proposed system. Comparative result with [2] cannot be considered as an objective comparison because we used a different method and a corpus. At any rate, the average length of MNP is similar in both corpora; hence, we expect that it can also provide useful information.

6 Conclusion and Future Work

The proposed method shows a promising result of the 2-phase hybrid approach, which works well with effective features such as expanded chunks and classified punctuations.

Expanded chunks make it easy to reference wider context information by considering expanded chunks as one unit. However, it also has latent problem of error propagation. The results indicate that the rule-based system of expanded chunks has 97% of precision and 65% of recall. Therefore, a more efficient method is needed to improve the recall without reducing the precision. Classified punctuations are also proved to be a powerful feature. We have grouped the Chinese punctuations into five groups in consideration of their linguistic functions and usages in Penn Chinese Treebank.

From the error analysis, we can see that the noun is the main causing factor which produces classification errors, since it is possible that noun could be tagged as one of the candidate tags such as MNP_B, MNP_E, MNP_I, and MNP_S. As a topic-prominent and isolating language, topic and subject could be MNPs without any inflections or functional words between them. These problems should be resolved by referencing more complicated linguistic knowledge such as valency information and so on. Also, more elaborate post-processing is necessary to guarantee that the result can be used in other applications which adopt MNP identification as a pre-process. We will also apply different ML methods such as Conditional Random Field (CRF) in our system. Further research should be carried out to resolve these problems.

Acknowledgments. This work was supported by the Korea Science and Engineering Foundation (KOSEF) through the Advanced Information Technology Research Center (AITrc), and also partially by the BK 21 Project in 2006.

References

1. Steven P. Abney: Parsing by Chunks, In: Principle-Based Parsing, Kluwer Academic Publishers, Dordrecht (1991) 257--278
2. Zhou Qiang, Sun Maosong and Huang Changning: Automatically Identify Chinese Maximal Noun Phrase, Technical Report 99001, State Key Lab. of Intelligent Technology and Systems, Dept. of Computer Science and Technology, Tsinghua University (1998)
3. Didier Bourigault: Surface Grammatical Analysis for the Extraction of Terminological Noun Phrases, In: Christian Boitet ed. Proceedings of the 15th International Conference on Computational Linguistics (COLING 92), Nantes, France (1992) 977—981
4. Kuang-hua Chen, Hsin-His Chen: Extracting Noun Phrases from Large-Scale Texts: A Hybrid Approach and Its Automatic Evaluation, In: Proceedings of 32nd Annual Meeting of Association of Computational Linguistics, New York (1994) 234--241
5. Wenjie Li, Haihua Pan, Ming Zhou, Kam-Fai Wong and Vincent Lum: Corpus-based Maximal-length Chinese Noun Phrase Extraction, In: Key-Sun Choi ed. Proceedings of Natural Language Processing Pacific Rim Symposium (NLPRS'95), Korea (1995) 246--251

6. Angel S. Y. Tse, Kam-Fai Wong, & al.: Effectiveness Analysis of Linguistics- and Corpus-based Noun Phrase Partial Parsers, In: Key-Sun Choi ed. Proceedings of Natural Language Processing Pacific Rim Symposium (NLPRS'95), Korea (1995) 252--257
7. Changhao Yin: Identification of Maximal Noun Phrase in Chinese: Using the Head of Base Phrases, Master Dissertation, POSTECH, Korea (2005) (in Korean)
8. Erik F. Tjong Kim Sang, Sabine Buchholz: Introduction to the CoNLL-2000 Shared Task: Chunking, In: Proceedings of CoNLL-2000 and LLL-2000 (2000)127--132
9. Yongmei Tan, Tianshun Yao, Qing Chen and Jongbo Zhu: Applying Conditional Random Fields to Chinese Shallow Parsing, In: The Sixth International Conference on Intelligent Text Processing and Computational Linguistics (CICLing-2005), LNCS, Vol.3406, Springer (2005) 167--176
10. Erik F. Tjong Kim Sang, Walter Daelemans, Herve Dejean, Rob Koeling, Yuval Krymolowski, Vasin Punyakanok, and Dan Roth: Applying system combination to base noun phrase identification, In: proceedings of CoNLL-2000 (2000) 857--863
11. Taku Kudo and Yuji Matsumoto: Chunking with Support Vector Machines, In: Proceedings of Second Meeting of North American Chapter of the Association for Computational Linguistics (NAACL) (2001) 192--199
12. WEKA machine learning toolkit http://www.cs.waikato.ac.nz/~ml/
13. LIBSVM: Multi-Class Support Vector Machine Learning Toolkit http://www.csie.ntu.edu.tw/~cjlin/libsvm/index.html
14. Shui-Fang Lin: Study and Application of Punctuation (标点符号的学习和应用), People's Pulisher, P.R.China (in Chinese)
15. Penn Chinese TreeBank 4.0 http://www.cis.upenn.edu/~chinese
16. Ming Zhou: A Block-based Robust Dependency Parser for Unrestricted Chinese Text, In: Proceedings of the Second Chinese Language Processing Workshop (2000) 78--84

A Hybrid Approach to Chinese Abbreviation Expansion

Guohong Fu[1,2], Kang-Kuong Luke[2], Min Zhang[3], and GuoDong Zhou[3,4]

[1] Dept of Chinese, Translation and Linguistics, City University of Hong Kong, Hong Kong
[2] Department of Linguistics, The University of Hong Kong, Hong Kong
[3] Institute for Infocomm Research, Singapore 119613
[4] School of Computer Science and Technology, Suzhou University, China 215006
ghfu@hotmail.com, kkluke@hkusua.hku.hk,
mzhang@i2r.a-star.edu.sg, gdzhou@suda.edu.cn

Abstract. This paper presents a hybrid approach to Chinese abbreviation expansion. In this study, each short-form in Chinese text is assumed to be created by the method of reduction and the method of elimination or generalization, respectively. A mapping table between short words and long words and a dictionary of non-reduced short-form/full-form pairs are thus applied to generate the respective expansion candidates. Then, a hidden Markov model (HMM) based disambiguation is employed to rank these candidates and select a proper expansion for each ambiguous abbreviation. In order to improve expansion accuracy, some linguistic knowledge like discourse information and abbreviation patterns are further employed to double-check the expanded results and revise some error expansions if any. The proposed approach was evaluated on an abbreviation-expanded corpus built from the Peking University Corpus. The results showed that a recall of 83.8% and a precision of 86.3% can be achieved on average for different types of Chinese abbreviations.

Keywords: Chinese abbreviation expansion, hidden Markov models (HMMs), abbreviation disambiguation.

1 Introduction

Abbreviations (also referred to as short-forms of words or phrases) are widely used in current articles. Expanding abbreviation to their original full-forms plays an important role in improving information extraction and retrieval systems [1][2][3][4].

Over the past years, great progress has been achieved in English abbreviation resolution, and various methods have been proposed for the identification and expansion of abbreviation in English, including rule-based methods [1][2], statistically-based methods [3] and machine learning methods [4][5][6]. Yu et al (2002) presented a solution for mapping abbreviations in biomedical articles to their full-forms, which is based on a set of pattern-matching rules [1]. This method achieved an average recall of 70% and an average precision of 95% for the defined abbreviations in a test on four public abbreviation databases, namely GenBank Locuslink, SWISSPROT, LRABR of the UMLS Speciailist Lexiocn and BioBACUS. Toole (2000) described another rule-based method for expansion resolution [2]. In this study, different linguistic morphological features of the

Y. Matsumoto et al. (Eds.): ICCPOL 2006, LNAI 4285, pp. 277 – 287, 2006.

abbreviations and their expansions were taken into account and further incorporated for expansion resolution, including initial character match, final character match, word length, difference in length, consonant/vowel ratio and cluster code. It is found that these features help capture abbreviation candidates in English. More recently, some statistically-based methods and machine learning techniques have been applied to resolve English abbreviations. Terada et al (2004) presented a corpus-based method to automatic abbreviation expansion by using context and character information [3]. Their experiments on some 10,000 documents in the field of aviation showed that their method improved the accuracy of expansion by about 10% in compassion with previous rule or dictionary based methods. Gaudan et al (2005) applied support vector machines (SVMs) to disambiguate global abbreviations in Medline [4]. This disambiguation strategy achieved a precision of 98.9% and a recall of 98.2%. Yu et al (2003) also demonstrated that the incorporation of SVMs and the majority voting method is a promising technique for abbreviation disambiguation [5]. Pakhomov (2002) put forward a semi-supervised maximum entropy approach to abbreviation expansion in medical text and an accuracy of 89% was reported on a sample of 10,000 rheumatology notes [6].

However, the study of Chinese abbreviation expansion is still at its early stage. Only in recent years, have some studies been reported on the expansion of abbreviations in Chinese. Chang and Lai (2004) presented a HMM-based generation model for Chinese abbreviation generation and expansion [7]. Although their method achieved a reasonable performance their preliminary experiments, it is not effective for non-reduced abbreviations in Chinese. Lee (2005) proposed a rule-based approach to automatic Chinese abbreviation expansion [8]. In this study, this method can incorporate both contextual information and the underlying principals of abbreviation formations are combined to expand different types of abbreviations in Chinese. However, this method is not tested on a large-scale corpus. Moreover, the performance, particularly the recall of expansion is highly dependent on the coverage of rules for expansion.

In comparison with abbreviation resolution in English, it is relatively more difficult for a computer system to deal with abbreviations in Chinese. First, Chinese text has no explicit delimiters like space in English text to mark word boundaries. Therefore, word segmentation is usually the first step for Chinese abbreviation processing. Second, abbreviation is an important word formation in Chinese. It is difficult to distinguish an abbreviation from a normal word in Chinese. Third, Chinese abbreviations are produced with some complicated methods, including reduction, elimination and generalization [8][9]. Finally, the Chinese language lacks exterior morphological hints like capitalization for abbreviation resolution.

In this paper, we propose a hybrid approach to Chinese abbreviation expansion. In order to improve expansion recall, each abbreviation under expansion is assumed to be created by reduction, elimination or generalization, respectively. Based on this assumption, a mapping table of short-words and long-words and a dictionary of short-form/full-from pairs are applied to generate the respective expansion. Then, a hidden Markov model (HMM) based disambiguation is employed to rank theses candidates and select a proper expansion for each abbreviation. To improve expansion accuracy, some linguistic knowledge like discourse information and abbreviation patterns are further applied to double-check the expanded results and revise some error expansions if any. The proposed approach is also evaluated using an abbreviation-expanded corpus built from the Peking University (PKU) corpus [10]. The results show that most Chinese abbreviations can be correctly expanded by our system.

The rest of the paper is organized as follows: Section 2 presents a typology of abbreviations in Chinese text. Section 3 details the proposed method for Chinese abbreviation expansion. Section 4 introduces an abbreviation-expanded corpus built from the PKU corpus [10]. Section 5 gives the experimental results. Finally, we draw our conclusions in Section 6.

2 Abbreviations in Chinese

2.1 Methods for Creating Chinese Abbreviations

In general, Chinese abbreviations or short-forms are created using three major methods, namely reduction, elimination and generalization [8][9].

Reduction is the most popular method for creating Chinese abbreviations [9], which produces abbreviations by selecting one or more key morphemes from each constituent word of the original full-forms. For example, 香港 大学 (*The University of Hong Kong*) is abbreviated to 港大 by selecting the second morpheme of the first word 香港 (*Hong Kong*) and the first morpheme of the second word 大学 (*university*), respectively.

By the method of elimination, one or more constituent words of the original full-form are eliminated and the rest parts remain as an abbreviation. The eliminated parts are usually the modifiers or representing the natures of the original forms. For example, 清华 大学 (*Tsinghua University*) is conventionally abbreviated to 清华 (*Tsinghua*) by eliminating the second word 大学 (*university*) which indicates the nature of the full-form.

In the way of generation, abbreviations are created by generalizing parallel parts of their corresponding full-forms. For example, 三防 (*three preventions*) is an abbreviation for the combined phrase 防火、防盗、防交通事故 (*fire prevention, theft prevention and traffic accident prevention*). The idea of *prevention* is common among the three parts of the original expression, so it is being generalized.

2.2 Types of Abbreviations in Chinese

According to the above three methods, Chinese abbreviations can be classified into three types, namely reduced abbreviations, eliminated abbreviations and generalized abbreviations.

Given a full-form $F = f_1 f_2 \cdots f_m$ consisting of m constituent words and its corresponding abbreviation $S = s_1 s_2 \cdots s_n$ consisting of n component words, then the three types of Chinese abbreviations can be formally defined as follows:

If $n = m$ and s_i is the corresponding short-form of the constituent word f_i (namely for $i = 1$ to n, $s_i \in f_i$), then S is a reduced abbreviation. This means that each constituent word of a full-form should have remains in its short-form if it is abbreviated with reduction.

If $n < m$ and $\forall s_j \in F(1 \le j \le n)$, then S is an eliminated abbreviation. This implies that each component of an eliminated abbreviation should be remains of its original full-form after abbreviation.

If $n < m$ and $\exists s_j \notin F(1 \le j \le n)$, S is a generalized abbreviation. This means that some additional morphemes or words are usually needed to abbreviate an expression with generalization.

For convenience, the latter two types of abbreviations are called by a joint name in this paper, namely non-reduced abbreviations. Furthermore, the component words of a short-form are designated as short-words while the constituent words in a full-form are referred as long-words in this paper. With the above formal definitions, we can distinguish reduced abbreviations from non-reduced abbreviations.

3 The Method

3.1 Overview

Fig 1 illustrates an overview of our system for Chinese abbreviation expansion. As can be seen from this figure, our system expands a given abbreviation in context to its original full-form in three main steps:

(1) Generating expansion candidates: In this step, a set of expansion candidates are generated for each abbreviation under expansion. As discussed above, abbreviations in Chinese text may be created using three possible methods. However, we do not know exactly how a given abbreviation is created before expansion. To ensure every abbreviation can be expanded by our system, we assume that a given abbreviation could be created either by the method of reduction or by the method of non-reduction (viz. elimination or generalization). If the abbreviation is assumed to be produced by

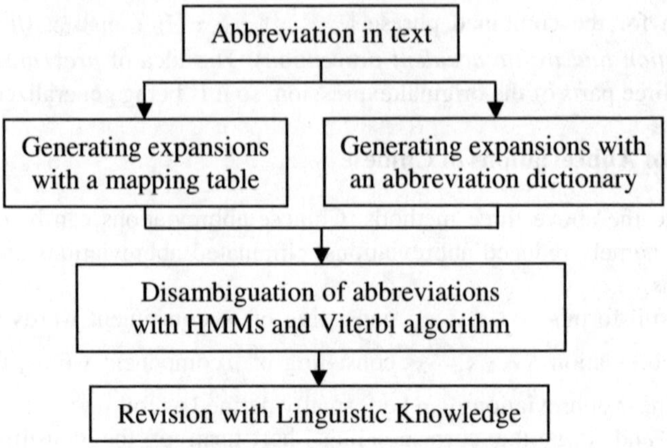

Fig. 1. Overview of the system for Chinese abbreviation expansion

reduction, a mapping table between short-words and long-words is used to generate expansion candidates. If the abbreviation is assumed to be created by elimination or generalization, then its expansion candidates will be generated by consulting a dictionary of short-form/full-form pairs.

(2) Disambiguation of abbreviations: The first step may come out with multiple expansion candidates for a given abbreviation in the first step. But only one is correct in a specific context. The task of the second step is therefore to find a proper expansion for an abbreviation in context. In view of expansion efficiency, a hidden Markov model disambiguation is developed in this paper to perform this task.

(3) Revision with linguistic knowledge: The third step is actually a post-processing of the second step. Actually, HMM-based disambiguation can only handle local contextual information for disambiguation, which is not effective for some ambiguous abbreviations. To address this problem, some further linguistic knowledge like discourse information is applied in the third step to double-check the results yielded by HMM-based disambiguation and revise the error expansion if any.

3.2 Generation of Expansion Candidates

3.2.1 Generating Expansion Candidates with a Mapping Table

The idea of generating expansion candidates using a mapping table is based on the characteristics of a reduced abbreviation. According to the formal definition in Section 2.2, each short-word within a reduced abbreviation must match a long-word in its original full-form. Given a reduced abbreviation, if the matching long-words of each short-word in it are known, its expansion candidates can be thus determined by combing exhaustively the relevant long-words. In this study, a mapping table like Table 1 is designed to map each short-word in reduced abbreviations to a set of long-words. Such a mapping table can be acquired automatically from a dictionary of reduced abbreviation/full-form pairs or an abbreviation-expanded corpus, in which all the abbreviations have been expanded to their respective full-forms.

The detailed process of generating expansions for a reduced abbreviation is as follows: Firstly, the reduced abbreviation is segmented into a sequence of short-words using a mapping table between short-words and long-words and a dictionary of

Table 1. A mapping table between short-words and long-words

Short-words	Long-words	English translation
阿	阿拉伯 | 阿尔巴尼亚 | 阿根廷 | …	Arabia | Albania | Argentina | …
工	工业 | 工程 | 工会 | 工作 | …	industry | engineering | labor union | work | …
大	大会 | 大学 | 大学生 | 大型 | …	meeting | university | undergraduate | large-scale | …
委	委员 | 委员会 | 委内瑞拉 | …	commissioner | committee | Venezuela | …
联	联邦 | 联合会 | 联合国 | 联盟 | …	federation | union | the United Nations | alliance | …
…	…	…

normal Chinese words. Then, each segmented short-word is mapped to a set of long-words by consulting the mapping table. In our current system, all the generated long-words and their matching short-words are stored in a lattice structure. Obviously, any combination of the relevant long-words forms an expansion candidate.

3.2.2 Generating Expansions with a Dictionary of Abbreviations

As defined in Section 2.2, if a full-form is abbreviated by the method of elimination or generalization, some words will be eliminated from it or some additional morphemes or words will be added to its short-form. Therefore, a one-to-one mapping relationship no longer exists between constituent words of a non-reduced abbreviation and the component words within its full-form. Due to this reason, the above mapping table based approach is not feasible for non-reduced abbreviations though it has a strong capacity of expansion generation.

In order to address this problem, a dictionary of abbreviation /full-form pairs is applied in our system, in which each non-reduced abbreviation is mapped to a set of full-forms. This dictionary can be manually compiled or automatically compiled from an abbreviation-expanded corpus. Table 2 presents some pairs of non-reduced abbreviations and their full-forms used in our system.

Table 2. A dictionary of non-reduced abbreviation / full-form pairs

Short-forms	Full-forms	English translation
清华	清华大学	Tsinghua Univeristy
三资企业	外商独资企业、中外合资企业、合作经营企业	Wholly foreign-owned enterprise, Chinese-foreign equity joint venture and cooperative joint venture in China
三通	通邮、通商、通航	the direct links in mail, transport and trade across the Taiwan Straits (viz. the Three Direct Links)
上交所	上海证券交易所	Shanghai Stock Exchange (SSE)
世贸	世界贸易组织	World Trade Organization (WTO)
…	…	…

3.3 Disambiguation of Abbreviations with HMMs

Given an abbreviation $S = s_1 s_2 \cdots s_m$ in context, let $F = f_1 f_2 \cdots f_n$ denote one of its expansion candidates. Where, $s_i (1 \le i \le m)$ stands for a short-word in the abbreviation and $f_i (1 \le i \le n)$ denotes a long-word in an expansion candidate. Based on hidden Markov models (HMMs), the task of abbreviation disambiguation is to find a proper expansion $\hat{F}$ that maximizes the following score

$$\hat{F} = \arg\max_F P(F \mid S) \approx \arg\max_F P(S \mid F) P(F) \quad (1)$$

Equation (1) presents a general HMM for abbreviation expansion, which consists of two parts: the abbreviation model $P(S \mid F)$ and the full-form model $P(F)$.

The abbreviation model reflects the conditional probability of a short-form S given a full-form F. If the short-form is a non-reduced abbreviation, and its expansion candidates are generated with the above dictionary of short-form/full-form pairs, the probability can be estimated directly from an abbreviation-expanded corpus using the maximum likelihood estimation technique. If the short-form is a reduced abbreviation and the relevant full-form is generated with the above mapping table, then the abbreviation model can be approximately calculated with the following formula

$$P(S \mid F) = P(s_1 s_2 \cdots s_m \mid f_1 f_2 \cdots f_m) = \prod_{i=1}^{m} P(s_i \mid f_i) \tag{2}$$

The full-form model reflects the probability that a full-form occurs. In this study, the n-gram language model is applied to approximate the full-form model as follows:

$$P(F) = P(Lf_1 f_2 \cdots f_n R) \approx P(f_1 \mid L) \times P(R \mid f_{n-N+2,n}) \times \prod_{i=1}^{n} P(f_i \mid f_{i-N+1,i-1}) \tag{3}$$

Where $P(f_i \mid f_{i-N+1,i-1})$ denotes an n-gram probability and N denotes the number of contextual words. It should be noted that the left context L and the right context R around the abbreviation are also taken into account in the full-form model for disambiguation. With a view to the serious data sparseness for high order models, we employ a bigram LM in this study, namely $N=2$. Let f_0 and f_{n+1} denote the respective words on the left and right of the abbreviation, we have $P(f_1 \mid L) = P(f_1 \mid f_0)$ and $P(R \mid f_{n-N+1,n}) = P(f_{n+1} \mid f_n)$. Thus, Equation (3) can be rewritten as

$$P(F) = P(Lf_1 f_2 \cdots f_n R) \approx \prod_{i=1}^{n+1} P(f_i \mid f_{i-1}) \tag{4}$$

3.4 Revising Error Expansions Using Linguistic Knowledge

In this study, the following two types of linguistic knowledge are applied to check whether the expansion yielded by the above procedure is correct for a given abbreviation in a specific context:

(1) The hypothesis of one sense per discourse.

The hypothesis of one sense per discourse is first introduced in the study of word sense disambiguation [11] and is further applied to the expansion of abbreviations in MEDLINE Abstract [5]. In this study, this hypothesis is employed to double-check results yielded by the above statistical expansion procedure and revise some error expansions if any.

By this hypothesis, a candidate will be selected as the resulting expansion for the abbreviation if it occurs in the text. For example, 伊 has two possible expansions, namely 伊朗 (*Iran*) and 伊拉克 (*Iraq*). But in a text like 安南/批准/伊拉克/食品/分配/计划/并/建议/增加/伊/石油/出口量/ (*Annan authorized the plan of Iraqi food distribution and suggested increasing the exportation of Iraqi oil*), the short-form 伊 is more likely to be abbreviated from 伊拉克 (*Iraq*), rather than 伊朗 (*Iran*).

(2) The format for defining abbreviations.

The formation for defining abbreviation has been widely used in the study of English abbreviation expansion [1]. In Chinese text, an abbreviation is commonly defined in the format "*<full-form>*（简称*<short-form>*）" or "*<full-form>*（以下简称*<short-form>*）, as illustrated in the text "...*全国科学技术名词审定委员会*（简称*名词委*）..." (*China National Committee for Terms in Sciences and Technologies*, CNCTST). Based on the format of abbreviation definition in Chinese text, we can find the full-form of a given abbreviation directly from the relevant article.

4 Building an Abbreviation-Expanded Corpus

In order to acquire the expansion knowledge bases such as abbreviation dictionary, mapping table and HMMs, we built an abbreviation-expanded corpus from the Peking University (PKU) Corpus [10], in which each abbreviation is mapped to its original full-form. Furthermore, each abbreviation and its full-form are segmented to a sequence of short-words and a sequence of full-words, respectively. It should be noted that the segmentation of a short-form is different from the normal lexical word segmentation, which is aiming at recognizing the remaining parts of each long-word of the full-form in the short-form after abbreviation. For example, the short form 常委会 (*standing committee*) is usually segmented to one word in lexical word segmentation. But, it is segmented to 常/委会/ in short-form segmentation because its full-form, i.e. 常务委员会 (*standing committee*) consists of two words 常务(*standing*) and 委员会 (*committee*). In this way, the mapping relationships between short-words and full-words can be easily identified.

The original PKU Corpus contains six month (viz. January to June in 1998) of news text from the People's Daily, which has been manually segmented and tagged with part-of-speech by the Institute of Computational Linguistics of the Peking University [10]. In this study, the first two month (viz. January and February) is chosen as our experimental corpus, in which all the abbreviations are manually expanded to their respective full-forms. It should noted that a special tag, i.e. "*j*" is specified to label abbreviations in the original PKU corpus. For convenience, we only consider these explicitly labeled abbreviations in our current evaluation.

Table 3 presents the distribution of different abbreviations in the experimental corpora. It is observed that Chinese news text is not rich in abbreviations. Among a total of more than one million words in January or February of the PKU corpus, there are only about ten thousand abbreviations. However, these abbreviations are widely distributed in different sentences. As shown in Table 3, more than 14% of sentences have abbreviation(s). Moreover, about 60% of abbreviations in the selected source data are observed to be reduced abbreviations. This demonstrates in a sense that reduction is a relatively more popular method for producing abbreviations in Chinese news text, in comparison with the other two methods, i.e. elimination and generalization.

Table 3. Number of abbreviations in the first two month of texts from the PKU corpus

Corpus	No. words	No. sentences	No. abbreviations		Total	No. sentences with abbreviations
			Reduced (%)	Non-reduced (%)		
For training (January)	1.12M	47,288	6,505 (62.6%)	3,883 (37.4%)	10,388	6,729 (14.2%)
For testing (February)	1.15M	48,095	6,655 (59.7%)	4,498 (40.3%)	11,153	7,137 (14.8%)

5 Evaluation Results and Discussions

We evaluate our system on the abbreviation-expanded corpus shown in Table 3. In particular, the first month is used for training while the second month is for open test. To measure the expansion performance, we calculate *recall* and *precision* in our evaluation. Here, recall (R) is defined as the number of correctly-expanded abbreviations divided by the total number of abbreviations under evaluation; Precision (P) is defined as the number of correctly-expanded abbreviations divided by the total number of abbreviations expanded automatically by the system.

Table 4. Experimental results for different methods

Measures		Unigram	Bigram	HMMs+Revision
Reduced abbreviation	R (%)	69.2	86.6	87.2
	P (%)	63.5	83.3	83.5
Non-reduced abbreviation	R (%)	82.1	82.3	83.1
	P (%)	83.2	83.9	84.5
Overall	R (%)	71.0	82.9	**83.8**
	P (%)	75.1	85.5	**86.3**

Table 4 presents the experimental results for different methods. In this experiment, two baseline methods, namely the unigram language model and the bigram language model, are introduced for comparison. It should be noted that both the mapping table and the abbreviation dictionary used in this experiment are extracted from the training corpus only. As can be seen from Table 4, our system can achieve a recall of 83.8% and a precision of 86.3% on average for different types of Chinese abbreviations if a bigram LM trained with the expanded corpus is applied. This indicates that our system is effective for a majority of abbreviations in Chinese text. The results in Table 4 also show that the proposed method performs better than the unigram LM and bigram LM, which indicates that using HMMs and linguistic knowledge like discourse information helps improving expansion performance.

Table 5. Experimental results for different mapping tables and abbreviation dictionaries

	Reduced abbr.		Non-reduced abbr.		Overall	
	R (%)	P (%)	R (%)	P (%)	R (%)	P (%)
The mapping table and the abbreviation dictionary are from training data only	87.2	83.5	83.1	84.5	83.8	86.3
The mapping table and the abbreviation dictionary are from both training and test data	87.5	84.7	93.4	94.3	89.1	88.6

Table 5 gives the experimental results for different mapping tables and abbreviation dictionaries of different sizes. In this experiment, the system is trained and tested using the expanded corpora for training and test, respectively. It is observed that the respective overall recall and precision are improved from 83.8% and 86.3% to 89.1% and 88.6% if all the abbreviations being tested are covered by the mapping table and the abbreviation dictionary. This indicates that a broad-coverage mapping table or abbreviation dictionary is of great value to correct expansion of Chinese abbreviations. However, acquiring such types of knowledge bases is still a challenge because abbreviations are dynamically produced in Chinese texts and their identification remains unresolved at present.

6 Conclusion

In this paper, we presented a hybrid approach to Chinese abbreviation expansion using HMMs and linguistic knowledge. Given an abbreviation in a Chinese text, it is assumed that this short-form may be created either by reduction or by non-reduction (viz. elimination or generalization). Based on this assumption, a mapping table between short words and long words and a dictionary of non-reduced short-form/full-form pairs are first applied to generate the respective expansion candidates. Then, a HMM-based disambiguation is employed to rank these candidates and select a proper expansion for each ambiguous abbreviation. In order to improve expansion accuracy, some linguistic knowledge like discourse information and abbreviation patterns are finally applied to double-check the expanded results and revise some error expansions if any. We evaluated our approach with an abbreviation-expanded corpus built from the Peking University corpus. The results showed our system achieved a recall of 83.8% and a precision of 86.3% on average for different abbreviations in Chinese text. In future, we hope to improve our system by exploring more features and acquiring dynamically an abbreviation dictionary for expansion disambiguation.

Acknowledgments. We would like to thank the Institute of Computational Linguistics of the Peking University for their corpus. This study was supported in part by Hong Kong Research Grants Council Competitive Earmarked Research Grant 7271/03H.

References

1. Yu, H., G. Hripcsak, and C. Friedman: Mapping abbreviations to full-forms in biomedical articles. Journal of American Medical Information Association, Vol. 9, No.3 (2002) 262-272
2. Toole, J.: A hybrid approach to the identification and expansion of abbreviations. In: Proceedings of RIAO'2000, (2000) 725-736
3. Terada, A., T. Tokunaga, and H. Tanaka: Automatic expansion of abbreviations by using context and character information. Information Processing and Management, Vol.40, No.1 (2004) 31-45
4. Gaudan, S., H. Kirsch, and D. Rebholz-Schuhmann: Resolving abbreviations to their senses in Medline. Bioinformatics, Vol. 21, No.18, (2005) 3658-3664
5. Yu, Z., Y. Tsuruoka, and J. Tsujii: Automatic resolution of ambiguous abbreviations in biomedical texts using support vector machines and one sense per discourse hypothesis. In: Proceedings of the 26th ACM SIGIR, Toronto, Canada (2003) 57-62
6. Pakhomov, S.: Semi-supervised maximum entropy based approach to acronym and abbreviation normalization in medical texts. In: Proceedings of the 40th Annual Meeting of the Association for Computational Linguistics, Philadelphia, USA (2002) 160-167
7. Chang, J.-S., and Y.-T. Lai: A preliminary study on probabilistic models for Chinese abbreviations. In: Proceedings of the 3rd SIGHAN Workshop on Chinese Language Processing, Barcelona, Spain (2004) 9-16
8. Lee, H.-W.: A study of automatic expansion of Chinese abbreviations. MA Thesis, The University of Hong Kong, (2005)
9. Yin, Z.: Methodologies and principles of Chinese abbreviation formation. Language Teaching and Study, No.2 (1999) 73-82
10. Yu, S., H. Duan, S. Zhu, B. Swen, and B. Chang: Specification for corpus processing at Peking University: Word segmentation, POS tagging and phonetic notation. Journal of Chinese Language and Computing, Vol. 13, No.2 (2003) 121-158
11. Gale, W.A., K.W. Church, and D. Yarowsky: One sense per discourse. In: Proceedings of the ARPA Workshop on Speech and Natural Language Processing, (1992) 233-237

Category-Pattern-Based Korean Word-Spacing

Mi-young Kang[1,2], Sung-won Jung[1,2], and Hyuk-chul Kwon[1,2]

[1] Pusan National University, Korean Language Processing Laboratory, Department of Computer Science Engineering
[2] Pusan National University, Center for U-Port IT Research and Education, Jangjeon-dong, Geumjeong-gu, 609-735, Busan, Korea
{kmyoung, swjung, hckwon}@pusan.ac.kr

Abstract. It is difficult to cope with data sparseness, unless augmenting the size of the dictionary in a stochastic-based word-spacing model is an option. To resolve both data sparseness and the dictionary memory size problem, this paper describes the process of dynamically providing candidate words to detect correct words using morpheme unigrams and their categories. Each candidate word's probability was estimated from the morpheme probability, which was weighted according to its category. The category weights were trained to minimize the mean of the errors between the observed probability of a word and that estimated by the word's individual morpheme probability weighted by its category power in a category pattern for producing the given word.

Keywords: Word-spacing, Category pattern, Morpheme unigram, Agglutinative morphology.

1 Introduction

Korean, like Western European languages, includes orthographic-words [8], which are defined by writing convention and separated by delimiters (spaces), in contrast to many Asian languages, such as Chinese and Japanese. The boundary of an orthographic word (hereafter referred to as a "word") offers clues for tokenization into syntactic words. Erroneous recognition of a word in Korean induces linguistic errors and ambiguities in lexical part-of-speech (POS) interpretation. Many rule-/knowledge- or stochastic-based approaches have been used in investigating automatic word-spacing systems. Among those, the stochastic approach has advantages in set-up time and cost savings, as well as the capability of coping with unregistered words. However, data sparseness is a challenge to the stochastic method. In order to cope with data sparseness, most of the previous stochastic studies on Korean word-spacing have adopted the syllable-N-gram-based method, though, in the epistemological view, it is rather unnatural compared to the word-N-gram-based method. Further, resolving data sparseness often induces an increase in stochastic dictionary size. Therefore, the present study aimed (a) to develop a stochastic word-spacing model based on the word recognition approach and (b) to resolve both data sparseness and increased memory size while considering Korean morphological and orthographical characteristics.

Y. Matsumoto et al. (Eds.): ICCPOL 2006, LNAI 4285, pp. 288 – 298, 2006.

This paper is organized as follows. Section 2 discusses previous studies undertaken to develop a word-spacing algorithm. Section 3 predicts a word's probability, used to detect the optimal word-spacing points, based on the morpheme unigram and its power weight in appearing in the category pattern for constituting a word, Section 4 discusses the experiments, and Section 5 offers concluding comments.

2 Related Studies and Discussion of Various N-Gram Approaches

In general, higher-order N-gram models are not used when there is not sufficient data [4]. On extracting enough word N-grams and applying the N-gram model, which uses the preceding word(s) to predict the next word, we can estimate a correct sentence using the contiguousness between words. However, this method is not easy in Korean due to the data sparseness produced by the typological characteristics of Korean: the predominance of agglutinative morphology (Suffixes can be attached successively to the ends of stems) and the remarkable productivity of the Korean word. Automatic word-spacing in Korean is more intricate than in Western languages, in which fewer morphemes for inflection simultaneously encode several meanings, because Korean words can be analyzed into several different morpheme strings that include different parts of speech.

2.1 Syllable-N-Gram-Based Approaches

Many Korean stochastic word-spacing studies have used syllable-N-gram-based models, because the syllable-N-gram approach easily augments the order of the N-gram model, thanks to the fact that there are a finite number of syllables in a language. The model in [2] used syllable-bigram information extracted from the raw corpus to estimate the space insertion possibility by considering the left-, right-, and inside-space probabilities. According to experiments carried out by [3], the model in [2] showed a word-unit precision of 76.71. And [3] obtained the best results using a syllable trigram-based approach. Word-spacing problems were treated using a hidden Markov model. Using the trigram model, which considered the current tag and previous two syllables, a word-unit spacing precision of 93.06% was obtained. These representative syllable-N-gram models can resolve Korean word-spacing problems to a certain degree, because syllable-N-gram statistics reflect the characteristics of the Korean writing system[1], its morphophonotactic restrictions, and its local syntactic dependency.

However, a meaningful performance can only be obtained by augmenting the N-gram order until a trigram model value is reached, which has important consequences for memory size (see Table 1). Moreover, using syllable-N-grams to predict "words" is less epistemologically acceptable than using word-N-grams, because word-spacing can be conceived as detecting word boundaries and not syllable boundaries, even if in some cases we find frequent syllable patterns due to phonotactic restrictions and local syntactic dependence.

---

[1] The Korean language has an alphabet composed of 24 letters and a syllabary composed of 19 consonants and 22 vowels arranged into squares as initial consonant(s), medial vowel(s), and final consonant(s), whereas other languages are arranged in horizontal or vertical rows.

2.2 Word-N-Gram-Based Approaches

As far as we are aware, the word recognition approach has not been adopted in any Korean stochastic word-spacing model except [1], whereas the word recognition approach has been widely used in rule-/knowledge-based approaches [6]. The word-spacing model in [1] was constructed using a word-unigram model while minimizing the N-gram order, to avoid a drastic augmentation of the parameters count, in spite of the drawback of using the word-N-gram-based approach. The probability of the words was estimated by the relative frequencies of the word unigrams extracted from the training data. In the model, the occurrence of a word is independent from that of another. Although the Korean language shows a free-word-order predominance, local syntactic dependencies do exist. The syllable patterns reflect these local dependencies. Thus, the odds favoring an inner-spacing probability of a syllable bigram at the boundary of the kth word in a sentence (w_k) compared with a no-inner-spacing probability, were included in the model to compensate for the lack of context with word unigrams. The value of the odds ($\sigma_{k,\lambda}$, $\sigma_{k+1,1}$) was assigned to the link between two consecutive words, w_k and w_{k+1}, which were composed of the syllable sequences $\sigma_{k,1}... \sigma_{k,\lambda}$ and $\sigma_{k+1,1}... \sigma_{k+1,\lambda}$, respectively. Thus, the optimum sentence is obtained by maximizing the probability of w_k and the odds favoring the inner-spacing probability of the syllable bigram $\sigma_{k,\lambda}$ and $\sigma_{k+1,1}$.

$$\textit{The optimum sentence} = \arg\max_{S} \prod_{k=1}^{n} p(w_k) \cdot odds(\sigma_{k,\lambda}, \sigma_{k+1,1}) \qquad (1)$$

Applying (1), [1] obtained a word-unit precision of 93.39% using 29.2 MB (word unigram + syllable bigram = 25.1 + 4.1 MB, respectively), which is a smaller memory size than the trigram model, which used 63.7 MB of memory, and obtained a word precision of 93.06% (see Table 1).

2.3 Resolving Data Sparseness and Memory Increase in Word-Unigram Model

Although the word-unigram-based model in [1] obtained a high performance with only smoothing using a syllable bigram, the unseen-words problem remained. Moreover, the word-unigram-based model still demands a significant memory size, though that memory size is smaller than that of the syllable trigram model. A combined model compensated for the data sparseness of the word-unigram-based model using the Korean Morphological Analyzer (KMA) implemented at Pusan National University in Korea.[1] This dynamically expanded the number of candidate words among the possible morphemes, using a longest-radix search strategy, and assigned them to a fixed heuristic probability value. Thus, the system obtained a similar performance using internal and external test data, with a precision word-unit correction of 98.39% and 97.51%, respectively, being observed. Data-sparseness-resolving methods exhibit highly efficient performance when dealing with unseen words. However, smoothing using the KMA technique could not prevent an increase in the dictionary memory size by 33.5 MB (word unigram + syllable bigram + KMA = 25.1 + 4.1 MB + 4.3 MB, respectively) (see Table 1).

Table 1. Statistics extracted from the training corpora in a previous study[2] and in the KMA

		Total No. of unities	Total memory size	
The training corpora in [1]	Word unigrams	33,643,884		
	Different word unigrams	1,950,068		25.1 MB
	Syllable bigrams	90,235,529		
	Different syllable bigrams	391,732	+ spacing tags	4.1 MB
	Syllable trigrams	84,239,729		
	Different syllable trigrams	5,116,404	+ spacing tags	63.7 MB
	Morpheme unigrams	68,771,225		
	Different morpheme unigrams	500,920		5.4 MB
KMA	Different morpheme entries	381,443	+ morphotactic rules	4.3 MB

3 A Category-Pattern-Based Word-Spacing Model

Our alternative method for resolving both (a) data sparseness and (b) the increase in memory size is based on the following definition and assumption.

Definition 1. Morphemes are the immediate constituents of a word.

Assumption 1. The word probability can be estimated from the individual morpheme(s)' probability in the word.

Starting with Definition 1 and Assumption 1, we can remove the stochastic word dictionary while introducing the morpheme N-gram in our word recognition model. If we can show that Assumption 1 is true, then (a) we can resolve the data sparseness because the system can more productively propose candidate words combining morphemes, and (b) we can reduce the dictionary memory size because the number of morphemes in a language is considerably smaller than the number of words. This section discusses how to predict a word using the factors that control the combination of individual morphemes in a real word, while not neglecting our ultimate aim, which is *to resolve both data sparseness and increased memory size while maintaining an epistemological approach, that is, word recognition for word-spacing*.

3.1 The Correlation Between the Observed Word and Its Category Pattern

In contrast to the syllable-N-gram approach, the morpheme-N-gram order is not easy to augment, because morphemes, although finite in number, are much more numerous than the number of syllables. Our alternative model was constructed by considering that *the occurrences of morphemes are independent of each other.* The probability of a morpheme was estimated by its *relative frequency*, which was the same as that of

[2] These statistics are drawn from two different newspaper articles and three years' worth of news broadcasting scripts. (See Ref. [1] for further description on the training corpora)

the word. Holding Definition 1 and Assumptions 1 and 2 to be true, the occurrence probability of a word, w_k, is proportional to the product of the probabilities of all of the individual morphemes in the word, w_k, as in (2).

$$p(w_k) \approx \prod_{i=1}^{\mu} p(m_{k,i}) \approx \prod_{i=1}^{\mu} \frac{f(m_{k,i})}{Tm} \tag{2}$$

μ = Total N° of morphemes in w_k.
$m_{k,i}$ = i^{th} morpheme in w_k.
$f(m_{k,i})$ = Frequency of the ith morpheme in w_k.
Tm = Total N° of morphemes extracted from the training data set

However, (2) does not reflect language reality. The combination of morphemes in a word performs under certain constraints. Each morpheme, as part of a word, refers to a grammatical category (i.e., the POS, see Table 2). The morphemes are strictly ordered according to their categories in a particular pattern in a word.

Definition 1 is reformulated as Definition 2 for considering the correlation among individual morpheme categories in a word, w_k. To fulfill our objective, we use an indirect method based on Assumption 2, while excluding category patterns in which only a single morpheme appears.

Definition 2. A morpheme constitutes a particular category pattern in a word alone, or with other morpheme(s).

Assumption 2. A morpheme participates as a category in a category pattern with a particular power for constituting a word.

When considering Definition 2 and Assumption 2, we can predict a word's probability based on the morpheme unigram and its power weight in appearing in the category pattern for constituting a word. In this work, we trained each morpheme's category power weight parameters within the category pattern.

3.2 Basic Concepts Involved in the Category-Pattern-Based Model Construction

As a prerequisite to a discussion on parameter fitting, we need to define some concepts involved in model construction. The exhaustive elements list of morpheme categories used in the model construction are listed in Table 2.

Remark 1. The categories that are generally accepted in Korean normative grammar are mostly used in our training as they are, but with some exceptions, which are grouped in a generalized category or not considered in model construction according to their linguistic characteristics and heuristics, as follows.

- All inflectional endings, except pre-endings, are grouped into one category, whereas derivational endings, including ETN and ETM, are considered as generalized categories.
- Not all prefixes and suffixes, except derivational suffixes including STV and STA, are considered as a generalized category, but they are included in a major category because they derive new words that are only semantically different and are less productive than derivational suffixes and inflectional suffixes.
- Verbs and adjectival verbs that mostly show the same morphosyntactic characteristics are not grouped into a general category, as there exists a clear morphological criterion for classifying them.

Table 2. The categories of Korean morphemes used in training

Category	Tag	Category	Tag
Noun, Cardinal number	N	Exclamation	III
Bound noun (General, Numeral)	BN	Postposition(Subjective, Objective, Modifier, Adverb, Vocative, Connective Adjunctive)	P
Pronoun(Personal, Demonstrative)	PN		
Ordinal number	NO	Ending(General, Conjunctive, Quotation)	END
Verb	V	Nominalization ending	ETN
Adjectival verb	VA	Modifier ending	ETM
Auxiliary verb	VX	Pre-ending	EPP
General modifier	MD	Verbalization suffix	STV
Numeral modifier	MDQ	Adjectivation suffix	STA
Adverb	ADV	Copula	COP

Let *CP* be the set of possible category patterns we could find within a Korean word. The word w_k is composed of $mc_{k,1} \ldots mc_{k,\mu}$, related to cp_j, an element of *CP*, as shown in (3).

$$
\begin{aligned}
& CP=\{cp_1,\cdots,cp_\beta\}, cp_j = cp_{j,1}\cdots cp_{j,\varepsilon}, \\
& if\,(w_k = m_{k,1}\ldots m_{k,\mu} \text{ and } mc_{k,1}\cdots mc_{k,\mu} = cp_j) \\
& \quad then\, mc_{k,i} = cp_{j,i}
\end{aligned}
\tag{3}
$$

$mc_{k,i}$ = Category of the ith morpheme in w_k
cp_j = Category pattern of $mc_{k,1} \ldots mc_{k,\mu}$, that constitute w_k

According to Assumption 2, the morphemes constitute a word under a particular category pattern, cp_j, with a correlation between those morphemes in the word. Each morpheme appears in a category pattern with a particular power. The estimated word probability based on the morpheme weighted according to its category power in the category pattern should be in direct proportion to the probability of the word.

$$
p(w_k) \approx \prod_{i>1}^{\mu} p(m_{k,i})^{wcp_{j,i}}
\tag{4}
$$

$wcp_{j,i}$ = Weight (i.e. Power) of the individual category, $cp_{j,i}$, in category pattern cp_j.

3.3 Parameter Fitting Using Simulated Annealing

This subsection discusses the training of parameter $wcp_{j,i}$ necessary to satisfy (4). To fit the parameter $wcp_{j,i}$, *real-word lists according to each* cp_j (CPWL) were extracted from the corpora in [1], which are tagged with POS tags. A total of 383 types of CPWL were extracted. Each CPWL was sorted by word-frequency rank, and sample

Table 3. The optimum parameters for the different patterns containing more than one morpheme, according to frequency rank[3]

Category Pattern (cp_j)		Parameters				Error Mean	Error Standard Deviation
		$wcp_{j,1}$	$wcp_{j,2}$	$wcp_{j,3}$	$wcp_{j,4}$		
cp_1	N + PP	0.81	0.87			2.94e-07	3.42e-06
cp_2	V + END	0.73	0.81			7.02e-07	1.16e-05
cp_3	V + ETM	0.77	0.62			1.68e-06	2.06e-05
cp_4	MDQ + BN	0.51	0.73			4.12e-06	8.10e-05
cp_5	N + STV + END	0.64	0.38	0.70		2.58e-07	1.80e-06
...	...	...	...	...	...	...	...

data sets for training (SCPWLs) were constructed by selecting one word every 10 words from each CPWL in order to save time during training. When the number of words in the CPWL was less than 1,000 words (about 82 CPWLs in all), they were all sampled in SCPWLs. (For an illustration of the process of parameter fitting, see Fig. 3). The optimum parameters according to each pattern cp_j were fitted by applying a simulated annealing algorithm to these training-sample data sets. The *hill-climbing algorithm* was adopted in order to fit the parameters. However, the *hill-climbing algorithm* was less economical than *simulated annealing* because it always chose the parameters with the best value, and thus could possibly become trapped in a local maximum, thus requiring many restarts. In contrast, the simulated annealing algorithm tolerated the choosing of the worst value and was thus protected from any effect of *local maxima*.

The parameter-fitting training algorithm shown in (5) adjusted $wcp_{j,(1...n)}$ so as to minimize the mean of the error between the observed probability of a word, as in (1), and the word probability estimated from the morpheme probability, which was weighted with a category weight, $wcp_{j,(1...n)}$,in the category pattern cp_j.

$$ErrorMean(cp_j) = \left[\sum_{\gamma=1}^{Tcp_j} \left| p(w_\gamma) - \prod_{i=1}^{\mu} p(m_{\gamma,i})^{wcp_{j,i}} \right| \right] \Big/ Tcp_j$$

$$\text{if } mc_{\gamma,1} \cdots mc_{\gamma,\mu} = cp_j, \arg\min_{wmcp_{j,i}} \left(\left[\sum_{\gamma=1}^{Tcp_j} \left| p(w_\gamma) - \prod_{i=1}^{\mu} p(m_{\gamma,i})^{wcp_{j,i}} \right| \right] \Big/ Tcp_j \right) \quad (5)$$

Tcp_j = Total number of training category patterns (cp_j).

[3] Patterns containing a single category, such as *N, ADV, BN, MDQ, MD, PN*, and *III*, corresponded exactly to a word. Thus, they were not considered in our training of the category weights of the morpheme, while assigning to them a relative frequency as a word probability without a weighted probability of a morpheme.

3.4 Word Prediction with Morpheme Unigram and Category Power

Equation (1) determining the word probability can be replaced by the morpheme probability with a category weight $wcp_{j,i}$ as in (6). Thus, the word-unigram-based word-spacing model in (1) can be rewritten as a category-pattern-based word-spacing model, as in (7).

$$EP(cw_k) = \begin{cases} if\ mc_{k,1} \cdots mc_{k,\mu} = cp_j & \prod_{i=1}^{\mu} p(m_{k,i})^{wcp_{j,i}} \\ otherwise & 1/Tw \end{cases} \quad (6)$$

$$The\ optimal\ sentence =$$

$$\arg\max_{S} \sum_{k=1}^{n} \{\log(EP(cw_k)) + \log(odds(\sigma_{k,\lambda}, \sigma_{k+1,1}))\} \quad (7)$$

Remark 2. Figure 1 illustrates the mean of the error and the standard deviation of each category pattern obtained from parameter-fitting training. For effective visualization, the standard deviation is plotted in units of 10. Based on the data shown in Fig. 1 in which the two results follow the same flowing, decreasing curve, we can understand that the higher the mean of the error and the standard deviation, the greater the average frequency of different words. For example, the average frequency of different words (total number of words/total number of different words) was 1069.33 occurrences for the *cp16* <VX+ETM>, which had the highest standard deviation, and 13.48 occurrences for the *cp1* <N + PP> in SCPWLs.

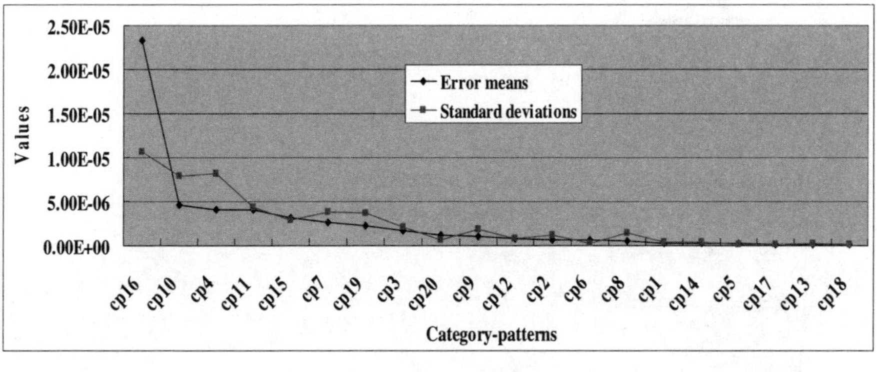

Fig. 1. The mean of the error and the standard deviation per category pattern

4 Experimentation

In this section, the category-based model is tested on data sets provided in a previous study. [1]

Table 4. The test data suite

	No. of sentences	No. of words	No. of syllables
Internal	2,000	25,020	103,196
External	2,000	17,191	52,688

Key: (a) Internal = test data extracted at the same distribution ratio as a given corpus over the whole training data; (b) External = external test data extracted from the POS-tagged corpus by the Electronics and Telecommunications Research Institute (ETRI) in Korea

Table 5. The F-measure of the category-pattern-based Model (in %)

	Model A: with SWD, *Odds ()*	**Model B** ($wcp_{j,l}$; *Odds ()* = 1)
Internal	95.45	**96.08**
External	89.96	**96.89**

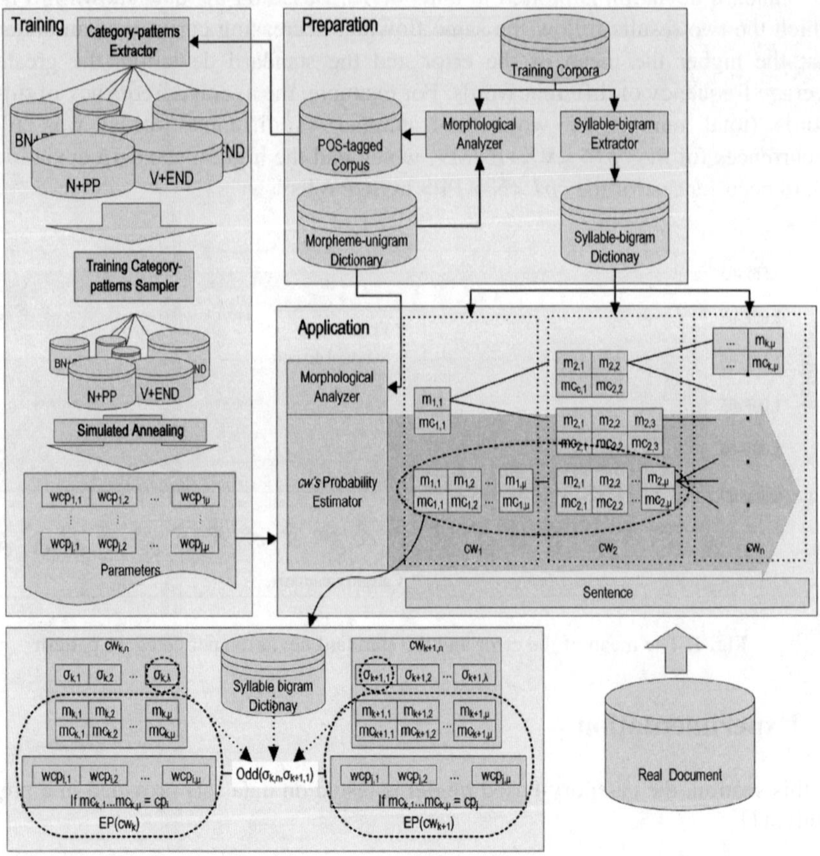

Fig. 2. The process of category-pattern-based word-spacing system implementation

Improving the recall and improving the precision are antagonistic. The more a system splits an input sentence into words, the more the recall value increases, whereas the less a system splits an input sentence into words, the higher the precision value becomes. Therefore, to obtain an optimum word-spacing system, we need to improve the F-measure of correctly spaced words, which improvement combines both recall and precision.

In Table 5, Model A represents the word-unigram-based model using SWD and *odds()*. Model B represents the word-unigram-based model with the morpheme-unigram dictionary, $wcp_{j,i}$ and *odds()* assigned by 1, which represents a single application of the category-patterns-based algorithm. According to the results, the system shows a comparably higher performance with SWD for internal data than for external data. However, it fared less well without SWD, and again showed a similar performance using external data with a trained category weight, $wcp_{j,i}$.

5 Conclusions and Further Work

This study resolved data sparseness by producing abundant candidate words based on morpheme unigrams and category weights, and reduced the dictionary memory size by using a morpheme-unigram dictionary instead of a word-unigram dictionary. We obtained a similar F-measure for external data and with trained category weights. Thus, in this paper, we showed that a morpheme participates as a category in a category pattern with a particular power for constituting a word in Korean. Furthermore, this approach still maintains an epistemological approach because word recognition proceeds through category patterns composed of the immediate constituent(s) of a word. Our work trained the weight for a category in a category pattern, and thus the system was robust toward unseen words; however, the opposite effect sometimes occurred because the weight was not trained for each morpheme. The same trained category weight value was applied to each morpheme probability when it appeared in the same category. Thus, exceptions in the category patterns appeared. We intend to investigate this point further in future work by treating the morphemes producing those exceptions as a particular category.

Acknowledgements. This work was supported by the National Research Program under Grant M10400000332-06J0000-33210.

References

1. Kang, M.Y., Yoon, A.S., Kwon, H.C.: Combined Word-Spacing Method for Disambiguating Korean Texts. Lecture Notes in Computer Science, Vol.3339(2004), 562–573
2. Kang, S.S., Woo, C.W.: Automatic Segmentation of Words Using Syllable Bigram Statistics. Proceedings of the 6th Natural Language Processing Pacific Rim Symposium (2001), 729–732
3. Lee, D.G., Lee, S.Z., Lim, H.S., Rim, H.CH.: Two Statistical Models for Automatic Word spacing of Korean Sentences. Journal of KISS(B): Software and Applications, Vol. 30(4)(2003), 358–370

4. Manning, C.D., Schütze, H.: Foundations of Statistical Natural Language Processing. MIT Press, Cambridge, MA, USA (2001)
5. Shim, K.S.: Automated Word-Segmentation for Korean using Mutual Information of Syllables. Journal of KISS(B), Vol. 23(1996), 991–1000
6. Sin, H.CH., :A Study of Word-spacing using Morphological Analysis. Korean Linguistic 12, Vol. 12(2000), 167–185
7. Sproat, R, Shih, c., Gale, W., Chang, N.: A Stochastic Finite-State Word-Segmentation Algorithm for Chinese. Computational Linguistics, 22(3)(1996), 377-404
8. Tsai, C.-H.: Word identification and eye movements in reading Chinese: A modeling approach. Doctoral thesis, University of Illinois at Urbana-Champaign, IL, USA, (2001)

An Integrated Approach to Chinese Word Segmentation and Part-of-Speech Tagging

Maosong Sun, Dongliang Xu, Benjamin K. Tsou[1], and Huaming Lu[2]

National Lab. of Intelligent Tech. & Systems, Tsinghua University, Beijing 100084, China
[1] Language Information Sciences Research Centre, City University of Hong Kong
[2] Beijing Information Science and Technology University, Beijing 100085, China
sms@tsinghua.edu.cn, rlbtsou@cityu.edu.hk, huaminglu@sohu.com

Abstract. This paper discusses and compares various integration schemes of Chinese word segmentation and part-of-speech tagging in the framework of true-integration and pseudo-integration. A true-integration approach, named 'the divide-and-conquer integration', is presented. The experiments based on a manually word-segmented and part-of-speech tagged corpus with about 5.8 million words show that this true integration achieves 98.61% F-measure in word segmentation, 95.18% F-measure in part-of-speech tagging, and 93.86% F-measure in word segmentation and part-of-speech tagging, outperforming all other kinds of combinations to some extent. The experimental results demonstrate the potential for further improving the performance of Chinese word segmentation and part-of-speech tagging.

Keywords: Chinese word segmentation, part-of-speech tagging, integration, smoothing.

1 Introduction

In the last decades, enormous efforts have been made to Chinese word segmentation and part-of-speech tagging. However, the state-of-the-art system performance for the both is still not very satisfactory in general. For example, in the First International Chinese Word Segmentation Bakeoff in 2003 organized by SIGHAN [1], the highest F-measures for word segmentation in the open test on four small-scale corpora were 95.9%, 95.6%, 90.4% and 91.2%, respectively. In the Second SIGHAN International Chinese Word Segmentation Bakeoff [2], the situation remains unchanged essentially, despite the minor increase in performance of word segmentation.

It has been observed that, Chinese word segmentation (including segmentation ambiguity resolution and unknown word processing) in some cases needs the assistance of higher level linguistic processing to some extent, at least at the morph-syntactic level [3, 4, 5]. Since part-of-speech tagging techniques have been relatively mature, the idea of combining word segmentation with part-of-speech tagging comes with apparent ease. Researchers have found that these two processes can be

Y. Matsumoto et al. (Eds.): ICCPOL 2006, LNAI 4285, pp. 299 – 309, 2006.

integrated, leading to an improved performance in both word segmentation and part-of-speech tagging [5, 6].

In general, there are two different ways of integration. [5] presented a method which consists of three main steps: (1) generate the N-best word sequences for an input sentence in terms of word frequencies and sentence length; (2) perform part-of-speech tagging for each of the N-best word sequences, achieving the N-best tag sequences accordingly; and (3) use a weighted score that mixes the factors of (1) and (2) to choose the 'best' solution from the N-best word sequences, along with its 'best' tag sequence. Note that step (1) and (2) are in fact carried out sequentially: this is therefore sort of pseudo-integration. A similar approach was exploited in [7]. [6] suggested a true-integration: (1) take into account all possible word sequences of a given sentence; (2) continue to expand each word sequence into all of its possible tag sequences, producing a 'unified' candidate space of word segmentation and part-of-speech tagging for the sentence; and (3) find the optimal path over such a space using dynamic programming in the framework of Hidden Markov Model (HMM), yielding a solution for word segmentation and part-of-speech tagging at the same time. It is worth stressing that in this approach, step (1) and (2) are done simultaneously.

[8] demonstrated the distinction between these two strategies by the following example:

(1) 他俩儿谈恋爱是从头年元月开始的。

There are two possible segmentations:

a. --- 是 | 从头 | 年 | 元月 | ---
V ADV TIME-CLASSIFIER TIME-N

b. --- 是 | 从 | 头年 | 元月 | ---
V PREP TIME-N TIME-N

The strategy of pseudo-integration would prefer segmentation (1a) because the probability of word sequence '从头+年' is greater than that of '从+头年' in terms of word frequencies, whereas the strategy of true-integration would prefer segmentation (1b) because the probability of tag sequence 'V + PREP + TIME-N + TIME-N' is likely to be greater than that of 'V + ADV + TIME-CLASSIFIER + TIME-N'.

[9] went ever farther by integrating word segmentation with full syntactic parsing of whole sentences. They believed that this strategy will be beneficial for resolving some segmentation ambiguities. For example:

(2) 在这些企业中国有企业有十个。

Two possible segmentations can be found for it:

a. 在 | 这些 | 企业 | 中 | 国有 | 企业 | 有 | 十 | 个 | 。
PREP DET N D ADJ N V NUM CLASSIFIER

b. 在 | 这些 | 企业 | 中国 | 有 | 企业 | 有 | 十 | 个 | 。
PREP DET N N V N V NUM CLASSIFIER

The probability of the tag sequence '…+N+V+N+V+…' for segmentation (2b) is likely to be greater than that of '…+D+ADJ+N+V+…' for segmentation (2a) according to part-of-speech N-grams in Chinese. So the strategy of [6, 8] may fail to handle this sentence. However, segmentation (2b) could be rejected when it fails to yield a full parse tree.

Generally speaking, the higher the level of linguistic analysis adopted in a computational model, the more powerful the disambiguation capability of this model, and the more expensive the realization of the model. The strategy of [9] does not seem to be feasible due to the fact that a parser capable of dealing with unrestricted Chinese texts is impossible in the near future.

Our basic idea on the issue of integration is fundamentally in line with [6, 8]. Under this framework, the integration is obtained at the morpho-syntactic level, a level adequate for the task here considering computational effectiveness and feasibility.

[8] reported an experiment on 729 sentences with 10,734 Chinese characters, in the condition that the experiment is free of unknown words. The training set is a manually word-segmented and part-of-speech tagged corpus in the news domain, consisting of about 70 thousand Chinese characters. For the sake of description, we use FMM-Seg to stand for segmentation using 'forward maximal matching', AP to stand for 'all-possible-segmentations', POSBigram-Tagging to stand for 'part-of-speech tagging with a Bigram model', 'x || y' to stand for that x and y are performed simultaneously, and 'x → y' to stand for x and y are performed sequentially (This convention is effective throughout this paper). In their experimental results, compared with 'FMM-Seg → POSBigram-Tagging', 'AP || POSBigram-Tagging' has a 1.31% improvement in word segmentation, a 1.47% improvement in part-of-speech tagging, and a 2.69% improvement in total performance for the combined processes, showing the reasonableness of integration.

A limitation of the work of [8] is its experimental scale is too small. Furthermore, it is necessary to study various ways of integration, including but not limited to the methods discussed in [8], as well as other related key factors (for example, to do word segmentation using word N-grams, instead of merely using the simplest FMM in the pseudo-integration). [10] proposed a true-integration scheme named 'divide-and-conquer integration'. The scheme has achieved some promising results. But this work is preliminary, -- there still are some room for improvement. Here, we shall follow that scheme and try to improve it further. Experiments are carried out on PDA9801-06, a manually word-segmented and part-of-speech-tagged news corpus of the People's Daily from January to June 1998, constructed by Institute of Computational Linguistics, Peking University (By the way, PDA9801-06 is public available from that institute). Its size is over 100 times larger than the corpus used by [8]. About 80% of PDA9801-06 (with 5,826,338 words) is randomly selected as the training set to learn parameters for the related models, and the remaining 20% (with 1,460,487 words) as the test set.

2 Alternative Approaches to Integration and Experimental Comparisons

We define three sets of metrics for measuring the performance of the proposed schemes:

(1) The precision, recall and F-measure of word segmentation, denoted P-WS, R-WS and F-WS respectively.

$$P-WS = \frac{Number-of-words-correctly-segmented}{Number-of-words-segmented}$$

$$R-WS = \frac{Number-of-words-correctly-segmentated}{Total-number-of-words-in-the-test-set}$$

$$F-WS = \frac{2 \times P-WS \times R-WS}{P-WS + R-WS}$$

(2) The precision, recall and F-measure of part-of-speech tagging, denoted P-PT, R-PT and F-PT respectively.

$$P-PT = \frac{Number-of-words-correctly-segmented-and-tagged}{Number-of-wrods-correctly-segmented}$$

$$R-PT = \frac{Number-of-words-correctly-segmentated-and-tagged}{Number-of-words-correctly-segmentated} = P-PT$$

$$F-PT = \frac{2 \times P-PT \times R-PT}{P-PT + R-PT} = P-PT = R-PT$$

(3) The precision, recall and F-measure of word segmentation and part-of-speech tagging combined, denoted P-WSPT, R-WSPT and F-WSPT respectively.

$$P-WSPT = \frac{Number-of-words-correctly-segmented-and-tagged}{Number-of-words-segmented}$$

$$R-WSPT = \frac{Number-of-words-correctly-segmentated-and-tagged}{Total-number-of-words-in-the-test-set}$$

$$F-WSPT = \frac{2 \times P-WSPT \times R-WSPT}{P-WSPT + R-WSPT}$$

Obviously, the following equations hold:

$P\text{-}WS \times P\text{-}PT = P\text{-}WSPT$

$R\text{-}WS \times R\text{-}PT = R\text{-}WSPT$

$F\text{-}WS \times F\text{-}PT = F\text{-}WSPT$

In all the experiments throughout this paper, the following basic strategy is employed for almost all the proposed schemes: in the process of word segmentation, all possible word sequences will be generated by AP for an input sentence; and, in the process of part-of-speech tagging (if any), all possible tag sequences for all possible word sequences will be generated. In cases where we need to find the best path from either the space consisting of word sequences or the space consisting of tag sequences, dynamic programming will be used to accomplish the task, which is in polynomial time complexity.

2.1 Segmentation Using Word N-Grams

[8] adopted FMM for word segmentation in the pseudo-integration. However, this scheme is too simple. So we explore some more sophisticated schemes here.

(1) Segmentation using 'the forward maximal matching' (denoted FMM-Seg)

(2) Segmentation using 'all-possible-segmentations', and using word unigrams to find the best word sequence (denoted AP&WordUnigram-Seg)

$$W_0 = \arg\max_W P(W) \approx \arg\max_W \prod_i P(w_i) \tag{1}$$

(3) Segmentation using 'all-possible-segmentations', and using word bigrams to find the best word sequence (denoted AP&WordBigram-Seg) :

$$W_0 = \arg\max_W P(W) \approx \arg\max_W \prod_i P(w_i \mid w_{i-1}) \tag{2}$$

It is expected that word bigrams will be more powerful than word unigrams.

(4) Segmentation using 'all-possible-segmentations', and using word bigrams to find the best word sequence, along with word unigram smoothing (denoted AP&WordBigram-Seg+WordUnigram-Smoothing):

$$W_0 = \arg\max_W P(W)$$

$$\approx \arg\max_W \left(\prod_i (P(w_i \mid w_{i-1}) \times (1-a) + P(w_i) \times a)\right)$$

a=0.70 by experiment. (3)

The word bigram model will encounter a serious data sparseness problem. A word unigram back-off would be a natural choice for smoothing.

(5) Segmentation using 'all-possible-segmentations', and using word bigrams, along with part-of-speech bigram smoothing (denoted AP&WordBigram-Seg+POSBigram-Smoothing):

$$W_0 = \arg\max_{W} P(W)$$

$$\approx \arg\max_{W,T} \prod_i (P(w_i \mid w_{i-1}) \times (1-a) + P(t_i \mid t_{i-1}) P(w_i \mid t_i) \times a)$$

a=0.77 by experiment. (4)

Smoothing here is somehow unusual, working in a similar way to part-of-speech tagging based on the bigram model.

Experimental results for the above five word segmentation schemes are given in Table 1.

Table 1. Word segmentation with word N-grams

Segmentation schemes	P-WS(%)	R-WS(%)	F-WS(%)
FMM-Seg	93.90	95.65	94.77
AP&WordUnigram-Seg	97.23	98.45	97.83
AP&WordBigram-Seg	97.40	96.85	97.12
AP&WordBigram-Seg+ WordUnigram-Smoothing	98.21	98.76	98.48
AP&WordBigram-Seg+ POSBigram-Smoothing	98.25	98.80	98.53

We regard FMM-Seg as the baseline for word segmentation. As can be seen in Table 1, all WordUnigram-based and WordBigram-based schemes significantly outperform FMM-Seg. The performance of AP&WordBigram-Seg without smoothing is even worse than that of AP&WordUnigram-Seg, indicating that smoothing is definitely necessary for word bigrams. The two kinds of smoothing, AP&WordBigram-Seg+ WordUnigram-Smoothing and AP&WordBigram-Seg+POSBigram-Smoothing, show 3.71% and 3.76% improvement in F-WS, respectively, as compared to the baseline. AP&WordBigram-Seg+POSBigram-Smoothing is the best among the five schemes.

2.2 Integration Schemes in the Conventional Framework

In the experiments below, the following bigram part-of-speech tagging model is used (denoted POSBigram-Tagging):

$$T_0 = \arg\max_{T} P(T \mid W) \approx \arg\max_{T} \prod_i P(t_i \mid t_{i-1}) P(w_i \mid t_i) \quad (5)$$

We try six integration schemes under the conventional framework:

(1) FMM-Seg → POSBigram-Tagging

This simplest pseudo-integration will serve as the baseline.

(2) AP&WordUnigram-Seg → POSBigram-Tagging

This pseudo-integration is similar to the work of [5].

(3) AP || POSBigram-Tagging

$$(W_0, T_0) = \arg\max_{W,T} P(W,T) = \arg\max_{W,T} \prod_i P(t_i \mid t_{i-1}) P(w_i \mid t_i) \quad (6)$$

This is same as the best scheme given in [8].

(4) AP&WordBigram-Seg → POSBigram-Tagging

(5) AP&WordBigram-Seg + WordUnigram-Smoothing → POSBigram-Tagging

(6) AP&WordBigram-Seg + POSBigram-Smoothing → POSBigram-Tagging

Note that the left hand side of this scheme has the highest performance for word segmentation. It thus deserves special attention.

Experimental results are summarized in Table 2.

Table 2. Comparison of various integration schemes

	Integration scheme	P-WS (%)	R-WS (%)	F-WS (%)	P-PT= R-PT= FPT(%)	P-WSPT (%)	R-WSPT (%)	F-WSPT (%)
1	FMM-Seg → POSBigram-Tagging	93.90	95.65	94.77	95.18	89.37	91.04	90.20
2	AP&WordUnigram-Seg → POSBigram-Tagging	97.23	98.45	97.83	95.21	92.57	93.73	93.15
3	AP || POSBigram-Tagging	97.60	98.58	98.09	95.24	92.95	93.89	93.42
4	AP&WordBigram-Seg → POSBigram-Tagging	97.40	96.85	97.12	95.18	92.71	92.18	92.44
5	AP&WordBigram-Seg+ WordUnigram-Smoothing → POSBigram-Tagging	98.21	98.76	98.48	95.18	93.48	94.00	93.74
6	AP&WordBigram-Seg+ POSBigram-Smoothing → POSBigram-Tagging	98.25	98.80	98.53	95.20	93.53	94.05	93.79

AP&WordBigram-Seg + POSBigram-Smoothing→ POSBigram-Tagging (scheme 6) outperforms all the other schemes. Although it appears to be a sequential integration at the first glance (as signaled by '→'), it is actually a semi true-integration, because the smoothing here has incorporated part-of-speech tagging into the process of word segmentation. Compared with the baseline (scheme 1), scheme 6 has a 3.59% improvement in F-WSPT. Note also that AP-Seg || POSBigram-Tagging (scheme 3, a typical true-integration) is better than scheme 1 and scheme 2 (two typical pseudo-integrations) by 3.22% and 0.27% in F-WSPT respectively.

2.3 Divide-and-Conquer Integration

In principle, word bigrams are more precise and thus more powerful than part-of-speech bigrams, as long as the data sparseness problem is overcome. Formulae 3 and 4 provide two alternative ways for smoothing. Here, we propose a further alternative.

Let w_{i-1} and w_i be two adjacent words in a possible word sequence for a given sentence, and $f(w_{i-1})$ and $f(w_i)$ be frequencies of w_{i-1} and w_i in the training corpus. We distinguish five cases in terms of $f(w_{i-1})$ and $f(w_i)$:

(1) Both $f(w_{i-1})$ and $f(w_i)$ are high

The contribution of w_{i-1} and w_i to the probability estimation of the word sequence can be accounted for directly by the bigram of these two words, because the data is sufficient:

$$P(w_i \mid w_{i-1}) \tag{7}$$

We need to set a frequency threshold η_u for this (experimentally, η_u =4,750).

(2) Otherwise

(2.1) $f(w_{i-1}\ w_i)$ is quite high

The bigram of w_{i-1} and w_i is supposed to still be quite reliable. Formula 7 will also work in this case.

We need to set another threshold η_b (experimentally, η_b=3).

(2.2) Otherwise

(2.2.1) $f(w_{i-1})$ is high, but $f(w_i)$ is low (according to η_u)

The contribution of w_i will become unreliable if we continue to use formula 7.

Thus we replace w_i with its part-of-speech, i.e., back off from a pure word bigram to a bigram consisting of a word and a part-of-speech, so as to make the related transition probability more reliable:

$$P(t_i \mid w_{i-1})P(w_i \mid t_i) \tag{8}$$

(2.2.2) $f(w_{i-1})$ is low, but $f(w_i)$ is high (according to η_u)

The contribution of w_{i-1} will become unreliable if we continue to use formula 7. Thus we replace w_{i-1} with its part-of-speech, i.e., back off from a pure word bigram to a bigram consisting of a part-of-speech and a word:

$$P(w_i \mid t_{i-1}) \tag{9}$$

(2.2.3) Both $f(w_{i-1})$ and $f(w_i)$ are low (according to η_u)

We should completely back off from a word bigram to a part-of-speech bigram, because the data sparseness problem is very serious:

$$P(t_i \mid t_{i-1})P(w_i \mid t_i) \tag{10}$$

In essence, cases 2.2.1, 2.2.2 and 2.2.3 reflect some degree of smoothing. We name this strategy 'divide-and-conquer integration', as summarized in Table 3.

Table 3. The divide-and-conquer integration

<table>
<tr><th colspan="3">Case (η_u =4,750, η_b=3)</th><th>Contribution of w_{i-1} and w_i</th></tr>
<tr><td colspan="3">$f(w_{i-1}) \geq \eta_u$ and $f(w_i) \geq \eta_u$</td><td>$P(w_i \mid w_{i-1})$</td></tr>
<tr><td rowspan="4">Otherwise</td><td colspan="2">$f(w_{i-1}\ w_i) \geq \eta_b$</td><td>$P(w_i \mid w_{i-1})$</td></tr>
<tr><td rowspan="3">Otherwise</td><td>$f(w_{i-1}) \geq \eta_u$ and $f(w_i) < \eta_u$</td><td>$P(t_i \mid w_{i-1})P(w_i \mid t_i)$</td></tr>
<tr><td>$f(w_{i-1}) < \eta_u$ and $f(w_i) \geq \eta_u$</td><td>$P(w_i \mid t_{i-1})$</td></tr>
<tr><td>$f(w_{i-1}) < \eta_u$ and $f(w_i) < \eta_u$</td><td>$P(t_i \mid t_{i-1})P(w_i \mid t_i)$</td></tr>
</table>

A two-round process is needed for the divide-and-conquer integration. The first round will apply formulae 7, 8, 9 and 10 to $f(w_{i-1})$, $f(w_i)$ and $f(w_{i-1}\ w_i)$ for $i = 1, \ldots, n$ (n is the length of the input sentence). Some words in the sentence may not be assigned a part-of-speech in this round, as in cases 1, 2.1, 2.2.1 and 2.2.2. For these untagged words (note that these words may constitute some word spans separated by the tagged words which are given in the first round), part-of-speech tagging is invoked in the second round. We denote this strategy AP&DivideConquer-Seg-Tag→ POSBigram-Tagging (scheme 7). It becomes apparent from the above discussion that scheme 7 is indeed a true integration, due to the fact that the first round, i.e., AP&DivideConquer-Seg-Tag, is sort of true-integration itself, though '→' appears in the notation.

Results of the divide-and-conquer integration approach are listed in Table 4.

Table 4. Results of the divide-and-conquer integration

	Integration scheme	P-WS (%)	R-WS (%)	F-WS (%)	P-PT= R-PT= F-PT(%)	P-WSPT (%)	R-WSPT (%)	F-WSPT (%)
7	AP&DivideConquer-Seg-Tag → POSBigram-Tagging	98.56	98.66	98.61	95.18	93.81	93.91	93.86

The divide-and-conquer integration (scheme 7) shows a small improvement in F-WSPT (0.07%) as compared to the best scheme in Table 2 (scheme 6), i.e., the best scheme in the conventional framework.

We also compare the space and time complexities of all schemes through experiment, as listed in Table 5.

As can be seen from Table 5, scheme 7 significantly outperforms scheme 6 in both time complexity and space complexity: it only uses 86.11% of the time and 46.17% of the space that scheme 6 used to process the test set.

Table 5. Space and time costs of all the schemes

Cost	Scheme 1	Scheme 2	Scheme 3	Scheme 4	Scheme 5	Scheme 6	Scheme 7
Time (sec.)	295	533	518	556	576	785	676
Space (KB)	1,471	1,755	1,471	7,942	7,942	7,942	3,667

It is worth noting that, experimentally, the strategy of 'all-possible-segmentations' (AP) does not lead to an explosion of space and time complexities compared to the simplest scheme, scheme1, in which only one word sequence is under consideration before part-of-speech tagging.

To deepen our understanding of scheme 7, we randomly divide the training set into 6 sub-corpora with roughly equal size, namely, T1, T2, T3, T4, T5, T6, train the scheme with these sub-corpora incrementally, and observe its F-WSPT in the test set accordingly. Figure 1 demonstrates that the F-WSPT of scheme 7 is not overly sensitive to the size of the training corpus.

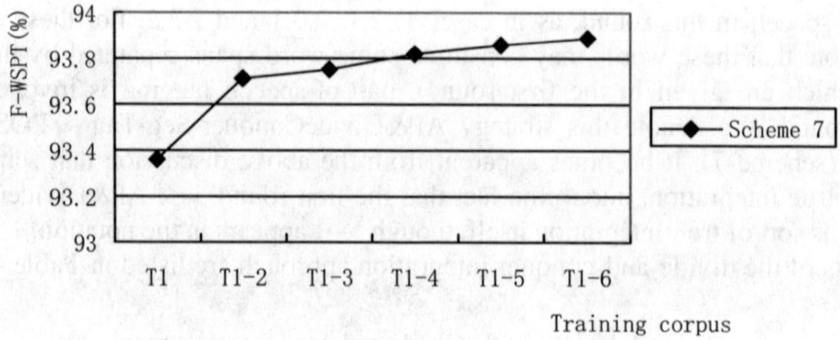

Fig. 1. F-WSPT of scheme 7 vs. corpus size

3 Conclusions

In this paper, various ways of integration of Chinese word segmentation and part-of-speech tagging, including the true-integration and the pseudo-integration, are tested and compared based on a large-scale test corpus. A novel true-integration approach, named 'the divide-and-conquer integration', is originally proposed. Preliminary experiments have demonstrated that this true integration outperforms all other kinds of combinations in some degree, showing the potential for further improving the performance of Chinese word segmentation and part-of-speech tagging.

Acknowledgments. The research is supported by the National Natural Science Foundation of China under grant number 60573187 and 60321002, and the Tsinghua-ALVIS Project co-sponsored by the National Natural Science Foundation of China under grant number 60520130299 and EU FP6.

References

1. Sproat R., Emerson T.: The First International Chinese Word Segmentation Bakeoff. Proceedings of the Second SIHAN Workshop on Chinese Language Processing. Sapporo, Japan, 133-143, (2003)
2. Emerson T.: The Second International Chinese Word Segmentation Bakeoff. Proceedings of the Third SIHAN Workshop on Chinese Language Processing. Jeju, Korea, (2005)
3. Liang N.Y.: Knowledge of Chinese Word Segmentation. Journal of Chinese Information Processing. Vol. 4, No. 2, 29-33, (1990)
4. Sun M.S., Lai B.Y. et al.: Some Issues on Statistical Approach to Chinese Word Identification. Proceedings of the 3rd International Conference on Chinese Information Processing. Beijing, 246-253, (1992)
5. Chang C.H., Chen C.D.: A Study on Integrating Chinese Word Segmentation and Part-of-speech Tagging. Communications of COLIPS. Vol. 3, No. 2, 69-77, (1993)
6. Lai B.Y., Sun M.S. et al.: Tagging-based First Order Markov Model Approach to Chinese Word Identification. Proceedings of 1992 International Conference on Computer Processing of Chinese and Oriental Languages. Florida, USA, (1992)
7. Bai S.H.: The Method of Integration of Word Segmentation and Part-of-speech Tagging in Chinese Texts. Advance and Application of Computational Linguistics. Tsinghua University Press, Beijing, 56-61, (1995)
8. Lai B.Y., Sun M.S. et al.: Chinese Word Segmentation and Part-of-speech Tagging in One Step. Proceedings of International Conference: 1997 Research on Computational Linguistics. Taipei, 229-236, (1997)
9. Wu A.D., Jiang Z.X.: Word Segmentation in Sentence Analysis. Proceedings of the 1998 International Conference on Chinese Information Processing. Beijing, 169-180, (1998)
10. Sun M.S., Xu D.L., Tsou B.K.: Integrated Chinese Word Segmentation and Part-of-speech Tagging Based on the Divide-and-Conquer Strategy. Proceedings of IEEE-NLPKE, Beijing, (2003)

Kansuke: A Kanji Look-Up System Based on a Few Stroke Prototypes

Kumiko Tanaka-Ishii and Julian Godon

Graduate School of Information Science and Technology
University of Tokyo

Abstract. We have developed a method that makes it easier for language beginners to look up Japanese kanji characters. Instead of using the arbitrary conventions of kanjis, this method is based on three simple prototypes: horizontal, vertical, and other strokes. For example, the code for the kanji 田 (*ta*, meaning rice field) is '3-3-0', indicating the kanji consists of three horizontal strokes and three vertical strokes. Such codes allow a beginner to look up kanjis even with no knowledge of the ideographic conventions used by native speakers. To make the search easier, a complex kanji can be looked up via the components making up the kanji. We conducted a user evaluation of this system and found that non-native speakers could look up kanjis more quickly and reliably, and with fewer failures, with our system than with conventional methods.

1 Introduction

Many people who are not natives of Japanese or Chinese find it very difficult to learn Japanese or Chinese characters. This is partially because their ability to look up those characters is often quite limited. Roughly, there are three conventional ways to look up kanjis in kanji dictionaries:

- by how they are read,
- by their traditional Chinese radicals, recorded in a shape index, and
- by the number of strokes.

All of these approaches follow the conventions of the ideographic system, which are not easily understood by beginners. Beginners do not know how to read kanjis, nor are they used to viewing kanji shapes and determining the special parts used to consult dictionaries. Moreover, looking up kanjis by stroke count can seem arbitrary and confusing to beginners, as what appear to be multiple strokes are often conventionally counted as one stroke (because one stroke is considered what can be continuously drawn with a brush in calligraphy).

An overview of previous work on kanji look-up for non-natives can be found in [1]. Most methods, however, are designed on the basis of the traditional conventions. For example, the SKIP code [2] is based on the traditional stroke counting method. One exception is a method allowing kanjis to be looked up by estimated readings according to the user's knowledge, even though such estimated

Y. Matsumoto et al. (Eds.): ICCPOL 2006, LNAI 4285, pp. 310–320, 2006.

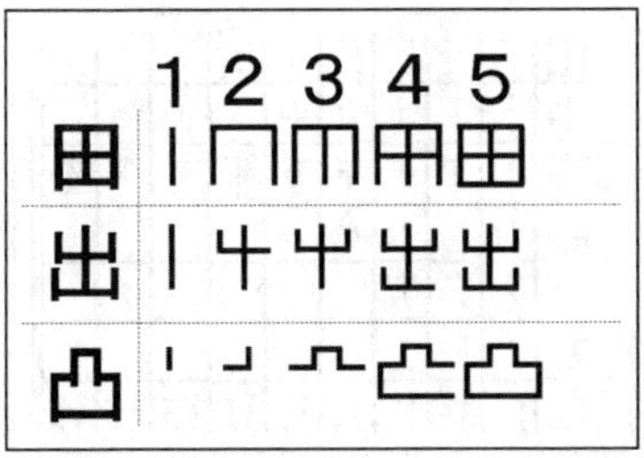

Fig. 1. Japanese kanji writing

readings could be incorrect [3]. This approach assumes that users have a certain basic knowledge of kanjis, enabling them to guess how an unfamiliar one should be read.

There are various shape-based entry systems for Chinese (well summarized in [5]), but these systems again assume that the user is well acquainted with kanjis. Methods based on decomposed kanji parts require the user to memorize 10 to 30 character shapes, as in methods such as the Cangjie method and the four-corner method. Stroke-based methods require that strokes be entered in the proper order, for example, by the five-stroke method. Still, this last method is probably the most similar to our contribution in that it only uses five stroke prototypes.

Given this situation, we have developed an alternative stroke-based kanji look-up method that does not rely on ideographic conventions. Our idea is to use three stroke prototypes which form the basis of kanjis. The three prototypes are vertical, horizontal, and other (diagonal or bent strokes). For example, in the case of the character 東, the code is '4-3-2', meaning four horizontal strokes, three vertical strokes, and two other strokes (which are diagonals). The stroke prototypes were selected through an analysis of kanjis written by several complete beginners. The codes are entered into our web-based kanji look-up system, called Kansuke, and the characters corresponding to a given code will be displayed.

We first look at how arbitrary the kanji system is, and then go on to explain how a non-native may look up kanjis.

2 Arbitrariness of the Kanji Writing System

Each kanji has a *correct* ordering of strokes for writing it down. All Japanese natives are educated in kanji writing from age 6 to age 17, learning the correct stroke order for writing each kanji. The order comes from calligraphy, and there are some general rules: from top to bottom, and from left to right.

There are also related rules for counting the number of strokes, where some connected vertexes are combined and regarded as one stroke. For example, Figure 1 shows how to correctly write three Japanese kanjis. The left-most column shows the three complete characters, with the stroke orders shown on the right. The stroke count is five for all three characters. In the first row, the

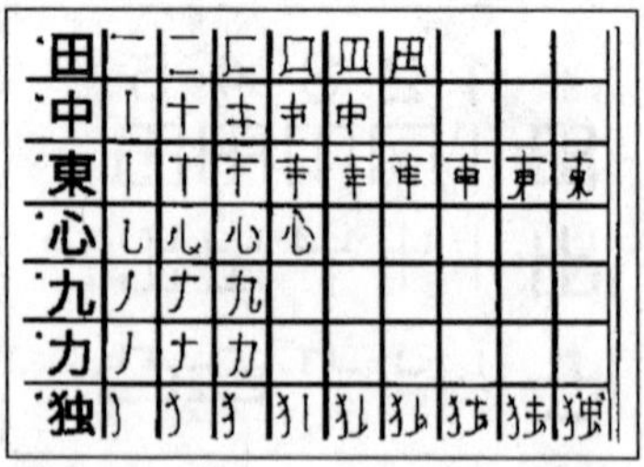

Fig. 2. Beginner's kanji writing

top-most and right-most lines are considered one stroke, while the left-most and bottom lines are counted as two different strokes. This is different from the second character, however, where the two left-most strokes are connected to horizontal strokes.

As seen from these two examples of relatively simple kanjis, the stroke order and stroke count are arbitrary. An extreme case is shown in the third row. Even though this character does not appear very complex, even many Japanese natives do not know how to write it in the correct order.

When a person does not know how to read a kanji, a kanji dictionary is consulted by examining the radicals (the individual parts of a character) of which it consists. This is also done by Japanese, using the *bushu* methodology. For those who are not familiar with the radicals, there is a multi-radical method [1], where people visually scan through parts consisting of 249 radical prototypes and choose those contained in the target kanji. When even this is difficult, the last option is to look up a kanji by its stroke count; e.g., by entering 'five' for the characters in Figure 2. However the conventions that determine stroke count are largely arbitrary. Consequently, non-natives must leap a large hurdle to become *initiated* in looking up Japanese kanjis.

One way to overcome this problem is to provide a universal way to describe kanjis. For this purpose, the way non-native beginners write kanjis will give us hints.

3 Designing Codes

3.1 A Beginner's View of Kanjis

Figure 2 shows the efforts of an anonymous beginner. The left-most column shows the printed kanjis, with the written stroke sequence to the right.

We can see how variant a subject can be in writing these kanjis. If we tested more subjects, this variety would become even greater. By viewing such results produced by several beginners, however, we found the subjects had some common ideas:

- The feature of a stroke being straight or bent is preserved.
- The gradients of strokes are mostly preserved. Especially, horizontals, verticals, and strokes of other gradients are distinguished.

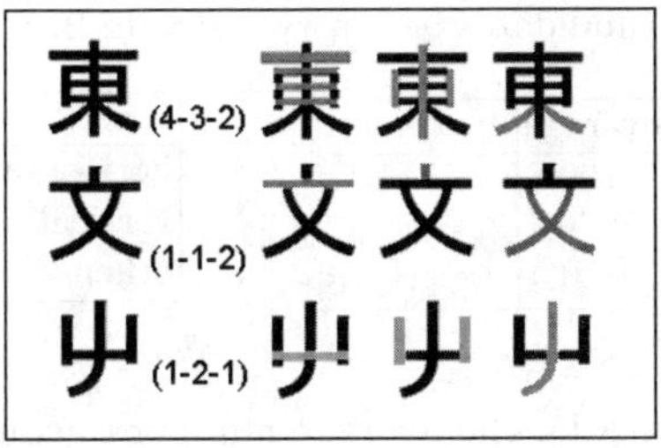

Fig. 3. Kansuke code examples

- Sharply curved bent lines are considered one stroke. For example, when we compare the results for the third, fourth, and fifth kanjis of Figure 2, which include sharply bent lines, the longest bent line in 心 (*kokoro*, meaning heart) and the long bent stroke on the right-hand side of 九 (*kyuu*, meaning nine) were each considered one stroke.

From these observations, we can see how few features the non-natives shared. Under such circumstances, predictive/ambiguous entry is one possibility for designing an entry method, where users enter a code sequence formed only of letters with a few codes, which is expanded into the corresponding actual data by a machine. We designed such prototypes based on the above common observations.

3.2 Kansuke Code

Non-natives are generally not used to browsing kanjis, and they can scan a list of kanjis only slowly. Therefore, the code should allow precise classification of a huge number of kanjis through their assigned codes.

As the notion of traditional strokes is quite arbitrary, we first redefined this notion based on the small analysis explained in the previous section: a stroke is any part of the kanji that does not contain a sharp angle. Kansuke code was then designed as three numbers representing the quantity of horizontal, vertical, and other strokes (re-defined), so that it can be used to search for a kanji. Strokes classified as 'other' can be diagonal, bent, or complex. Examples of Kansuke code can be seen in Figure 3. For example, in the case of 東, there are four horizontal strokes, three vertical strokes, and two other strokes, so the Kansuke code is '4-3-2'. Likewise, the code for the second character will be '1-1-2'. A slightly tricky case is shown in the third row, where the stroke in the middle is classified as a vertical stroke but can also be classified as an 'other' stroke. For such ambiguous cases, multiple codes are attached, as explained in more detail in §4.3.

We compared the use of these codes with other conventional methods. First, the average number of candidates per entry for the traditional stroke-count method, the system of kanji indexing by patterns (SKIP), and Kansuke are compared in Table 3.2. These results were obtained for 2965 kanjis, all within the JIS first standard. For the traditional stroke count (for example, a count of

Table 1. No.of candidates per entry

	average	maximum
stroke count	100.8	232
SKIP	9.7	67
Kansuke	4.1	28

Table 2. Stroke Distribution

type	total number
horizontal	10549
vertical	7499
other	10676

five for all kanjis in Figure 1), the average number of candidates was 100, meaning beginners often have to go through a painfully long list to find the kanjis they seek.

As an alternative, SKIP is a method applied to a kanji dictionary especially designed for beginners [2]. In this method, a kanji is typically classified into one of four initial patterns describing the division of the kanji into two portions, and a code is constructed using this initial pattern and the stroke count of each portion. Thus, 村 has a SKIP code of '1-4-3', indicating a vertical division (1), and the stroke counts for the left-hand (4) and right-hand sides (3). The SKIP method produced only 9.7 candidates on average, but the maximum number of candidates was as high as 67.

In contrast, Kansuke produced only four candidates on average, with a maximum of 28 candidates. Thus, our code classifies kanjis far better than the other conventional methods. This superior classification is due to the fairly uniform distribution of strokes as regards the code type, as shown in Table 2. The numbers of all strokes taken from 2965 kanjis were quite evenly distributed among our three prototypes (horizontal/vertical/other) with about 10,000 strokes in each category.

4 Kansuke System

4.1 Interface

Based on the Kansuke Code, the Kansuke system was designed for quick and efficient kanji look-up.

The Kansuke system consists of an interface and a database where the interface interactively filters candidate characters obtained from the database. The whole system is written as an HTML/CSS web page and can run on any Internet browser. Our database of kanjis is enclosed in text format, and accessed through JavaScript functions. This allows dynamic queries without connection to a server. The system is accessible on the web [4]. We are now preparing a free-download version for local machine use.

The Kansuke system can be used by entering Kansuke code for the target character. Kansuke code can be constructed for the whole character, for example, '4-3-2' for 東. An alternative, perhaps a better way for the novice is to look up the target via *components* included in the target character. We explain this here by looking up an example character 漢. The user first chooses a component that is included in 漢. Suppose that the user chose 艹 (at the

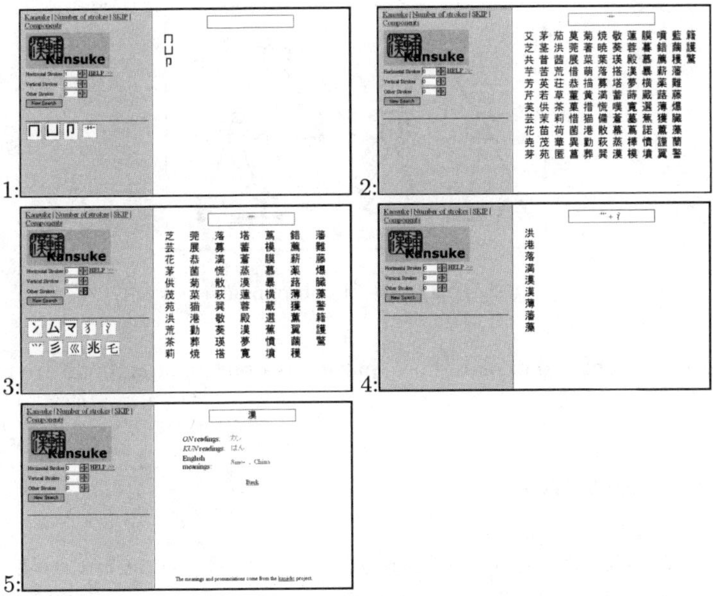

Fig. 4. Looking up a Kanji with Kansuke

top on the right-hand side of 漢). As 艹's Kansuke code is '1-2-0' (one horizontal, two vertical strokes), the user enters the code. The system display then looks like panel 1 of Figure 4, with corresponding kanjis shown on the right side of the page, while corresponding components with the code '1-2-0' appear in the lower-left part. When the user selects the target component 艹 from the left side of the page, the system shows all characters that contain 艹 (Figure 4-2). Note that the 艹 appears at the top of the right side of the page. As there are still many kanjis that appeared as candidates, suppose that the user chose to filter further by 氵 (the left part of 漢). As 氵 is formed of three *other* strokes, the code '0-0-3' is entered. The system shifts to the figure showing corresponding components on the left page and all kanjis that consists of *any* of those components in addition to 艹 are shown on the right-hand side (Figure 4-3). When the component 氵 is selected, the kanji candidates are filtered so that the kanjis that consist of 艹 and 氵 are shown (Figure 4-4). The user may now choose his target, the sixth from the top on the right side of the page, and the dictionary entry for the kanji is shown (Figure 4-5).

The list of candidates on the right side is sorted according to their stroke counts, which provides a good approximation of each candidate's complexity. Thus, simpler kanjis, often targeted by beginners, appear at the beginning of the list and can be retrieved more quickly. On the other hand, components shown on the left side are placed according to a manual grouping of similar components, so that similar components will be placed close to each other.

Consequently, kanji lookup is enhanced not only by Kansuke code, but also through the use of components. We will next show how this is designed within

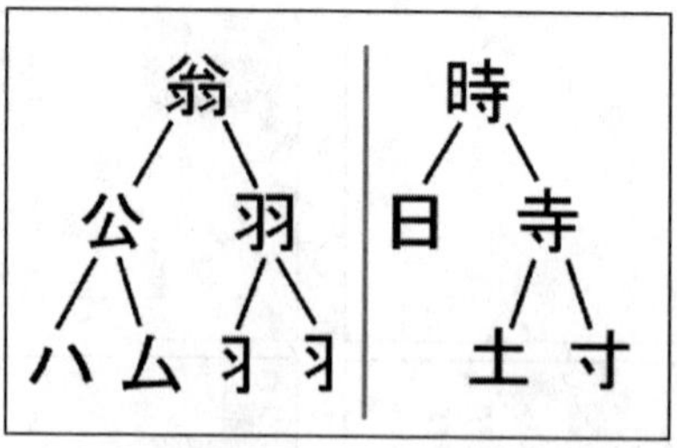

Fig. 5. Examples of component trees

the Kansuke system and explain how characters of ambiguous code are processed in the system.

4.2 Component Trees

Given a kanji, classifying all of its strokes to get the code is time-consuming and the probability of making a counting mistake is high. The Kansuke code of the character 躍, for example, is '13-8-3', for a total of 24 strokes. As almost all complex kanjis are composites of smaller kanji parts, one way to make the search easier is to search by *components*.

A component is a group of strokes that is visually isolated in the kanji. If we take, for instance, 時 (*toki*, meaning time), some of its components are 日 and 寺. Someone more knowledgeable of kanjis, however, might prefer to search by 日, 土, and 寸, which are the traditional Chinese radicals. Another example is the kanji 翁 (*okina*, meaning venerable man), which can be seen by a skilled reader as the combination of 公 and 羽, whereas a beginner could only distinguish ハ, ム, and two instances of 习.

Taking all such possibilities into account, every kanji in the database is represented as a *component tree*. Within this component tree, a node denotes a part of the kanji and its children nodes denote its components. Having the root node denote the whole kanji, the tree thus denotes a recursive decomposition into smaller components. Figure 5 shows an example, in this case component trees for the kanjis 翁 and 時 discussed above.

In the Kansuke system, any component corresponding to the node of the tree can be encoded by the Kansuke code and used to search the target kanji, as explained in the last section. Such flexibility is realized by this tree structure and is not provided by other kanji search methods based on a component-type look-up system, such as multi-radical methods [1].

The tree structure is also advantageous for maintaining the database, because the computer can dynamically calculate the Kansuke code of a node by summing up the codes of the children nodes. If we change a value at a node, the change is automatically reflected in the upper nodes. This function is especially powerful, because some of the components require the attachment of multiple Kansuke codes as mentioned; we explain this in more detail next.

4.3 Multiple Codes for Ambiguous Kanjis

As the number of stroke prototypes is very small, some stroke types could be difficult to uniquely classify into a type. For example, 心 (*kokoro*, heart), is represented in Kansuke code as '0-0-4', but some users might code it as '1-1-3'.

The easiest solution would be to display all kanjis corresponding to codes with small code differences from the user's entry. In our case, however, this can result in too many candidates, due to the small number of prototypes. For example, even for the smallest difference of one, an average of almost 24 additional kanjis would be added ((three stroke types) × (the increase or decrease in the total number of strokes; i.e., 2) × 4.1 (from Table 3.2)).

We therefore added all the probable codes to each kanji, and hard-coded them in our database. For example, 心 is represented by the codes '0-0-4' and '1-1-3'. The user can enter either to retrieve this kanji. With this solution, the average number of candidates per entry increases, but only slightly.

In fact, as most kanjis are just a group of components, we only had to consider the ambiguity for the component, and the rest of the calculation is done dynamically via the tree explained in the previous section. For now, of about 250 components, 50 have such multiple code definitions.

5 User Evaluation

5.1 Settings

To evaluate the effectiveness of the Kansuke system, we compared our method with three other methods: the SKIP code method, the traditional look-up method using only the number of strokes (denoted as 'stroke count'), and a multi-radical method (denoted as 'radicals'). There are other conventional look-up methods, but since most require some knowledge of kanji pronunciation they cannot be used by beginners.

A total of 16 subjects participated in the evaluation. Eight were native speakers of Japanese or Chinese, and the other eight had no knowledge of the Japanese language or kanjis. We refer to these groups as native speakers and non-native speakers. Note that the non-native speakers were familiar with none of the four methods, whereas the native speakers were familiar with the stroke count, radicals, and part of the look-up process used in SKIP.

As it took about 10 minutes to test a method on a subject, only three methods were tested on each subject. Three sets of five kanjis each (sets A, B, and C) were constructed randomly by computer (Table 3), under the condition that the nth kanji of each set had close to the same stroke count. The subjects looked up each set using one of the four systems (stroke count, SKIP, radicals, or Kansuke) in different combinations and in different orders.

The methods and sets assigned to each subject were chosen according to the Latin square, as shown in Table 4. The order of methods was decided randomly at the run time for each subject. Such experimental conditions ensured that the differences between the three sets (A, B, and C) and the ordering of the methods

Table 3. Sets of kanjis used in the evaluation

No.	1	2	3	4
set A	白	咲	款	酵
set B	王	承	射	駿
set C	夊	取	庸	察

Table 4. Experimental course: combinations of systems and kanji sets

	set A	set B	set C
Course 1,5	stroke count	SKIP	Kansuke
Course 2,6	SKIP	Kansuke	radicals
Course 3,7	Kansuke	radicals	stroke count
Course 4,8	radicals	stroke count	SKIP

did not influence the results. Moreover, the interface used to look up kanjis was exactly the same for all methods, except for the GUI buttons and labels that had to be changed to match the type of input data required by each method.

Every subject connected to our test website to participate in the experiment. For each method, each subject went through the following stages:

1. Read the instructions and the explanation of the system (one of the four methods).
2. Look up one kanji as a preliminary test under the same environment used in the real test.
3. Perform the real test for four kanjis using the assigned method. The target kanji was displayed on top of the page, and the subject used the interface to find the target. The time to look up each kanji and the number of trials were automatically recorded by the program. If the user did not find the kanji after several attempts, he could click on a 'skip' button to proceed to the next kanji. This was recorded as a failure.

Each subject was asked afterwards to make qualitative comments on each system.

5.2 Results

The evaluation results are shown in Table 5. The average time needed to look up one kanji and the standard deviation are shown in the second column in seconds. The average number of trials is shown in the third column, except for the SKIP method where this information was not available. The last column shows the percentage of kanjis that the subjects failed to find.

The results in Table 5 show that the performance of non-native speakers was best when using the Kansuke system. Kansuke enabled the shortest look-up time because the average number of trials was low. In addition, the low standard deviation shows how reliably a kanji could be found. Apart from time considerations, Kansuke also had the lowest percentage of failures. This is important

Table 5. Performance of non-native and native speakers

	non-natives			natives		
	time (secs) (avg./std. dev.)	trials (avg.)	failures (%)	time (secs) (avg./std. dev.)	trials (avg.)	failures (%)
stroke count	132.8 / 96.4	1.9	17.9%	37.1 / 24.5	1.2	0.0%
SKIP	112.5 / 85.0	-	20.0%	31.7 / 17.4	-	10.7%
radicals	141.6 / 104.5	2.1	31.2%	48.6 / 38.2	1.4	6.3%
Kansuke	92.7 / 71.9	1.3	7.1%	44.8 / 24.8	1.7	8.3%

since a method must work for all kanjis to be effective. When using SKIP, many non-native speakers could not find the target because they failed to select the correct initial pattern. The radical method produced the worst results, with an average time of more than two minutes and a failure rate close to 30%. This poor performance was due to the inability of non-native speakers to properly identify the radicals — the arbitrarily defined components — of some kanjis.

In contrast, the native speakers were much faster with the stroke count and SKIP methods. With the traditional method, they only required an average of about 37 seconds to look up a kanji, whereas the non-native speakers required more than 2 minutes. These results reflect the time needed to scan through the displayed kanji candidates as well as the time needed to count the number of strokes. As the native speakers were accustomed to reading kanjis, their scanning speed was much faster, and even the display of many kanjis did not seriously slow them down. The SKIP method also produced a good look-up time, but some failures occurred due to incorrect selection of the initial pattern. Curiously, the average number of trials with Kansuke was higher than for non-native speakers. We attribute this low performance of the Kansuke system for native speakers to the fact that for these subjects the Kansuke system was the only completely new method, which required some degree of learning overhead. In addition, as native speakers are used to the traditional way of counting strokes, this knowledge probably prevented these subjects from properly counting the Kansuke strokes.

We also had the subjects complete a questionnaire regarding the pros and cons of the different methods. Their opinions are summarized here:

- The way of counting strokes with Kansuke is easier for beginners.
- Both beginners and natives liked the decomposition of kanjis in the Kansuke method.
- Even with multiple codes attached, some strokes could not be uniquely classified to obtain the Kansuke code. For example, some strokes could be counted as one vertical stroke and one horizontal stroke, or just as one other stroke.
- The traditional method displays too many candidates.
- The stroke counts of the traditional method and SKIP are ambiguous.
- The native speakers were used to the traditional method.
- The SKIP codes were liked by those who were used to counting the number of traditional strokes.

- The SKIP codes were ambiguous when selecting the initial pattern.
- The radicals method is not suitable for some kanjis with no obvious radicals; i.e., the principal component indexed for look-up.

No method proved to be totally free of ambiguity, making this a problem with all methods. For our method, we may alleviate this problem by attaching more multiple codes as explained in §4.3. Overall, though, both beginners and native speakers showed great interest in our system and enjoyed the experiment.

6 Conclusion

We have developed a method for looking up Japanese kanjis using only a small number of stroke prototypes. Our method differs from previous methods in that the stroke classification does not require the user to have any preliminary knowledge of kanjis. Given a kanji, the user constructs a Kansuke code, which is based on the counts of three stroke prototypes: horizontal, vertical, and other strokes. For example, the code for 東 is '4-3-2', because there are four horizontal strokes, three vertical strokes and two other strokes. After this code is entered into a web-based Kansuke interface we have developed, a list of the corresponding candidates is displayed. Even though the number of prototypes is very small, the average number of candidates for a given code is about 4, whereas that for the traditional method is about 100. Further refinement enables a kanji to be looked up using different codes. This feature is enhanced by a tree-structure decomposition of each kanji.

We conducted a user evaluation with two groups of subjects: Japanese or Chinese native speakers and non-native speakers. The non-native speakers could look up kanjis more quickly and reliably with our Kansuke system, and with fewer failures, than with other existing methods. This demonstrates that our method can support complete beginners who want to access information in East Asian languages.

References

1. J. Breen. Jim Breen's WWWDic Site, 2005. available at `http://www.csse.monash.edu.au/ jwb/cgi-bin/wwwjdic.cgi`.
2. J. Halpern. *Kanji Learner's Dictionary.* Kodansha, 1999.
3. B. Slaven, T. Baldwin, and H. Tanaka. Bringing the dictionary to the user: the foks system. In *In Proceedings of the 19th International Conference on Computational Linguistics*, pages 84–91, 2002.
4. K. Tanaka-Ishii and J. Godon. Kansuke : Japanese kanji lookup site, 2006. available at http://www.ish.ci.i.u-tokyo.ac.jp/kansuke.html.
5. Wikipedia. Chinese character entry on keygoard and its evaluation and test technology, 2005. available at `http://en.wikipedia.org/wiki/-Chinese_input_method`.

Modelling the Orthographic Neighbourhood for Japanese Kanji

Lars Yencken and Timothy Baldwin

Computer Science and Software Engineering
University of Melbourne, Victoria 3010 Australia
NICTA Victoria Research Lab
University of Melbourne, Victoria 3010 Australia
{lljy, tim}@csse.unimelb.edu.au

Abstract. Japanese kanji recognition experiments are typically narrowly focused, and feature only native speakers as participants. It remains unclear how to apply their results to kanji similarity applications, especially when learners are much more likely to make similarity-based confusion errors. We describe an experiment to collect authentic human similarity judgements from participants of all levels of Japanese proficiency, from non-speaker to native. The data was used to construct simple similarity models for kanji based on pixel difference and radical cosine similarity, in order to work towards genuine confusability data. The latter model proved the best predictor of human responses.

1 Introduction

In everyday reading tasks, humans distinguish effortlessly between written words. This is despite languages often seeming ill-suited to error-free word recognition, through a combination of inter-*character* similarity (i.e. the existence of graphically-similar character pairs such as 士 [*shi*] and 土 [*tsuchi*]) and inter-*word* similarity (i.e. the existence of orthographically-similar word pairs such as *bottle* and *battle*). While native speakers of a language tend to be oblivious to such similarities, language learners are often forced to consciously adapt their mental model of a language in order to cope with the effects of similarity. Additionally, native speakers of a language may perceive the same character pair significantly differently to language learners, and there may be radical differences between language learners at different levels of proficiency or from different language backgrounds.

This paper is focused on the similarity and confusability of Japanese kanji characters. This research is novel in that it analyses the effects of kanji confusability across the full spectrum of Japanese proficiency, from complete kanji novices to native speakers of the language. Also, unlike conventional psycholinguistic research on kanji confusability, it draws on large-scale data to construct and validate computational models of similarity and confusability. This data set was collected for the purposes of this research via a web experiment, and consists of a selection of both control pairs aimed at targeted phenomena, and also

Y. Matsumoto et al. (Eds.): ICCPOL 2006, LNAI 4285, pp. 321–332, 2006.

random-selected character pairs. The research builds on psycholinguistic studies of the visual recognition of both Chinese hanzi and Japanese kanji.

The paper is structured as follows. We begin by discussing the background to this research (Section 2), then follow with a description of our web experiment and its basic results (Section 3). We construct some simple models of the similarity data (Section 4), then evaluate the models using the experimental data (Section 5). Finally, we lay out our plans for future work (Section 6).

2 Background

2.1 Types of Similarity

This paper chiefly concerns itself with orthographic similarity of individual Japanese characters, that is, graphical similarity in the way the characters are written. Note that within the scope of this paper, we do not concern ourselves directly with the question of orthographic similarity of multi-character words. That is, we focus exclusively on inter-character similarity and confusability.

Other than simple orthography, character similarity can also be quantised *semantically* or *phonetically*; identically-pronounced characters are termed *homophones*. In Chinese and Japanese, radicals[1] are often semantic or phonetic cues of varying reliability in determining character similarity. When radicals are shared between kanji, more than orthographic similarity may be shared. If a radical is reliably semantic, then two kanji sharing it are likely to be semantically related in some way (e.g. kanji containing the radical 月, such as 胸 [*mune*] “chest” and 腕 [*ude*] “arm” are reliably body parts). If reliably phonetic (e.g. 同, as in 銅[*dō*] “copper” and 胴 [*dō*] “body”), the two kanji will share a Chinese or *on* reading, and will hence be homophones. It may thus not be impossible for skilled readers to give purely orthographic similarity judgements in the presence of shared radicals, since evidence shows these cues are crucial to reading, as to be discussed in Section 2.2.

2.2 Lexical Processing of Kanji and Hanzi

In considering potential effects to control for, and types of similarity effects, we draw on the many psycholinguistic studies of kanji or hanzi recognition, with the basic assumption that human reading processes for the two scripts are fundamentally similar. It is beyond the scope of this paper to give these studies full treatment, but we refer the interested reader to some pertinent results.

There is much support for a form of hierarchical activation model for the recognition of kanji (Taft and Zhu 1997, Taft, Zhu, and Peng 1999). In such a model, firstly strokes are visually activated, which in turn activate the radicals they form, which then activate entire kanji. Evidence for such a model includes

[1] Here, we use the term *radical* liberally to refer to any consistent stroke group, rather than in its narrow sense as either the dictionary index stroke group or the main semantic stroke group.

Fig. 1. Example stimulus pair for the similarity experiment. This pair contains a shared radical on the left.

experiments which showed stroke count effects (summarized by Taft and Zhu (1997)), and numerous radical effects, including radical frequency with semantic radical interference (Feldman and Siok 1997, Feldman and Siok 1999), and homophony effects which only occurred with shared phonetic radicals (Saito, Inoue, and Nomura 1995, Saito, Masuda, and Kawakami 1998). There is also evidence that structure may be important for orthographic similarity in general, and for radical-based effects (Taft and Zhu 1997, Yeh and Li 2002).

3 Similarity Experiment

3.1 Experiment Outline

A short web-based experiment was run to obtain a set of gold-standard orthographic similarity judgements. Participants were first asked to state their first-language background, and level of kanji knowledge, pegged to one of the levels of either the Japanese Kanji Aptitude Test (日本漢字能力検定試験)[2] or the Japanese Language Proficiency Test (日本語能力試験).[3] Participants were then exposed to pairs of kanji, in a manner shown in Figure 1, and asked to rate each pair on a five point graded similarity scale. The number of similarity grades chosen represents a trade-off between rater agreement, which is highest with only two grades, and discrimination, which is highest with a large number of grades.

Although participants included both first and second language readers of Chinese, only Japanese kanji were included in the stimulus. Chinese hanzi and Japanese hiragana and katakana were not used for stimulus, in order to avoid potential confounding effects of character variants and of differing scripts. The pairs were also shuffled for each participant, with the ordering of kanji within a pair also random, in order to reduce any effects caused by participants shifting their judgements part-way through the experiment.

Each participant was exposed to a common set of control pairs, to be discussed in Section 3.2 below. Further, a remaining 100 random kanji pairs were shown where both kanji were within the user's specified level of kanji knowledge

[2] A Japanese government test which is tied to Japanese grade school levels initially, but culminates at a level well above that expected in high-school graduates. `http://www.kanken.or.jp/`

[3] The standard general-purpose Japanese aptitude test taken by non-native Japanese learners of all first-language backgrounds. `http://www.jees.or.jp/jlpt/en/index.htm`

Effect type	Example	Description
Frequency (independent)	会 店	Frequency of occurrence of each kanji individually. Both kanji in the example pair are high-frequency.
Co-occurrence	法 考	Both kanji occur with high frequency with some third kanji. For example, 法 [*hō*] "Act (law)" occurs in 法案 [*hōaN*] "bill (law)", and 考 [*kaNga(e)*] "thought" occurs in 考案 [*kōaN*] "plan, idea".
Homophones	弘 博	Both kanji share a reading. In the example, both 弘 [*hiro(i)*] "spacious" and 博 [*haku*] "doctor" share a reading [*hiro*]. For 博 this is a name reading.
Stroke overlap	策 英	Both kanji share many similar strokes, although no radicals are shared.
Shared graphemes	働 動	Both kanji share one or more graphical elements. These elements might occur in any position.
Shared structure	幣 哲	Both kanji share the same structural break-down into subcomponents, although the sub-components differ.
Stroke count	奮 撃	Pairs comparing and contrasting stroke counts. Both examples here have a very high stroke count.
Part of speech/function	方 事	Both kanji have a common syntactic function.
Semantic similarity	千 万	Both kanji are semantically similar. In the example, they are both numbering units.

Fig. 2. Groups of control pairs used, with an example for each. Parts of readings in brackets indicate *okurigana*, necessary suffixes before the given kanji forms a word.

(where possible), and 100 were shown where one or both kanji were outside the user's level of knowledge. This was in order to determine any effects caused by knowing a kanji's meaning, its frequency, its readings, or any other potentially confounding properties.

Web-based experiments are known to provide access to large numbers of participants and a high degree of voluntariness, at the cost of self-selection (Reips 2002). Although participants of all language backgrounds and all levels of kanji knowledge were solicited, the nature of the experiment and the lists advertised to biased participants to be mainly of an English, Chinese or Japanese first-language background.

3.2 Control Pairs

There are many possible influences on orthographic similarity judgements which we hoped to detect in order to determine whether the data could be taken at face value. A sample pair and a description of each control effect is given in Figure 2. Since the number of potential effects considered was quite large, the aim was not statistical significance for the presence or absence of any effect, but rather guidance in similarity modelling should any individual effect seem strong. All frequency and co-occurrence counts were taken from 1990–1999 Nikkei Shinbun corpus data.

3.3 Results

The experiment had 236 participants, with a dropout rate of 24%. The participants who did not complete the experiment, and those who gave no positive responses, were filtered from the data set. The remaining 179 participants are

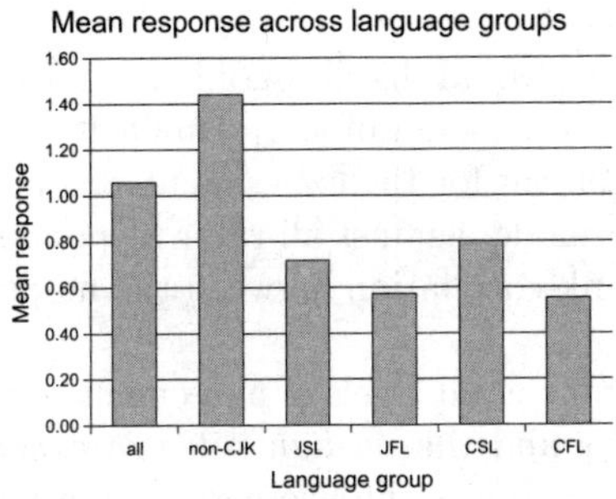

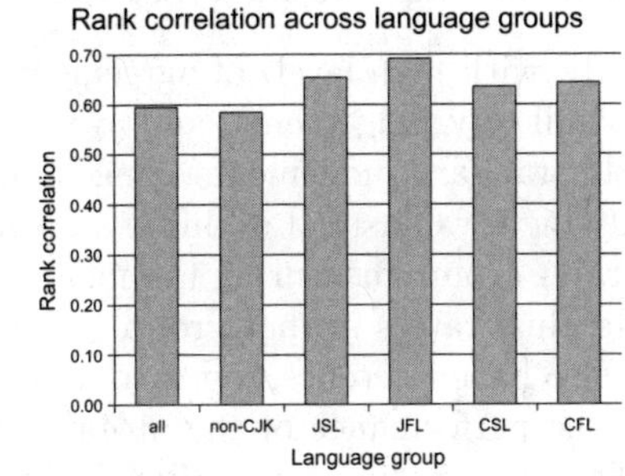

Fig. 3. Mean responses and rank correlation when broken up into participant groups, measured over the control set stimulus

spread across 20 different first languages. Mapping the responses from "Very different" as 0 to "Very similar" as 4, the mean response over the whole data set was 1.06, with an average standard deviation for each stimulus across raters of 0.98.

To measure the inter-rater agreement, we consider the mean rank-correlation across all pairs of raters. Although the kappa statistic is often used (Eugenio and Glass 2004), it underestimates agreement over data with graded responses. The mean rank correlation for all participants over the control set was strong at 0.60. However, it is still lower than that for many tasks, suggesting that many raters lack strong intuitions about what makes one kanji similar to another.

Since many of the first language backgrounds had too few raters to do significant analysis on, they were reduced to larger groupings of backgrounds, with the assumption that all alphabetic backgrounds were equivalent. Firstly, we group first-language speakers of Chinese (CFL) and Japanese (JFL). Secondly, we divide the remaining participants from alphabetic backgrounds into second language learners of Japanese (JSL), second language learners of Chinese (CSL), and the remainder (non-CJK). Participants who studied both languages were put into their dominant language based on their comments, or into the JSL group in borderline cases.[4]

Figure 3 shows mean responses and agreement data within these participant groups. This grouping of raters is validated by the marked difference in mean responses across these groups. The *non-CJK* group shows high mean responses, which are then halved for second language learners, and lowered still for first language speakers. Agreement is higher for the first-language groups (JFL and CFL) than the second-language groups (JSL and CSL), which in turn have higher agreement than the non-speakers. Both of these results together suggest that with increasing experience, participants were more discerning about what they found to be similar, and more consistent in their judgements.

[4] Many alternative groupings were considered. Here we restrict ourselves to the most interesting one.

3.4 Evaluating Similarity Models

Normally, with high levels of agreement, we would distill a gold standard dataset of similarity judgements, and evaluate any model of kanji similarity against our gold-standard judgements. Since agreement for the experiment was not sufficiently high, we instead evaluate a given model against all rater responses in a given rater group, measuring the mean rank-correlation between the model and all individual raters in that group.

We also have reference points to determine good levels of agreement, by measuring the performance of the *mean rating* and the *median rater response* this way. The mean rating for a stimulus pair is simply the average response across all raters to that pair. The median rater response is the response of the best performing rater within each stimulus set (i.e. the most "agreeable" rater for each ability level), calculated using the above measure.

4 Similarity Models

4.1 Pixel Difference Model

In Section 2.2, we briefly discussed evidence for stroke level processing in visual character recognition. Indeed, confusability data for Japanese learners taken from the logs of the FOKS (Forgiving Online Kanji Search) error-correcting dictionary interface suggests that stroke-level similarity is a source of error for Japanese learners. The example 基 [*ki, moto*] "basis" and 墓 [*bo, haka*] "grave / tomb" was taken from FOKS dictionary error logs (Bilac, Baldwin, and Tanaka 2003), and is one of the pairs in our "Stroke overlap" control subgroup (Figure 2).

This example shows that learners mistake very similar looking kanji, even when there are no shared radicals, if there are sufficient similar looking strokes between the two kanji. Ideally, with a sufficiently rich data set for kanji strokes, we could model the stroke similarity directly. As an approximation, we instead attempt to measure the amount that strokes overlap by rendering both kanji to an image, and then determine the pixel difference d_{pixel} between the two rendered kanji. We can easily move from this distance metric to a similarity measure, as below:

$$d_{\text{pixel}}(k_a, k_b) = \frac{1}{L} \sum_{L}^{i=0} |\text{Image}(k_a)(i) - \text{Image}(k_b)(i)| \quad (1)$$

$$s_{\text{pixel}}(k_a, k_b) = 1 - d_{\text{pixel}}(k_a, k_b) \quad (2)$$

This calculation is potentially sensitive both to the size of the rendered images, and the font used for rendering. For our purposes, we considered an image size of 100×100 pixels to be sufficiently detailed, and used this in all experiments described here. To attempt to attain reasonable font independence, the same calculation was done using 5 commonly available fonts, then averaged between them. The fonts used were: Kochi Gothic (medium gothic), Kochi Mincho (thin mincho), Mikachan (handwriting), MS Gothic (thick gothic), and MS Mincho (thin mincho). The graphics program Inkscape[5] was used to render them non-interactively.

This method of calculating similarity is brittle. Suppose two characters share a significant number of similar strokes. If the font renders the characters in such a way that the similar strokes are unaligned or overly scaled, then they will count as differences rather than similarities in the calculation. Further robustness could be added by using more sophisticated algorithms for scale and translation invariant image similarity.

Consider the minimum pixel difference of a pair over all possible offsets. This defines a translation invariant similarity measure. Since the current method calculates only one alignment, it is an underestimate of the true translation invariant similarity. Since characters are rendered in equal-sized square blocks, and radical frequency is position-dependent, the best alignment usually features a low offset between images. The current approximation is thus a close estimate on average, and is considerably less expensive to compute.

Pixel difference is also likely to underestimate the perceptual salience that repeated stroke units (i.e. radicals) have, and thus underestimate radical-level similarity, except where identical radicals are aligned. Nevertheless, we expect it to correlate well with human responses where stroke-level similarity is present. Pairs scored as highly similar by this method should thus also be rated as highly similar by human judgements.

4.2 Bag of Radicals Model

Just as the pixel model aimed to capture similarity effects at the stroke-level, we now try to capture them at the radical level. Fortunately, at the radical-level there is an existing data-set which indexes kanji by all of their contained radicals, the *radkfile*.[6] It was designed to aid dictionary look-up, and serves as a simple method of determining all the unique radicals used by a kanji. Two examples of kanji decomposed into their component kanji are given in Figure 4.

Using all the potential radicals as dimensions, we can map each kanji onto a vector space of radicals, giving it a boolean entry in each dimension determining whether or not the kanji contains that radical. On this vector space, we can calculate the cosine similarity between the two kanji vectors, to achieve a simple similarity measure based on radical composition:

$$s_{\text{radical}}(k_a, k_b) = \frac{\text{radicals}(k_a) \bullet \text{radicals}(k_b)}{|\text{radicals}(k_a)||\text{radicals}(k_b)|} \tag{3}$$

Comparing high-similarity examples from the different methods (Figure 5), we can immediately see some drawbacks to this model. Example (g) shows that the number of each radical present is discarded, hence 火 and 炎 are considered identical with this method. Example (h) shows that position is also discarded, yet there is evidence that radical effects are position specific (Taft and Zhu 1997,

[5] `http://www.inkscape.org`

[6] `http://ftp.monash.edu.au/pub/nihongo/radkfile.gz`

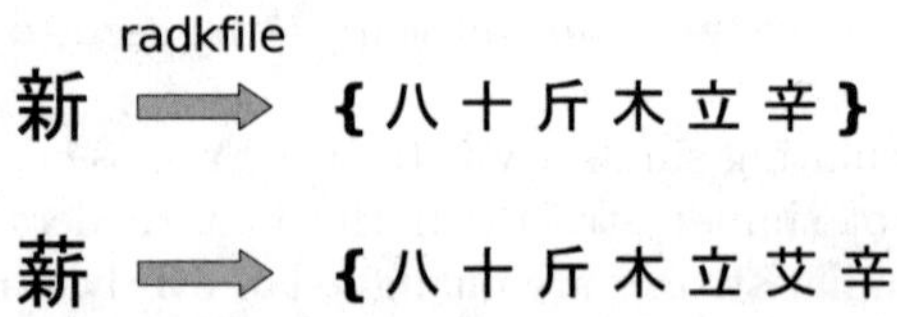

Fig. 4. Kanji are decomposed into their radicals using the radkfile. Each set of radicals can be considered a boolean vector over all radicals, where only the radicals which are present are stored. Note that the character used to represent each single radical is not always identical to that radical.

Predictor	Example	Similarity
Mean rating	a. 漢 菓	0.938
	b. 飫 餘	0.833
	c. 基 墓	0.830
Pixel	d. 土 士	0.878
	e. 人 入	0.844
	f. 官 宮	0.771
Bag of radicals	g. 火 炎	1.000
	h. 累 細	1.000
	i. 馬 駅	0.500

Fig. 5. Examples of high similarity pairs according to mean rating, the pixel model and the bag of radicals model. The mean rating pairs were taken from the experimental data, whereas the other pairs were taken from general use kanji.

1999). This model also ignores similarity due to stroke data, yet the existence of high-similarity examples such as (d) and (e) which do not share radicals indicates that stroke overlap can also be a significant contributor to similarity. Larger structure such as layout may also be important for similarity (Yeh and Li 2002), and it too is discarded here.

Nonetheless, radicals are clearly significant in the perception of kanji. If the presence or absence of shared radicals is the main way that individuals perceive similarity, then this model should agree well with human judgements, whether or not they make use of the additional semantic or phonetic information these radicals can encode.

5 Model Evaluation

The pixel and radical models were evaluated against human judgements in various participant groups, as shown in Figure 6, and can be compared to the mean rating and median raters. The pixel based similarity method exhibits weak rank correlation across the board, but increasing in correlation with increasing kanji knowledge. The radical model however shows strong rank correlation for all groups but the non-CJK, and better improvements in the other groups.

These results match our predictions with the pixel-based approach, in that it performs reasonably, but remains only an approximation. The radical method's results however are of a comparable level of agreement within the CFL and JFL

Group	*Mean*	*Median*	*Pixel*	*Radical*
Non-CJK	0.69	0.55	0.34	0.47
CSL	0.60	0.65	0.38	0.56
CFL	0.51	0.62	0.44	0.66
JSL	0.64	0.70	0.43	0.59
JFL	0.56	0.69	0.46	0.68
All	0.65	0.62	0.39	0.54

Fig. 6. Rank correlation of pixel and radical models against raters in given participant groups. Mean and median raters provided as reference scores.

Band	*Mean*	*Median*	*Pixel*	*Radical*
$[0, 1)$	0.69	0.55	0.34	0.47
$[1, 200)$	0.62	0.60	0.38	0.53
$[200, 600)$	0.64	0.69	0.41	0.61
$[600, 1000)$	0.69	0.72	0.46	0.52
$[1000, 2000)$	0.56	0.70	0.46	0.65
$[2000, \ldots)$	0.58	0.73	0.48	0.70

Fig. 7. Rank correlation of pixel and radical models against raters in across bands of kanji knowledge. Each band contains raters whose number of known kanji falls within that band's range.

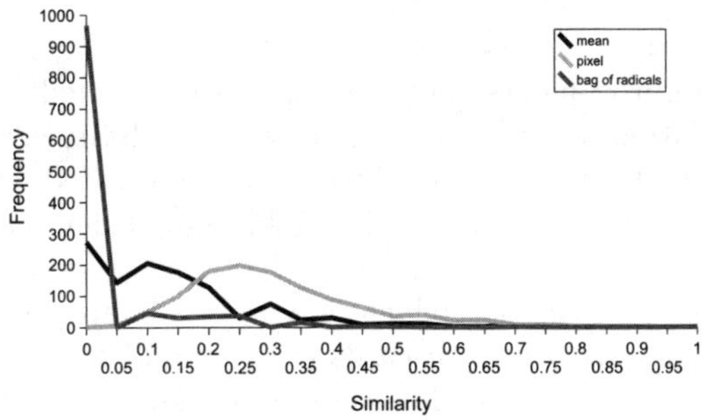

Fig. 8. Histograms of scaled responses across all experimental stimulus pairs, taken from mean rating, pixel and bag of radical models. Responses were scaled into the range $[0, 1]$.

groups to the median rater, a very strong result. It suggests that native speakers, when asked to assess the similarity of two characters, make their judgements primarily based either on the radicals which are shared between the two characters, or on some other measure which correlates well to identification of shared radicals. Intuitively, this makes sense. Native speakers have a great knowledge of the radicals, their meaning and their semantic or phonetic reliability. They also have the most experience in decomposing kanji into radicals for learning, writing and dictionary lookup.

The radical model still has poor correlation with the non-CJK group, but this is not an issue for applications, since similarity applications primarily target either native speakers or learners, who either already have or will pick up the skill of decomposing characters into radicals. To attempt to determine when such a skill gets picked up, Figure 7 shows agreement when raters are instead grouped by the number of kanji they claimed to know, based on their proficiency level. Aside from the $[600, 1000)$ band, there are consistent increases in agreement with the radical method as more kanji are learned, suggesting that the change is gradual, rather than sudden. Indeed, learners may start by focusing on strokes, only to shift towards using radicals more as their knowledge of radicals improves.

If we compare the histograms of the responses in Figure 8, we can see stark differences between human responses and the two models. The radical model considers the majority of stimuli to be completely dissimilar. Once it reaches stimulus pairs with at least one shared radical, its responses are highly quantised. The pixel model in comparison always finds some similarities and some differences, and exhibits a normal style bell curve. Human responses lie somewhere in between the pixel and radical models, featuring a much smaller number of stimuli which are completely dissimilar, and a shorter tail of high similarity than found with the pixel model.

6 Applications and Future Research

6.1 Similarity

Several potential improvements could be made to our similarity modelling. In particular, a translation invariant version of pixel similarity could be easily constructed and tested. On the other hand, the data-set created by Apel and Quint (2004) provides rich stroke data, which would allow a holistic model combining strokes, radicals and layout into a unified similarity metric. This should be superior to both the pixel model, which only approximates stroke-level similarity, and the radical model, which discards position and stroke information. The data-set created here allows fast and simple evaluation of any new similarity models, which should help foster further experimentation.

Kanji similarity metrics have many potential uses. A similarity or distance metric defines an orthographic space across kanji, which we can in turn use in novel ways. Our interest lies in dictionary lookup, and indeed a user could browse across this space from some seed point to quickly and intuitively arrive at a target kanji whose pronunciation is unknown. Particularly dense regions of this space will yield easily confusable pairs or clusters of high similarity. Presenting these to learners during study or testing could help these learners to differentiate between similar characters, but also to better structure their mental lexicon. Depending on the level of similarity the application is concerned with, the high amount of quantisation of responses may be a disadvantage for thresholding to only high-similarity responses. This remains one advantage of the pixel model over the radical model.

6.2 Confusability

From a similarity metric, the next step would be to determine confusability probabilities across pairs of kanji. Since confusability need not be symmetric, there may be other effects such as frequency which also play a role. Several studies of character perception at the word-level have found evidence of asymmetric interference effects for low frequency words with high frequency neighbours (van Heuven, Dijkstra, and Grainger 1998).

Similarity provides a means to bootstrap collection of confusability data, useful since authentic confusability data is difficult to find or construct. The available data mainly comes from controlled experiments in artificial environments, for example in explorations of illusory conjunctions (Fang and Wu 1989). Hand analysed logs for the FOKS dictionary detected a few accidentally corrected orthographic confusability examples, suggesting genuine occurrence of these errors (Bilac, Baldwin, and Tanaka 2004). The FOKS system also provides a method of turning a basic confusability model into a source of genuine confusability data. By adding the confusability model to the FOKS error model, any errors successfully corrected using the model will indicate genuine confusion pairs. We thus can create an informed confusability model, which bootstraps a cycle of confusability data collection and model validation.

6.3 Perception

There remain many open questions in orthographic similarity effects. Since the control pairs were not numerous enough to statistically determine similarity effects from the various effect types, further experimentation in this area is needed. In particular, it is unconfirmed as to whether semantic or phonetic similarity contributed to the similarity judgements analysed here. It could be tested by comparing pairs that share the same number of radicals, where the shared radicals for one pair were reliable semantic or phonetic cues, but the shared radicals for the other pair were not. We have discussed positional specificity of shared radicals as shown by Taft and Zhu (1997, 1999); the same specificity may also occur in radical-based similarity effects, and should be further investigated, as should stroke level effects.

7 Conclusion

We carried out an experiment seeking graphical similarity judgements, intending to form a similarity dataset to use for modelling. Since agreement between raters was moderate, we instead used the human judgements directly as our dataset, evaluating models against it. Two models were proposed, a stroke-based model and a radical based model. The stroke-based model was approximated using pixel differencing, rather than created directly. The pixel model showed medium agreement, but the radical model showed agreement as strong as the best individual rater for native speakers.

Although the radical-based model's performance may be adequate for some applications, there is much promise for a holistic model taking into account stroke, radical and positional effects. The data-set created here provides a means for quick and effective evaluation of new similarity models for kanji, thus allowing much experimentation. As well as seeding new dictionary lookup methods, the similarity models considered provide a basis for confusability models, which are themselves useful for error-correcting lookup, and in turn generating confusion data. Such confusion data, along with the similarity judgments collected here, will provide important evidence for understanding the perceptual process for kanji.

References

Apel, U. and Quint, J.: Building a graphetic dictionary for Japanese kanji – character look up based on brush strokes or stroke groups, and the display of kanji as path data, in Proceedings of the 20th International Conference on Computational Linguistics (2004)

Bilac, S., Baldwin, T., and Tanaka, H.: Modeling learners' cognitive processes for improved dictionary accessibility, in Proceedings of the 10th International Conference of the European Association for Japanese Studies, Warsaw, Poland (2003)

Bilac, S., Baldwin, T., and Tanaka, H.: Evaluating the FOKS error model, in Proc. of the 4th International Conference on Language Resources and Evaluation, Lisbon, Portugal (2004) 2119–2122

Eugenio, B. D. and Glass, M.: The kappa statistic: A second look. Computational Linguistics **30** (2004)(1) 95–101

Fang, S.-P. and Wu, P.: Illusory conjunctions in the preception of Chinese characters. Journal of Experimental Psychology: Human Perception and Performance **15** (1989) 434–447

Feldman, L. and Siok, W.: Semantic radicals contribute to the visual identification of chinese characters. Journal of Memory and Language **40** (1999) 559–576

Feldman, L. B. and Siok, W. W. T.: The role of component function in visual recognition of Chinese characters. Journal of Experimental Psychology: Learning, Memory and Cognition **23** (1997) 776–781

Reips, U.-D.: Standards for internet-based experimenting. Experimental Psychology **49** (2002)(4) 243–256

Saito, H., Inoue, M., and Nomura, Y.: Information processing of Kanji (Chinese characters) and Kana (Japanese characters). Psychologia **22** (1995) 195–206

Saito, H., Masuda, H., and Kawakami, M.: Form and sound similarity effects in kanji recognition. Reading and Writing **10** (1998)(3 - 5) 323–357

Taft, M. and Zhu, X.: Submorphemic processing in reading Chinese. Journal of Experimental Psychology: Learning, Memory and Cognition **23** (1997) 761–775

Taft, M., Zhu, X., and Peng, D.: Positional specificity of radicals in Chinese character recognition. Journal of Memory and Language **40** (1999) 498–519

van Heuven, W. J. B., Dijkstra, T., and Grainger, J.: Orthographic neighborhood effects in bilingual word recognition. Journal of Memory and Language **39** (1998) 458–483

Yeh, S.-L. and Li, J.-L.: Role of structure and component in judgments of visual similarity of Chinese characters. Journal of Experimental Psychology: Human Perception and Performance **28** (2002)(4) 933–947

Reconstructing the Correct Writing Sequence from a Set of Chinese Character Strokes

Kai-Tai Tang and Howard Leung

Department of Computer Science, City University of Hong Kong, Hong Kong
{itjeff, howard}@cityu.edu.hk

Abstract. A Chinese character is composed of several strokes ordered in a particular sequence. The stroke sequence contains useful online information for handwriting recognition and handwriting education. Although there exist some general heuristic stroke sequence rules, sometimes these rules can be inconsistent making it difficult to apply them to determine the standard writing sequence given a set of strokes. In this paper, we proposed a method to estimate the standard writing sequence given the strokes of a Chinese character. The strokes are modeled as discrete states with the state transition costs determined by the result of the classification into forward/backward order of each stroke pair using the positional features. Candidate sequences are found by shortest path algorithm and the final decision of the stroke sequence is made according to the total handwriting energy in case there is more than one candidate sequence. Experiments show that our results provide better performance than existing approaches.

Keywords: Pattern Recognition, Chinese character, handwriting, stroke sequence estimation.

1 Introduction

Each Chinese character is a logograph composed of strokes in a particular sequence. A generally accepted (standard) stroke sequence for each Chinese character is defined in Chinese dictionary. It is designed by taking the writing habit of people into consideration, e.g., from top to bottom and from left to right. People proposed some heuristic rules to generalize the standard stroke sequence [1], for example, "the upper strokes should be written first" and "the right strokes should be written last". However, the rules are not always accurate since there are many variations among the set of thousands of Chinese characters. Given a static Chinese handwriting image, after extracting the strokes, the heuristic rules may be ambiguous for predicting the standard stroke sequence for writing. This motivates us to develop a method to automatically determine the standard stroke sequence given a static Chinese character image. The stroke sequence information is useful in handwriting recognition and handwriting education applications.

In handwriting recognition researches, the stroke sequence contains important online information that can improve the recognition accuracy. The authors in [2] have surveyed the latest technologies in online handwriting recognition, such as

Y. Matsumoto et al. (Eds.): ICCPOL 2006, LNAI 4285, pp. 333 – 344, 2006.

classification using stroke sequence in the preliminary stage of character recognition. The handwriting recognition systems in [3] and [4] used the Hidden Markov Model (HMM) to identify the character with known stroke sequence. Chen *et al.* [5] suggested the input character is coarsely classified according to a set of features as well as the stroke sequence that is known in advance, and then followed by fine classification that performs detail comparisons. Liu *et al.* [6] model Chinese characters as code sequences of stroke segments defined by their shapes, line movement etc. A handwriting Chinese character (HCC) library is built and the recognition is done by matching the input code with the templates in database. Similarly, Hung *et al.* [7] identify characters by matching the input box code sequence with the templates in database. The box code is the label assigned to each stroke according to their shapes e.g. 1=horizontal stroke, 3=diagonal stroke. All the above work made use of the stroke sequence for performing online handwriting recognition.

For improving the accuracy of offline handwriting recognition, one can first extract the online information from the offline data and then apply online handwriting recognition techniques. The handwriting system proposed by Lin *et al.* [8] used a video camera to capture the writing of Chinese characters on ordinary paper. The hand movement showing the stroke sequence is recorded. The input strokes are then extracted and aligned with the corresponding instances on the stroke sequence. They are then parsed to the online character recognizer. The authors in [9] proposed to trace the offline strokes after line thinning to obtain a sequence of points as the online data. The sequence of stroke segments is then used to recognize Chinese characters and alphabets. This application motivates us to develop a method for determining the sequence of extracted strokes from characters on an image.

Besides handwriting recognition, the stroke sequence information is also important for handwriting education systems to train users to write the characters with the correct stroke sequence [10][11][12]. In [10], the threshold values of each stroke feature are manually defined for each character, for example, the straightness of the stroke. The system proposed in [11] and [12] compared the input handwriting with the template that is the standard handwriting by a skilled teacher.

A few researchers have targeted on the problem of recovering the stroke sequence given a set of strokes from a Chinese handwriting character. Shimomura [13] proposed a solution by minimizing the energy of hand movement. This method is only valid for characters with few strokes because the structure is more complex for characters with many strokes so that the standard stroke sequence may not necessary yield the minimum hand movement. Lau *et al.* considered the inter-stroke relationship and the position of the starting pixel of each stroke [14][15] in the cost and applied dynamic programming to determine the stroke sequence. It is not applicable in non-cursive handwriting because the algorithm considers the relationship between continuous strokes. Lee *et al.* [16] proposed a method similar to the one proposed by Shimomura [13], but with the application in Hangul (Korean character) recognition.

In this paper, we propose a method to estimate the standard sequence given a set of strokes from a Chinese character. Instead of solving for the minimum hand movement, we first perform classification to identify whether the order is forward or backward for every pair of strokes and use the classification result to define the state transition cost. Shortest path algorithm is then used to select the best candidate stroke sequence(s). Afterwards, energy minimization is applied for the final selection of the stroke sequence if there is more than one candidate from the shortest path algorithm.

2 The Proposed Method

We propose a method to estimate the stroke sequence given a set of strokes from a handwriting Chinese character image. In this paper, we focus on the method of stroke sequence determination so we assume all the strokes in each character sample have been correctly extracted from the image. The analysis of our proposed algorithm will thus not be affected by any errors from the stroke extraction. The extracted strokes are in the online form without the sequence information and they serve as the input data of our proposed study.

Fig. 1(a) shows our coordinate system. The top-left corner is considered as the origin (0, 0). The Chinese character "王" has four strokes with labels S_1, S_2, S_3 and S_4. We model the strokes as discrete states s_1, s_2, s_3, and s_4. The corresponding state diagram for the strokes of the Chinese character "王" is shown in Fig. 1(b). It can be seen that there are altogether 4 × 3 = 12 possible transitions among these four states. We formulate the shortest path problem to recover the stroke sequence in a Chinese character with online strokes but unknown sequence. To complete the route, each state should be visited once only without returning from the last state to the first state. Fig. 1(c) shows the expected route ($s_1 \rightarrow s_2 \rightarrow s_3 \rightarrow s_4$) with the associated state transition costs and it corresponds to the standard stroke sequence.

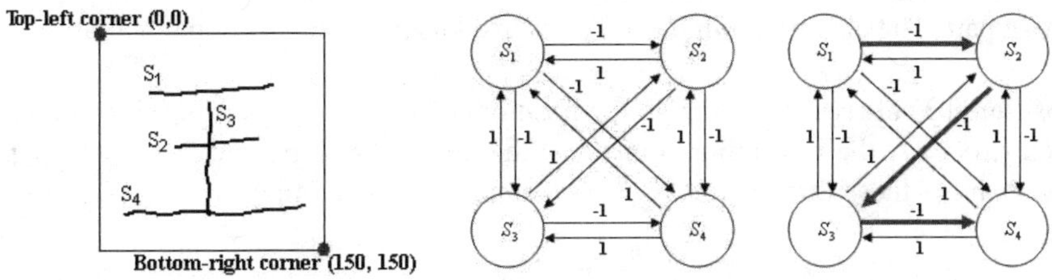

(a) The coordinate system (b) Possible transitions among the states (c) The expected route (the standard stroke sequence)

Fig. 1. Mapping between the strokes of the Chinese character "王" and the corresponding states

Existing methods ([13] and [14]) consider mainly the handwriting energy in order to determine the stroke sequence of Chinese characters. These methods ignored the relation (e.g. the coordinate difference) of stroke pairs that may reflect their relative orders. Hence, in our proposed method, we try to identify the relative order (forward/backward) of every stroke pair and then make use of this information to determine the stroke sequence.

Fig. 2 illustrates the block diagram of our proposed method. As the first step, we perform classification to determine in each stroke pair (s_i, s_j) whether they are in forward or backward order. The stroke pair is in forward order if s_i appears *before* s_j in the stroke sequence and it is in backward order if s_i appears *after* s_j in the stroke sequence. The classification result is treated as the traveling cost between states. In the next step, the genetic algorithm is used to solve for the shortest path(s) that correspond to candidate stroke sequences. The classification result (cost) is a discrete value (-1 for forward or 1 for backward) hence it may result in more than one shortest path (stroke sequence). The third step is applied if and only if the solution from the last step is not unique. In this case, the total handwriting energy along each candidate sequence is considered. The candidate sequence with the minimal hand movement is the best estimated sequence.

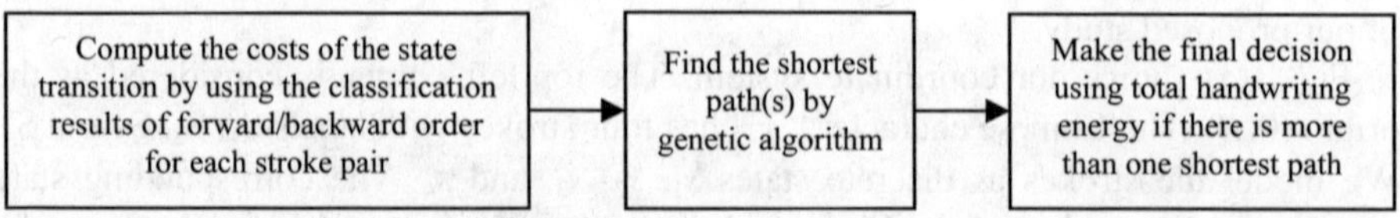

Fig. 2. Block diagram of our proposed method

2.1 State Transition Cost Computation

The state transition cost is set to be the result of the classification into forward/backward order for every stroke pair. We extract positional features by first selecting some points on each stroke and measuring some offsets between every stroke pair. The classification is then performed using these positional features.

Positional Features. To compute the positional features, we consider three points for each stroke: 1) the beginning point *b*, 2) the middle point *m*, and 3) the end point *e*. Fig. 3 shows these three points for two example strokes s_i and s_j.

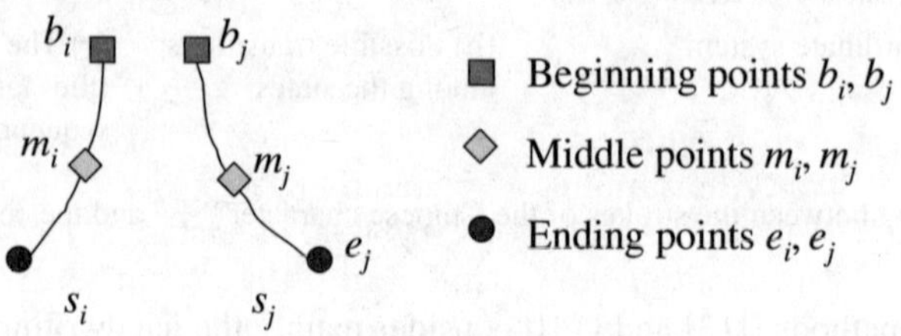

Fig. 3. Beginning, middle and ending points for two example strokes s_i and s_j

We apply several measures to compute the geometric difference between each pair of strokes. In particular, we compute the vertical offset (*V*), horizontal offset (*H*), and radial offset (*R*) between two points with one point from each stroke. Fig. 4 shows an illustration of these offsets.

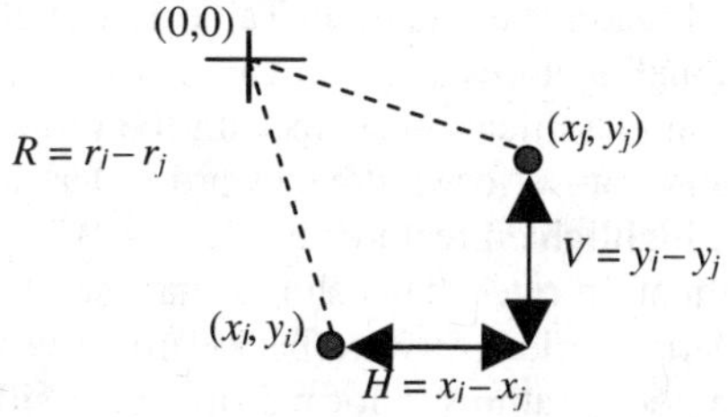

Fig. 4. Illustration of horizontal offset *H*, vertical offset *V* and radial offset *R*

(a) Vertical Offset

The vertical offset (*V*) is defined as $V = y_i - y_j$. The sign of vertical offset can help classify the relative order of the stroke pair. If the vertical offset is negative, then the stroke pair is likely in forward order, otherwise it is likely in backward order.

(b) Horizontal Offset

Similarly, the horizontal offset (*H*) is defined as $H = x_i - x_j$. If the horizontal offset is negative, then the stroke pair is likely in forward order, otherwise it is likely in backward order.

(c) Radial Offset

The radial offset (*R*) is the difference in radial component when the points are expressed in polar coordinates. It is defined as $R = r_i - r_j$ where. $r_i = \sqrt{x_i^2 + y_i^2}$ and $r_j = \sqrt{x_j^2 + y_j^2}$. If the radial offset is negative, then the stroke pair is likely in forward order.

Classification into Forward/Backward Order. We build a classifier to determine whether a stroke pair (s_i, s_j) is in forward/backward order, i.e., whether s_i appears before/after s_j in the standard stroke sequence. We have collected some Chinese handwriting characters written according to the standard stroke sequence thus the ground truth stroke sequence information is available. This means that given any two arbitrary strokes from a character, we know from the ground truth whether they are in forward or backward order. The positional features are taken as input to the classifier. We randomly choose one-tenth of the collected data for training the classifier, and the other nine-tenth for testing.

Table 1 shows the positional features that were examined. Each feature corresponds to a particular offset between various reference points on the strokes (i.e. beginning, middle or ending points) for a stroke pair. For instance, the feature $V(b_i, b_j)$ represents the vertical offset between the beginning points of a particular stroke pair (s_i, s_j). We compute the offsets for the same reference points on the two strokes, e.g., beginning point of stroke s_i versus beginning point of stroke s_j. As a result, the forward and backward classification rates are symmetric. We have altogether examined nine positional features as stated in Table 1. The classification results using each of these 9 positional features are shown in Table 1. They are obtained by comparing the sign of the classification score with the ground truth. It represents the percentage of stroke pairs whose forward/backward orders are correctly identified in the testing data. The six highlighted features labeled as "1", "2", "3", "7", "8" and "9" yield the highest classification rates. This shows that they are good features that can well distinguish the relative order (forward/backward) given a stroke pair. The next step is to combine these good features to form a single classifier.

Table 1. Results of classification into forward/backward order with various positional features

Feature Label	Feature	Classification Rate
1	$V(b_i, b_j)$	**77.56%**
2	$V(m_i, m_j)$	**85.19%**
3	$V(e_i, e_j)$	**79.73%**
4	$H(b_i, b_j)$	60.78%
5	$H(m_i, m_j)$	60.13%
6	$H(e_i, e_j)$	61.22%
7	$R(b_i, b_j)$	**87.15%**
8	$R(m_i, m_j)$	**88.24%**
9	$R(e_i, e_j)$	**82.35%**

Feature Description

$V(\bullet,\bullet)$: vertical offset
$H(\bullet,\bullet)$: horizontal offset
$R(\bullet,\bullet)$: radial offset

b_i, b_j: beginning points of strokes s_i and s_j
m_i, m_j: middle points of strokes s_i and s_j
e_i, e_j: ending points of strokes s_i and s_j

We combine the component feature values by a linear method to find the weight associated with each feature. Assume that there are n feature vectors in the training set and each feature vector contains m elements. We form an $n \times m$ feature matrix **A** and formulate our classification problem by constructing a linear classifier using matrix pseudoinverse [17]. The margins are either -1 or 1 for the classes "forward" and "backward" respectively. The weights of the features form the decision boundary and can be obtained by multiplying the pseudoinverse of the feature matrix **A** with the margin vector. Finally, the classification score is given by $c_f = \sum_{k=1}^{m} w_i f_k$. We have considered all 63 possible combinations of the six selected features. It is found that the combination of the features labeled "2", "7", "8" and "9" gives the highest classification rate of 93.46%.

2.2 Determination of the Shortest Path(s)

For each pair of stroke, the classification score is obtained by multiplying the feature vector with the weight vector determined in the previous step. When the classification score is negative, the classification result is set to be -1. When the classification score is positive, the classification result is set to be +1. The classification result is zero when the classification score is also zero. These classification results between every pair of strokes form a matrix and are used as the state transition costs. Next, we would like to find the path with the minimum cost such that all states are visited once.

This path finding problem is equivalent to the Traveling Salesman Problem (TSP). TSP belongs to the category of combinatorial optimization that has many variations and implementations [18]. Traditional TSP can be solved by branch-and-bound algorithms and linear programming. However, the complexity of TSP is large because it is a NP-hard problem. As a result, people developed methods to improve the performance of TSP, such as Markov Chain and genetic algorithm (GA). In this paper, we have adapted the genetic algorithm [19] that treats the TSP proposition as a biological gene and enhances the searching and optimization by randomization path selection. The shortest path is solved for finding possible stroke sequences. Since the classification result (cost) is a discrete value (-1 for forward or 1 for backward) hence it may result in more than one shortest path (stroke sequence).

2.3 Final Decision

If a unique minimal cost path is resulted after the step described in section 2.2, it will be the final stroke sequence result. However, it is possible to have several paths with the same minimal cost because the state transition costs are discrete. In this case, the total handwriting energy in each sequence can help us determine the final solution from the set of paths (sequences) with the minimum cost. The total handwriting energy is defined as the total distance moved by the hand during both pen-up and pen-down in writing the character. During the pen-up at the ending point of the stroke, it is assumed that the hand moves in a straight line toward the beginning point of the next stroke.

It can be expected that many erroneous cases would have been resulted if only handwriting energy were considered. For instance, sometimes neighbor strokes that are not consecutive strokes could be mistaken as consecutive strokes if we only consider the handwriting energy without considering other features that may give us more clues about the relative stroke order. On the other hand, the shortest path algorithm in section 2.2 is independent of the handwriting energy. After considering the features for the relative stroke order to come up with a reduced set of candidate stroke sequences by the shortest path algorithm, it is more likely that these candidate stroke sequences can be ranked properly according to the total handwriting energy. As the standard stroke sequence is generally designed for saving handwriting energy, the sequence with the minimal handwriting energy is chosen as the final solution.

3 Experiment and Results

Table 2 shows the sample data set used in the experiment. We have invited some public users to input Chinese characters using a tablet and we have collected 360 online Chinese handwriting character samples. These sample characters are free of production errors such that there are no missing, extra, concatenated or broken strokes. The captured stroke sequences of each sample character may not be the same as the standard stroke sequence but we have manually inspected each of them to obtain the ground truth standard stroke sequence. The number of strokes of these characters varies from 2 (very simple characters) to 14 (quite complicated characters).

These characters such as "率" and "裏" are non-cursive but some contain complicated structure consisted of a few smaller radicals.

In the testing, we have analyzed altogether 9804 stroke pairs from the sample characters. We compared our proposed algorithm with two existing methods for the identification of the standard stroke sequence. The first method (Method I) is proposed by Shimomura [13] with the idea that the pen-path (the movement of the pen) should conserve the minimal energy law. This method does not work well for characters with more strokes. The second method (Method II) is proposed by Lau *et al.* [14] and it applies dynamic programming with the costs formed by the position of the starting point of each stroke and the inter-stroke distances. This method is focused on the recovery of the stroke sequence in cursive signature. In non-cursive handwriting, the strokes are separated so that the direction tendency of the previous stroke, which is one of the features used in their method, is not applicable.

Table 2. The Chinese characters in our sample data set

Number of strokes	Sample count	Character samples
2	11	九丁十七八力二
3	59	刃三弓上山下已
4	70	卅片巨牙木巴廿斗凶五匹丑六中王比水火犬
5	79	吉本四必丙古目可叩叵卡卉甲丘白皮甘生出田
6	90	衣臣至光共兆企舟丟考臼早自回式米
7	7	男告阜辰角門
8	5	虎雨非者
9	17	重面革美皇疫虐甚
10	7	衷衰病骨舀鬥疲
>10	15	率區畢爽虛啓黑貳華雋裏褭

We have compared our proposed method with existing methods I [13] and II [14] in terms of percentage of fully recovered stroke sequence and the rank distance. The percentage of fully recovered stroke sequence means the percentage of characters with the entire stroke sequences correctly estimated using the stated methods. The rank distance is a measure of sequence similarity between the experimental result and the ground truth. We use the Kendall distance [20] as our rank distance. It measures the similarity of two sequences by counting the minimum number of swaps of any pair of strokes on the estimated sequence which can restore to the standard (ground truth) sequence. As a result, the smaller the rank distance, the more similar the estimated sequence and the standard sequence are.

Fig. 5(a)-(b) shows the comparison results among all algorithms in terms of the percentage of fully recovered stroke sequence and rank distance versus the number of strokes. Fig. 5(a) shows that our proposed method can correctly estimate stroke sequence in a relatively higher percentage. Fig. 5(b) shows that the rank distance of our proposed method is relatively low as compared with the existing methods.

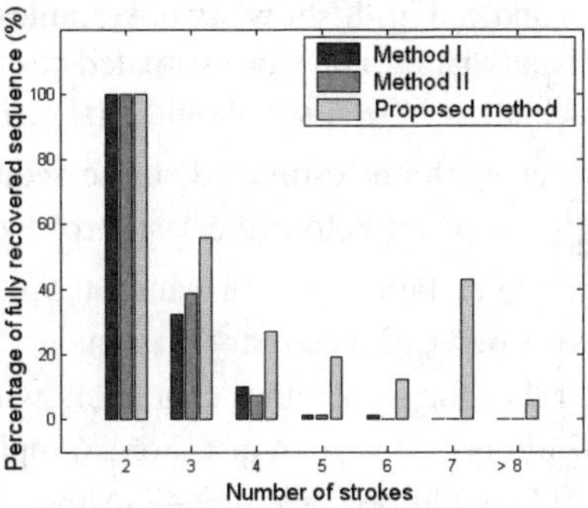

(a) Comparison of the percentage of fully recovered stroke sequence

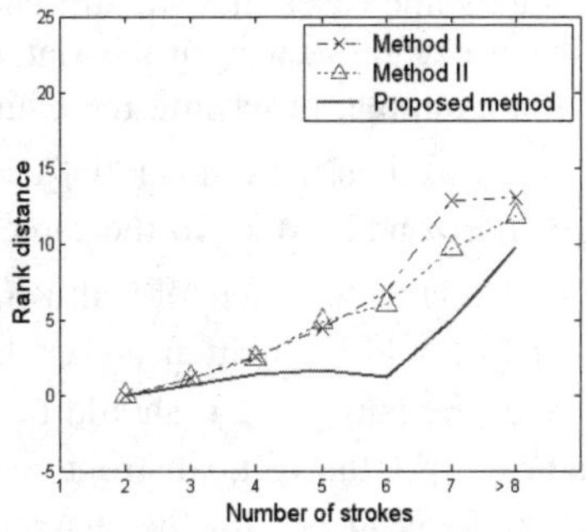

(b) Comparison of the rank distances

Fig. 5. Comparison among our proposed algorithm and two existing methods

The Chinese character “角” is one challenging case with which our proposed method can identify the standard stroke sequence correctly but not with the other two methods. Fig. 6 shows the standard stroke sequence from the Chinese dictionary. Fig. 7 illustrates the stroke-by-stroke sequences deduced by our method as well as the other two methods. For method I, the pen-up to pen-down positions of successive strokes are proven dominant in the stroke sequence determination as the successive strokes are near each other as shown in Fig. 7(b). The result of Method II looks better. However, since it considers the direction tendency of successive cursive strokes, it is not directly applicable to some non-cursive strokes. This explains why it gives a wrong sub-sequence in the radical “土” inside the character “角”. Our proposed method is more robust than the other methods that only consider the stroke proximity or direction tendency. In fact we have considered the relative order of each stroke pair by examining various distinguishable positional features described in Table 1.

Fig. 6. Standard stroke sequence of the character “角”

Methods	Output stroke sequence
(a) Proposed Method	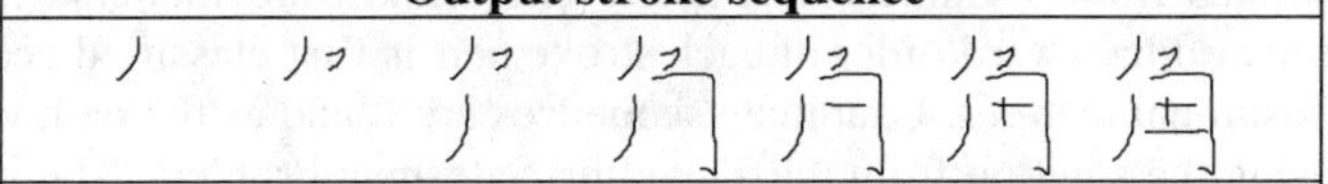
(b) Method I [13]	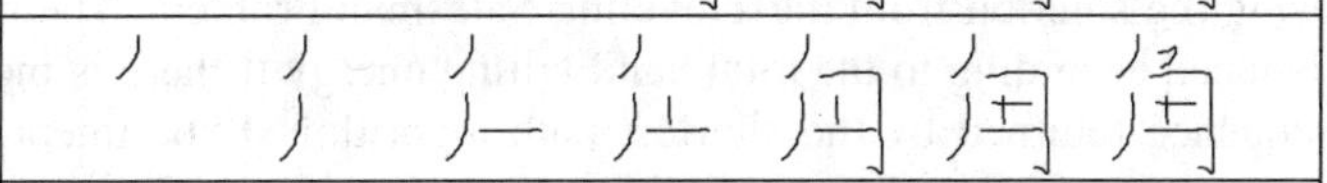
(c) Method II [14]	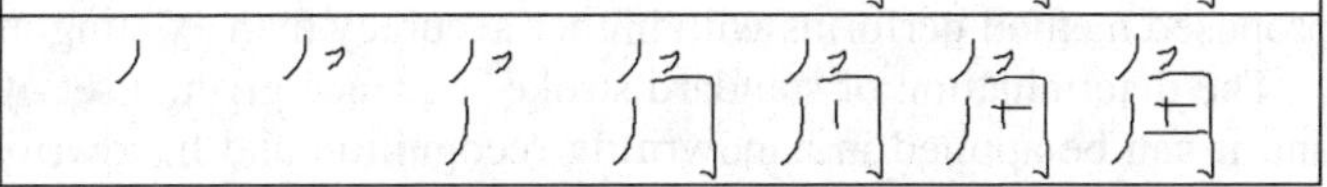

Fig. 7. Stroke sequence of the character “角” determined by various algorithms

There are some cases that we still cannot handle. Fig. 8 shows two examples with which the stroke sequences of the standard handwriting and our estimated result are shown. For example, in writing the Chinese character "病", we should first write the radical "疒"and then the radical "丙". However, with our estimated stroke sequence, the upper horizontal stroke in the radical "丙" appears before the last stroke of the radical "疒". The character "畢" illustrates another failed case in which the vertical stroke "丨" should be written at last but with our estimated stroke sequence, it is written after the part "曰". It should be noted that for some other characters with the same radical "甲", the vertical stroke "丨" would not be written last. For example, the standard stroke sequence for the character "里" should be 曰→甲→里. Although our algorithm is not 100% accurate, experimental results show that in general we can identify the standard stroke sequence correctly in more cases than existing methods. As future work, we can break down the character into smaller radicals, and then determine the stroke sequence by first considering stroke subsequence for each radical and then consider the sequence of the radicals.

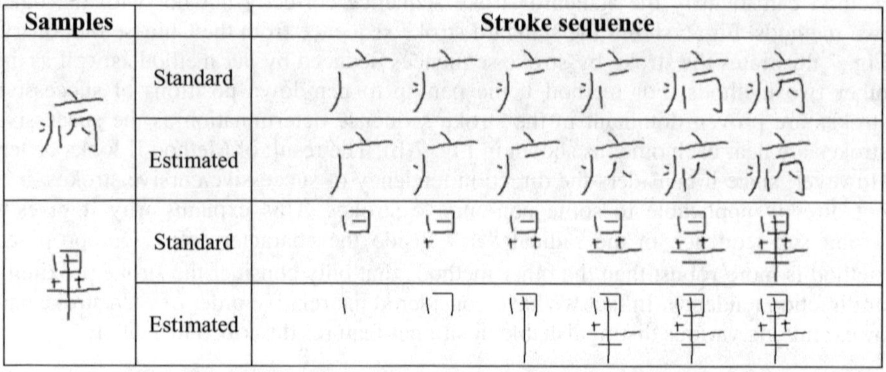

Fig. 8. Cases in which our proposed method failed to handle

4 Conclusion and Future Works

We proposed a method to determine the standard stroke sequence given a set of strokes from a Chinese character. The strokes are modeled as discrete states. The forward/backward order of each stroke pair is first classified according to the selected positional features. Candidate sequences are found as the path with the minimum cost using the solution from the Traveling Salesman Problem. The best stroke sequence is deduced according to the total handwriting energy if there is more than one candidate sequence returned by the shortest path algorithm. Experiment result shows that our proposed method performs with higher accuracy than existing methods.

The determination of standard stroke sequence given a set of strokes is very useful and it can be applied in handwriting recognition and handwriting education. A more accurate stroke sequence estimation enhances the extraction of online information from offline handwriting data and improve the performance of handwriting recognition. In handwriting education, it is common for students to write Chinese characters in the wrong stroke sequence. Our algorithm enables the computer to check the stroke sequence of the student's handwriting automatically. It benefits both the student and teacher as it could guide the student how to write the correct sequence and hence the teacher can save lot of effort in teaching and correcting student's handwriting homework.

In our experiment, we found that it is quite difficult to estimate with high accuracy the entire stroke sequence of characters with many strokes (i.e. more than 9 strokes). Some failed cases have been investigated. These cases are due to the complex/special structure of some Chinese characters such that the stroke sequence cannot be generalized solely in terms of positional features and handwriting energy.

As future work, we will explore more features to make the approach more general. We will also study the effect of breaking down a complex character into smaller parts/radicals, or assigning different weights for different stroke types. We may apply our proposed algorithm in other oriental character sets such as Japanese Hiragana/Katakana and Korean Hangul since these character sets are also logographs that look similar to Chinese characters.

Acknowledgments. The work described in this paper was supported by a grant from City University of Hong Kong (Project No. 9360092).

References

1. The structures and stroke sequence rules of Hanzi (In Chinese), The Commercial Press (HK) Ltd. http://www.cp-edu.com/TW/CIKU/free_html/fl_hzjglx.asp
2. Cheng-Lin Liu, Stefan Jaegerm and Masaki Nakagawa, "Online Recognition of Chinese Characters: The State-of-the-Art", IEEE Trans. Pattern Analysis and Machine Intelligence, vol. 26, no. 2, pp. 198–213, February 2004.
3. Mitsuru Nakai, Naoto Akira, Hiroshi Shimodaira, and Shigeki Sagayama, "Substroke Approach to HMM-based On-line Kanji Handwriting Recognition", 6th Intl. Conf. on Document Analysis and Recognition, pp. 491–495, Sep. 2001.
4. Thierry Artières and Patrick Gallinari, "Stroke level HMMs for on-line handwriting recognition", In 8th Intl. Workshop on Frontiers in Handwriting Recognition, Niagara, pp.227–232, August 2002.
5. Zen Chen, Chi-Wei Lee, and Rei-Hen Cheng, "Handwritten Chinese character analysis and preclassification using stroke structural sequence", Proc. of 13th Intl. Conf. on Pattern Recognition, vol. 3, pp. 89–93, 1996.
6. Ying-Jian Liu, Li-Qin Zhang, and Ju-Wei Tai, "A new approach to on-line handwritten Chinese character recognition", Proc. of the 2nd Intl. Conf. on Document Analysis and Recognition, pp. 192–195, 1993.
7. Kwok-Wah Hung, Wing-Nin Leung and Yau-Chuen Lai, "Boxing code for stroke-order free handprinted Chinese character recognition", IEEE Intl. Conf. on Systems, Man, and Cybernetics, vol. 4, pp. 2721–2724, 8–11 October 2000.
8. Feng Lin and Xiaoou Tang, "Dynamic stroke information analysis for video-based handwritten Chinese character recognition", Proceedings of the 9th IEEE Intl. Conf. on Computer Vision, vol. 1, pp. 695–700, 2003.

9. Charles C. Tappert, Ching Y. Suen, and Toru Wakahara, “The state of the art in on-line handwriting recognition”, IEEE Trans. Pattern Analysis and Machine Intelligence, vol. 12, pp. 787–808, 1990.
10. Chwee-Keng Tan, “An algorithm for online strokes verification of Chinese characters using discrete features”, 8th Intl. Workshop on Frontiers in Handwriting Recognition, pp. 339–344, 2002.
11. Kerry Tsang and Howard Leung, "Teaching Stroke Order for Chinese Characters by Using Minimal Feedback", Intl. Conf. on Web-based Learning (ICWL 2005), Hong Kong, August 2005.
12. Kai-Tai Tang, Ka-Ki Li and Howard Leung, "A Web-based Chinese Handwriting Education System with Automatic Feedback and Analysis", 5th Intl. Conf. on Web-based Learning (ICWL 2006), Malaysia, July 2006.
13. Takeshi Shimomura, “Informatics: input and output: Science of the stroke sequence of Kanji”, 8th Intl. Conf. on Computational Linguistics, pp. 270–273, 1980.
14. Kai-Kwong Lau, Pong-Chi Yuen and Yuan Yan Tang, “Stroke extraction and stroke sequence estimation on signatures”, 16th Intl. Conf. on Pattern Recognition, vol. 3, pp. 119–122, Aug 2002.
15. Kai-Kwong Lau, Pong-Chi Yuen and Yuan-Yan Tang, “Universal Writing Model for Recovery of Writing Sequence of Static Handwriting Images”, Intl. Journal of Pattern Recognition and Artificial Intelligence, vol. 19, no.5, pp. 1–27, 2005.
16. Moon-Jeung Joe, Huen-Joo Lee, “A combined method on the handwritten character recognition”, Proc. of the 3rd Intl. Conf. on Document Analysis and Recognition, vol. 1, pp. 112–115, August 1995.
17. Richard O. Duda, Peter E. Hart and David G. Stork, Pattern Classification (2nd edition), Wiley Interscience, 2000.
18. D. S. Johnson and L. A. McGeoch, The Traveling Salesman Problem: A Case Study in Local Optimization, Local Search in Combinatorial Optimization, E. H. L. Aarts and J.K. Lenstra (ed), John Wiley and Sons Ltd, 1997, pp 215-310.
19. J. Grefenstette, R. Gopal, R. Rosmaita, and D. Gucht, “Genetic algorithms for the traveling salesman problem”, Proceedings of the 2nd Intl. Conf. on Genetic Algorithms, Lawrence Eribaum Associates, Mahwah, NJ, 1985.
20. Teknomo Kardi. Similarity Measurement. http://people.revoledu.com/kardi/tutorial/Similarity/

Expansion of Machine Translation Bilingual Dictionaries by Using Existing Dictionaries and Thesauruses

Takeshi Kutsumi[1], Takehiko Yoshimi[2], Katsunori Kotani[3], Ichiko Sata[1], and Hitoshi Isahara[3]

[1] SHARP Corporation
Minosho-cho, Yamatokoriyama, Nara, Japan
{kutsumi.takeshi, sata.ichiko}@sharp.co.jp
[2] Ryukoku University
Seta-oe, Otsu, Shiga, Japan
yoshimi@rins.ryukoku.ac.jp
[3] National Institute of Information and Communications Technology
Seika, Soraku-gun, Kyoto, Japan
kat@khn.nict.go.jp, isahara@nict.go.jp

Abstract. This paper gives a method of expanding bilingual dictionaries by creating a new multi-word entry (MWE) and its possible translation previously unregistered in bilingual dictionaries by replacing one of the components of a registered MWE with its semantically similar words, and then selecting appropriate lexical entries from the pairs of new MWEs and their possible translations according to a prioritizing method. In the proposed method, the pairs of new nominal MWEs and their possible translations are prioritized by referring to more than one thesaurus and considering the number of original MWEs from which a single new MWE is created. As a result, the pairs which are effective for improving translation quality, if registered in bilingual dictionaries, are acquired with an improvement of 55.0% for the top 500 prioritized pairs. This accuracy rate exceeds the one marked with the baseline method.

1 Introduction

Bilingual dictionaries have such a great influence on the performance of machine translation (MT) systems and cross-lingual information retrieval applications, etc., that the expansion of these dictionaries is one of the principal objectives in MT system development. Above all, adding multi-word lexical entries is a promising area for dictionary improvement. There are many recurrent multi-word expressions, the proper translation of which is often different from the simple combination of each individually translated component of the expression. In this paper, such a bilingually fixed expression is defined as a "multi-word expression (MWE)". Since it is hard to say at this stage that MWEs are registered comprehensively enough in the dictionaries, it is necessary to continue to research and collect them.

Y. Matsumoto et al. (Eds.): ICCPOL 2006, LNAI 4285, pp. 345–354, 2006.

Generally speaking, a variance can be seen in the degree to which the proper translation for a nominal MWE matches with the sequence of its individually translated components. This study classifies two-word MWEs into three types as described in Section 2 below, and focuses especially on the two classes of MWEs which match comparatively better in the proper translation and the individual translation, such as "salt shaker".

There are two methods of acquiring expressions to be registered in bilingual dictionaries: one is using raw data in corpora, etc. (Kaji 2005; Shibata et al. 2005; Utsuro et al. 2005, etc.) and the other is using language resources created by people. (Ogawa et al. 2004; Fujita et al. 2005, etc.) The proposed method in this paper falls into the latter category since bilingual dictionaries and thesauruses constructed by people are used as resources.

The basic concept of this study is as follows: in the MWE "salt shaker" and its proper translation "*shio-ire*" , "shaker" corresponds to "*ire*" and this correspondence is highly likely to be accurate in the case of the MWE "spice shaker", since "spice" is a word semantically similar to "salt." Thus, the MWE "spice shaker" and its possible translation "*supaisu-ire*" are expected to be associatively acquired as a new lexical entry. Based on this assumption, this paper proposes a method of creating a new MWE and its possible translation previously unregistered in bilingual dictionaries by replacing one of the components of a registered MWE with its semantically similar words.

Some of the new multi-word lexical entries acquired with this method are almost certain to improve the translation quality, but others are not. In order to select appropriate entries, this method prioritizes the pairs of newly acquired MWEs and their possible translations by the criteria described in Section 5 below and outputs the pairs in order of highest priority.

2 Classifying MWEs

In this paper, from the standpoint of how the MWE's proper translation matches with the individual translation, two-word English MWEs are classified into the following three types.

Compounds with First Matched and Second Unmatched components: When the translation of the first component of an MWE matches with the first component of its proper translation while the translation of its second component does not match with the second component of the proper translation, it is classified as "First Matched and Second Unmatched".

For example, in the bilingual dictionary of the MT system used in this experiment, the proper translation "*shio-ire*" is given for the MWE "salt shaker". When the MT system translates them individually, "*shio*" is given for "salt" and "*sheeka*" for "shaker". Therefore, the first component's translation "*shio*" of "salt shaker" matches with the first component of its proper translation "*shio-ire*" while the second component's translation "*sheeka*" does not match with the second component of the proper translation.

Compounds with First Unmatched and Second Matched components: When the translation of the first component of an MWE does not match with the first component of its proper translation while the translation of its second component matches with the second component of the proper translation, this MWE is classified as "First Unmatched and Second Matched".

For example, the proper translation "*kaiin-kaisha*" is given for the MWE "member company". When the MT system translates them individually "*menbaa*" is given for "member" and "*kaisha*" for "company". Therefore, the first component's translation "*menbaa*" for "member company" does not match with the first component of its proper translation "*kaiin-kaisha*" while the second component's translation "*kaisha*" for "company"' matches with the second component of the proper translation.

Both Unmatched: When the translation of the first component of an MWE does not match with the first component of its proper translation and also the translation of its second component does not match with the second component of the proper translation, this MWE is classified as "Both Unmatched". For example, the proper translation "*seikuu-ken*" is given for the MWE "air superiority." Neither of the components' translations "*kuuki*" for "air" or "*yuuetsu*" for "superiority" individually matches with the proper translation "*seikuu-ken*" for "air superiority" at all.

This study proposes a method of dealing with the MWE types "First Matched and Second Unmatched" and "First Unmatched and Second Matched" out of the three types described above. This paper, however, refers only to the process of the MWE type "First Matched and Second Unmatched."

3 Outline of the Method Proposed

For the "First Matched and Second Unmatched" MWEs focused on in this paper, the translation of the second component does not match with the second component of its proper translation. It can be assumed that the same thing happens with replacing the first component of the original MWE with a semantically similar words. In other words, it is highly possible that the appropriate possible translation of a new MWE is acquired not by just joining each translation of its first and second components together, but by replacing the first component of the proper translation of the original MWE with the translation of a word semantically similar to the first component of the original MWE.

For example, as described in Section 1 above, since "*ire*" corresponds to "shaker" of "salt shaker", the translation for "shaker" must be "*ire*", not "*sheeka*" even in the case of the MWE "spice shaker," which is created by replacing "salt" in "salt shaker" with a word semantically similar to "salt" ("spice", etc.)

Based on this assumption, this paper proposes a method of acquiring a new MWE and its possible previously unregistered translation in bilingual dictionaries. The method proposed consists of three major steps as follows:

1. Extract the "First Matched and Second Unmatched" MWEs from bilingual dictionaries.
2. Extract words semantically similar to the first component of the original MWEs from a thesaurus. Create new MWEs by joining one of the semantically similar words and the second component of the original MWE together and then generate the possible translations for the new MWEs.
3. Prioritize the pairs of the new MWEs and their possible translations.

4 Creating a New MWE and Its Possible Translation

The most exhaustive method of acquiring new MWEs is to create all the possible combinations of the words fitting the rules of English syntax as candidates to be registered in a dictionary. It is, however, not desirable to take this method, since it will create a large number of incompatible combinations or compositional expressions, which are unnecessary to register in the dictionary.

In the method proposed, based on the assumption described in Section 3 above, new MWEs are created based on existing bilingual dictionaries and thesauruses. More specifically, synonyms for the first component of the original MWE are extracted from a thesaurus. Replacing the first component with its synonym creates the new MWEs.

WordNet Version 2.0(Miller 1998)[1] is used as a thesaurus. For the given word X, its synonym Y is extracted from WordNet only when it satisfies the following condition; Similarity $SIM(X, Y)$ (Kurohashi et al. 1996)[2] calculated in equation (1) below, is more than or equal to a certain threshold. The threshold is tentatively set at 0.7.

$$SIM(X,Y) = \frac{2 \times d_c}{d_x + d_y} \tag{1}$$

The values d_X and d_Y in the equation (1) are the depths from the root node to X and Y in WordNet, respectively, and d_C is the depth from the root node to the common node of both X and Y. From here on after, asemantically similar word Y refers to a word which has a similarity $SIM(X, Y)$ is more than or equal to the threshold.

A node in WordNet is represented as a word sense not as a word. Therefore when a word has more than one sense, it appears in WordNet more than once. There may exist more than one path from the root node to a given node. In order to extract words semantically similar to a word, all the possible paths for each word sense are followed.

By joining together the translation by an MT system of a word semantically similar to the first component of the original MWE, and the second component of the proper translation which corresponds to the second component of the original MWE, a possible translation for the new MWE is created.

[1] `http://wordnet.princeton.edu/online/`

[2] This degree of similarity is also used in the reference (Ogawa et al. 2004 etc.).

As shown in Table 1 below, in the case of the original MWE "salt shaker", "spice" is extracted from WordNet as a word semantically similar to the component "salt" more than or equal to the threshold. As a possible translation for "spice shaker", "*supaisu-ire*" is acquired by joining "*supaisu*" as the translation for "spice" from the MT system and "*ire*" corresponding to "shaker" in "salt shaker".

Table 1. Example of a new MWE and its possible translation

Original MWE:	salt shaker
Original proper translation:	*shio ire*
New MWE:	spice shaker
New possible translation:	*supaisu ire*

5 Prioritizing the Pairs of New MWE and Its Possible Translation

It is important to select appropriate entries for an MT dictionary from the pairs of new MWEs and their possible translations, which are acquired with the method described in Section 4 above. This study proposes a method of prioritizing each pair and outputting the pairs in order of highest priority from the standpoint of the contribution to the better translation quality.

5.1 Prioritizing by Similarity in an English Thesaurus

Replacing the first component of the original MWE with its semantically similar word creates a new MWE. Regarding this, it is assumed that the higher word sense similarity between the first component of the original MWE and its semantically similar word, the higher the probability of the new MWE as a correct English noun phrase.

Based on this assumption, the similarity $SIM(W_E, SimW_E)$ for the first component of the original MWE, WE, and its semantically similar word, $SimW_E$, is calculated in equation (1) described in Section 4 above. The similarity score is employed as priority score $Score_{SeedPair}(NewPair)$ for the pair of new MWE and its possible translation, $NewPair$, created from the pair of the original MWE and its proper translation, $SeedPair$.

$$Score_{SeedPair}(NewPair) = SIM(W_E, SimW_E) \tag{2}$$

5.2 Prioritizing by Similarity in a Japanese Thesaurus

This section describes countermeasures for the two significant problems found from the observation of the new MWEs created and prioritized through the methods described in Section 4 and Subsection 5.1 above. The problems are:

inappropriate selection of the word senses in WordNet and the inappropriate translation by the MT system. It is possible to acquire an appropriate possible translation for the new MWE, only when the selection of the word sense in WordNet and its translation by the MT system are both appropriate.

As described in Section 4 above, all the paths for each word sense are followed in order to extract semantically similar words from WordNet. As a result of this process, when the first component of the original MWE is replaced with a semantically similar words, not only appropriate MWEs but also inappropriate ones may be created. This is because some of the words extracted from WordNet cannot co-occur with the second component of the original MWE. For example, in the case of the original MWE "pop vocal" ("*poppu-kashu*") registered in the bilingual dictionary, "*tansansui*" can be selected, as well as "*ryuukouka*" , as a word sense for the first component "pop". When the path is followed with the word sense of drink "*tansansui*" in WordNet, the semantically similar word "soda" is acquired. However, the new MWE "soda vocal" ("*sooda-kashu*"), replacing "pop" with "soda", is an incorrect selection.

Also, the inappropriate translation by the MT system of a word semantically similar to the first component of the original MWE causes problems. For example, "state" is extracted from WordNet as a word semantically similar to the first component "government" of the original MWE "government authority" ("*seifu-toukyoku*"). The appropriate translations for "state" as the first component of the new MWE are "*kokka*", "*shuu*", etc. The MT system used in this experiment, however, gives the translation "*joutai*" "condition" so that the inappropriate possible translation "*joutai-toukyoku*" is created.

When either or both of the problems described above occur, it is highly possible that the similarity is low between the translation for the first component of the original MWE and the translation for the word semantically similar to the first component of the original MWE. Therefore, as a countermeasure for these two problems, it is useful to introduce a new criterion to lower the priority when the similarity of the translations is low. For such a criterion, the similarity in a Japanese thesaurus is used between the translation for the first component of the original MWE, W_J, and the translation for its semantically similar component , $SimW_J$. The similarity in the Japanese thesaurus is also calculated in equation (1) in Section 4 above. The EDR Electronic Dictionary[3] is used here for a Japanese thesaurus. As an appropriate example, in Table 2, the similarity between the translation "*shio*" for the first component of the original MWE "salt" and the translation "*supaisu*" for its semantically similar word "spice" is 0.667, which is comparatively high. On the other hand, as an inappropriate example, the similarity between the translation "*poppu*" for the first component of the original MWE "pop" and the translation "*sooda*" for the semantically similar word "soda" is 0.222, which is very low. Consider also the similarity between the translation "*seifu*" for the first component of the original MWE "government" and the translation "*jotai*" "condition" for the semantically similar word "state" for "government": 0.0.

[3] http://www2.nict.go.jp/kk/e416/EDR/J_index.html

The priority score for the pair of new MWE and its possible translation, $NewPair$, acquired from the pair of original MWE and its proper translation, $SeedPair$, is determined by both the similarity score in the English thesaurus $SIM(W_E, SimW_E)$ and the similarity score in the Japanese thesaurus $SIM(W_J, SimW_J)$. Thus, the adjusted score $Score_{SeedPair}(NewPair)$ calculated in equation (3) below is given to each pair $NewPair$.

$$Score_{SeedPair}(NewPair) = SIM(W_E, SimW_E) \times SIM(W_J, SimW_J) \quad (3)$$

5.3 Prioritizing Considered with a Number of Original MWEs

Original MWEs in bilingual dictionaries have been judged by their developers as having a positive effect on translation quality. Consequently, a new MWE created from a larger number of original MWEs is regarded as more plausible than one created from a smaller number of original MWEs; i.e. the former contributes more to the improvement of translation quality than the latter. Since this paper focuses on two-word MWEs of the type 'First Matched and Second Unmatched', the original MWEs, from which a single new MWE is created, have the same second component in common, such as "salt shaker" and "pepper shaker". While "spice" is extracted from WordNet as a word semantically similar to both "salt" and "pepper", "carbonate (*tansan'en*)" is extracted as a word semantically similar to only "salt". In this case, when calculating priority of newly created MWEs, "spice shaker" takes precedence over "carbonate shaker".

Based on this idea, when a single pair of a new MWE and its possible translation, $NewPair$, is created from the pairs of its original MWEs and their proper translations, $SeedPair_1$, ..., $SeedPair_n$,, the final priority for the new pair $AggScore(NewPair)$ is given as the sum of each priority for the new pair $Score_{SeedPair}(NewPair)$:

$$AggScore(NewPair) = \sum_{i=1}^{n} Score_{SeedPair_i}(NewPair) \quad (4)$$

For example, as shown in Table 2 below, the priority for new pair (spice shaker, *supaisu-ire*) is calculated to be 1.204 by adding priorities 0.556 and 0.648 respectively for the original pairs, when created from original pairs (salt shaker, *shio-ire*) and (pepper shaker, *koshou-ire*), while the priority for new pair (carbonate shaker, *tansan'en-ire*) is 0.762 when created from original pair (salt shaker, *shio-ire*).

Table 2. Example of new MWE created from multiple original MWEs

Original MWE:	salt shaker	pepper shaker
Original proper translation:	*shio ire*	*koshou ire*
New MWE:	spice shaker	
New possible translation:	*supaisu ire*	
priority	1.204 = 0.556 + 0.648	

6 Evaluation

6.1 Experimental Procedure

In this experiment, a version of our English-Japanese MT system[4] was used. Two-word lexical entries were extracted from part of the bilingual dictionary in the MT system. A total of 25,351 entries were extracted for the experiment. They were classified into four types, "First Matched and Second Unmatched", "First Unmatched and Second Matched", "Both Unmatched" as described in Section 2, and "Both Matched". There were some cases in which the combined translations of the first and second components of a new MWE completely match with the proper translation of an MWE registered in the bilingual dictionaries; therefore, they were classified as "Both Matched". The total percentage of the entries in types "First Matched and Second Unmatched" and "First Unmatched and Second Matched", which are focused on in this study, is 31.1%. The pairs of new MWEs and their possible translations were created from original MWEs in types "First Matched and Second Unmatched" and "First Unmatched and Second Matched", and output in order of highest priority.

The newly acquired pairs were evaluated as follows: First, English native speakers evaluated the new MWEs to judge whether or not they were appropriate as English noun phrases. Second, as for the new MWEs judged as appropriate, Japanese native speakers compared their possible translations acquired through our proposed method with their individual translations generated by the MT system, and then evaluated them as "Good", "Bad" or "Same". "Good" signifies that the new possible translation is better than the individual translation and adding this lexical entry is expected to improve the translation quality. "Bad" signifies that the new possible translation is worse than the individual translation and adding this lexical entry would deteriorate the translation quality. "Same" signifies that both or neither of the possible translation and the individual translation are appropriate and adding this lexical entry would not affect the translation quality. In Table 5 below, the new MWEs evaluated as inappropriate English noun phrases by the English native speakers are grouped with the ones with "Same" under "Safe". This tallying up was conducted because such inappropriate MWEs rarely appear in actual text and there would be little possibility of their hampering the translation quality.

6.2 Experimental Results and Discussion

A total of 759,704 pairs of new MWEs and their possible translations were acquired from the 2,148 pairs of original MWEs and their proper translations in type "First Matched and Second Unmatched" with the method described in Section 4 and 5 above. The top 500 prioritized pairs were evaluated.

In order to verify how much each criterion proposed in Section 5 above contributes to the improvement of prioritization performance, the results of the following four prioritizing methods are compared:

[4] `http://www.sharp.co.jp/ej/`

(a) Prioritize by the similarity in an English thesaurus (baseline method).
(b) Prioritize using the similarity in both English and Japanese thesauruses.
(c) Prioritize considering the similarity in an English thesaurus and the number of original MWEs.
(d) Prioritize considering the similarity in both English and Japanese thesauruses and the number of original MWEs (proposed method).

Table 3. Performances of prioritizing methods for top 500 pairs

	(a)	(b)	(c)	(d)
"Good"	85(17.0%)	165(33.0%)	186(37.2%)	297(59.4%)
"Safe" (Inappropriate)	353(70.6%)	247(49.4%)	233(46.6%)	150(30.0%)
"Safe" (Same)	50(10.0%)	67(13.4%)	64(12.8%)	31(6.2%)
"Bad"	12(2.4%)	21(4.2%)	17(3.4%)	22(4.4%)
Improvement rate	73(14.6%)	144(28.8%)	169(33.8%)	275(55.0%)

The performances of the prioritizing methods are shown in Table 3 above. The following points are found in the table, and it can be concluded that the effectiveness of the proposed method is verified:

1. In the baseline method (a), the percentage of "Good" minus that of "Bad", which is referred to as "improvement rate", is only 14.6%, and the accuracy achieved with this method is not enough. Also the percentage of "Safe (Inappropriate)" is 70.6%, which is very high.
2. In the method employing the similarity in a Japanese thesaurus as well as in an English thesaurus (b), the improvement rate increased to 28.8%, compared to the baseline method (a), and the percentage for "Safe (Inappropriate)" decreased to 49.4%. This shows that the consideration of similarity in a Japanese thesaurus prevented incorrect selection of word senses in an English thesaurus.
3. The method considering the number of original MWEs (c)is almost as effective as method (b).
4. In the proposed method (d), compared to the baseline method (a), although the percentage of "Bad" slightly increased from 2.4% to 4.4%, the improvement rate greatly increased from 14.6% to 55.0%. The percentages of "Safe (Same)" and "Safe (Inappropriate)" decreased from 10.0% and 70.6% to 6.2% and 30.0% respectively. Additionally, the proposed method (d) demonstrates high performance compared to methods (b) and (c). It is verified that it is highly effective to introduce into the baseline method (a) both the consideration of similarity in a Japanese thesaurus and reference to a number of original MWEs.

A group of the "Good" pairs of new MWEs and their possible translations acquired by the proposed method include such pairs as "hockey equipment" and "*hokkee-yougu*" and "car plant" and "*jidousha-koujou*", which are not listed even in the large-scale bilingual dictionary "*eijiro*"[5] .

7 Conclusion

This paper gives a method of expanding bilingual dictionaries in MT systems by creating a new MWE and its possible translation which had previously been unregistered in bilingual dictionaries, by replacing one of the components of a registered MWE with a semantically similar words, and then selecting appropriate lexical entries from the pairs of new MWEs and their possible translations according to a prioritizing method. In the proposed method, the pairs of new MWEs and their possible translations are prioritized by referring to more than one thesaurus and considering the number of original MWEs from which a single new MWE is created. As a result, the pairs which are effective for improving translation quality if registered in bilingual dictionaries are acquired with an improvement rate of 55.0% for the top 500 prioritized pairs. This accuracy rate exceeds the one marked with the baseline method.

References

1. Fujita, S., Bond, F. (2005). "An investigation into the nature of verbal alternations and their use in the creation of bilingual valency entries." Journal of Natural Language Processing, 12(3), 67-89. (in Japanese).
2. Kaji, H. (2005). "Extracting Translation Equivalents from Bilingual Comparable Corpora." IEICE Transactions on information and systems, E88-D(2), 313-323.
3. Kurohashi, S., Nagao, M.(editor) (1996). "Jisho to koopasu: Dictionary and Corpus." In Shizen-gengo-shori: Natural Language Processing, pp.231-264, Iwanami Shoten, Publishers. (in Japanese)
4. Mandala, R., Tokunaga, T., and Tanaka, H. (2000), "The Exploration and Analysis of Using Multiple Thesaurus Types for Query Expansion in Information Retrieval." Journal of Natural Language Processing, 7(2), 117-140.
5. Miller, G. (1998). "Nouns in WordNet." In Fellbaum, C. (Ed.), WordNet: An Electronic Lexical Database, pp.23-46. The MIT Press.
6. Ogawa, Y., Kamatani, S., Mahsut, M., Inagaki, Y. (2004). "Expansion of a Japanese-Uighur Bilingual Dictionary by Paraphrasing." Journal of Natural Language Processing, 11(5), 39-62. (in Japanese).
7. Shibata, M., Tomiura, Y., Tanaka, S (2005). "Assisting with Translating Collocations Based on the Word Co-occurrence on the Web Texts." Transactions of Information Processing Society of Japan, 46(6), 1480-1491.(in Japanese).
8. Utsuro, T., Hino, K., Horiuchi, T. and Nakagawa, S. (2005). "Estimating Bilingual Term Correspondences from Relevant Japanese-English News Articles." Journal of Natural Language Processing, 12(5), 43-69. (in Japanese).

[5] Searched at `http://www.alc.co.jp/index.html` on December 5, 2005.

Feature Rich Translation Model for Example-Based Machine Translation*

Yin Chen, Muyun Yang, Sheng Li, and Hongfei Jiang

MOE-MS Key Laboratory of Natural Language Processing and Speech,
Harbin Institute of Technology,
No. 92, West Dazhi Street, NanGang, Harbin, China, 150001
{chenyin, ymy, lisheng, hfjiang}@mtlab.hit.edu.cn

Abstract. Most EBMT systems select the best example scored by the similarity between the input sentence and existing examples. However, there is still much matching and mutual-translation information unexplored from examples. This paper introduces log-linear translation model into EBMT in order to adequately incorporate different kinds of features inherited in the translation examples. Instead of designing translation model by human intuition, this paper formally constructs a multi-dimensional feature space to include various features of different aspects. In the experiments, the proposed model shows significantly better result.

Keywords: EBMT, log-linear translation model, feature space.

1 Introduction

Nowadays, much attention has been given to data-driven (or corpus-based) machine translation, such as example-based machine translation [1] and statistical machine translation [2]. This paper focuses on EBMT approach.

The basic idea of EBMT is that translation examples similar to a part of an input sentence are retrieved and combined to produce a translation based on some heuristic criterion/measures. Most EBMT systems select the best example scored by the similarity between the input sentence and existing examples [3, 4, 5, 6, 7]. However, there is still much matching and mutual-translation information unexplored from examples. In this paper, log-linear translation model, which can be formally derived from the maximum entropy (ME) framework [8] is introduced into EBMT in order to adequately incorporate different kinds of information that can be explored from examples.

Most EBMT systems handle their translation examples using some heuristic meatures/criterion based on human intuition [9, 10, 11, 12]. For example, EBMT basically prefers larger translation examples, because the larger the translation is, the wider context is taken into account. Sometimes, translation model is designed for a specific domain which gives poor performance after it is transplanted to another

* Sponsored by the National Natural Science Foundation of China(60375019).

Y. Matsumoto et al. (Eds.): ICCPOL 2006, LNAI 4285, pp. 355 – 362, 2006.

domain. In order to alleviate this problem, this paper formally constructs a multi-dimensional feature space to include general features of different aspects.

The rest of this paper is organized as follows: Section 2 presents the basic idea of our approach, Section 3 discusses the design of feature functions. Section 4 presents the training method, Section 5 reports the experimental results and conclusion is drawn in Section 6.

2 A Maximum Entropy Approach to EBMT Model Training

The process of EBMT could be described as: Given an input sentence, match fragments against existing examples in the example base (EB) and identify the corresponding translation fragments, then, recombine them to give the target text.

The overall architecture of our approach is summarized in Figure 1. The first step is to search the optimal translations from all the possible candidate translations according to automatic evaluation metrics such as NIST and BLEU. Then, machine learning approach is used to explore what features they have as optimal translations, and therefore an optimized translation model is trained.

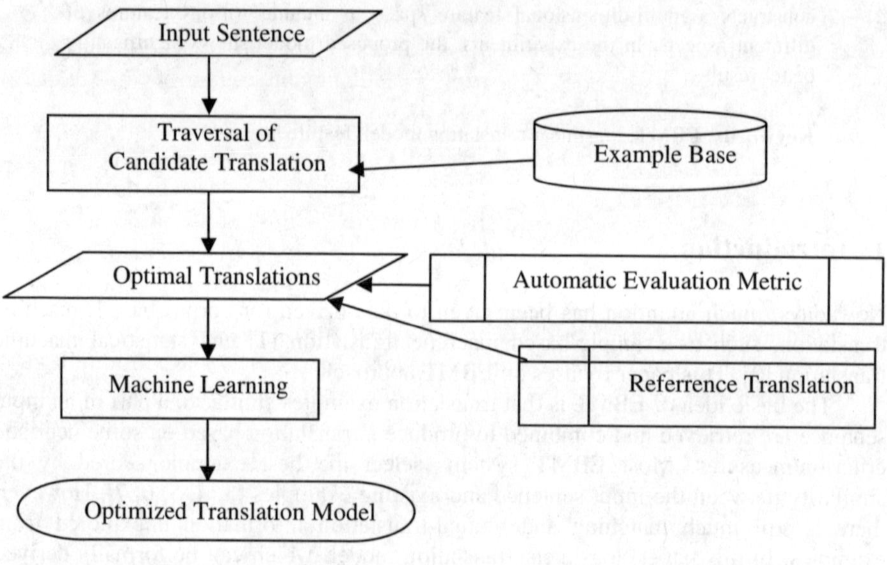

Fig. 1. Architecture of the translation approach

In this paper, the translation model is constructed under the maximum entropy (ME) framework which is very suitable to incorporate additional dependences. In this framework, we have a set of M feature functions $h_m(\mathrm{S},\mathrm{T}), m = 1,...,M$. For each feature function, there exists a model parameter $\lambda_m, m = 1,...,M$. The direct translation probability is given by:

$$p(\mathrm{T} \mid \mathrm{S}) = p_{\lambda_1^M}(\mathrm{T} \mid \mathrm{S}) = \frac{\exp[\sum_{m=1}^{M} \lambda_m h_m(\mathrm{S},\mathrm{T})]}{\sum_{\mathrm{T}'} \exp[\sum_{m=1}^{M} \lambda_m h_m(\mathrm{S},\mathrm{T}')]} \tag{1}$$

This approach has been suggested by [13, 14] for a natural language understanding task and successfully applied to statistical machine translation by [15].

We obtain the following decision rule:

$$\hat{\mathrm{T}} = \arg\max_{\mathrm{T}} \{ p_{\lambda_1^M}(\mathrm{T} \mid \mathrm{S}) \} = \arg\max_{\mathrm{T}} \{ \sum_{m=1}^{M} \lambda_m h_m(\mathrm{S},\mathrm{T}) \} \tag{2}$$

Hence, the time-consuming renormalization in Eq. 1 is not needed in search.

To train the model parameters λ_1^M of the log-linear models according to Eq. 1, we use the GIS (Generalized Iterative Scaling) algorithm [16]. It should be noted that, as was already shown by [16], by applying suitable transformations, the GIS algorithm is able to handle any type of real-valued features. In practice, we use YASMET[1] written by Franz J. Och for performing training.

3 Feature Functions

The design of features for natural language processing tasks is, in general, a critical problem. The inherent complexity of linguistic phenomena, often characterized by structured data, makes difficult to find effective linear feature representations for the target learning models. Therefore, we are considering to formally construct a high dimensional feature space to include different kinds of features. The following essential factors motivate our definition for features:

- The construct ion of feature space should take into account linguistic aspects as well as non-linguistic aspects. Non-linguistic feature are mainly based on the words matching and frequency statistics. These features only capture the surface-based information for the proper translation. However, linguistic features can provide the deeper understanding of the sentences.
- The linguistic features should be selected from multiple linguistic levels, i.e., lexics, syntax and semantics [17].
- The features should be selected from source language, target language and bilingual viewpoint respectively.

The feature space which is empirically defined according to the above motivations is shown in Table 1.

---

[1] Available at http://www.fjoch.com/YASMET.html

Table 1. Basic features

		Source Language	Target Language	Bilingual
Non-Linguistic	Word	• Length • Digit format ……	• Length • Digit format ……	
	Phrase (example)	• Length • Frequency • # of words ……	• Length • Frequency • # of words ……	• # of aligned words ……
	Sentence	• Length • # of words ……	• Length • # of words ……	• # of aligned words • # of examples ……
Linguistic	Lexical	• POS • Content word • Function word ……	• POS • Content word • Function word • Language model ……	
	Syntactic	• Sentence type • Voice ……	• Sentence type • Voice ……	
	Semantic	• WordNet class ……	• WordNet class ……	

In fact, features in Table 1 are only basic features. From the viewpoint of feature engineering, complex features should be further generated from these basic features by some mathematical techniques to represent specific meanings in deeper levels. The following lists the equations for the key features adopted in experiments:

• Sentence length ratio

Given a bilingual corpus that has an average sentence length ratio r, it is reasonable to believe that a pair of sentences whose length ratio is close to this r are more likely to match each other than two sentences whose length ratio is far different.

$$h(S,T) = \frac{\text{length of target sentence}}{\text{length of source sentence}} \tag{3}$$

• Word alignment rate

The more words are aligned, the more accurate translation is. word number of source sentence

$$h(S,T) = \frac{\sum_{\text{all examples}} \text{number of aligned word}}{\text{word number of source sentence}} \tag{4}$$

• Covering rate

This feature measures how much the input sentence is covered by examples.

$$h(S,T) = \frac{\sum_{\text{all examples}} \text{example length}}{\text{length of source sentence}} \quad (5)$$

- Average frequency of examples

This feature measures how often the examples are used.

$$h(S,T) = \frac{\sum_{\text{all examples}} \text{frequency of example}}{\text{number of examples}} \quad (6)$$

In this way, we can create numerous new features from the basic features to deal with specific problems of the baseline EBMT system.

4 Experiments

4.1 Experimental Setting

The corpus used in the experiment comes from the Basic Travel Expression Corpus (BTEC) which is provided in the IWSLT2004. The experiment consists of two parts: close test and open test. In the close test, example base is built using the development set of BTEC which consists of 506 Chinese sentences and their English references (506 × 16). In the open test, example base is built using the training set of BTEC which consists of 20000 English-Chinese sentence pairs. We build translation examples by using the alignment method mentioned in [18].

In the experiment, the performance of translation is evaluated in terms of BLEU score [19]. The baseline system is based on the work presented in [18]. It works in three steps. First, the input sentence is decomposed into fragments according to the example base, and the optimal decomposition is searched by an evaluation function based on some heuristic measures. Second, the optimal translation is searched for each example according to a translation selection model. The last step is translation generation in which examples' orders are adjusted according to N-Gram model.

The translation model proposed in this paper totally uses 21 features in the experiment. 6 of them are features at word level such as word alignment rate, 10 of them are features at phrase (example) level such as average frequency of examples, another 5 are features at sentence level such as sentence length ratio and covering rate. As an initial experiment, we are only testing the 302 affirmative sentences in the development set of BTEC. A comprehensive test and comparison of the other somewhat complex sentences such as negative sentences is going on. Further result will be reported elsewhere.

4.2 Results

The result is shown in Table 2. We can see that the proposed method outperform the baseline method significantly both in closed test and open test. The results demonstrate that more features really help to improve translation performance. Maximum entropy translation modeling provides an effective way to combine all these features.

Table 2. Experiment result

		BASELINE	PROPOSED
close	BLEU4	0.79	0.94
	BLEU3	0.83	0.95
	BLEU1	0.93	0.98
open	BLEU4	0.57	0.64
	BLEU3	0.62	0.71
	BLEU1	0.81	0.91

Feature selection is an important part of maximum entropy translation modeling. Table 3 evaluates the effect of each kind of feature adopted in our translation model. In the first row, all features are taken into consideration. In the second rows, features at word level are omitted; In the third row, features at phrase level are omitted; And in the last row, features at sentence level are omitted.

From the results, we see that all three kinds of features are very helpful in improving translation performance since omitting features at any level will lead to worse performance. In the close test, features at phrase level are the most important, it just validates the importance of examples in EBMT. In the open test, however, features at word level are the most important, this is because some examples can not be found in example base, so, features at word level play a more important role.

Table 3. Evaluation for each kind of features

		All features	w/o word level	w/o phrase level	w/o sentence level
close	BLEU4	0.94	0.87	0.70	0.83
	BLEU3	0.95	0.90	0.75	0.83
	BLEU1	0.98	0.95	0.92	0.91
open	BLEU4	0.64	0.51	0.52	0.62
	BLEU3	0.71	0.58	0.61	0.70
	BLEU1	0.91	0.79	0.82	0.90

For more concrete analysis, we randomly selected 60 proposed translations and checked them by hand. The hand check determined that 28 outputs are correct and the other 32 outputs are incorrect. The errors of translation can be classified as follows and the numbers of each error are listed in Table 4.

- Data sparseness

Data sparseness is the error caused by lack of translation examples. In such a case, the proposed method sometimes generates wrong translations by using a translation dictionary.

- Alignment error

Alignment error is the error caused by incorrect alignment results

- Word order

Word order refers to the case where the word order is ungrammatical.

- Selection error

Selection error is the error caused by unsuitable translation examples.

- Others

Others is a case that multiple errors occur ,and we could not classify it into the above error types.

Table 4. Error analysis

Error	Number
Data Sparseness	13
Alignment Error	4
Word Order	3
Selection Error	3
Others	9

Among them, data sparseness is the most outstanding problem. Therefore, we can believe that the system will achieve a higher performance if we obtain more corpora.

5 Conclusion

In order to adequately incorporate different kinds of information that can be explored from examples, this paper introduces log-linear translation model into EBMT. In addition, a high dimensional feature space is formally constructed to include general features of different aspects.

In the experiments, the proposed model shows significantly better result. The result demonstrated the validity of the proposed model.

In future work, we will consider of incorporating context similarity of translation examples as proposed in [20] to alleviate the data sparseness problem. On the other hand, we need to further improve the quality of example base since many errors are caused by incorrect alignment results. At the same time, we will consider of introducing some features to reflect the word orders of translations.

References

1. Makoto, N.: A framework of a mechanical translation between Japanese and English by analogy principle. In: Elithorn, A. and Banerji, R. (eds.): Artificial and Human Intelligence (1984) 173–180.
2. Brown, P. F., Stephen, A. D. P., Vicent, J. D. P., Robert, L. M.: The mathematics of statistical machine translation: Parameter estimation. Computational Linguistics, 19(2) (1993)

3. Nirenburg, S., Beale, S., Domashnev, C.: A full-text experiment in example-based machine translation. Int'l Conf. on New Methods in Language Processing Manchester. (1994) 78-87
4. Brown, R. D.: Adding Linguistic Knowledge to a Lexical Example-Based Translation Ayatem, in Proceedings of the 8th International Conference on Theoretical and Methodological Issues in Machine Translation (1999)
5. Macklovitch, E., Russell, G.: What's been Forgotten in Translation Memory, in Proceedings of the Conference of Association for the Machine Translation in Americas (2000)
6. Watanabe, H., Kurohashi, S., Aramaki, E.: Finding Steuctural Correspondences from Bilingual Parsed Corpus for Corpus-based Translation, in Proceedings of the 18th International Conference on Computational Linguistics (2000)
7. Liu, Z., Wang, H., Wu, H.: Example-based Machine Translation Based on TSC and Statistical Generation, in Proceedings of MT Summit X (2005)
8. Adam, L. B., Stephen, A. D. P., Vincent, J. D. P.: A maximum entropy approach to natural language processing. Computational Linguistics, 22(1), March (1996) 39-72
9. Eiji, A., Sadao, K., Hideki, K., Hideki, T.: Word selection for ebmt based on monolingual similarity and translation confidence. In Proceedings of the HLT-NAACL 2003 Workshop on Building and Using Parallel Texts: Data Driven Machine Translation and Beyond (2003) 57–64
10. Osamu, F., Hitoshi, I.: Constituent boundary parsing for example-based machine translation. In Proceedings of the 15th COLING (1994) 105–111
11. Kenji, I.: Application of translation knowledgeacquired by hierarchical phrase alignment for pattern-based mt. In Proceedings of TMI-2002 (2002) 74–84
12. Stephen, D. R., William, B. D., Arul, M., Monica, C. O.: Overcoming the customization bottleneck using example-based mt. In Proceedings of the ACL 2001 Workshop on Data-Driven Methods in Machine Translation (2001) 9–16
13. Papineni, K. A., Roukos, S., Ward, R. T.: Feature-based language understanding. In European Conf. on Speech Communication and Technology, Rhodes, Greece, September (1997) 1435–1438
14. Papineni, K. A., Roukos, S., Ward, R. T.: Maximum likelihood and discriminative training of direct translation models. In Proc. Int. Conf. on Acoustics, Speech, and Signal Processing, Seattle, WA, May (1998) 189–192
15. Franz J. O., Hermann, N.: Discriminative training and maximum entropy models for statistical machine translation. In Proceedings of the 40th Annual Meeting of the Association for Computational Linguistics (ACL), Philadelphia, PA, July (2002) 295-302
16. Darroch, J. N., Ratcliff, D.: Generalized iterative scaling for log-linear models. Annals of Mathematical Statistics (1972) 43:1470–1480
17. Cardie, C.: Automating Feature Set Selection for Case-Based Learning of Linguistic Knowledge. In Proc. of the Conference on Empirical Methods in Natural Language Processing. University of Pennsylvania, Philadelphia, USA (1996)
18. Liu, Z.: Research and Realization of Example Based Machine Translationon. Master Thesis of Harbin Institute of Technology (2003)
19. Papineni, K. Roukos, S. Ward, T. Zhu, W.: BLEU: a Method for Automatic Evaluation of machine translation. In *Proc. of the 40th Annual Conf. of the Association for Computational Linguistics (ACL 02)* Philadelphia, PA(2002)311–318
20. Eiji, A., Sadao, K., Hideki, K., Naoto, K.: Probabilistic Model for Example-based Machine Translation, in Proceedings of MT Summit X (2005)

Dictionaries for English-Vietnamese Machine Translation

Le Manh Hai, Nguyen Chanh Thanh, Nguyen Chi Hieu, and Phan Thi Tuoi

Ho Chi Minh City University of technology, 268 Ly Thuong Kiet Street, District 10, HoChiMinh City, Vietnam
{lmhai, ncthanh, nchieu, tuoi}@dit.hcmut.edu.vn

Abstract. Dictionary has an important role in Rule-Based Machine Translation. Many efforts have been concentrated on building machine-readable dictionaries. However, researchers have a long debate about structure and entry of these dictionaries. We develop a syntactic-semantic structure for English-Vietnamese dictionary as first measure to solve lexical gap problem, then use extend feature to improve Vietnamese dictionary to get grammatical target sentences. This work describes dictionaries used in English-Vietnamese Machine Translation (EVMT) at Ho Chi Minh City University of technology. There are three dictionaries: The English dictionary, the bilingual English-Vietnamese and the Vietnamese dictionary.

Keywords: Machine Translation, machine readable dictionary.

1 Introduction

In 2003 the English-Vietnamese Machine Translation (EVMT) has been found in Ho Chi Minh City's University of Technology. Due to limitation of budget and labor, the project has been developed in three phases. The first phase establishes framework and tools. This phase should end in 2007. The second phase will translate text in medical field from English to Vietnamese as stand-alone system. The second phase will lasts for 3 years. The third stage will combine this rule-based machine translation with other statistic machine translation to get unique powerful system. To the moment of this writing, the model for the EVMT has been established, including grammar rules for English, Vietnamese and small dictionaries for both languages. The model for EVMT is phrasal transfer model [see 5], which consist of analyzing, transferring and generating steps.

Phrasal transfer model for EVMT allows system to transfer a source lexical entry to a phrase in destination language. This feature of model could solve lexical gap – the problem of EVMT. However, it makes dictionary more complex and slows down real application where dictionary includes more than 10000 entries. To speed up lookup process, three dictionaries have built. First dictionary is English dictionary. It consists of English lexicon with syntactic-semantic information. This dictionary helps EVMT to analyze English sentences. The second dictionary is bilingual English-Vietnamese dictionary. The bilingual dictionary is word - to – phrase translation,

Y. Matsumoto et al. (Eds.): ICCPOL 2006, LNAI 4285, pp. 363 – 369, 2006.

which needed for EVMT to solve lexical gap. The third dictionary is Vietnamese dictionary, which is required for generating grammatical Vietnamese sentence.

Next section describes word - to – phrase transfer model for EVMT. These dictionaries are discussed in third section before some results has been reported.

2 Word-to-Phrase Transfer Model for EVMT

Word-to-Phrase Transfer model [5] was developed from transfer system of machine translation [6]. The special feature of word-to-phrase transfer model involves some syntactic structure changes in target text (see figure 1).

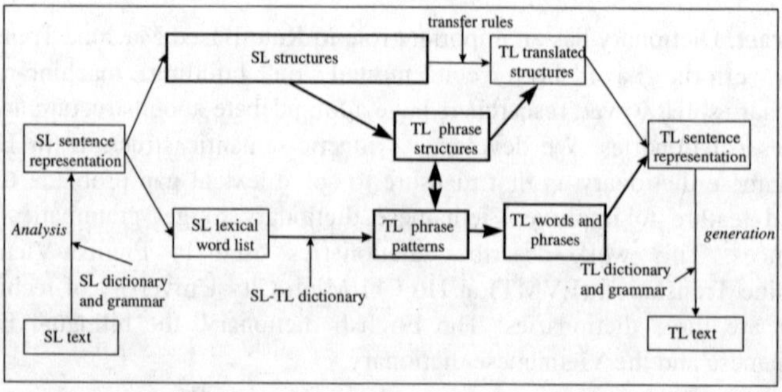

Fig. 1. Word-to-Phrase Transfer model

Important components of the model are Target language Phrase pattern and Target language Phrase structures that affect lexicons and structure of target text.

This model has two differences to the standard model: first, it allows transferring source lexicons to destination phrases, and second, it requires correcting structure of target sentence after lexicon mapping. Figure 2 represents workflow of transfer stage.

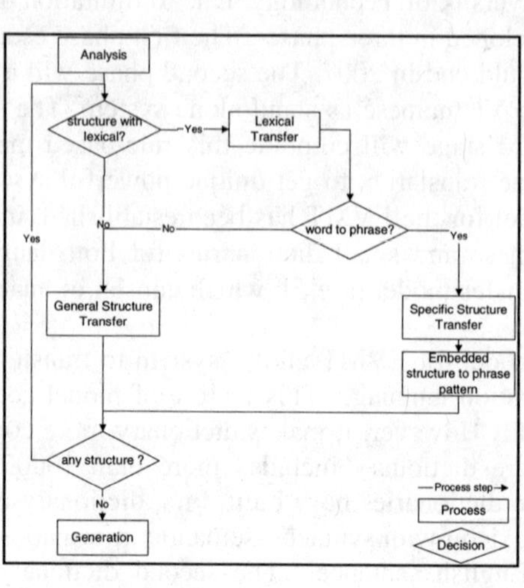

Fig. 2. Workflow of transfer stage of word-to-phrase transfer model

The structure and lexicon after analysis stage has transfer to lexicon and structure in destination text. However, in case of lexical gap, a source word should be translated to phrase in pattern of target language, then specific structure transfer is required. The analysis and generation phases depend on grammar rules of each language.

3 Dictionaries for EVMT

This section describes three dictionaries used in EVMT.

3.1 English Dictionary

English dictionary is the first data source involved in analysis process. English dictionary. The English dictionary covers only root words, instead of inflectional and derivational words. This feature speeds up looking process and simplifies parser for English. For this reason the Englex database [9] is used as input of English dictionary Englex supports root words in following format:

```
\lf `better
\lx AJ-AV
\alt Suffix
\fea comp
\gl1 `good
\gl2
```

Fig. 3. Entry in Englex 2.0 - a morphological description of English

```
<text>
    <body>
      <entry>
        <form>
          <orth>A</orth>
          <orth>a</orth>
        </form>
        <gramGrp>
          <pos>n</pos>
          <gen>m</gen>
        </gramGrp>
        <sense>
          <def>the first letter of the Language1
Alphabet</def>
        </sense>
      </entry>

    </body>
  </text>
```

Fig. 4. TEI XML format for English dictionary from [12] for lexicon “A”

The advantages of this approach are simplicity and therefore, speed. Unfortunately, many inflectional and derivational words are not covered in this dictionary. To map root lexicons to surface words, some morphological parsers like PC-KIMMO are applied.

Additionally, English dictionary applies XML format to TEI XML template [11] as shown in Fig. 4.

3.2 Bilingual English-Vietnamese Dictionary

The bilingual dictionary has central role in transfer model. Unfortunately, there is a little resource for English-Vietnamese dictionary. One of the most complete dictionaries has been developed by HoNgocDuc in [11].

Sample of entries is shown in figure 5.

Abuser

<ə'bju:zə>

-danh từ

/noun/

+ người lạm dụng

/man abuse/

+ người lăng mạ, người sỉ nhục, người chửi rủa

/man revile, man dishonour, man scold/

+ người nói xấu, kẻ gièm pha

/man backbite/

+ người đánh lừa, người lừa gạt

/man deceive/

Fig. 5. Entry in English –Vietnamese dictionary [11]

At first glance, this dictionary is not suitable for machine translation. Some tools are needed to change the paper-based dictionary to XML machine readable format like figure 6.

The special feature introduced here is <schema > which defines pattern of phrase. In this example "NP" mean "noun phrase" and number_of_schema =3 means this phrase has noun-verb structure. There are some templates for noun phrases such as noun-noun, noun –adjective, noun-verb, noun – adverb and so on [2].

```
<text>
    <body>
      <entry>
        <form>
          <orth> Abuser </orth>
          <orth> abuser </orth>
        </form>
        <gramGrp>
          <pos>n</pos>
          <gen></gen>
        </gramGrp>
        <vietnamese_sense sensenumber =1 >
          <def> người lạm dụng </def>
          <gramGrp>
                <pos>np</pos>
                <schema type = "NP" number_of_schema = 3/>
          </gramGrp>

        </vietnamese_sense >
        <vietnamese_sense sensenumber =2 >
          <def> người lăng mạ </def>
          <gramGrp>
                <pos>np</pos>
                <schema type = "NP" number_of_schema = 3/>
          </gramGrp>

        </vietnamese_sense >
        <vietnamese_sense sensenumber =3 >
          <def> người sỉ nhục </def>
          <gramGrp>
                <pos>np</pos>
                <schema type = "NP" number_of_schema = 3/>
          </gramGrp>


        </vietnamese_sense >

      </entry>

    </body>
  </text>
```

Fig. 6. Entry in English –Vietnamese dictionary in XML

3.3 Vietnamese Dictionary

Vietnamese dictionary helps choosing correct lexicons and reordering them in target text. The format of Vietnamese dictionary is the same for English dictionary discussed in section 3.1. For example, a noun “ác cảm” (“dislike”) has following form (figure 7).

@ác cảm
- d. Cảm giác không ưa thích đối với ai. Có ác cảm. Gây ác cảm.

```
<entry>
        <form>
          <orth> Ác cảm </orth>
          <orth> ác cảm </orth>
        </form>
        <gramGrp>
          <pos>n</pos>
        </gramGrp>
        <sense sensenumber =1 >
          <def> Cảm giác không ưa thích đối với ai </def>
        </sense>
        <sense sensenumber =2 >
          <def> Có ác cảm </def>
        </sense>
        <sense sensenumber =3 >
          <def> Gây ác cảm </def>
        </sense>

      </entry>
```

Fig. 7. Entry in Vietnamese dictionary in XML

This dictionary works well with Vietnamese grammar [2].

4 Results and Discussion

To the moment of this writing, there are more than 20,000 entries have been added to English dictionary, 109,000 words for English-vietnamese dictionary and about 30,000 lexicons for Vietnamese dictionary. Unfortuatelly, the final result of translation does not correlate to numbers of dictionary entries. From 1000 test sentences only 189 were successful transferred. Among these transferred text, only 112 sentences were acceptable. There are some reasons of the unsuccess: first, analyzed structure too complex, and too ambiguous; second, generation is simple and need more sophisticated to adapt Vietnamese grammar.

References

1. Le Manh Hai, Phan Thi Tuoi Nguyen Chi Hieu. English - Vietnamese dictionary with lexical conceptual structure for machine translation. FAIR’05. (2005)
2. Ban D.Q. ,Thung H.V. Vietnamese Grammar. Education publisher. (2000)

3. Binh D.T. , Chuong T.V., Du D.V. English Vietnamese dictionary. Social science publisher. (1990)
4. Dorr, B., C. R. Voss, E. Peterson and M. Kiker. Concept Based Lexical Selection. Proceedings of the AAAI-94 fall symposium on Knowledge Representation for Natural Language Processing in Implemented Systems, New Orleans. (2): 256-272. (1994)
5. Le Manh Hai, Asanee Kawtrakul, Yuen Poovorawan. Phrasal Transfer Model for Vietnamese-English Machine Translation. NLPRS '97. (1997)
6. Arnold D., Balkan L, Meijer S, R. Lee Humphreys, Sadler L. MACHINE TRANSLATION: An Introductory Guide. NCC Blackwell, London, (1994.)
7. Nguyen Thanh Bon, Nguyen Thi MinhHuyen, Lauren Romary, Vu Xuan Luong. Lexical descriptions for Vietnamese language processing. ALRWS2004. (2004)
8. John Hutchins. Towards a definition of example-based machine translation. Machine Translation Summit 2005. (2005)
9. Englex http://www.sil.org/pckimmo/v2/englex.html
10. Wordnet- a lexical database for the English language. http://wordnet.princeton.edu
11. The Free Vietnamese Dictionary Project. http://www.ifis.uni-luebeck.de/~duc/Dict
12. FreeDict - free bilingual dictionaries. http://www.freedict.org/en

Translation Selection Through Machine Learning with Language Resources

Hyun Ah Lee

School of Computer and Software Engineering,
Kumoh National Institute of Technology,
1, Yangho-dong, Gumi, Gyeongbuk, 730-701, Republic of Korea
halee@kumoh.ac.kr

Abstract. Knowledge acquisition is a critical problem for machine translation and translation selection. In this paper, I propose a tranlsation selection method that combines variable features from multiple language resources using machine learning. I introduce multiple measures for sense disambiguation and word selection that are based on language resources, and apply machine learning to combine those measures for translation selection. In evaluation, precision of translation selection improves even though a small-sized bilingual corpus is used as training data.

Keywords: Translation Selection, Knowledge Acquisition Problem, Machine Readable Dictionary, Target Language Corpus, Machine Learning.

1 Introduction

Translation selection is a process to select, from a set of target language words corresponding to a source language word, one that conveys the correct sense of a source word and makes more fluent target language sentences. Translation selection is a key problem in machine translation (MT) since the quality of translation varies significantly according to results of translation selection.

The difficulty of translation selection and machine translation is that they link two different languages thus requiring more complex knowledge than other problems concerning only one language. So, knowledge acquisition is a critical problem for many MT systems and translation selection methods including rule-based, knowledge-based and statistical methods since hand-crafting knowledge or bilingual corpora used for them are not easily available.

Most of recent researches on translation selection are based on statistical methods, which utilize various text resources such as a parallel corpus, a non-parallel corpus, a monolingual corpus or web documents. They focus on statistical information and empirical patterns in those resources and are usually unconcerned with other linguistic knowledge resources like a dictionary or a thesaurus. In this paper, I propose a method for translation selection that combines variable features from multiple language resources through machine learning. A mono-bilingual dictionary, WordNet,

Y. Matsumoto et al. (Eds.): ICCPOL 2006, LNAI 4285, pp. 370 – 377, 2006.

and a target language monolingual corpus are utilized to extract features and a small-sized bilingual corpus is used to train machine learning programs.

Machine learning has not been preferred for translation selection because it is hard to obtain an aligned bilingual corpus that provides enough training data for each source and target word. In this paper, machine learning is applied not to select a target word but to decide whether combination of employed features is appropriate for translation selection or not, so it generates binary classifiers that decide a target word is appropriate to a source word as translation based on features for translation selection. Base features for translation selection are extracted from a dictionary and a target language corpus, and then some of them are generalized into numerical scores by relevant measures to make features for translation selection not to depend on a specific word.

2 Previous Work

As masses of language resources have become available, a lot of statistical methods have been attempted for translation selection. Approaches based on statistical machine translation extract lexical information or word-class information from a bilingual corpus and calculate the probability for translation with extracted information. Along with them, much research has been devoted to extract translation equivalence to align sentences or words and to refine statistical models [1, 2]. Aligned sentences can be acquired from a bilingual corpus, therefore those approaches were generally regarded as a solution for the knowledge acquisition problem. However, a bilingual corpus is hard to acquire in itself and even it does not provide sufficient information for translation.

Dagan and Itai have proposed a new method for sense disambiguation and translation selection that uses word co-occurrence in a target language corpus [3]. Based on this method, some latest approaches have exploited word co-occurrence that is extracted from target and source monolingual corpora [4, 5, 6]. They extract clusters of target language words by using their co-occurrence in target language corpora. Since those clusters contain words that show similar distribution of contextual words, they are expected to have similar meaning and to server to reduce data sparseness of word co-occurrence. Those target language based methods could relieve the knowledge acquisition problem since they need only a monolingual corpus and simple mapping information between a source word and its target words. However, they are apt to select an incorrect translation because of ambiguity of target word senses for individual source words as shown in Lee et al. [7].

To overcome the difficulty of knowledge acquisition, some studies have attempted to extract rules or knowledge automatically from existing resources like a machine readable dictionary (MRD). Early researches on MRDs were concerned with limited types of words or applicable only when extra knowledge sources including a multilingual large knowledge base or a bilingual corpus already exist. But, recent researches exploit diverse information in MRDs and show practical result by combining such information with easily obtainable resources like a monolingual corpus. Lee et al. have proposed a hybrid method that combines sense disambiguation and word selection for translation selection [7]. They exploit a bilingual dictionary to extract knowledge for sense disambiguation and word selection, and introduce multiple measures for translation selection that combine extracted knowledge. They use linear combination to join multiple measures, and result shows that their method selects better translation than previous methods based on monolingual corpora even it conducts translation between languages in difference linguistics families.

3 Feature Extraction from Knowledge Resources

In this paper, I propose a translation selection method that refines the work of Lee and et al., which combines multiple measures for translation selection through an arbitrary way of linear combination, by adopting machine learning. With employing a new combining method, new features are introduced for sense disambiguation and word selection.

As indicated in Lee et al. [7], a bilingual dictionary classifies senses of a word into several sense divisions and groups its translations by each sense division. Along with their finding of 'word-to sense and sense-to-word' relationship between a source word and its translations, features for translation selection are extracted on the basis of a sense division of a bilingual dictionary. Features are extracted by exploiting a dictionary and a target language corpus and they are categorized into two groups - some for sense disambiguation and the others for word selection. Two features for sense disambiguation (*dPOS, dORD*) and one feature for word selection (*tPOS*) are directly extracted from a bilingual dictionary. A feature *tFREQ* is extracted from a target language monolingual corpus. Scores for four features for sense disambiguation (*simDEF, simEX, SYN* and *TYP*) are calculated by utilizing WordNet and a partial parser based on information from a dictionary. Finally, scores for three features for sense disambiguation (*sCOOCp, sCASEp* and *sRCASEp*) and six feature for word selection (*tCOOCp, tCASEp, sRCASEp, stCOOCp, stCASEp*, and *stRCASEp*) are computed using word co-occurrence in a target language corpus.

3.1 Features for Sense Disambiguation from a Mono-bilingual Dictionary

As shown in Fig. 1, a mono-bilingual dictionary provides various information including a POS, syntactic codes, sense definition sentences, and example sentences for each sense division, which can be used as clues for sense disambiguation. A

an·swer [ænsər, ɑ:n-] *vt.* 1 (P6,11,13,14) reply; respond to (a question) …라고 (대)답하다; …에 회답하다. ¶ ~ *my question* 내 질문에 답하다 / ~ *a speech* 답사를 하다/ ~ *a letter* 편지에 회답하다 / ~ *a teacher* 선생님께 대답하다 / *She answered that she knew something about it.* 그녀는 그것에 대해 무언가를 . … 2. (P6) act in answer for, say or do in return, repay. …에 응하다; (말, 행동으로) 되갚다. ¶ ~ *a knock [the door]* 노크 소리에 응하여 나오다. / ~ *the bell* 초인종 소리에 나오다[전화를 받다]…
— *vi.* (P1) 1 (P1,3) ((to)) reply. 대답하다; 회답하다. ¶ ~ *to a question* 질문에 답하다 / ~ *in the affirmative [negative]* 긍정[부정]의 대답을 하다 / …2 (P1) respond, act in reply to 응하다…
— *n.* © 1 something said or written in return; a replay. 대답; 회답. ¶ *an ~ to a question* 질문에 …

Fig. 1. A part of a English-English-Korean dictionary

feature *dPOS* is a POS of a given entry word. *dORD* is a feature that reflects sense distribution in real text. Let us suppose that the *i*-th word in an input sentence is s_i, and the *k*-th sense of s_i is s_i^k. An integer value *k* is the order of a sense s_i^k and it generally reflects the frequency of each sense because senses in a dictionary are ordered by significance in most cases. *dORD* has an integer value of *k* in the supposed case.

SYN has a value of 0 or 1. A dictionary provides syntactic codes like '©' that means an entry word is an uncountable noun or like 'P6' that means a verb entry word must has an object. We can test whether a word in an input sentence satisfies a given syntactic code and score 1 for *SYN* if an input word satisfies it or 0 if not. *TYP* also has a binary value. For a verb and an adjective, a sense definition sentence contains a typical subject or a typical object in brackets like '(a question)' in the first line of Fig. 1. *TYP* scores 1 if a subject or an object in an input sentence satisfies a given condition. WordNet is used to test compatibility of typical words.

As shown in Fig. 1, example sentences and definition sentences in a bilingual dictionary provide contexts for each sense of an entry word, thus they can work as good indicators for a sense. Given the *i*-th word in an input sentence s_i and its *k*-th sense s_i^k, we calculate *simDEF* and *simEX* of each sense s_i^k by the equation (1). *simDEF* calculates similarity between context words and words in senses definition sentences, and *simEX* calculate similarity between context words and words in example sentences. In the equation, *SNT* is a set of all content words except s_i in an input sentence, and $DEF_{s_i^k}$ and $EX_{s_i^k}$ are the sets of words in sense definitions and examples respectively. $\left|EX_{s_i^k}\right|$ is the number of elements in $EX_{s_i^k}$. $sim(s_j, w_d)$ returns a value for semantic similarity between two words s_j and w_d, which is also calculated based on WordNet.

$$simDEF(s_i^k) = \frac{\sum_{s_j \in SNT} \max_{w_d \in DEF_{s_i^k}} sim(s_j, w_d)}{|DEF_{s_i^k}|} \qquad simEX(s_i^k) = \frac{\sum_{s_j \in SNT} \max_{w_e \in EX_{s_i^k}} sim(s_j, w_e)}{|EX_{s_i^k}|} \tag{1}$$

3.2 Features for Sense Disambiguation from a Target Language Corpus

sCOOCp, sCASEp and *sRCASEp* are features for sense disambiguation that correspond to sense probability of Lee et al. In a bilingual dictionary, each sense division of a source word is mapped into a set of target words. Words in an individual set have the same sense or similar usage, so they can be replaced with each other in a translated sentence. *sCOOCp, sCASEp* and *sRCASEp* represent how likely target words with the same sense co-occur with translations of other words in an input sentence, which are calculated using the equation (2).

$$n(t_{iq}^k) = \sum_{(s_j,m,c)\in\Theta(s_i)} \sum_{p=1}^{m} \frac{f(t_{iq}^k, t_{jp}, c)}{f(t_{iq}^k) + f(t_{jp})} \qquad sCASEp(s_i^k) = \hat{p}_{sense}(s_i^k) = \frac{\sum_q n(t_{iq}^k)}{\sum_x \sum_y n(t_{iy}^x)} \tag{2}$$

In the equation, $\Theta(s_i)$ signifies a set of co-occurrences of a word s_i on a syntactic relation. In an element (s_j, *m*, c) of $\Theta(s_i)$, s_j is a word that co-occurs with s_i in an input sentences, c is a syntactic relation between s_i and s_j, and m is the number of translations of s_j. Given the set of translation of a sense s_i^k is T(s_i^k) and a member of T(s_i^k) is t_{iq}^k, the frequency of co-occurring t_{iq}^k and t_{jp} with a syntactic relation c is denoted by f(t_{iq}^k, t_{jp}, c). When all of the n(t_{iq}^k) are 0, it is smoothed using frequency of a target word with a weighting factor σ *i.e.* $n(t_{iq}^k) = \sigma \cdot f(t_{iq}^k) / \sum_p f(t_{ip}^k)$.

A syntactic relation between words is changed in the process of translation in some cases. For example, 'question' in "answer my question" is an object to a verb 'answer' but, in its translation "질문에 답하다(*jilmum-e dapha-da*)", '질문(*jilmun*)' that is a translation of 'question' is not an object of a verb translation '답하다(*dapha-da*)'. A score for *sRCASEp* is obtained by replacing c by a refined syntactic relation c' in the equation (2) when $c'$is a syntactic relation in a target language that corresponds to a syntactic relation c in a source language. A score for *sCOOCp,* which is a measure ignoring syntactic dependency of context words in an input sentence, is computed by using all words in a input sentence as context words in the equation (2),.

3.3 Features for Word Selection from a Dictionary and a Corpus

tPOS, tFREQ, tCOOCp, tCASEp, sRCASEp, stCOOCp, stCASEp, and *stRCASEp* are features for word selection and they are extracted or calculated based on a dictionary and a target language corpus. A feature *tPOS* is a tagged POS of a target word that is defined as a translation of a source word in a dictionary. A feature *tFREQ* has an integer value of target word frequency in a target language corpus.

tCASEp is calculated using a similar way of calculating *sCASEp* above. *tCASEp* represents how frequently a target word in a sense division co-occurs in a corpus with translations of other words of an input sentence. It is a probability of selecting t_{iq}^k from $T(s_i^k)$ and obtained using $n(t_{iq}^k)$ the equation (2) with the equation (3). *tCOOCp* and *tRCASEp* are calculated by ignoring syntactic dependency like *sCOOCp* and considering syntactic relations in a target language like *sRCASEp* respectively.

$$tCASEp(t_{iq}^k) = \hat{p}_{word}(t_{iq}^k) = \frac{n(t_{iq}^k)}{\sum_x n(t_{ix}^k)} \qquad stCASEp(t_{iq}^k) = \frac{tCASEp(t_{iq}^k)}{\max_j(tCASEp(t_{ij}^k))} \tag{3}$$

stCOOCp, stCASEp, and *stRCASEp* are obtained by normalizing *tCOOCp, tCASEp* and *sRCASEp* for each sense. *tCASEp* is a probability of selecting t_{iq}^k from $T(s_i^k)$ thus *tCASEp* becomes 1 when $T(s_i^k)$ has only one element. To make the maximum *stCASEp* for each sense to 1 not to discount the score of a word the sense of which has many corresponding target words, *tCASEp* of target words in a sense division is divided with its maximum value.

4 Combining Features Using Machine Learning

To combining variable features explained in the previous section, machine learning is adopted. Comparing with previous methods based on machine learning or a statistical method that try to choose a target word directly from a source word thus need huge training data, a machine learning program in this paper is utilized to decide only whether a word with given features is appropriate as translation or not. That means machine learning for translation selection is employed not to depend on features that relate on a specific word (like contextual word information) by generalizing those into numerical values as explained in the previous section.

Fig. 2 shows example feature vectors in training data that is extracted from an English sentence "Shares in all three banks are suspended on the Paris Bourse." and its translation, in which 'bank' is translated into '은행(*eunhaeng*)'. Each line has 22 values. First four values are an input word, its tagged POS in a input sentence, and two features - *dPOS* and *dORD*. And then, values for seven features are listed – *simDEF, simEX, TYP, SYN, sCOOC, sCASE* and *sRCASE*. The next value represents whether a given sense is appropriate in an input sentence, and it has 0 if a sense is inappropriate and has 1 if a sense is appropriate for an input sentence. In the figure, the first twelve values for '비탈' and '사면' are equal because they are included in the same sense division. A target word and eight features for word selection (*tPOS, tFREQ, tCOOC, tCASE, tRCASE, stCOOC, stCASE* and *stRCASE*) follow after sense appropriateness. The last value represents whether a given target word is a translation word in an aligned bilingual sentence, which also has a binary value of 0 or 1.

Training data is utilized to make three types of binary classifiers: the first one that decides sense appropriateness with first ten features for sense disambiguation (a tagged POS, *dPOS, dORD, simDEF, simEX, TYP, SYN, sCOOC, sCASE* and *sRCASE*), the second one for word selection with sense appropriateness and the eight features for word selection, and the last one that uses simultaneously all features for sense disambiguation and word selection with bypassing sense appropriateness. A source word and its target word are included in feature vectors only for readability and they are completely ignored in training.

Translation selection is processed in two ways – stepwise combining and simultaneous combining of features for sense disambiguation and word selection, and each utilizes three classifiers listed above. In stepwise combining, the first classifier decides appropriateness of a given sense, and then using result of the first classifier the second classifier decides appropriateness of a target word as translation. Simultaneous combining is done by the third classifier. If no target word is decided to be appropriate as translation, the first target word of the first sense of a source word is selected as translation, and if it decides more than two words are appropriate as translation, a word that has a smaller *sORD* value is selected as translation[1].

```
bank N n 1 0.9161 0.7617 1 1 0.00 0.0373 0.0373 0 둑 N 142 0.0000 0.0373 0.0373 0.0000 1.0000 1.0000 0
bank N n 2 0.8365 0.0000 1 1 0.00 0.0699 0.0699 0 비탈 N 98 0.0000 0.0258 0.0258 0.0000 0.5833 0.5833 0
bank N n 2 0.8365 0.0000 1 1 0.00 0.0699 0.0699 0 사면 N 168 0.0000 0.0442 0.0442 0.0000 1.0000 1.0000 0
bank N n 5 0.8887 0.0000 1 1 0.00 0.0055 0.0055 0 건반 N 21 0.0000 0.0055 0.0055 0.0000 1.0000 1.0000 0
bank N n 6 0.8641 0.8181 1 1 0.00 0.0292 0.0292 0 퇴적 N 21 0.0000 0.0055 0.0055 0.0000 0.2333 0.2333 0
bank N n 6 0.8641 0.8181 1 1 0.00 0.0292 0.0292 0 층 N 654 0.0000 0.0000 0.0000 0.0000 0.0000 0.0000 0
bank N n 1 0.8875 0.9126 1 1 1.00 0.7713 0.7713 1 은행 N 2934 1.0000 0.7713 0.7713 1.0000 1.0000 1.0000 1
bank N n 2 0.8799 0.7093 1 1 0.00 0.0000 0.0000 1 저장소 N 0 0.0000 0.0000 0.0000 0.0000 0.0000 0.0000 0
```

Fig. 2. Example feature vectors in training data

[1] As shown in Lee et al.[7], precision of selecting the first target word of the first sense is about 13%, so it is unlikely that this heuristic is a major factor for improved precision.

5 Evaluation

The proposed method was automatically evaluated with the same environment and the same evaluation method of Lee et al. [7]. Training data was extracted from test data of Lee et al. From 3,462 content words in 1,304 example sentences of a Korean-to-English bilingual dictionary, 70,142 feature vectors were extracted as training data. Test data is composed of remaining test data of Lee et al., which is extracted from reference books for the study of English, and sentences from an evaluation set for English-to-Korean machine translation. In total, 2,375 sentences with 7,580 words were used as test data, which are relatively longer than test sentences of Lee et al. Two machine learning algorithms of C4.5 and Tilburg Memory Based Learner (TiMBL)[8] were trained on 3,462 words. With the same test data, the translation selection system of Lee et al. was evaluated

Experiment was conducted while altering the combination of features. In cross-validation, a classifier for sense disambiguation showed highest precision when using a tagged POS, *dPOS, sORD, simEX, SYN* and *sRCASE.* In word selection, the combination of a tagged POS, *tPOS, tFREQ* and *stRCASE* showed the highest precision. Although it is clear that *sRCASE* and *stRCASE* are refined features for *sCASE* and *stCASE* and a feature *SYN* is an important linguistic clue for sense disambiguation, they decrease precision of translation selection in the system of Lee et al. It leads us to believe that adoption of machine learning for feature combining prevents wrong or excessive application of features. Features that result the highest precision in sense disambiguation and word selection were used to evaluate precision of translation selection.

In the proposed method, features for sense disambiguation and words selection are combined two ways of simultaneous combining and stepwise combining for translation selection. Table 1 shows precision of each combining method with precision of Lee et al. Result shows that feature combining through machine learning gets better precision than linear combination of Lee et al. for all cases. Two types of feature combining exert different effects for word classes, and it may come from lower precision of sense disambiguation for verbs and adverbs. In a decision tree generated by C4.5, many irrelevant decision nodes are detected, which seems to come from noisy training data.

To compare precisions of various researches, Prescher et al. [8] proposed to use standardized precision and their best result for noun was 79.7%. In test data used in this work, the average number of translations per noun word is 12.3, thus a

Table 1. Precision of Translation Selection

		Noun	Verb	Adj.	Adv.	All
Translation Preference of Lee et al.		52.67%	48.40%	55.79%	45.56%	51.32%
Simultaneous Feature Combining	C4.5	58.27%	54.46%	63.16%	52.86%	57.47%
	TiMBL	56.75%	54.16%	59.89%	48.78%	55.78%
Stepwise Feature Combining	C4.5	59.12%	51.31%	63.43%	55.14%	57.28%
	TiMBL	58.92%	52.23%	60.93%	49.45%	56.60%

standardized precision of the proposed method is 85.5%~86.5%. This is better than result of previous researches although the proposed method used automatically extracted knowledge with excluding word-dependent features in machine learning and was evaluated with languages of different linguistic families like English and Korean. Better precision would be obtained by manual evaluation since in automatic evaluation a selected translation is treated as incorrect if it is not exactly the same as a word in an aligned target language sentence even though it is also a good translation of a given source word.

6 Conclusion

In this paper, I proposed a machine learning based translation selection method. Appling machine learning after generalizing features for translation selection, the proposed method increases precision of translation selection even though it uses just a small-sized bilingual corpus with 3,462 content words as training data. Though result of the proposed method with a small-sized bilingual corpus shows better performance than that of previous methods, additional experiment should be conducted to analyze effect of increasing the size of training data. And I expect to improve the proposed method by introducing other knowledge resources like a target language thesaurus.

Acknowledgments. This paper was supported by Research Fund, Kumoh National Institute of Technology.

References

1. Peter F. Brown, John Cocke, Vincent Della Pietra, Stephen Della Pietra, Frederick Jelinek, John D. Lafferty, Robert L. Mercer and Paul S. Roossin: A Statistical Approach to Machine Translation. Computational Linguistics, Vol. 16, No. 2. (1990)
2. Dragos Munteanu and Daniel Marcu: Improving Machine Translation Performance by Exploiting Comparable Corpora. Computational Linguistics, Vol. 31, No. 4. (2005)
3. Ido Dagan and Alon Itai: Word Sense Disambiguation Using a Second Language Monolingual Corpus. Computational Linguistics, Vol. 20, No. 4. (1994)
4. Detlef Prescher, Stefan Riezler and Mats Rooth: Using a Probabilistic Class-Based Lexicon for Lexical Ambiguity Resolution. Proceedings of the 18th International Conference on Computational Linguistics. (2000)
5. Philipp Koehn and Kevin Knight: Estimating Word Translation Probabilities from Unrelated Monolingual Corpora Using the EM Algorithm. Proceedings of National Conference on Artificial Intelligence. (2000)
6. Eric Gaussier, Jean-Michel Renders, Irina Matveeva, Cyril Goutte and Hervé Déjean; A Geometric view on bilingual lexicon extraction from comparable corpora. Proceedings of the 42nd Annual Meeting of the Association for Computational Linguistics. (2004)
7. Hyun Ah Lee, Juntae Yoon and Gil Chang Kim. : Translation Selection by Combining Multiple Measures for Sense Disambiguation and Word Selection. International Journal of Computer Processing of Oriental Languages, Vol. 16, No. 3. (2003)
8. Daelemans, W., Zavrel, J., Van der Sloot, K., and Van den Bosch, A. : TiMBL: Tilburg Memory Based Learner, version 5.1, Reference Guide. ILK Technical Report Series 04-02. (2004)

Acquiring Translational Equivalence from a Japanese-Chinese Parallel Corpus

Yujie Zhang[1], Qing Ma[2], Qun Liu[3], Wenliang Chen[1], and Hitoshi Isahara[1]

[1] Computational Linguistics Group
National Institute of Information and Communications Technology
3-5 Hikari-dai, Seika-cho, Soraku-gun, Kyoto, 619-0289, Japan
{yujie, chenwl, isahara}@nict.go.jp
[2] Department of Applied Mathematics and Informatics
Ryukoku University, Seta, Otsu, 520-2194, Japan
qma@math.ryukoku.ac.jp
[3] Institute of Computing Technology
Chinese Academy of Sciences, Beijing, China
liuqun@ict.ac.cn

Abstract. This paper presents our work on acquiring translational equivalence from a Japanese-Chinese parallel corpus. We follow and extend existing word alignment techniques, including statistical model and heuristic model, in order to achieve a high performance. In addition to the statistics of the parallel corpus, the lexical knowledge of the language pair, such as orthographic cognates and bilingual dictionary are exploited. The implemented aligner is applied to the annotation of word alignment in the parallel corpus and the evaluation is conducted also. The experimental results prove the usability of the aligner in our task.

1 Introduction

Acquiring translational equivalence from parallel corpus is needed in corpus-based machine translation, whether in statistical machine translation or in example-based machine translation. In such systems, the quality of the machine translation output directly depends on the quality of the initial word alignment [1]. There are numerous researches on word alignment [1,2,3,4,5], which can be classified into two types, statistical models and heuristic models, according to the used models. Statistical models are based on statistical estimation theory and therefore are more coherent [1,3]. However, statistical models incorrectly align less frequently occurring words when statistically significant evidence is not available. Heuristic models use a function of the similarity between the types of the two languages and therefore are language dependent. It is easy for heuristic models to integrate linguistic knowledge for improving performance given a language pair or task [6,7].

Although the research on general model (language independent) for word alignment is important, the studies of how to apply the existing techniques to a given language pair or task also have special significance. So far reported researches on word alignment, whether using statistical models or heuristic models, involved many language pairs, such as English-French, English-Japanese, Chinese-English, and Chinese-Korean. To

Y. Matsumoto et al. (Eds.): ICCPOL 2006, LNAI 4285, pp. 378–386, 2006.

our knowledge, there is no report on Japanese-Chinese language pair. Word alignment between Japanese-Chinese is thought to be easier because of a similarity in orthography. However, no quantitative result has been reported.

This paper presents our work on word alignment for a Japanese-Chinese parallel corpus. We propose a new word alignment method by combining statistical model and heuristic model. For statistical model, we use GIZA++ software [2] and examine its performance on the Japanese-Chinese language pair. For heuristic model, we extend the method introduced by Ker [4] and implement a lexical-knowledge based aligner. We conduct a comparison between GIZA++ and the lexical-knowledge based aligner. The experimental results show that each aligner has respective advantages. We, therefore, develop a multi-aligner by combining the different aligners and confirm its effectiveness. The experimental results obtained in this work gave us insights on aligning words between Japanese and Chinese parallel texts.

2 NICT Japanese-Chinese Corpus

The Japanese-Chinese parallel corpus we used is developed at NICT (National Institute of Information and Communications Technology) of Japan [8]. Hereafter, we call it NICT Japanese-Chinese Corpus. The corpus consists of original Japanese sentences from Mainichi Newspaper and their Chinese translations. The corpus is already sentence aligned. In Japanese side, morphological and syntactic structures are annotated following the specification of the Corpus of Spontaneous Japanese [9]. In Chinese side, word segmentation and part-of-speech are annotated following the specification of Peking University [10]. The detail of the corpus is listed in Table 1.

Table 1. Characteristics of NICT Japanese-Chinese Parallel Corpus

	Japanese	Chinese
Sentences	38,383	38,383
Words	947,066	877,859
Vocabulary	36,657	33,425
Singletons	15,036	13,238
Aver. Sentence length	24.7	22.9

3 Lexical-Knowledge Based Aligner

This section describes the implementation of a lexical-knowledge based aligner. The aligner consists of two components. The first one is to establish reliable alignments and the second one is to extend alignments based on the alignments established in the first component. For a given Japanese sentence J and its Chinese translation C, let W_J and W_C denote their word sequences. (j, c) denotes a word alignment between the word j in W_J and the word c in W_C. j_i denotes a specified word at the position i in W_J and c_k^l denotes a specified word sequence with length l in W_C that starts at the position k.

3.1 Component for Establishing Reliable Alignment

In this component, we consider one-to-many alignment, the case of a Japanese word j being aligned with a sequence of Chinese words c_k^l $(1 \leq k \leq |W_C|, 1 \leq l \leq L)$. We set $L = 4$ in this paper. Hereafter, we use $\ddot{c}$ to express any word sequence within the length of 4. One-to-one alignment is a special case when $l = 1$. Actually, the case of many-to-one is also considered in the study. For simplicity of description, however, only the case of one-to-many is described here.

In measuring the degree of similarity between two strings, Dice coefficient defined in formula (1) is used. Based on this measure, we define the score of $(j, \ddot{c})$. The higher the score is, the more likely j is aligned with $\ddot{c}$.

$$Sim(x, y) = \frac{2 \times |x \cap y|}{|x| + |y|} \tag{1}$$

Three kinds of lexical resources we exploit are described below.

Bilingual Dictionary. A bilingual dictionary can help to identify the translation relations. Let C_j denote the Chinese translation set of j. We can define the score of $(j, \ddot{c})$ using the following formula [4].

$$Score_{dic}(j, \ddot{c}) = \max Sim_{c' \in C_j}(c', \ddot{c}) \tag{2}$$

$Score_{dic}$ expresses the score calculated by using a bilingual dictionary. In Section 4, we will describe how to automatically build a Japanese-Chinese dictionary.

Because of the deficiency of the bilingual dictionary, the other two kinds of lexical-knowledge are exploited as follows.

Orthography. About a half of Japanese words contain kanji, the Chinese characters used in Japanese writing. Japanese words may also contain hiragana or katakana, which are phonetic characters. Because some kanji words were adapted directly from China, their Chinese translations are the same as the words themselves. We then define the following score calculating formula. $Score_{ort}$ expresses the score of $(j, \ddot{c})$ that is calculated by using orthography information.

$$Score_{ort}(j, \ddot{c}) = Sim(j, \ddot{c}) \tag{3}$$

Simple and Traditional Chinese Characters. In Chinese, the simplified Chinese characters are used which were simplified from the traditional Chinese characters. At the same time, many Japanese kanji words maintain the form of the traditional Chinese characters as they were when they were introduced from China. The Chinese translations of such kanji words are usually the simplified character of the traditional characters. For example, the Chinese translation of the Japanese word 故郷 (hometown) is 故乡 in which 乡 is the simplified character of 郷. Let $S(j)$ denote the simplified form of j by converting each traditional character of j into a simplified character. We then define the following score calculating formula. $Score_{t-s}$ expresses the score estimated by using the correspondence between the traditional and the simplified Chinese characters.

$$Score_{t-s}(j, \ddot{c}) = Sim(S(j), \ddot{c}) \tag{4}$$

We then combine the three scores as described above in the following way, where $Score_{lex}$ expresses the score calculated by utilizing the three kinds of lexical resources.

$$Score_{lex}(j, \ddot{c}) = \max\{Score_{dic}(j, \ddot{c}), Score_{ort}(j, \ddot{c}), Score_{t-s}(j, \ddot{c})\} \quad (5)$$

The Algorithm of establishing reliable alignment is described as follows.

Algorithm 1. Align j in W_J with $\ddot{c}$ in W_C using lexical-knowledge.
Input W_J and W_C
Output Reliable alignment A_{rel}
Step 1 For each j in W_J , get $S(j)$ and obtain C_j from the bilingual dictionary. Initialize a $matrix[1 : |W_J|, 1 : |W_C|]$.
Step 2 For each candidate $(j, \ddot{c})$, compute $Score_{dic}(j, \ddot{c})$ using formula (2), $Score_{ort}$ $(j, \ddot{c})$ using formula (3), $Score_{t-s}(j, \ddot{c})$ using formula (4), and $Score_{lex}(j, \ddot{c})$ using formula (5). Store $Score_{lex}(j, \ddot{c})$ into $matrix$.

$$matrix[i, k] = \max_{1 \le l \le min\{|W_C|-k,4\}} Score_{lex}(j_i, \ddot{c}_k^l).$$

$\hat{l} = \arg\max_{1 \le l \le min\{|W_C|-k,4\}} Score_{lex}(j_i, c_k^l)$ is also stored into $matrix[i, k]$.
Step 3 Loop.
If $\max_{1 \le i \le |W_J|, 1 \le k \le |W_C|}\{matrix[i, k]\} \ge \theta_{lex}$, output $(j_{\hat{i}}, c_{\hat{k}}^{\hat{l}})$ to A_{rel},
$(\hat{i}, \hat{k}) = \max_{1 \le i \le |W_J|, 1 \le k \le |W_c|}\{matrix[i, k]\}$, where θ_{lex} is a preset threshold.
Set $matrix[i, k]_{i=\hat{i} \ or \ \hat{k} \le k \le \hat{k}+\hat{l}-1} = 0$, i.e., not consider $j_{\hat{i}}$ and $c_{\hat{k}}^{\hat{l}}$ again at the next step of the loop.
Remove $j_{\hat{i}}$ from W_J and $c_{\hat{k}}^{\hat{l}}$ from W_c .

3.2 Component for Extending Alignment

It is observed that words within one syntactic structure are often to be translated into words that belong to the same syntactic structure in the target sentence [4]. For example, when j_1 and j_2 belong to the same syntactic structure and j_1 has been aligned with Chinese word c_1, we can use this result to infer that j_2 will be aligned to a word which is near to c_1. For this purpose, we use the reliable alignments obtained in the first component. For an candidate $(\tilde{j}, \tilde{c})$, we take four reliable alignments into account: the two alignments that are the nearest to $\tilde{j}$ on the left and right and the two alignments that are the nearest to $\tilde{c}$ on the left and right. We estimate the score of $(\tilde{j}, \tilde{c})$ as follows. In this algorithm we only consider one to one alignment.

First, add $(Null_0, Null_0)$ and $(Null_{|W_J|+1}, Null_{|W_C|+1})$ to A_{rel} as the leftmost and the rightmost reliable alignments.

Second, search four alignments from A_{rel} as follows.

(1) $a_{left\_of\_\tilde{j}} = (j_{left\_of\_\tilde{j}}, \ddot{c}_{left\_of\_\tilde{j}})$, in which $j_{left\_of\_\tilde{j}}$ is the nearest word to $\tilde{j}$ from the left side, of all j that have been aligned in the first component.

(2) $a_{right\_of\_\tilde{j}} = (j_{right\_of\_\tilde{j}}, \ddot{c}_{right\_of\_\tilde{j}})$, in which $j_{right\_of\_\tilde{j}}$ is the nearest word to $\tilde{j}$ from the right side, of all j that have been aligned in the first component.

(3) $a_{left\_of\_\tilde{c}} = (j_{left\_of\_\tilde{c}}, \ddot{c}_{left\_of\_\tilde{c}})$, in which the last word of $\ddot{c}_{left\_of\_\tilde{c}}$ is the nearest word to $\tilde{c}$ from the left side, of all $\ddot{c}$ that have been aligned in the first component.

(4) $a_{right\_of\_\tilde{c}} = (j_{right\_of\_\tilde{c}}, \ddot{c}_{right\_of\_\tilde{c}})$, in which the first word of $\ddot{c}_{right\_of\_\tilde{c}}$ is the nearest word to $\tilde{c}$ from the right side, of all $\ddot{c}$ that have been aligned in the first component.

We illustrate this searching in Fig. 1. There are five reliable alignments, expressed in lines. For the considered candidate $(\tilde{j}, \tilde{c}) = (5, 5)$, $a_{left\_of\_\tilde{j}} = (4, 2)$, $a_{right\_of\_\tilde{j}} = (6, 6)$, $a_{left\_of\_\tilde{c}} = (3, 3)$, $a_{right\_of\_\tilde{c}} = (6, 6)$ are selected, because the Japanese words 4 and 6 are the nearest words to $\tilde{j} = 5$, the Chinese words 3 and 6 are the nearest words to $\tilde{c} = 5$, from the left side and right side, respectively, as expressed in black ball.

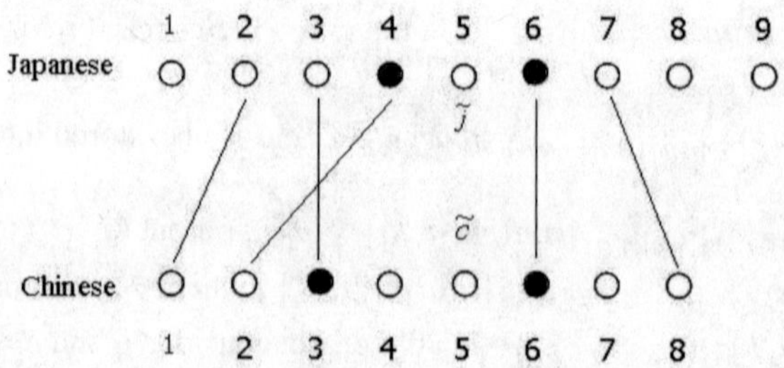

Fig. 1. Illustration of how to select four reliable alignments for the considered alignment candidate $(\tilde{j}, \tilde{c})$

Regarding to $a_{left\_of\_\tilde{j}}$, the degree at which the candidate $(\tilde{j}, \tilde{c})$ dislocate from it can be measured using the following quantitative variables.

$$\Delta i = \tilde{i} - i, \Delta k = \tilde{k} - k, \tag{6}$$

in which $\tilde{i}$ and $\tilde{k}$ are the positions of $\tilde{j}$ and $\tilde{c}$, i and k are the positions of $j_{left\_of\_\tilde{j}}$ and the last word of $\ddot{c}_{left\_of\_\tilde{j}}$, respectively.

Third, estimate the score of $(\tilde{j}, \tilde{c})$ by regarding to $a_{left\_of\_\tilde{j}}$ as follows [7].

$$Score_{left\_of\_\tilde{j}}(\tilde{j}, \tilde{c}) = \frac{2}{(|\Delta i| + |\Delta k|)e^{|\Delta i - \Delta k|}} \tag{7}$$

We assume that a candidate $(\tilde{j}, \tilde{c})$ is likely to be true if it nears a reliable alignment. The item $(|\Delta i| + |\Delta k|)$ imposes penalty on the score using the degree at which $(\tilde{j}, \tilde{c})$ dislocate from the reliable alignment $a_{left\_of\_\tilde{j}}$. The smaller the sum of them is, the larger the score of the alignment is. The item $e^{|\Delta i - \Delta k|}$ also imposes penalty on the score. When $\tilde{j}$ and $\tilde{c}$ dislocate from the reliable alignment in the same direction, i.e., $\tilde{j}$ and $\tilde{c}$ are on the left or right side of the reliable alignment at the same time, Δi and Δk

are positive or negative at the same time. In this case, $|\Delta i - \Delta k| = ||\Delta i| - |\Delta k|| \leq \min\{|\Delta i|, |\Delta k|\}$. When $\tilde{j}$ and $\tilde{c}$ dislocate from a reliable alignment in an opposite direction, i.e., $\tilde{j}$ is on the left side and $\tilde{c}$ on the right side, Δi is negative and Δk is positive, or vice versa. In this case, $|\Delta i - \Delta k| = ||\Delta i| + |\Delta k|| \geq \max\{|\Delta i|, |\Delta k|\}$. As a result, the case of $\tilde{j}$ and $\tilde{c}$ dislocating from a reliable alignment in the same direction will be imposed by a smaller penalty while the case in an opposite direction will be imposed by a larger penalty. The exponential function is used because the fact that $\tilde{j}$ and $\tilde{c}$ dislocate from a reliable alignment in the same direction or not is regarded more important.

In the same way, we can calculate the score of $(\tilde{j}, \tilde{c})$ by regarding to $a_{right\_of\_\tilde{j}}$, $a_{left\_of\_\tilde{c}}$ and $a_{right\_of\_\tilde{c}}$, respectively. Note that the position of the first word of $\tilde{c}$ will be taken in computing Δk when referring to $a_{right\_of\_\tilde{j}}$ and $a_{right\_of\_\tilde{c}}$.

Finally, select the score with the largest value.

$$\begin{aligned} Score_{dis}(\tilde{j},\tilde{c}) = \max\{\ & Score_{left\_of\_\tilde{j}}(\tilde{j},\tilde{c}), Score_{right\_of\_\tilde{j}}(\tilde{j},\tilde{c}), \\ & Score_{left\_of\_\tilde{c}}(\tilde{j},\tilde{c}), Score_{right\_of\_\tilde{c}}(\tilde{j},\tilde{c})\} \end{aligned} \tag{8}$$

$Score_{dis}$ expresses the score calculated by using dislocation information. The algorithm for extending alignment is described as follows.

Algorithm 2. Align $\tilde{j}$ in W_J with $\tilde{c}$ in W_C by referring to the reliable alignments.
Input W_J and W_C, from which j and $\ddot{c}$ aligned in the first component have been removed, respectively. A_{rel}.
Output Extended alignment A_{aug}.
Step 1 For each candidate $(\tilde{j}, \tilde{c})$, search for $a_{left\_of\_\tilde{j}}$, $a_{right\_of\_\tilde{j}}$, $a_{left\_of\_\tilde{c}}$ and $a_{right\_of\_\tilde{c}}$ in A_{rel}.
Step 2 For each candidate $(\tilde{j}, \tilde{c})$, compute $Score_{left\_of\_\tilde{j}}(\tilde{j},\tilde{c})$, $Score_{right\_of\_\tilde{j}}(\tilde{j},\tilde{c})$, $Score_{left\_of\_\tilde{c}}(\tilde{j},\tilde{c})$,$Score_{right\_of\_\tilde{c}}(\tilde{j},\tilde{c})$ using formula (6) and (7), and then $Score_{dis}(\tilde{j},\tilde{c})$ using formula (8).
Step 3 Loop.
If $\max_{\tilde{j}\ in\ W_J, \tilde{c}\ in\ W_C} Score_{dis}(\tilde{j},\tilde{c}) > \theta_{dis}$ and $Score_{lex}(\hat{j},\hat{c}) > \theta'_{lex}$, where $(\hat{j},\hat{c}) = \arg\max_{\tilde{j}\ in\ W_J, \tilde{c}\ in\ W_C} Score_{dis}(\tilde{j},\tilde{c}) > \theta_{dis}$, output $(\hat{j},\hat{c})$ to A_{aug}. Remove $\hat{j}$ from W_J and $\hat{c}$ from W_c.

θ_{dis} and $\theta'_{lex} (< \theta_{lex})$ are preset thresholds. In Step 3, $Score_{lex}(\hat{j},\hat{c}) > \theta'_{lex}$ means that the lexical-knowledge is also used in deciding $(\hat{j},\hat{c})$. Finally, A_{rel} and A_{aug} are output as alignment results.

4 Automatically Building a Japanese-Chinese Dictionary

In this work, we automatically built a Japanese-Chinese dictionary by using English as an intermediary. We then use the built dictionary in the word alignment and verify the dictionary. Two used machine-readable dictionaries are as follows.

EDR Japanese-English Dictionary [11]
It contains 364,430 records, each of which consists of Japanese word, part-of-speech, English translations, etc.

LDC English-Chinese Dictionary [12]

It contains 110,834 records, each of which consists of English word and Chinese translations.

We ranked the Chinese translation candidates utilizing three types of heuristic information: the number of English translations in common [13], the part of speech, and Japanese kanji information [14]. We then took the results that were ranked within top 5. As a result, we obtained a Japanese-Chinese dictionary that contains 144,002 Japanese entries.

5 Experiment

We evaluated the performance of the implemented lexical-knowledge based aligner and GIZA++ tool on the NICT corpus. We randomly selected 1,127 sentence pairs from the corpus and manually annotated them with word alignments. There are totally 7,332 reference alignments. The results were evaluated in terms of three measures, Precision, Recall and F-measure. In the lexical-knowledge based aligner, the thresholds are set as $\theta_{lex} = 0.85$, $\theta_{lex}^{'} = 0.4$ and $\theta_{dis} = 0.8$ [4][7]. In the application of GIZA++, two directions are tested: the Japanese is used as source language and the Chinese as target language, and vice versa. For training data, we used the whole corpus, 38,383 sentence pairs. In addition, we used the Japanese-Chinese dictionary. The comparison result is listed in Table 2.

Och used post-processing step that combines the alignment results of GIZA++ in both translation directions [1]. Based on this idea, we develop a multi-aligner by utilizing three groups of alignment results which are produced by the lexical-knowledge based aligner, J⟶C of GIZA++, and C⟶J of GIZA++, respectively. Then a majority decision is used do decide the final result. If an alignment appears at the two or three result groups, the alignment is accepted. Otherwise, it is abandoned. In this way, we aim at increasing recall rate and avoiding a large loss in precision at the same time. The evaluation result on the same date is also listed in Table 2.

Table 2. Comparison of performances between the lexical-knowledge based aligner, GIZA++ and a multi-aligner

aligner	Precision(%)	Recall(%)	F-measure
Lexical-knowledge based aligner	73.4	55.2	63.0
GIZA++(J⟶C)	52.4	64.8	57.9
GIZA++(C⟶J)	50.7	61.8	55.7
Multi-aligner	85.6	61.1	71.3

The lexical-knowledge based aligner obtained a higher precision (73.4%) and GIZA++ obtained a higher recall rate (64.8%). The former could correctly align the less frequently occurring words by using lexical knowledge, while the latter could not because statistically significant evidence was not available. On the other hand, the latter could correctly align the often occurring words, for some of which the former could not because of the deficiency of the current bilingual dictionary.

Compared with the lexical-knowledge based aligner, the multi-aligner achieved an improvement of 12.2% in precision, 5.9% in recall rate and 8.3% in F-measure. Compared with J ⟶C of GIZA++, the multi-aligner achieved an improvement of 33.3% in precision and 13.4 in F-measure, while a 3.7% loss in recall rate. We think that the performance of the multi-aligner, i.e. 61.1% recall rate with 85.6% precision, is applicable to our task of assisting manual annotation of word alignment.

Two results obtained by the lexical-knowledge based aligner are shown in Figure 2 and 3. The upper are Chinese sentences and the lower are Japanese Sentences. In Figure 2, the one-to-many alignment example is one Japanese word "クウスメート" (classmate) being aligned with two Chinese word "同班" (same class) and "同学" (classmate). In Figure 3, the many-to-one alignment example is two Japanese words "文化" (culture) and "財" (asset) being aligned with one Chinese word "文物" (culture asset).

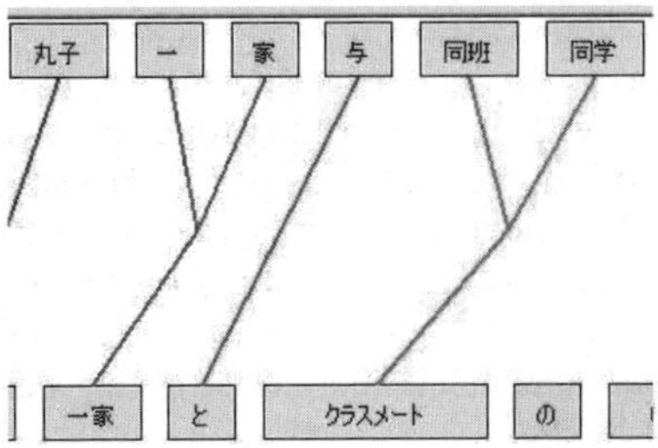

Fig. 2. Examples of the obtained one-to-many alignments

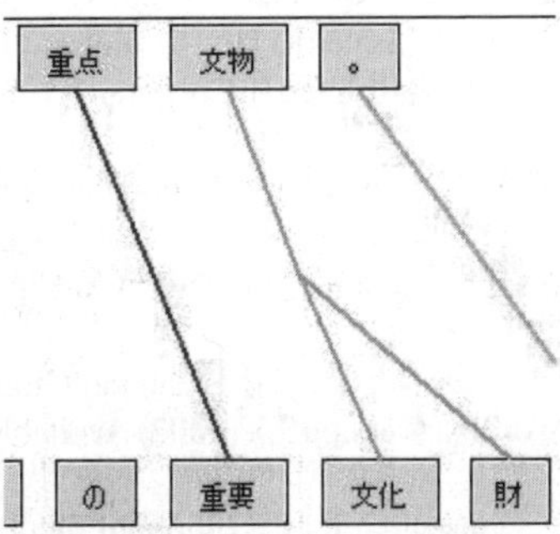

Fig. 3. Examples of the obtained many-to-one alignment

6 Conclusion

This paper presents the implementation of a lexical-knowledge based aligner, which consists of two components. The first components obtain reliable alignments by using three kinds of heuristics: an automatically built Japanese-Chinese dictionary, orthography and the correspondence between the traditional and the simplified Chinese characters. The second component extends alignments by using dislocation information. The performance of the implemented aligned is evaluated and the comparison with GIZA++ is conducted on the NICT Japanese-Chinese parallel corpus. Furthermore, a multi-aligner is developed based on the lexical-knowledge based aligner and GIZA++. The experimental results show that the multi-aligner achieved a satisfied results and it is applicable in assisting manual annotation of word alignment.

In the future research, we will improve the lexical-knowledge based approach to increase recall rate further. We will also attempt the method of using a bilingual dictionary in GIZA++, proposed by Och.

References

1. Och, F.J., Ney, H.: A systematic comparison of various statistical alignment models. Computational Linguistics **29** (2003) 19–51
2. Och, F.J., Ney, H.: Giza++: Training of statistical translation models. (2000) Available at http://www-i6.informatik.rwthaachen.de/ och/software/GIZA++.html.
3. Brown, P.F., Pietra, S.D., Pietra, V.J.D., Mercer, R.L.: The mathematic of statistical machine translation: Parameter estimation. Computational Linguistics **19** (1993) 263–311
4. Ker, S.J., Chang, J.S.: A class-based approach to word alignment. Computational Linguistics **23** (1997) 313–343
5. Melamed, I.D.: Models of translational equivalence among words. Computational Linguistics **26** (2000) 221–249
6. Huang, J.X., Choi, K.S.: Chinese-korean word alignment based on linguistic comparison. In: ACL. (2000)
7. Deng, D.: Research on Chinese-English word alignment. Master's thesis, Institute of Computing Technology, Chinese Academy of Sciences (2004)
8. Zhang, Y., Uchimoto, K., Ma, Q., Isahara, H.: Building an annotated Japanese-Chinese parallel corpus - a part of NICT Multilingual Corpora. In: the Tenth Machine Translation Summit. (2005) 71–78
9. Maekawa, K., Koiso, H., Furui, F., Isahara, H.: Spontaneous speech corpus of Japanese. In: LREC2000. (2000) 947–952
10. Zhou, Q., Yu, S.: Blending segmentation with tagging in Chinese language corpus processing. In: COLING. (1994) 1274–1278
11. NICT: EDR Electronic Dictionary Version 2.0 Technical Guide. (2002)
12. LDC: English-to-Chinese Wordlist (version 2.). (2002) Available at http://www.ldc.upenn.edu/Projects/Chinese/.
13. Tanaka, K., Umemura, K.: Construction of a bilingual dictionary intermediated by a third language. In: COLING. (1994) 297–303
14. Zhang, Y., Ma, Q., Isahara, H.: Automatic construction of Japanese-Chinese translation dictionary using English as intermediary. Journal of Natural Language Processing **12** (2005) 63–85

Deep Processing of Korean Floating Quantifier Constructions

Jong-Bok Kim[1] and Jaehyung Yang[2]

[1] School of English, Kyung Hee University, Seoul, Korea 130-701
[2] School of Computer Engineering, Kangnam University, Kyunggi, Korea, 449-702

Abstract. The so-called floating quantifier constructions in languages like Korean display intriguing properties whose successful processing can prove the robustness of a parsing system.[1] This paper shows that a constraint-based analysis, in particular couched upon the framework of HPSG, can offer us an efficient way of parsing these constructions together with proper semantic representations. It also shows how the analysis has been successfully implemented in the LKB (Linguistic Knowledge Building) system.

1 Processing Difficulties

One of the most salient features in languages like Korean is the complex behavior of numeral classifiers (Num-CL) linked to an NP they classify. Among several types of Num-CL constructions, the most complicated type includes the one where the Num-CL floats away from its antecedent:[2]

(1) pemin-i cengmal sey myeng-i/*-ul te iss-ta
criminal-NOM really three CL-NOM/ACC more exist-DECL
'There are three more criminals.'

There also exist constraints on which arguments can 'launch' floating quantifiers (FQ). Literature (cf. [1]) has proposed that the antecedent of the FQ needs to have the identical case marking as in (1). However, issues become more complicated with raising and causative constructions where the two do not agree in the case value:

(2) a. haksayng-tul-**ul** sey myeng-**i/ul** chencay-i-lako mit-ess-ta.
student-PL-ACC three-CL-NOM/*ACC genius-COP-COMP believed
'(We) believed three students to be genius.'

---

[1] We thank three anonymous reviewers for the constructive comments and suggestions. This work was supported by the Korea Research Foundation Grant funded by the Korean Government (KRF-2005-042-A00056).

[2] The following are the abbreviations used for the glosses and attributes in this paper: CL (CLASSIFIER), CONJ (CONJUNCTION), COP (COPULA), COMP (COMPLEMENTIZER), DECL (DECLARATIVE), GEN (GENITIVE), LBL (LABEL), LTOP (LOCAL TOP), NOM (NOMINATIVE), PNE (PRENOMINAL ENDING), PST (PAST), RELS (RELATIONS), SEM (SEMANTICS), SPR (SPECIFIER), SYN (SYNTAX), TOP (TOPIC), etc.

Y. Matsumoto et al. (Eds.): ICCPOL 2006, LNAI 4285, pp. 387–394, 2006.

b. haksayng-tul-**ul** sey-myeng-**i/ul/*eykey** ttena-key hayessta
student-PL-ACC three-CL-NOM/ACC/*DAT leave-COMP did
'(We) made three students to leave.'

As given in the raising (2a) and causative (2b), the Num-CL *sey myeng* 'three CL' can have a different case marking from its antecedent, functioning as the matrix object. In a sense, it is linked to the original grammatical function of the raised object and the causee, respectively.

Central issues in deep-parsing numeral classifier constructions thus concern how to generate such FQ constructions and link the FQ with its remote antecedent together with appropriate semantics. This paper shows that a typed feature structure grammar, HPSG, together with Minimal Recursion Semantics (MRS), is well-suited in providing the syntax and semantics of these constructions for computational implementations.[3]

2 Data Distribution

We have inspected the Sejong Treebank Corpus to figure out the distributional frequency of Korean numeral classifiers in real texts. From the corpus of total 378,689 words (33,953 sentences), we identified 694 occurrences of numeral classifier expressions. Of these 694 examples, we identified 36 FQ examples, some of which are given in the following:

(3) a. ... salam-i cengmal **han salam-to** epsessta.
person-NOM really one CL-also not.exist
'Nobody was really there.'

b. ... kkoma-lul **han myeng** pwuthcapassta
...little.boy-ACC one CL caught
'(We) grasped one little boy.'

The FQ type is relatively rare partly because the Sejong Corpus we inspected consists mainly of written texts. However, the statistics clearly show that these FQ constructions are legitimate constructions and should be taken into consideration if we want to build a robust grammar for Korean numeral classifiers.[4]

3 Implementing an Analysis

3.1 Forming a Numeral-Classifier Sequence and Its Semantics

The starting point of our analysis is forming the well-formed Num-CL expressions. Syntactically, numeral classifiers are a subclass of nouns (for Japanese see

[3] Minimal Recursion Semantics, developed by [2], is a framework of computational semantics designed to enable semantic composition using only the unification of type feature structures. See [2] and [3]. The value of the attribute SEM(ANTICS) in our system represents a simplified MRS.

[4] In addition to the FQ type, the Num-Cl can appear with a genitive case marking (GC type) or can combine with a noun without any particle (NI type). For an analysis of the GC and NI types, see [4].

[5]). However, unlike common nouns, they cannot stand alone and must combine with a numeral or a limited set of determiners:[5]

(4) a. *(twu) kay 'two CL' (Numeral)
  b. *(myech) kay 'how many' (Interrogative)

Semantically, there are tight sortal constraints between the classifiers and the nouns (or NPs) they modify. For example, *pen* can classify only events, *tay* machinery, and *kwuen* just books. Such sortal constraints block classifiers like *tay* from modifying thin entities like books as in **chayk twu tay* 'book two-CL'. Reflecting these syntactic and semantic properties, we can assign the following lexical information to numerals (*num-det*) and classifiers (*cl-n*) within the feature structure system of HPSG and MRS.[6]

(5)

a.
$$\begin{bmatrix} \textit{num-det} \\ \text{ORTH } \langle \text{sey `'} \rangle \\ \text{SYN}\,|\,\text{HEAD} \begin{bmatrix}\text{POS } \textit{det} \\ \text{NUM } + \end{bmatrix} \\ \text{SEM} \begin{bmatrix} \text{HOOK} \begin{bmatrix} \text{INDEX } i \\ \text{LTOP } h2 \end{bmatrix} \\ \text{RELS} \left\langle \begin{bmatrix} \text{PRED } \textit{card\_rel} \\ \text{LBL } h2 \\ \text{ARG0 } i \\ \text{CARG } 3 \end{bmatrix} \right\rangle \end{bmatrix} \end{bmatrix}$$

b.
$$\begin{bmatrix} \textit{cl-n} \\ \text{ORTH } \langle \text{myeng `'} \rangle \\ \text{SYN} \begin{bmatrix} \text{HEAD} \begin{bmatrix} \text{POS } \textit{noun} \\ \text{CLTYPE } + \end{bmatrix} \\ \text{VAL}\,|\,\text{SPR } \left\langle \begin{bmatrix} \text{NUM } + \\ \text{INDEX } i \end{bmatrix} \right\rangle \end{bmatrix} \\ \text{SEM} \begin{bmatrix} \text{HOOK} \begin{bmatrix} \text{INDEX } i \\ \text{LTOP } h1 \end{bmatrix} \\ \text{RELS} \left\langle \begin{bmatrix} \text{PRED } \textit{person\_rel} \\ \text{LBL } h1 \\ \text{ARG0 } i \end{bmatrix} \right\rangle \end{bmatrix} \end{bmatrix}$$

The feature structure in (5a) represents that there exists an individual x whose CARG (constant argument) value is "3". The feature NUM is assigned to the numerals as well as to determiners like *yele* 'several' and *myech* 'some' which combine with classifiers. Meanwhile, (5b) indicates that syntactically a classifier selects a NUM element through the SPR, whereas semantically it belongs to the ontological category *person_rel*. The feature CLTYPE differentiates classifiers from common nouns. An independent grammar rule then ensures that only [NUM +] elements can combine with the [CLTYPE +] expression, ruling out unwanted forms such as **ku myeng* 'the CL'.

3.2 Syntax and Semantics of the Floating Quantifier Constructions

As noted earlier, the Num-CL can float away from the NP it classifies. There exist several supporting phenomena indicating that the FQ modifies the

[5] A limited set of common nouns such as *salam* 'person', *kulus* 'vessel', *can* 'cup', *khep* 'cup', and *thong* 'bucket' can also function as classifiers.

[6] The value of LBL is a token to a given EP (elementary predicate). The feature HOOK includes externally visible attributes of the atomic predications in RELS. The value of LTOP is the local top handle, the handle of the relations with the widest scope within the constituent. See [2] for the exact functions of each attribute.

following verbal expression. One phenomenon is the substitution by the proverb *kule-* 'do so'. As noted in (6), unlike the NI type, only in the NC type, an FQ and the following main verb can be together substituted by the proverb *kulay-ss-ta*:

(6) a. namca-ka [sey myeng o-ass-ko], yeca-to kulay-ss-ta
man-NOM three CL come-PST-CONJ woman-also do-PST-DECL.
'As for man, three came, and as for woman, the same number came.'
b. *[namca sey myeng-i] o-ass-ko, yeca-to [kulay-ss-ta]

This means that the FQ in the NC type is a VP modifier, though it is linked to a preceding NP.

Coordination data also support a VP modifier analysis:

(7) [namhaksayng-kwa] kuliko [yehaksayng-i] [sey myeng-i] oassta
boy student-and and girl student-NOM three CL-NOM came
'The total 3 of boys and girls came.'

The FQ 'three-CL' cannot refer to only the second conjunct 'girl students': its antecedent must be the total number of boys and girls together. This means the FQ refers to the whole NP constituent as its reference. This implies that an analysis in which the FQ forms a constituent with the preceding NP then cannot ensure the reading such that the number of boys and girls is in total three.

Given this VP-modifier treatment, the following question then is how to link an FQ with its appropriate antecedent. There exist several constraints in identifying the antecedents. When the floating quantifier is case-marked, it seems to be linked to an argument with the same case marking. However, further complication arises from examples in which either the antecedent NP or the FQ are marked not with a case marker, but a marker like a TOP:

(8) a. haksayng-tul-i/un sakwa-lul sey kay-lul mekessta
student-PL-NOM/TOP apple-ACC three CL-ACC eat
'As for the students, they ate three apples.'
b. sakwa-lul haksayng-tul-i/un sey kay-lul mekessta

The data suggest that a surface case marking cannot be a sole indicator for the linking relation, and that we need to refer to grammatical functions. What we can observe is that, regardless of the location, the NOM-marked FQ is linked to the subject whereas the ACC-marked FQ is linked to the object. This observation is reflected in the following lexical information:[7]

[7] When the FQ has a delimiter marker (rather than a case marker) or no marker at all, it will refer to one of the elements in the ARG-ST (argument structure). Its antecedent will be determined in context.

(9)
$$\begin{bmatrix} \textit{num-cl-mw} \\ \text{ORTH } \langle \text{sey myeng-i} \rangle \\ \text{HEAD} \begin{bmatrix} \text{POS } \textit{noun} \\ \text{CASE} \mid \text{GCASE } \textit{nom} \\ \text{MOD} \left\langle \begin{bmatrix} \text{POS } \textit{verb} \\ \text{SUBJ} \left\langle \text{NP}_i \right\rangle \end{bmatrix} \right\rangle \end{bmatrix} \\ \text{SEM} \mid \text{HOOK} \mid \text{INDEX } i \end{bmatrix} \begin{bmatrix} \textit{num-cl-mw} \\ \text{ORTH } \langle \text{sey myeng-ul} \rangle \\ \text{HEAD} \begin{bmatrix} \text{POS } \textit{noun} \\ \text{CASE} \mid \text{GCASE } \textit{acc} \\ \text{MOD} \left\langle \begin{bmatrix} \text{POS } \textit{verb} \\ \text{COMPS} \left\langle \text{NP}_{i,\ldots} \right\rangle \end{bmatrix} \right\rangle \end{bmatrix} \\ \text{SEM} \mid \text{HOOK} \mid \text{INDEX } i \end{bmatrix}$$

As given in (9), the NOM-marked *num-cl-mw* modifies a verbal element whose SUBJ has the same index value, whereas the ACC-marked *num-cl-mw* modifies a verbal element which has at least one unsaturated COMPS element whose INDEX value is identical with its own INDEX value. What this means is that the NOM or ACC marked *num-cl-mw* is semantically linked to the SUBJ or COMPS element through the INDEX value. Our system yields the following parsing results for (8b):[8]

(10)

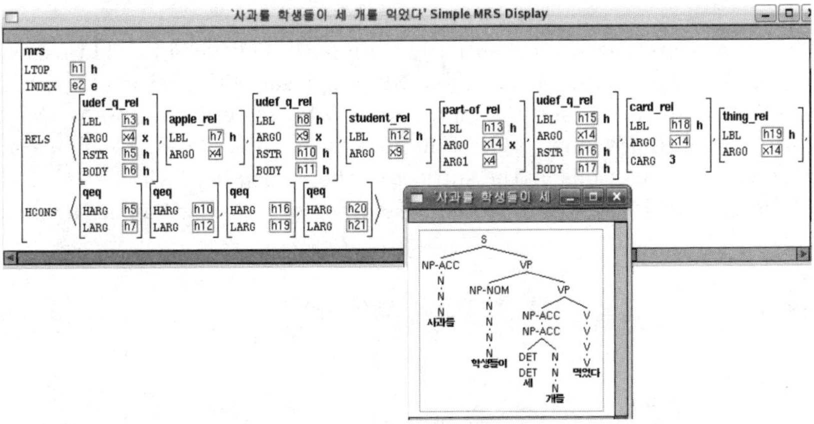

As seen from the parsed syntactic structure, the FQ *sey kay-lul* 'three CL-ACC' (NP-ACC) modifies the verbal expression *mek-ess-ta* 'eat-PST-DECL'. However, as noted from the output MRS, this modifying FQ is linked with its antecedent *sakwa-lul* 'apple-ACC' through the relation *part-of_rel*. Leaving aside the irrelevant semantic relations, let's see *card_rel* and *apple_rel*. As noted, the ARG0 value (x14) of *part-of_rel* is identified with that of *card_rel* whereas its ARG1 value (x4) is identified with the ARG0 value of the *apple_rel*. We thus can have the interpretation that there are three individuals x14s which belongs to the set x4.

4 Case Mismatches

Further complication in parsing FQ constructions comes from raising, causatives, and topicalization where the FQ and its antecedent have different case values.

[8] The attribute HCONS is to represent quantificational information. See [3].

In such examples, the two need not have an identical case value. For example, as given in (11b), the ACC-marked raised object can function as the antecedent of either the NOM-marked or ACC-marked FQ:

(11) a. namcatul-i [yecatul-**i** sey myeng-**i/*ul** chakhata-ko] mitessta.
men-NOM women-NOM three-CL-NOM/*ACC honest-COMP thought
'Men thought that three women are honest.'
b. namcatul-i yecatul-**ul** sey myeng-**i** chakhata-ko mitessta.

In the present analysis in which the case-marked FQ is linked to either the SUBJ or a COMPS element as given in (12), we can expect these variations. Let us consider the lexical entry for the raising verb *mitessta* 'believed':

(12) a. $\begin{bmatrix} \text{HEAD} \mid \text{POS } \textit{verb} \\ \text{SUBJ } \langle \boxed{1}\text{NP} \rangle \\ \text{COMPS } \langle \boxed{2}\text{S} \rangle \\ \text{ARG-ST } \langle \boxed{1}, \boxed{2} \rangle \end{bmatrix}$ b. $\begin{bmatrix} \text{HEAD} \mid \text{POS } \textit{verb} \\ \text{SUBJ } \langle \boxed{1}\text{NP} \rangle \\ \text{COMPS } \langle \boxed{2}\text{NP}_i, \boxed{3}\text{VP[SUBJ } \langle \text{NP}_i \rangle] \rangle \\ \text{ARG-ST } \langle \boxed{1}, \boxed{2}, \boxed{3} \rangle \end{bmatrix}$

(12a) represents the lexical entry for *mitessta* 'believed' in (11a) selecting a sentential complement. Meanwhile, (12b) represents the raising verb 'thought' in (11b) in which the subject of the embedded clause is raised as the object. That is, *yecatul-ul* 'women-ACC' functions as its object even though it originally (semantically) functions as the subject of the embedded clause.

Equipped with these, our grammar generates the following parsing results for (11a):

(13) 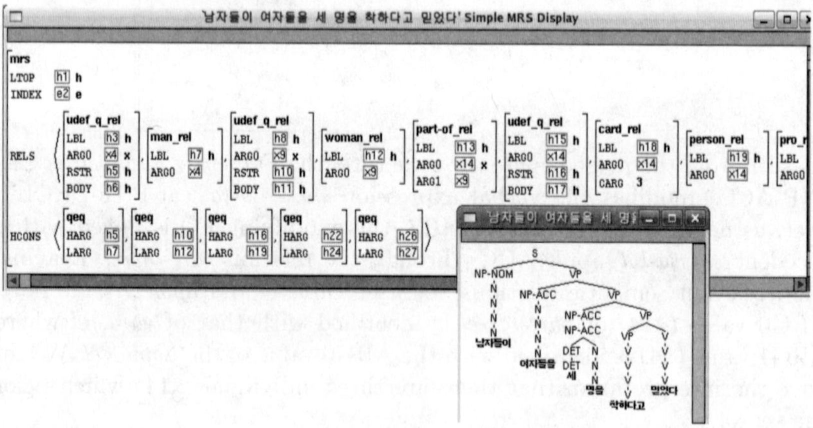

Syntactically, as noted from the parsed structure, the ACC-marked FQ *sey myeng-ul* 'three CL-ACC' (NP-ACC) modifies the VP *chakhata-ko mitessta* 'honest-COMP believed'.[9] Meanwhile, semantically, the ACC-marked FQ is linked to the ACC-marked object *yecatul-ul* 'woman-ACC'. This is because in

[9] Our grammar allows only binary structures for the language. One strong advantage of assuming binary structures comes from scrambling facts. See [7].

our grammar the antecedent of the ACC-marked FQ must be an unsaturated complement of the VP it modifies. As noted from the semantic relations *part-of_rel*, *card_rel* and *woman_rel* in the parsed MRS, this linking relation is attested. That is, the ARG0 value (x9) of *woman_rel* is identified with the ARG1 value of *part-of_rel* whereas the ARG0 value of *card_rel* is identical with the ARG0 value of *part-of_rel*. Thus, the semantic output correctly indicates that the individuals denoted by the FQ is a subset of the individuals denoted by the antecedent.

For the raising example (11b), our grammar correctly produces two structures. Let's see (14) first. As seen from the parsed syntactic structure here, the FQ *sey myeng-i* 'three CL-NOM' (NP-NOM) modifies the complex VP *chakhata-ko mitessta* 'honest-COMP believed'. However, in terms of semantics, the FQ is linked to the subject of the VP that it modifies.[10] This linking relation is once again attested by the MRS structure here. As noted here, the two semantic arguments of *part-of_rel*, ARG0 and ARG1, have identical values with the ARG0 value of *card_rel* (x14) and *man_rel* (x4), respectively.

(14)

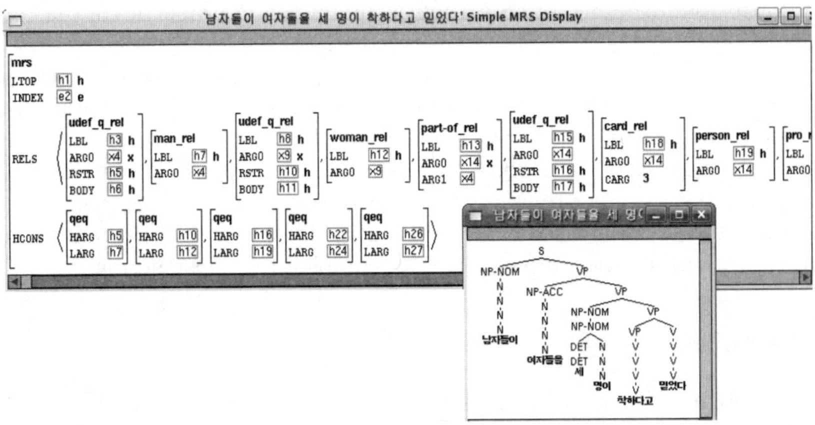

(15)

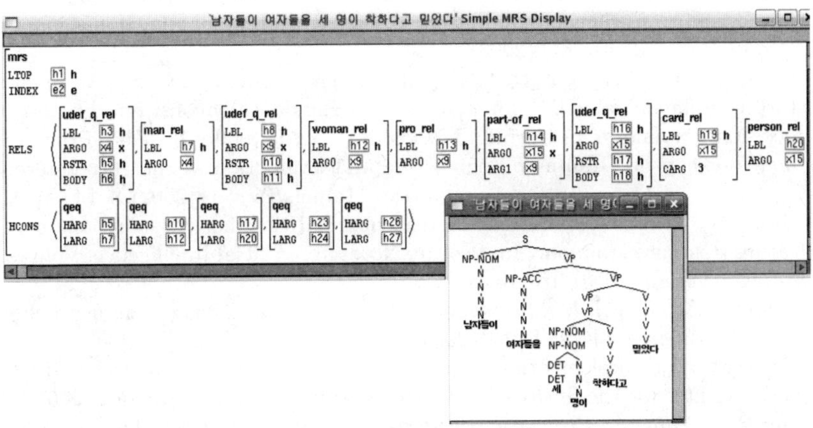

[10] As another semantic constraint, the FQ can be linked only to a sentential internal argument.

Meanwhile, as given in the second parsing result (15), the FQ *sey myeng-i* 'three CL-NOM' modifies the simple VP *chakhata-ko* 'honest-COMP' only. Since the VP that the FQ modifies has only its SUBJ unsaturated, the SUBJ is the only possible antecedent. The output MRS reflects this raising property: The ARG0 value of *part-of_rel* identified with that of *card_rel* whereas its ARG1 value is identified with the ARG0 value of *woman_rel*. Our system thus correctly links the NOM-marked FQ with the ACC-marked antecedent even though they have different case values.

5 Future Work and Conclusion

One of the complicated issues in building a robust parsing system is whether to cover empirical as well as psychological (intuition-based) data. Even though examples like the case mismatches in FQ occur not often in the corpus data we inquired, we need to deal with such legitimate constructions if we want to develop a system aiming for reflecting the fundamental properties of the language in question.

The grammar we have built within the typed-feature structure system and well-defined constraints, eventually aiming at working with real-world data, has been implemented in the HPSG for Korean. We have shown that the grammar can parse the appropriate syntactic and semantic aspects of the FQ constructions. The test results provide a promising indication that the grammar, built upon the typed feature structure system, is efficient enough to build semantic representations for the simple as well as complex FQ constructions.

References

1. Kang, B.M.: Categories and meanings of Korean floating quantifiers-with some reference to Japanese. Journal of East Asian Linguistics **11** (2002) 375–398
2. Copestake, A., Flickenger, D., Sag, I., Pollard, C.: Minimal recursion semantics: An introduction. Manuscript (2003)
3. Bender, E.M., Flickinger, D.P., Oepen, S.: The grammar matrix: An open-source starter-kit for the rapid development of cross-linguistically consistent broad-coverage precision grammars. In Carroll, J., Oostdijk, N., Sutcliffe, R., eds.: Proceedings of the Workshop on Grammar Engineering and Evaluation at the 19th International Conference on Computational Linguistics, Taipei, Taiwan (2002) 8–14
4. Kim, J.B., Yang, J.: Processing Korean numeral classifier constructions in a typed feature structure grammar. In: Lecture Notes in Artificial Intelligence. Volume 2945. Springer-Verlag (2006) To appear
5. Bender, E.M., Siegel, M.: Implementing the syntax of Japanese numeral classifiers. In: Proceedings of IJCNLP-04. (2004)
6. Copestake, A.: Implementing Typed Feature Structure Grammars. CSLI Lecture Notes. Center for the Study of Language and Information, Stanford (2001)
7. Kim, J.B., Yang, J.: Projections from morphology to syntax in the Korean resource grammar: implementing typed feature structures. In: Lecture Notes in Computer Science. Volume 2945. Springer-Verlag (2004) 13–24

Compilation of a Dictionary of Japanese Functional Expressions with Hierarchical Organization

Suguru Matsuyoshi[1], Satoshi Sato[1], and Takehito Utsuro[2]

[1] Graduate School of Engineering, Nagoya University,
Chikusa, Nagoya, 464-8603, Japan

[2] Graduate School of Systems and Information Engineering, University of Tsukuba,
Tennodai, Tsukuba, 305-8573, Japan

Abstract. The Japanese language has a lot of functional expressions, which consist of more than one word and behave like a single functional word. A remarkable characteristic of Japanese functional expressions is that each functional expression has many different surface forms. This paper proposes a methodology for compilation of a dictionary of Japanese functional expressions with hierarchical organization. We use a hierarchy with nine abstraction levels: the root node is a dummy node that governs all entries; a node in the first level is a headword in the dictionary; a leaf node corresponds to a surface form of a functional expression. Two or more lists of functional expressions can be integrated into this hierarchy. This hierarchy also provides a way of systematic generation of all different surface forms. We have compiled the dictionary with 292 headwords and 13,958 surface forms, which covers almost all of major functional expressions.

1 Introduction

Some languages have *functional expressions*, which consist of more than one word and behave like a single functional word. In English, "in spite of" is a typical example, which behaves like a single preposition. In natural language processing (NLP), correct detection of functional expressions is crucial because they determine sentence structures and meanings. Implementation of a detector of functional expressions requires a dictionary of functional expressions, which provides lexical knowledge of every functional expression.

The Japanese language has many functional expressions. They are classified into three types according to the classification of functional words: particle, auxiliary verb, and conjunction. The particle type is sub-divided into five sub-types: case-marking particle, conjunctive particle, adnominal particle, focus particle, and topic-marking particle. A remarkable characteristic of Japanese functional expressions is that each functional expression has many different surface forms; they include *derivations*, *expression variants* produced by particle alternation and insertion, *conjugation forms* produced by the final conjugation component, and *spelling variants*.

Y. Matsumoto et al. (Eds.): ICCPOL 2006, LNAI 4285, pp. 395–402, 2006.

Compilation of a dictionary of Japanese functional expressions for natural language processing requires two lists. The first is a list of headwords of the dictionary; the second is the complete list of surface forms of entries in the first list.

Although there are several lists of Japanese functional expressions such as [2] and [3], compilation of the first list is not straightforward because there is no concrete agreement on the selection guideline of headwords of Japanese functional expressions. For example, [2] and [3] follow different selection guidelines: both of "にたいして (ni-taishi-te)" and "にたいする (ni-taisuru)" are headwords in [2]; only the former is a headword and the latter is its derivation in [3]. We need a way of resolving this type of contradiction to merge different lists of headwords.

The second list is required because NLP systems have to process functional expressions in surface forms that appear in actual texts. Because native speakers easily identify functional expressions in surface forms, there is no explicit list that enumerates all surface forms in dictionaries for human use. We need a systematic way of generating the complete list of surface forms for machine use.

This paper proposes a methodology for compilation of a dictionary of Japanese functional expressions with hierarchical organization. We design a hierarchy with nine abstraction levels. By using this hierarchy, we can merge different lists of headwords, which are compiled according to different guidelines. This hierarchy also provides a way of systematic generation of all different surface forms.

2 Hierarchical Organization of Functional Expressions

2.1 Various Surface Forms of Japanese Functional Expressions

Several different language phenomena are related to the production of various surface forms of Japanese functional expressions. We classify these surface-form variants into four categories: *derivations*, *expression variants*, *conjugation forms*, and *spelling variants*.

In case two forms that have different grammatical functions are closely related to each other, we classify them into *derivations*. For example, "にたいする (ni-taisuru)" and "にたいして (ni-taishi-te)" are closely related to each other because "たいする (taisuru)" and "たいして (taishi-te)" are different conjugation forms of the same verb. They have different grammatical functions: the former behaves like an adnominal particle and the latter behaves like a case-marking particle. Therefore we classify them into derivations. This view comes from the fact that several case-marking particles can be used as adnominal particles with slightly different forms.

In case two forms have slightly different morpheme sequences with the same grammatical function and meaning except style (formal or informal), we classify them into *expression variants*. Language phenomena that are related to production of expression variants are:

1. Alternation of functional words (particles and auxiliary verbs)
   In a functional expression, a component functional word may be replaced by another functional word with the same meaning. For example, "からすれば (kara-sure-ba)" is produced from "からすると (kara-suru-to)" by substitution of "ば (ba)" for "と (to)," where these two particles have the same meaning (assumption).
2. Phonetic phenomena
   (a) Phonetic contraction
   For example, "なけりゃならない (nakerya-nara-nai)" is produced from "なければならない (nakere-ba-nara-nai)," where "りゃ (rya)" is a shorter form of "れば (re-ba)," which is produced by phonetic contraction.
   (b) Ellipsis
   In case a component word has an informal (ellipsis) form, it may be replaced by the informal form. For example, "とこだった (toko-daQ-ta)" is produced from "ところだった (tokoro-daQ-ta)," where "とこ (toko)" is an informal form of "ところ (tokoro)," which is produced by omission of "ろ (ro)."
   (c) Voicing
   The initial consonant "*t*" of a functional expression may change to "*d*," depending on the previous word. For example, "ていい (te-ii)" changes into "でいい (de-ii)" when it occurs just after "読ん (yoN)."
3. Insertion of a focus particle
   A focus particle such as "は (ha)" and "も (mo)" [4] can be inserted just after a case-marking particle. For example, "とはいっても (to-ha-iQ-te-mo)" is produced from "といっても (to-iQ-te-mo)" by insertion of "は (ha)" just after "と (to)."

The third category of surface-form variants is *conjugation forms.* In case the last component of a functional expression is a conjugation word, the functional expression may have conjugation forms in addition to the base form. For example, a functional expression "ことにする (koto-ni-suru)" has conjugation forms such as "ことにし (koto-ni-shi)" and "ことにすれ (koto-ni-sure)," because the last component "する (suru)" is a conjugation word.

Some conjugation forms have two different forms: the normal conjugation form and the *desu/masu* (polite) conjugation form. For example, a variant "ことにします (koto-ni-shi-masu)" is the *desu/masu* conjugation form of "ことにする (koto-ni-suru)," where "します (shi-masu)" is the *desu/masu* form of "する (suru)."

The last category of surface-form variants is *spelling variants.* In Japanese, most words have *kanji* spelling in addition to *hiragana* spelling. For example, both of "にあたって (ni-ataQ-te)" (hiragana spelling) and "に当たって (ni-ataQ-te)" (kanji spelling) are used in practice.

2.2 Hierarchy with Nine Abstraction Levels

In order to organize functional expressions with various surface forms described in the previous subsection, we design a hierarchy with nine abstraction levels. Figure 1 shows a part of the hierarchy. In this hierarchy, the root node (in L^0) is a dummy node that governs all entries in the dictionary. A node in L^1 is an entry (headword) in the dictionary; the most generalized form of a functional expression. A leaf node (in L^9) corresponds to a surface form (completely-instantiated form) of a functional expression. An intermediate node corresponds to a partially-abstracted (partially-instantiated) form of a functional expression.

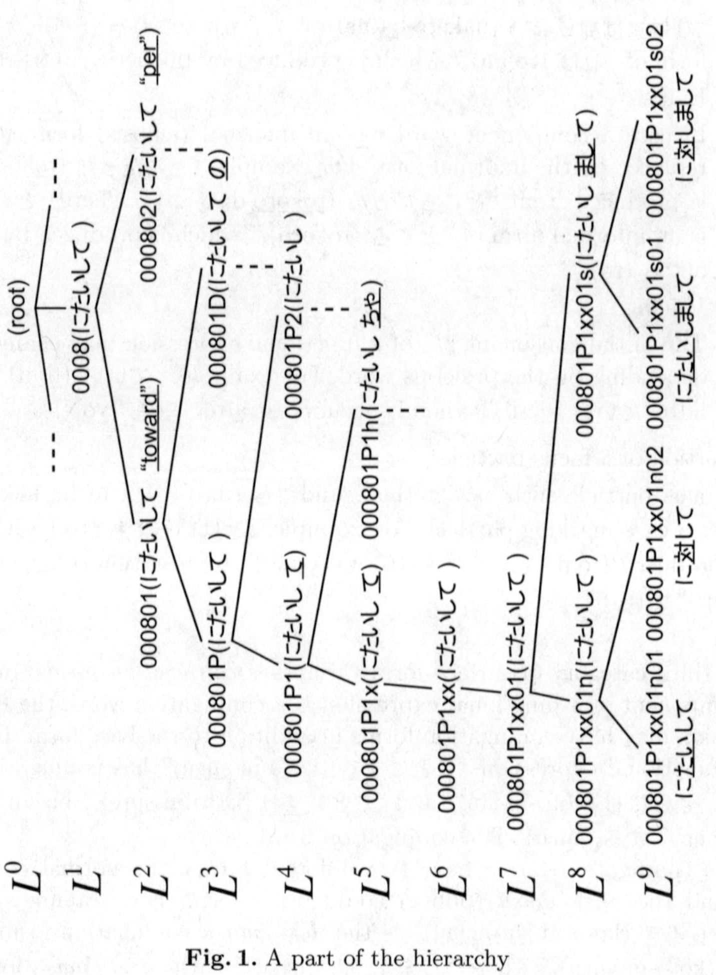

Fig. 1. A part of the hierarchy

Table 1 overviews the nine abstraction levels of the hierarchy. From L^3 to L^9 correspond to the phenomena described in the previous subsection. First, we have defined the following order according to the significance of categories of surface-form variants:

derivations (L^3) > expression variants (L^4, L^5, L^6)
> conjugation forms (L^7, L^8) > spelling variants (L^9) .

Then, we have defined the order L^4–L^6 and L^7–L^8 in order to make a simple hierarchy.

Table 1. Nine abstraction levels

Abstraction Levels		ID Character Type	ID Length	Number of Nodes
L^1	Headword	digit	4	292
L^2	Meaning categories	digit	2	354
L^3	Grammatical functions	8 alphabets	1	470
L^4	Alternations of functional words	digit	1	682
L^5	Phonetic variations	32 alphabets	1	1,032
L^6	Optional focus particles	17 alphabets	1	1,628
L^7	Conjugation forms	digit	2	6,190
L^8	Normal or *desu/masu* forms	2 alphabets	1	8,462
L^9	Spelling variations	digit	2	13,958

In addition to these seven levels, we define the following levels.

L^2 Meaning categories
Several functional expressions take more than one meaning. For example, "にたいして (ni-taishi-te)" takes two different meanings. The first meaning is "toward"; e.g., "彼は私にたいして親切だ" (He is kind toward me). The second meaning is "per"; e.g., "一人にたいして５つ" (five per one person). This level is introduced to distinguish such ambiguities.

L^1 Headword
A node of this level corresponds to a headword of the dictionary.

Because the hierarchy covers from the most generalized form (in L^1) of a functional expression to the completely-instantiated forms (in L^9) of it, any form of a functional expression can be inserted in some position in the hierarchy.

From this hierarchy, multiple lists of headwords can be generated. Our list of headwords is nodes in L^1. In case you follow the guideline that each headword has the unique meaning, which roughly corresponds to the guideline used by the book [3], nodes in L^2 become headwords. In case you follow the guideline that each headword has the unique grammatical function, nodes in L^3 become headwords.

We design an ID system in which the structure of hierarchy can be encoded; an ID consists of nine parts, each of which corresponds to one of nine levels of the hierarchy (in Fig. 2). We assign a unique ID to each surface form. Because an ID represents the position of the hierarchy, we easily obtain the relation between two surface forms by comparing their IDs. Table 2 shows three surface forms of functional expressions. By comparing IDs of (1) and (2), we obtain that the leftmost difference is "x" and "h" at the ninth character; it corresponds to L^5 so they are phonetic variants of the same functional expression. In contrast, the first 4 digits are different between (1) and (3); from this, we obtain that they are completely different functional expressions.

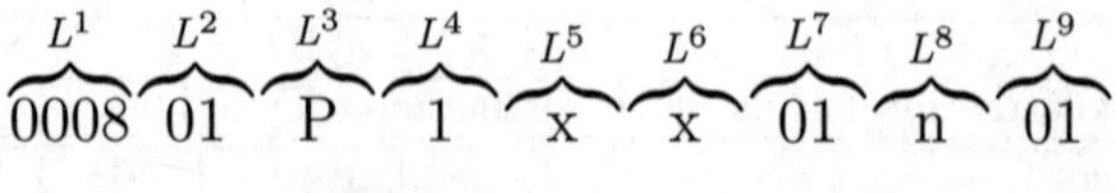

Fig. 2. An ID consists of nine parts

Table 2. Functional expressions with similar IDs

	ID	Functional Expression
(1)	000801P1xx01n01	にたいして (ni-taishi-te)
(2)	000801P1hx01n01	にたいしちゃ (ni-taishi-cha)
(3)	000901P1xx01n01	について (ni-tsui-te)

3 Compilation of a Dictionary of Functional Expressions

3.1 Compilation Procedure

We have compiled a dictionary of Japanese functional expressions, which has the hierarchy described in the previous section. The compilation process is incremental generation of the hierarchy, because we have neither the complete list of headwords nor the list of all possible surface forms in advance.

The compilation procedure of an incremental step is:

1. Pick up a functional expression from [3].
2. Create a node that corresponds to the given expression and insert it at the appropriate position of the hierarchy.
3. Create the lower subtree under the node.

Most of headwords in [3] correspond to nodes in L^2. Some exceptions correspond to nodes in L^4 or L^5. In order to insert such nodes into the hierarchy, we create the additional upper nodes if necessary.

In step 3, we create the lower subtree under the inserted node, which means enumeration of all possible surface forms of the functional expression. Most of surface forms can be generated automatically by applying generation templates to the inserted node. We manually remove overgenerated (incorrect) forms from the generated subtree. Several exceptional forms are not included in the generated subtree. We manually add such exceptional forms into the subtree.

We have already inserted 412 functional expressions, which are all functional expressions described in [3], into the hierarchy. The number of nodes in each level is shown in Table 1. The number of nodes in L^1 (headwords) is 292, and the number of leaf nodes (surface forms) is 13,958.

3.2 Description of Functional Expressions

When we create a leaf node in the hierarchy, we assign the following eight properties to the node.

1. ID (described in Sect. 2.2)
2. Meaning category
   We employ 103 meaning categories to describe meanings of functional expressions and to distinguish ambiguities in meaning (in L^2). We specify one of them in this slot.
3. Readability
   Some functional expressions are basic, i.e., everyone knows them; some are not. In this slot, we specify one of readability levels of A1, A2, B, C, and F, where A1 is the most basic level and F is the most advanced level.
4. Style
   We specify one of four styles: normal, polite, colloquial, and stiff.
5. Negative expressions
   We specify expressions that have an opposite meaning against the functional expression. These are required because literally negative forms of functional expressions may be ungrammatical.
6. Idiomatic expressions that include the functional expression
7. Example sentences
8. Reference

In practice, we specify the above properties at the appropriate intermediate nodes in the hierarchy, not at leaf nodes. For example, we specify meaning categories at nodes in L^2; we specify styles at nodes in L^8. A standard inheritance mechanism automatically fills all slots in the leaf nodes. This way of specification clarifies the relation between properties and forms of functional expressions; e.g., the style property is independent of spelling variants.

4 Related Work

There is no large electronic dictionary of Japanese functional expressions that is available in public.

Shudo et al. have collected 2,500 functional expressions in Japanese (1,000 of particle type and 1,500 of auxiliary-verb type) and classified them according to meaning [5,6]. In the list, the selection of headwords is not consistent, i.e., headwords of different abstraction levels exist; they correspond to the nodes at L^3, L^4, and L^5 in our dictionary. This list has no explicit organization structure except alphabetic order.

Hyodo et al. have proposed a dictionary of Japanese functional expressions with two layers [1]. This dictionary has 375 entries in the first layer: from these entries, 13,882 surface forms (in the second layer) are generated automatically. This dictionary does not provide precise classification between two surface forms, such as phonetic variants and spelling variants, which our dictionary provides.

5 Conclusion and Future Work

We have proposed a methodology for compilation of a dictionary of Japanese functional expressions with hierarchical organization. By using this methodology, we have compiled the dictionary with 292 headwords and 13,958 surface forms. It covers all functional expressions described in [3]. The compilation process of integrating additional functional expressions, which are described in [2], not in [3], is planned in the next step.

Our dictionary can be used for various NLP tasks including parsing, generation, and paraphrasing of Japanese sentences. For example, the use of our dictionary will improve the coverage of the detection method of functional expressions [7]. Experimental evaluation of application of this dictionary to actual NLP tasks is future work.

References

1. Yasuaki Hyodo, Yutaka Murakami, and Takashi Ikeda. A dictionary of long-unit functional words for analyzing *bunsetsu*. In *Proceedings of the 6th Annual Meeting of the Association for Natural Language Processing*, pages 407–410, 2000. (in Japanese).
2. Group Jamashii, editor. Nihongo Bunkei Jiten*(Dictionary of Sentence Patterns in Japanese)*. Kurosio Publisher, 1998. (in Japanese).
3. Yoshiyuki Morita and Masae Matsuki. Nihongo Hyougen Bunkei, *volume 5 of* NAFL Sensho *(Expression Patterns in Japanese)*. ALC Press Inc., 1989. (in Japanese).
4. Yoshiko Numata. Toritateshi (*Focus Particles*). In Keiichiro Okutsu, Yoshiko Numata, and Takeshi Sugimoto, editors, Iwayuru Nihongo Joshi no Kenkyu*(Studies on So-called Particles in Japanese)*, chapter 2. BONJINSHA, 1986. (In Japanese).
5. Kosho Shudo, Toshiko Narahara, and Sho Yoshida. Morphological aspect of Japanese language processing. In *Proceedings of the 8th COLING*, pages 1–8, 1980.
6. Kosho Shudo, Toshifumi Tanabe, Masahito Takahashi, and Kenji Yoshimura. MWEs as non-propositional content indicators. In *Proceedings of the 2nd ACL Workshop on Multiword Expressions: Integrating Processing (MWE-2004)*, pages 32–39, 2004.
7. Masatoshi Tsuchiya, Takao Shime, Toshihiro Takagi, Takehito Utsuro, Kiyotaka Uchimoto, Suguru Matsuyoshi, Satoshi Sato, and Seiichi Nakagawa. Chunking Japanese compound functional expressions by machine learning. In *Proceedings of the EACL 2006 Workshop on Multi-Word-Expressions in a Multilingual Context*, pages 25–32, 2006.

A System to Indicate Honorific Misuse in Spoken Japanese

Tamotsu Shirado[1], Satoko Marumoto[2], Masaki Murata[1], Kiyotaka Uchimoto[1], and Hitoshi Isahara[1]

[1] National Institute of Information and Communications Technology (NICT), Kyoto 619-0289, Japan
[2] The Institute of Behavioral Sciences (IBS), Tokyo 162-0845, Japan

Abstract. We developed a computational system to indicate the misuse of honorifics in word form and in performance of expressions in Japanese speech sentences. The misuse in word form was checked by constructing a list of expressions whose word form is bad in terms of honorifics. The misuse in performance was checked by constructing a consistency table. The consistency table defined the consistency between the honorific features of sentences and the social relationship among the people involved in the sentence. The social relationship was represented by combinations of [the number of people involved in the sentence] × [relative social position among the people] × [in-group/out-group relationship among the people]. The consistency table was obtained by using a machine learning technique. The proposed system was verified using test data prepared by the authors and also by third-party linguistic researchers. The results showed that the system was able to discriminate between the correct and the incorrect honorific sentences in all but a few cases. Furthermore, differences in the educational importance among the norms used in the system were revealed based on experiments using sentences written by people who are not linguistic experts.

1 Introduction

Politeness plays an important role in conversations, this is especially so in Japan. The correct use of honorific expressions is indispensable for maintaining an appropriate social distance between individuals. Recently, however, misuse of honorific Japanese expressions has increased. One of the origins for this misuse may be a lack of education regarding honorific conversations. As normative honorific expressions take a long time to learn, a computer-aided education system for learning honorific expressions would be useful. We have developed a computational system to indicate the misuse of honorifics in word form and in performance of expressions in Japanese speech sentences The proposed system was verified using test data prepared by the authors and also by third-party linguistic researchers. The tests showed that the system was able to discriminate between correct and incorrect honorific sentences in all but a few cases.

Y. Matsumoto et al. (Eds.): ICCPOL 2006, LNAI 4285, pp. 403–413, 2006.

2 Misuse of Honorific Expressions

2.1 Types of Honorific Expressions

The honorific expressions commonly used in spoken Japanese sentences can be divided into three types.

Subject honorific expressions ("sonkeigo" in Japanese) are used to show respect toward a person, who is usually the subject of a predicate in the sentence, by elevating the status of the person. For example, "osharu" is a subject honorific expression which means 'speak.'

Object honorific expressions ("kenjougo" in Japanese) are used to show respect toward a person, who is usually the object of a predicate in the sentence, by humbling the speaker or the subject of the predicate. Furthermore, object honorific expressions can be subdivided into expressions that elevate the status of the object and expressions that do not elevate the status of the object. We call these "object honorific-a" and "object honorific-b." For example, "itadaku" is an object honorific expression-a which means 'given,' and "mousu" is an object honorific expression-b which means 'speak.'

Polite expressions ("teineigo" in Japanese) are used to show politeness, but not necessarily respect, toward a person, who is usually the listener. Polite expressions include an auxiliary verb at the end of a sentence. For example, "soudesu" is a polite expression which means 'I hear that.'

2.2 Categories of Honorific Misuse

The misuse of honorific expressions can be divided into two kinds: misuse in word form and misuse in performance.

Misuse in word form

In a misuse in word form, the word form is bad; i.e., differs from the normative honorific forms. We constructed a list of expressions whose word form is bad in terms of honorifics by using traditional Japanese textbooks and articles written by Japanese linguistic researchers. For example, "itadakareru" is a misuse in word form of "itadaku."

Misuse in performance

In a misuse in performance, the word form is normative, but its honorific feature is inconsistent with the social relationships among the speakers, listeners, and individuals being referred to in the sentence. We took into account the relative social positions and in-group/out-group relationships among people to represent the social relationships because many textbooks of Japanese honorifics stated that these factors are important for choosing adequate honorific expressions in performance. Social distance (e.g., familiarity) among people may also affect the choice of honorific expressions in performance (Brown et al., 1987). However, this factor does not strongly relate to honorific norms because we do not need to follow honorific norms strictly when all the people involved are familiar with each other. We, therefore, ignored this factor in the proposed system and assumed all people involved in a sentence were "not so familiar each other."

3 System Description

3.1 Restrictions

The system deals with sentences that satisfy the following restrictions:

1. Only one predicate, with one subject and one object, is included.
2. Two to four people are involved; a speaker (denoted as "S") and a listener (denoted as "L") are involved in the case of two people, another person (denoted as "A") is also involved in the case of three people, and one more people (denoted as "B") is also involved in the case of four people.
3. Symbols indicating people ("S", "L", "A", or "B") must be shown in the sentence to indicate the subject or the object person of the predicate (pronouns can be used instead of "S" and "L" to indicate the speaker and the listener).

These restrictions require that sentences be simple so that the system does not need any high-precision parsing program because no parsing program is currently able to accurately identify the subject and object person for each predicate in complicated sentences. At this time, the following restrictions for the subject and object are assumed under the restrictions enumerated above: {subject, object}={S,L}, or {L,S} for two people, {subject, object}={S,A}, {L,A}, {A,S}, or {A,L} for three people, {subject, object}={A,B} for four people ({subject, object}={B,A} is equivalent to {subject, object}={A,B} in the system).

3.2 Honorific Features of Sentences

As we explained in Section 2.1, honorific expressions can be divided into subject honorific, object honorific (object honorific-a and object honorific-b), and polite expressions. These are concerned with the predicate, the honorific titles of the subject/object of the predicate, and the end of the sentence. The honorific features of the sentences that follow the restrictions stated in Section 3.1 can thus be represented by s, o, e, and p, whose value assignments are defined in Table 1. These are individually referred to as "honorific elements," and a set of them are referred to as "honorific pattern."

3.3 System Input and Output

Figure 1 shows an example of system input and output. The social relationship (i.e., [the number of people involved in the sentence] × [relative social position among the people] × [in-group/out-group relationship among the people]) among people named "Yamada," "Sato," and "Takahashi" (the fourth person's name was assigned as "Kimura," and these names were replaced by "S," "L," "A," and "B" in the system) are shown in the upper right portion of the figure. In this example, "Yamada (S)" and "Takahashi (A)" belong to the same organization, named "A," so they are in the "in-group." The social position of "Takahashi" is higher than that of "Yamada." "Sato" belongs to another organization, named

Table 1. Value assignment of honorific elements

Element	Condition
$s = 0$	No honorific title of *subj*
$s = 1$	Honorific title of *subj* used
$o = 0$	No honorific title of *obj*
$o = 1$	Honorific title of *obj* used
$e = 0$	Auxiliary verb at end of sentence is impolite / No auxiliary verb at end o sentence
$e = 1$	Auxiliary verb at end of sentence is polite
$p = 0$	Predicate is neither subject honorific nor object honorific
$p = 1$	Predicate is subject honorific
$p = 2$	Predicate is object honorific-a
$p = 3$	Predicate is subject honorific and also object honorific-a
$p = 4$	Predicate is object honorific-b

"B," so "Sato" is in the "out-group" from "Yamada" and "Takahashi." The social relationship among people involved in the sentence can be changed by using the buttons "△," "▽," "A," "B," "C," and "D," where "△" and "▽" are for changing the relative social position, and "A," "B," "C," or "D" are for changing the organization of each person.

The system checks the honorific validity of the input sentence and classifies any misuse as word form or performance and points out in the sentence.

The sample sentence shown in the upper portion of Figure 1 is "Takahashi" (a person's name) "ga" (a postpositional particle that indicates the subjective case) "Sato" (a person's name) "sama" (a honorific title that means 'Mr./Ms.') "ni" (a postpositional particle which means 'to') "gosetumei" (object honorific-a form of a verb that means 'explain') "si" (a conjugation of a verb, "suru," that means 'do') "masu" (a polite auxiliary verb the end of a sentence). The speech intention is 'Takahashi explains to Sato.' No misuse in word form was found in this sentence, but a misuse in performance was indicated because the existence of the honorific title for the subject ("san") is inconsistent with their social relationship. No misuses in word form and in performance were found in this sentence, So, the system output "input sentence is correct."

The sample sentence shown in the lower portion of Figure 1 is "Takahashi" (a person's name) "san" (a honorific title that means 'Mr./Ms.') "ga" (a postpositional particle that indicates the subjective case) "Sato" (a person's name) "sama" (a honorific title that means 'Mr./Ms.') "ni" (a postpositional particle which means 'to') "gosetumei" (object honorific-a form of a verb that means 'explain') "si" (a conjugation of a verb, "suru," that means 'do') "masu" (a polite auxiliary verb the end of a sentence). The speech intention is the same mentioned above. No misuse in word form was found in this sentence, but a misuse in performance was indicated because the existence of the honorific title for the subject ("san") is inconsistent with their social relationship.

3.4 Process Flow

Figure 2 shows a process flow chart of the system. The process consists of the following steps.

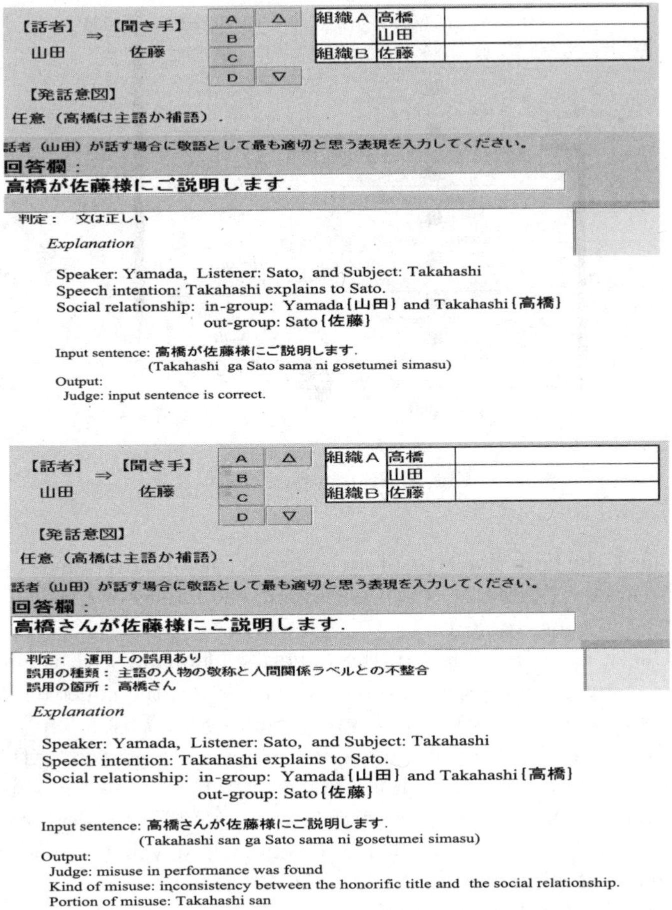

Fig. 1. System input and output

(Step 1) Replace the persons' names: "Yamada," "Sato," "Takahashi," and "Kimura" with "S," "L," "A," and "B," respectively.
(Step 2) Obtain a row of morphemes from the input sentence by using the Japanese morphological analysis program.
(Step 3) Check the row of morphemes for misuse in word form by using the list of words in bad form. If any misuses are found, proceed to Step 4; otherwise, proceed to Step 5.
(Step 4) Output "misuse in word form was found," along with all the portions of the input sentence corresponding to the partial rows that have misuse in word form. Then quit the process.
(Step 5) Identify the subject and the object of the predicate by using templates prepared by us.

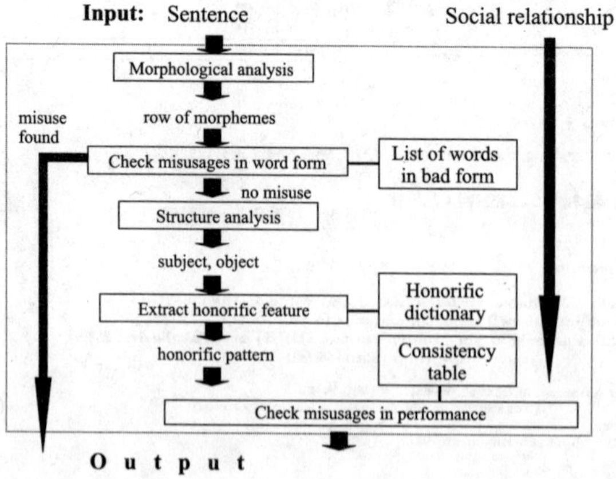

Fig. 2. System flow chart

(Step 6) Check the honorific type of each partial row of morphemes by using the honorific dictionary (Table 2). Then determine the values of honorific elements s, o, e, and p by using Table 1.
(Step 7) Check the consistency of the honorific elements with the social relationships by using the consistency table, which defines the consistency between them (details of the consistency table are explained in Section 3.5).
(Step 8) If all of the honorific elements are defined as consistent with the social relationships in the consistency table, output "the input sentence is correct." Otherwise, output "misuse in performance was found," the portions of the input sentence corresponding to honorific elements whose values are not consistent with the social relationship, and the kind of the inconsistency.

In the example shown in Figure 1, the sentence "Takahashi san ga Sato sama ni gosetumei simasu" has been replaced by "A san ga L sama ni gosetumei simasu" in Step 1. Then, a row of morphemes, "A" "san" "ga" "L" "sama" "ni" "go" "setumei" "suru" "masu," has been obtained in Step 2. No misuse in word form was identified in the row of morphemes, so Step 4 was skipped. In Step 5, the subject and the object were identified as "A" and "L," respectively. Then,

Table 2. A part of honorific dictionary

Partial row of morpheme	Honorific type
"L" "san"	honorific title of L
"A" "san"	honorific title of A
"o" / "go" *verb* "suru"	object honorific-a
"itadaku"	object honorific-a
"mousu"	object honorific-b
"o" / "go" *verb* "ni" "naru"	subject honorific
"ossharu"	subject honorific
"itadaku"+"rareru"	object honorific-a and also subject honorific
"desu"	polite
"da"	impolite

in Step 7, the values of honorific elements were assigned as $s = 1$, $o = 1$, $e = 1$, and $p = 2$ because there are honorific titles "san" and "sama" for persons A and L, the honorific type of the auxiliary verb at the end of the sentence is polite ("masu"), and the honorific type of the predicate ("go" *verb* "suru") is object honorific-a. Finally, the existence of the honorific title of person A ("san") is judged as misuse in performance because $s = 1$ is not defined as consistent with the social relationship among S, A, and L in Table 3.

3.5 Consistency Table

Table 3 shows a part of the consistency table for three people (S, L, and A). The consistency table defines the consistency between the honorific patterns of the sentences and the social relationships among the speakers, listeners, and individuals being referred to in the sentence, where the social relationship was represented by the combinations of [the number of persons involved in the sentence] × [relative social position among persons] × [in-group/out-group relationship among persons]. The conditions concerning the subject and the object of the predicate are also described with the social relationship. The symbols "S," "L," "A," and "B" indicate persons involved in the sentence, as described above. Symbols shown in same/other "()" means that the persons indicated by the symbols are in-group/out-group. Additionally, $(X > Y)$ means that the social position of the person corresponding to X is higher than that of the person corresponding to Y, where X and Y are in-group. The symbols "$\wedge$" and "$\vee$" mean logical "AND" and "OR," respectively. In the example shown in Figure 1, $s = 1$ (there is an honorific title for the subject) was judged to be inconsistent with the social relationship, (A>S)(L) and $\{subj{:}\mathrm{A}\}$, because such correspondence was not defined in Table 3.

The decision list (Rivest, 1987) was used to obtain Table 3 using a training set. The training set was comprised of 819 training data. Each training data is a pair of a social relationship among the people and a sentence that is considered to be honorifically normative on the social relationship. Three sentences were prepared for each condition of possible combinations of [the number of persons involved in the sentence] × [relative social position among persons] × [in-group/out-group relationship among persons] × [subject/object] (273 variations in total). That's why there were 819 (=3×273) training data as mentioned above. The sentences in the training set were written by us, following as closely as possible the honorific norms stated or suggested in traditional Japanese textbooks and articles written by Japanese linguistic researchers. The speech intention was set to "speak."

The procedure to obtain Table 3 is as follows:

(Step 1) Make a table for training. The table contains [the number of persons involved in the sentence], [relative social position among persons], [in-group/out-group relationship among persons], [subject/object], and a honorific pattern, of each training data in the training set.

(Step 2) Make a candidate list for consistency table. The candidate list is comprised of features. Each feature is one of the combinations of [the number of

Table 3. A Part of consistency table

Three persons (S,L,A)	
Elements	Conditions
$s = 0$	(S=L=A) ∨ (S>L=A) ∨ (S=L>A) ∨ (S>A>L) ∨ (S=A>L) ∨ (S>L>A) ∨ [(S=L)∧{subj:S}] ∨ [(S=L)∧{subj:L}] ∨ [(S=L)∧{obj:A}] ∨ [(S>L)∧{subj:S}] ∨ [(S>L)∧{subj:L}] ∨ [(S>L)∧{obj:A}] ∨ [(S=A)∧{subj:S}] ∨ [(S=A)∧{subj:A}] ∨ [(S=A)∧{obj:S}] ∨ [(S=A)∧{obj:L}] ∨ [(S>A)∧{subj:S}] ∨ [(S>A)∧{subj:A}] ∨ [(S>A)∧{obj:S}] ∨ [(S>A)∧{obj:L}] ∨ [(L=A)∧{subj:S}] ∨ [(L>A)∧{subj:S}] ∨ [(L>S)∧{subj:S}] ∨ [(A>S)∧{subj:S}] ∨ [(A>S)(L)∧{subj:A}] ∨ [(A>S)(L)∧{obj:S}] ∨ [(A>S)(L)∧{obj:L}] ∨ [(A>L)∧{subj:S}] ∨ [(S)(L)∧{subj:S}] ∨ [(S)(A)∧{subj:S}] ∨ [(L)(A)∧{subj:S}]
$s = 1$	[(L>S)∧{subj:L}] ∨ [(L>A>S)∧{subj:A}] ∨ [(L=A>S)∧{subj:A}] ∨ [(A>S=L)∧{subj:A}] ∨ [(A>L>S)∧{subj:A}] ∨ [(A>S>L)∧{subj:A}] ∨ [(L>A>S)∧{obj:S}] ∨ [(L>A>S)∧{obj:L}] ∨ [(L=A>S)∧{obj:L}] ∨ [(A>S=L)∧{obj:S}] ∨ [(A>S>L)∧{obj:S}] ∨ [(A>S>L)∧{obj:L}] ∨ [(A>S=L)∧{obj:L}] ∨ [(A>L>S)∧{obj:S}] ∨ [(A>L>S)∧{obj:L}] ∨ [(L=A>S)∧{obj:S}] ∨ [(S)(L)∧{subj:L}] ∨ [(S)(A)∧{subj:A}] ∨ [(S)(A)∧{obj:S}] ∨ [(S)(A)∧{obj:L}]
$o = 0$	(S>L=A) ∨ (S=L=A) ∨ (S=L>A) ∨ (S>L>A) ∨ (S=A>L) ∨ (S>A>L) ∨ [(S=A)∧{subj:S}] ∨ [(S=A)∧{subj:L}] ∨ [(S=L)∧{subj:A}] ∨ [(S>A)∧{subj:S}] ∨ [(S>A)∧{subj:L}] ∨ [(S>L)∧{subj:A}] ∨ [(S=L)∧{obj:S}] ∨ [(S=L)∧{obj:L}] ∨ [(S>L)∧{obj:S}] ∨ [(S>L)∧{obj:L}] ∨ [(S=A)∧{obj:S}] ∨ [(S=A)∧{obj:A}] ∨ [(S>A)∧{obj:S}] ∨ [(S>A)∧{obj:A}] ∨ [(L=A)∧{obj:S}] ∨ [(L>A)∧{obj:S}] ∨ [(L>S)∧{obj:S}] ∨ [(A>S)∧{obj:S}] ∨ [(A>L)∧{obj:S}] ∨ [(A>S)(L)∧{subj:S}] ∨ [(A>S)(L)∧{subj:L}] ∨ [(A>S)(L)∧{obj:A}] ∨ [(S)(L)∧{obj:S}] ∨ [(S)(A)∧{obj:S}] ∨ [(A)(L)∧{obj:S}]
$o = 1$	[(L>S)∧{obj:L}] ∨ [(L>A>S)∧{subj:S}] ∨ [(L=A>S)∧{subj:L}] ∨ [(L=A>S)∧{subj:S}] ∨ [(L>A>S)∧{subj:L}] ∨ [(L=A>S)∧{obj:A}] ∨ [(A>S>L)∧{subj:S}] ∨ [(A>S=L)∧{subj:L}] ∨ [(A>S=L)∧{obj:A}] ∨ [(A>L>S)∧{subj:S}] ∨ [(A>L>S)∧{subj:L}] ∨ [(A>S=L)∧{subj:S}] ∨ [(A>S>L)∧{subj:L}] ∨ [(A>S>L)∧{obj:A}] ∨ [(L>A>S)∧{obj:A}] ∨ [(A>L>S)∧{obj:A}] ∨ [(S)(A)∧{subj:L}] ∨ [(S)(A)∧{obj:A}] ∨ [(A)(S)∧{subj:S}] ∨ [(S)(L)∧{obj:L}]
$e = 0$	(S>L) ∨ (S=L)
$e = 1$	(L>S) ∨ (L)(S)
$p = 0$	(S=L=A) ∨ (S>L=A) ∨ (S=L>A) ∨ (S>L>A) ∨ (S=A>L) ∨ (S>A>L)
$p = 1$	[(L=A>S)∧{subj:L}] ∨ [(L=A>S)∧{subj:A}] ∨ [(L>A>S)∧{subj:L}] ∨ [(L>S>A)∧{subj:L}] ∨ [(L>S=A)∧{subj:L}] ∨ [(A>S=L)∧{subj:A}] ∨ [(A>S>L)∧{subj:A}] ∨ [(A>L>S)∧{subj:A}] ∨ [(L=A>S)∧{obj:S}] ∨ [(L>A>S)∧{obj:S}] ∨ [(L=A>S)∧{obj:L}] ∨ [(A>S=L)∧{obj:S}] ∨ [(A>S>L)∧{obj:S}] ∨ [(A>S>L)∧{obj:L}] ∨ [(A>S=L)∧{obj:L}] ∨ [(A>L>S)∧{obj:S}] ∨ [(A>L>S)∧{obj:L}] ∨ [(A)(L>S)∧{subj:L}]
$p = 2$	[(L>A>S)∧{subj:S}] ∨ [(L=A>S)∧{subj:S}] ∨ [(A>S>L)∧{subj:S}] ∨ [(A>S=L)∧{subj:S}] ∨ [(A>S=L)∧{subj:L}] ∨ [(A>S>L)∧{subj:L}] ∨ [(A>L>S)∧{subj:S}] ∨ [(A>S=L)∧{obj:A}] ∨ [(A>S>L)∧{obj:A}] ∨ [(L>S=A)∧{obj:L}] ∨ [(L>S>A)∧{obj:L}]
$p = 0 \vee 1$	[(S)(L)∧{subj:L}] ∨ [(S)(A)∧{obj:L}] ∨ [(A)(S)∧{subj:A}] ∨ [(A)(S)∧{obj:S}]
$p = 0 \vee 2$	[(S=L)(A)∧{subj:S}] ∨ [(S>L)(A)∧{subj:S}] ∨ [(S=L)(A)∧{subj:L}] ∨ [(S>L)(A)∧{subj:L}] ∨ [(S=L)(A)∧{obj:A}] ∨ [(S>L)(A)∧{obj:A}]
$p = 0 \vee 4$	[(L>S=A)∧{subj:S}] ∨ [(L>S>A)∧{subj:S}] ∨ [(L>S=A)∧{obj:S}] ∨ [(L>S>A)∧{obj:S}] ∨ [(S=A)(L)∧{obj:S}] ∨ [(S=A)(L)∧{subj:S}] ∨ [(S>A)(L)∧{subj:S}] ∨ [(S>A)(L)∧{obj:S}] ∨ [(A>S)(L)∧{subj:S}] ∨ [(A>S)(L)∧{obj:S}]
$p = 1 \vee 3$	[(A>L>S)∧{subj:L}]
$p = 0 \vee 2 \vee 4$	[(S=A)(L)∧{obj:L}] ∨ [(S>A)(L)∧{obj:L}] ∨ [(L>S)(A)∧{subj:S}] ∨ [(L=A)(S)∧{subj:S}] ∨ [(L>A)(S)∧{subj:S}] ∨ [(A>S)(L)∧{obj:L}] ∨ [(A>L)(S)∧{subj:S}] ∨ [(A)(S)(L)∧{subj:S}]
$p = 1 \vee 2 \vee 3$	(L>A>S)∧{obj:L}

persons involved in the sentence] × [relative social position among persons] × [in-group/out-group relationship among persons] × [subject/object] × [values of honorific elements], joined by logical "∧." The candidate list covers all the possible combination of them.

(Step 3) Maintain all the features in the candidate list which are consistent with the training table, the other features are deleted in the candidate list.

(Step 4) Delete features which are included in other features in the candidate list.

(Step 5) Join the remaining features in the candidate list by using logical "∨."

Each portion in Table 3 can be summarized to simpler conditions. For example, the conditions corresponding to $s = 0$ can be summarized to: (*subj*,S)(L) ∨ (S=*subj*) ∨ (S>*subj*) ∨ {*subj*:S}.

4 System Validity Check

The proposed system was verified by using test set-1 and test set-2. Both test sets were consisted of test data whose format was the same as that of the training data.

Validity Check using Test Set-1
We prepared test set-1 to contain correct and incorrect test sets. No test data included in the correct test set contained any misuses. All test data included in the incorrect test set contained some misuses. Both test sets covered all the conditions for each possible combination of [the number of people involved in the sentence] × [relative social position among the people] × [in-group/out-group relationship among the people] × [subject/object] (273 variations in total). The speech intention was set to "speak." The training set used to construct the consistency table was used as the correct test set. Both correct and incorrect test sets contained 819 test sets. The experimental results showed that the system accurately judged all of the test data in the correct test set to be "correct," and that it accurately indicated the misuses in all the test data from the incorrect test set.

Validity Check using Test Set-2
Test set-2 was prepared by third-party linguistic researchers. Five kinds of speech intentions, "speak," "phone," "explain," "visit," and "show," were assumed when preparing the test set. Other variations concerning social relationship ×[subject/object] were the same as those in test set-1. The total number of test sets included in both correct and incorrect test sets was 4,095. The experimental results showed that the system judged 99.4% of the correct test set to be "correct" but judged the rest of the data (0.6%) to be "misuse." The system accurately indicated 97.3% of misuses in the incorrect test set, but judged the rest of them (2.7%) to be "correct." Most cases of incorrect responses, i.e. the 0.6% in the former and 2.7% in the latter, were due to differences in honorific norm between the third-party linguistic researchers and us.

5 Experiments Using Sentences Described by Non-experts

Because the system was constructed to follow the honorific norms stated or suggested in traditional Japanese textbooks and articles written by Japanese linguistic researchers, the system may tend to regard some sentences as honorifically incorrect even though they are actually permissible in Japanese society. To reveal what kind and how many of the norms used in the system are perfectly acceptable honorific expressions for everyday use by non-experts (people who are not linguistic experts), we performed the following experiment.

5.1 Procedure

Forty subjects who are over 30, twenty males and females, participated in the experiments. We used an age requirement because we expected that people over 30 would have considerable experience in use of honorific expressions. Variation in speech intention× social relationship×[subject/object] were the same as those in test set-2 described in Section 4 (4,095 variations in total). The subjects were required to write at least one sentence for each variation under restrictions 1 to 3 described in Section 3.1. We obtained 54,600 sentences through this experiment. We prepared a test set that consisted of 54,600 data sets by using these sentences so that the data format was the same as that of the training data.

5.2 Experimental Results and Discussion

The experimental results showed that the system judged 70% of the test data sets to be "correct." All were considered to be valid. However, the remaining 30% of the test data sets were judged as "misuse." Among the test sets judged as "misuse", 20% of the test sets were judged to be "misuse in word form" and the remaining 80% of the test sets were judged to be "misuse in performance." Typical data of the misuse in performance are as follows.

(1) $[s = 1]$ is inconsistent with the situation where S and *subj* are in the in-group and S>(or =) *subj*. Example: "A san ga anata ni hanasitandesho": (S>A>L).
(2) $[s = 1]$ is inconsistent with the situation where S and *subj* are in the in-group, and L and them (i.e., S and *subj*) are in the out-group. Example: "A san ga watasi ni hanasimasita.": (A>S)(L).
(3) $[o = 1]$ is inconsistent with the situation where S and *obj* are in the in-group and S>(or =) *obj*. Example: "A san ni hanasitanda": (S>A>L).
(4) $[e = 1]$ is inconsistent with the situation where S and L are in the in-group and S>L. Example: "A kun ga anata ni ittandesune": (S>L>A).
(5) $[e = 0]$ is inconsistent with the situation where S and L are in the out-group. Example: "A san ga B san ni hanasita": (S)(L)(A)(B).
(6) $[p = 0]$ is inconsistent with the situation where S and L are in the in-group and *subj*>S. Example: "A san ga B kun ni hanasimasita": (L=A>S>B).

For educational purposes, cases (1), (3), (4), and (6) are not recommended to be indicated as serious misuse because these cases are considered to be permissible (Kokugoken 1992). However, cases (2) and (5) should be indicated as serious misuse because these cases are considered to be inappropriate for education (Kokugoken 1992). Such differences in educational importance should be reflected in a graphical user interface of the system (e.g., by changing the display scheme for an serious misuse).

6 Conclusion

We developed a computational system to indicate the misuse of honorifics in word form and in performance of expressions in Japanese speech sentences. The misuse in word form was checked by constructing the list of expressions whose word form is bad in terms of honorifics. The misuse in performance was checked by constructing a consistency table that defines consistency between the honorific features of sentences and the social relationship among the people involved in the sentences. The social relationship was represented by combinations of [the number of people involved in the sentence] × [relative social position among the people] × [in-group/out-group relationship among the people]. The proposed system was verified using test data prepared by the authors and also by third-party linguistic researchers. The results showed that the system was able to discriminate between the correct and the incorrect honorific sentences in all but a few cases. Furthermore, differences in the educational importance among the norms used in the system were revealed based on the experiments using sentences written by people who are not linguistic experts.

References

[Brown] Brown, P. and Levinson, S.: Politeness: - Some universals of language usage -. Cambridge (1987).

[Kokugoken] Kokugoken. Keigo-kyoiku-no-kihinmondai (in Japanese). M.O.F. Press Office, Tokyo(1972).

[Rivest] Ronald, L. Rivest. Learning decision lists. Machine Learning. **2** (1987): 229–246.

A Chinese Corpus with Word Sense Annotation

Yunfang Wu, Peng Jin, Yangsen Zhang, and Shiwen Yu

Institute of Computational Linguistics, Peking University
Beijing, China, 100871
{wuyf, jandp, yszhang, yusw}@pku.edu.cn

Abstract. This paper presents the construction of a Chinese word sense-tagged corpus. The resulting lexical resource includes mainly three components: 1) a corpus annotated with word senses; 2) a lexicon containing sense distinction and description in the feature-based formalism; 3) the linking between the sense entries in the lexicon and CCD synsets. A dynamic model is put forward to build the three knowledge bases simultaneously and interactively. The strategy to improve consistency is addressed since consistency is a thorny issue for constructing semantic resources. The inter-annotator agreement of the sense-tagged corpus is satisfied. The database will grow up to be a powerful lexical resource both for linguistic researches on Chinese lexical semantics and word sense disambiguation.

Keywords: sense, sense annotation, sense disambiguation, lexical semantics.

1 Introduction

There is a strong need for a large-scale Chinese corpus annotated with word senses both for word sense disambiguation (WSD) and linguistic research. Although much research has been carried out, there is still a long way to go for WSD techniques to meet the requirements of practical NLP programs such as machine translation and information retrieval. Although plenty of unsupervised learning algorithms have been put forward, SENSEVAL ([1]) evaluation results show that supervised leaning approaches, in general, are much better than unsupervised ones. Undoubtedly high accuracy WSD needs large-scale word sense tagged corpus as training material ([2]). It was argued that no fundamental progress in WSD could be made until large-scale lexical resources were built ([3]). The absence of a large-scale sense tagged corpus remains one of the most critical bottlenecks for successful WSD programs. In English a word sense annotated corpus SEMCOR (Semantic Concordances) ([4]) has been built, which was later trained and tested by many WSD systems and stimulated large amounts of WSD work. In the field of Chinese corpus construction, plenty of attention has been paid to POS tagging and syntactic structures bracketing, for instance the Penn Chinese Treebank ([5]) and Sinica Corpus ([6]), but very limited work has been done with semantic knowledge annotation. The semantic dependency knowledge has been annotated in a Chinese corpus ([7]), which is different from word

Y. Matsumoto et al. (Eds.): ICCPOL 2006, LNAI 4285, pp. 414–421, 2006.

sense tagging (WST) orientated towards WSD. [8] introduced the Sinica sense-based lexical knowledge base, but as everyone knows, Chinese pervasive in Taiwan is not the same as mandarin Chinese. SENSEVAL-3 ([1]) provides a Chinese word sense annotated corpus, which contains 20 words and 15 sentences per meaning for most words, but obviously the data is too limited to achieve wide coverage, high accuracy WSD systems.

This paper is devoted to building a large-scale Chinese corpus annotated with word senses, and the ambitious goal of the work is to build a comprehensive resource for Chinese lexical semantics. The resulting lexical knowledge base will contain three major components: 1) a corpus annotated with word senses; 2) a lexicon containing sense distinction and description; 3) the linking between the lexicon and the Chinese Concept Dictionary (CCD) ([9]). This paper is so organized as follows. The corpus, the lexicon, CCD as well as the interactive model are presented in detail in section 2 as 4 subsections. Section 3 is devoted to discuss the adapted strategy to improve consistency. Section 4 is the evaluation of the corpus. Finally in section 5 conclusions are drawn and future works are presented.

2 The Interactive Construction

2.1 The Corpus

At the present stage the sense tagged corpus mainly comes from three months' texts of People's Daily, and later we will move to other kinds of texts considering corpus balance. The input data for WST is POS tagged using Peking University's POS tagger. The high precision of Chinese POS tagging lays a sound foundation for researches on sense annotating. The emphasis of WST therefore falls on the ambiguous words with the same POS. All the ambiguous words in the corpus will be analyzed and annotated, which is different from most of the previous works that focus on limited specific words.

2.2 The Lexicon

Defining sense has long been one of the most heated-discussed topics in lexical semantics. The representation of word senses in the lexicon should be valid and efficient for WSD algorithms in the corpus.

Human beings make use of the context, communication surrounding and sometimes even world knowledge to disambiguate word senses. For machine understanding, the last resort for WSD is the context. In this paper the feature-based formalism is adopted to describe word senses. The features, which appear in the form "Attribute =Value", can incorporate extensive distributional information about a word sense. The many different features together constitute the representation of a sense, but the language definitions of meaning serve only as references for human readers. An example of feature-based description of meaning is shown in figure 1 as for some senses of verb "开/kai1".

Table 1. The feature-based description of some senses of verb "开/kai1"

word	senses	definition	subcategory	valence	subject	object	syntactic positions
开	1	open	[NP]	2	human		
开	2	unfold	[~]	1	~human		
开	3	pay	[NP]	2	human	asset	
开	4	drive	[NP]	2	human	vehicle	
开	5	establish	[NP]	2	human	building	
开	6	hold	[NP]	2	human	event	
开	7	write out	[NP]	2	human	bill	
开	8	used after verbs					V+～| V+de+～ | V+bu+～

Thus 开④, for example, can be defined in a set of features:

开④ {subcategory=[NP], valence=2, subject=human, object=vehicle, …}

The granularity of sense distinction has long been a thorny issue for WSD programs. Based on the features described in the lexicon, those features that are not specified will not be regarded as the distinguishing factors when discriminating word senses, such as the goal, the cause of the action of a verb. That is, finer sense distinctions without clear distributional indicators are ignored. As a result, the senses defining in our lexicon are somewhat coarse-grained compared with the senses in conventional dictionaries, and the inter-annotator agreement is more easily reached. [10] argued that the standard fine-grained division of senses for use by human reader may not be appropriate for the computational WSD task, and that the level of sense-discrimination that NLP needs corresponds roughly to homographs. The sense distinctions in our lexicon lie in between fine-grained senses and homographs.

With the feature-based description as the base, the computer can accordingly do the unification in WSD algorithms, and also the human annotator can correctly identify the meaning through reading the sentence context. What's more, using feature-based formalisms the syntax / semantics interface can be easily realized ([11]).

2.3 The Interactive Construction

To achieve the ambitious goal of constructing a comprehensive resource for Chinese lexical semantics, the lexicon containing sense descriptions and the corpus annotated with senses are built interactively, simultaneously and dynamically. On one hand, the sense distinctions are relying heavily on the corpus rather than on human's introspection. The annotator can add or delete a sense entry, and can also edit a sense description according to the real word uses in the corpus. The strategy adapted here conforms to the spirit of lexicon construction nowadays, that is, to commit to corpus evidence for semantic and syntactic generalization just as Berkeley FrameNet project did ([12]). On the other hand, using the sense information specified in the lexicon the human annotators assign semantic tags to all the instances of the word in a corpus. The knowledge base of lexical semantics built here can be viewed either as a corpus in which word senses have been tagged, or as a lexicon in which example sentences can be found for any sense entry. The lexicon and the corpus can also be split as separate lexical resource to study when needed. SEMCOR provides an important dynamic model ([4]) for sense-tagged programs, where the tagging process is used as a vehicle for improving WordNet coverage. Prior to SEMCOR WordNet has already existed and the tagging process served only as a way to improve the coverage. However, the Chinese lexicon containing sense descriptions oriented towards computer understanding has not yet been built, so the lexicon and the corpus are built in the same procedure. To some extent we can say that the interactive model adapted here is more fundamental than SEMCOR.

A software tool is developed in Java to be used as the word sense tagging interface (figure 1). The interface embodies the spirit of interactive construction properly. In the upper section there displays the context in the corpus, with the target ambiguous word highlighted. The range of the context can shift as required. The word senses with feature-based description from the lexicon are displayed in the bottom section. Through reading the context, the human annotator decides to add or delete a sense entry, and can also edit a sense's feature description. The annotator clicks on the appropriate entry to assign a sense tag to the word occurrence. A sample sentence can be selected from the corpus and added automatically to the lexicon in the corresponding sense entry.

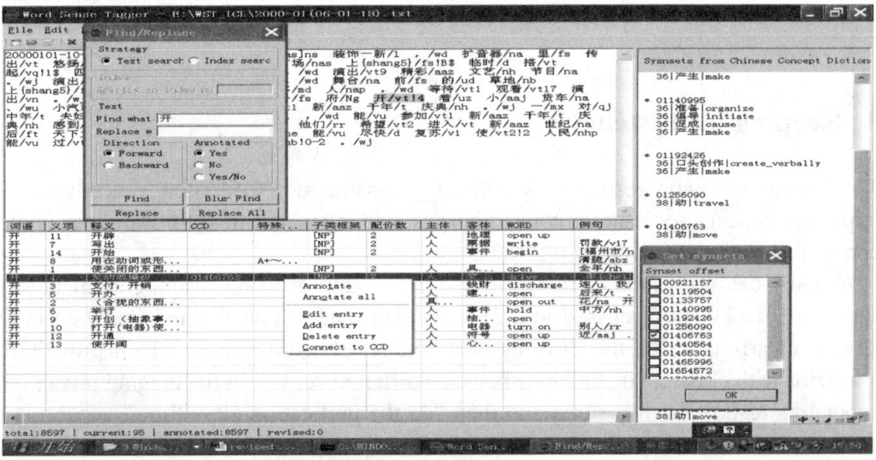

Fig. 1. The word sense tagging interface

2.4 Linking Senses to CCD Synsets

The feature-based description of word meaning as discussed in 2.2 describes mainly the syntagmatic information but cannot include the paradigmatic relations. A lexical knowledge base is well defined only when both the syntagmatic and paradigmatic information are included. WordNet has been widely experimented in WSD researches. We are trying to establish the linking between the sense entries in the lexicon and the synsets in WordNet. CCD is a WordNet-like Chinese lexicon ([9]), which carries the main relations defined in WordNet and is a bilingual concept lexicon with the parallel Chinese-English concepts to be simultaneously displayed. The offset number of the corresponding synset in CCD is used to convey the linking, which expresses the structural information of the tree. Through the linking to CCD, the paradigmatic relations (such as hypernym / hyponym, meronym / holonym) between word senses in the lexicon can be reasonably acquired. After the linking has been established, the many existing WSD approaches based on WordNet can be trained and tested on the Chinese sense tagged corpus.

The linking is now done manually. In the word sense tagging interface (figure 1) the different synsets of the word in CCD, along with the hypernyms of each sense (expressed by the first word in a synset), are displayed in the right section. A synset selection window (named Set synsets) contains the offset numbers of the synsets. The annotator clicks on the appropriate box(es) to assign a synset or synsets to the currently selected sense in the lexicon.

Table 2. An example of the linking between the sense entries in the lexicon and CCD synsets

word	senses	definition	CCD synset offset	subcategory	valence	subject
迷惑	1	puzzle	00419748, 00421101	[~]	1	human
迷惑	2	mislead	00361554, 00419112	[NP]	2	~human

3 Keeping Consistency

The lexical semantic resource is manually constructed. Consistency is always an important concern for hand-annotated corpus, and is even critical for the sense tagged corpus due to the subtle meanings to handle.

Actually the motivation of feature-based description of word meaning is intended to keep consistency (as discussed in 2.2). The "Attribute = Value" pairs specified in the lexicon clearly describe the distributional context of word senses, which provide the annotator with clear-cut distinctions between different senses. In the tagging process the annotators read through the text and then "do the unification algorithms" according to the features of senses, and then select the most appropriate sense tag. The operational guidelines for sense distinction can be set up based on the features, and thus the unified principle may be followed when distinguishing different words' senses.

Another observation is that the consistency is easier to keep when the annotator manages many different instances of the same word than handle many different words in a specific time frame, because the former method enables the annotator to establish an integrative knowledge of a specific word and its sense distinction. The word sense tagging interface as shown in figure 2 provides the tool (the window named Find/Replace), which allows the annotator to search for a specific word to finish tagging all its occurrences in the same period of time rather than move sequentially through the text as SEMCOR did ([4]).

Checking is of course a necessary procedure to keep the consistency. The annotators are also checkers, who check other annotator's work. A text generally is first tagged by one annotator and then verified by two checkers. There are five annotators together in the program, of which three are majored in linguistics and two are majored in computational linguistics. A heated discussion inevitably happens when there exist different views. After discussion, the disagreement between annotators will be greatly reduced. Checking all the instances of a word in a specific time frame will greatly improve the precision and accelerate the speed just as in the process of tagging. A software tool is designed to gather all the occurrences of a word in the corpus into a checking file with the sense KWIC (Key Word in Context) format in sense tags order. Figure 2 illustrates some example sentences containing different senses of verb "开/kai1". The checking file enables the checker to have a closer examination of how the senses are used and distributed, and to form a general view of how the sense distinctions are made. The inconsistency thus can be reached quickly and correctly.

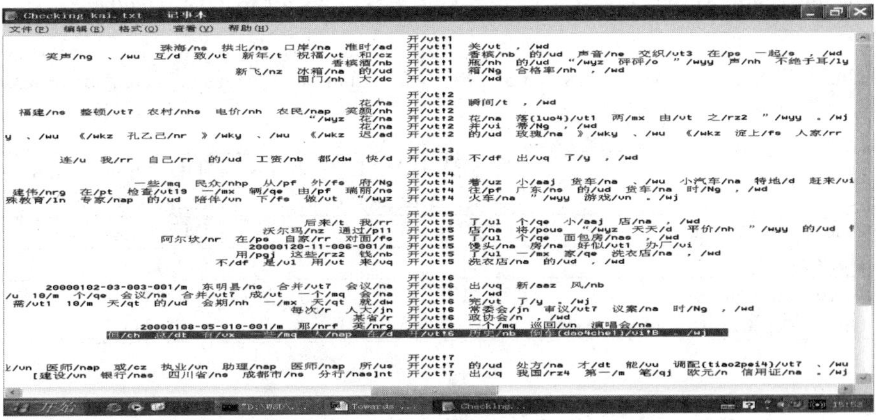

Fig. 2. Some example sentences of verb 开/kai1

4 Evaluation

4.1 Inter-annotator Agreement

The agreement rate between human annotators on word sense annotation is an important concern both for the evaluation of WSD algorithms and word sense tagged corpus. Suppose that there are *n* occurrences of ambiguous target words in the corpus. Let *m* be the number of tokens that are assigned identical sense by two human annotators (the annotator and the checker in this program). Then a simple measure to quantify the agreement rate between two human annotators is *p*, where $p = m/n$.

Table 3 summarizes the inter-annotator agreement in the sense tagged corpus. Combining nouns and verbs the agreement rate achieves 84.8%, which is comparable to the agreement figures reported in the literatures. [13] mentioned that for the

Table 3. Inter-annotator agreement

POS	Num of words	*n*	*m*	*p*
N	813	19,197	17,757	92.5%
V	132	41,698	33,900	81.3%
ALL	945	60,895	51,657	84.8%

SENSEVAL-3 lexical sample task there was a 67.3% agreement between the first two taggings. Table 3 also shows that the inter-annotator agreement for nouns is obviously higher than verbs.

4.2 The State of the Art

The project is now going on. Up to now, 813 nouns and 132 verbs have been analyzed and described in the lexicon with the feature-based formalism. In three-month People's Daily texts together 60,895 word occurrences have been sense tagged. By now this is almost the biggest scale sense tagged corpus for mandarin Chinese.

Table 4. The data comparing between SENSEVAL-3 Chinese sense-tagged data and the sense-tagged corpus of People's Daily

	SENSEVAL-3 Chinese sense-tagged data	The Chinese sense-tagged corpus
ambiguous words	20	945
senses	81	2268
word occurrences (including training and test data)	1172	60895

5 Conclusion and Future Works

This paper describes the construction of a sense-tagged Chinese corpus. The goal is to create a valuable resource both for word sense disambiguation and researches on Chinese lexical semantics. Actually some researches on Chinese word senses have been carried out based on the corpus, and some supervised learning approaches, such as SVM, ME, and Bayes algorithms have been trained and tested on the corpus. A small part of the sense-tagged corpus has been published in the website www.icl.pku.edu.cn. Later we will move to other kinds of texts on account of corpus balance and data sparseness. To analyze more ambiguous words and to describe more senses in the lexicon, and to annotate more word instances in the corpus are of course the next urgent task. To train algorithms and to develop software tools to realize semi-automatically sense tagging are also next undertakings.

Acknowledgments. This research is funded by National Basic Research Program of China (No. 2004CB318102).

References

1. SENSEVAL: http://ww.senseval.org
2. Ng, H. T.: Getting Serious about Word Sense Disambiguation. In Proceedings of ANLP-97 Workshop on Tagging Text with Lexical Semantics: Why, What, and How? (1997)
3. Veronis, J.: Sense Tagging: Does It Make Sense? In Wilson et al. (Eds). Corpus Linguistics by the Rule: a Festschrift for Geoffrey Leech. (2003)
4. Landes, S., Leacock, C. and Tengi, R.I.: Building Semantic Concordances. In Christiane Fellbaum (Ed.). WordNet: an Electronic Lexical Database. MIT Press, Cambridge (1999)
5. Xue, N., Chiou, F. D. and Palmer, M.: Building a Large-Scale Annotated Chinese Corpus. In Proceedings of COLING (2002)
6. Huang, Ch. R and Chen, K. J.: A Chinese Corpus for Linguistics Research. In Proceedings of COLING (1992)
7. Li, M.Q., Li, J. Z., Dong, Zh. D., Wang Z. Y. and Lu, D. J.: Building a Large Chinese Corpus Annotated with Semantic Dependency. In Proceedings of the 2nd SIGHAN Workshop on Chinese Language Processing (2003)
8. Huang, Ch. R., Chen, Ch. L., Weng C. X. and Chen. K. J.: The Sinica Sense Management System: Design and Implementation. In Recent advancement in Chinese lexical semantics (2004)
9. Liu, Y., Yu, S. W. and Yu, J.S.: Building a Bilingual WordNet-like Lexicon: the New Approach and Algorithms. In Proceedings of COLING (2002)
10. Ide N. and Wilks, Y.: Making Sense About Sense. In Agirre, E., Edmonds, P. (eds.):Word Sense Disambiguation: Algorithms and Applications. Springer (2006)
11. Nerbonne, J.: Computational Semantics – Linguistics and Processing. In Shalom Lappin. (Ed.) The Handbook of contemporary semantic theory. Foreign Language Teaching and Research Press and Blackwell Publishers Ltd (2001)
12. Colin F. B., Fillmore, C. J. and Lowe, J. B.: The Berkeley FrameNet Project. In Proceedings of COLING-ACL (1998)
13. Mihalcea, R., Chklovsky, T. and Kilgarriff, A.: The SENSEVAL-3 English Lexical Sample Task. In Third International Workshop on the Evaluation of Systems for the Semantic analysis of Text (2004)

Multilingual Machine Translation of Closed Captions for Digital Television with Dynamic Dictionary Adaptation

Sanghwa Yuh[1] and Jungyun Seo[2]

[1] School of Computer and Information, Kyungin Women's College, Gyesan 2-Dong, Gyeyang-Ku, Incheon, 407-740, Korea
shyuh@kic.ac.kr

[2] Department of Computer Science and Interdisciplinary Program of Integrated Biotechnology, Sogang University, Mapo-Ku, Sinsu Dong 1, Seoul, 121-742, Korea
seojy@sogang.ac.kr

Abstract. In this paper, we present a multilingual machine translation system for closed captions for digital television. To cope with frequent appearance of unregistered words and the articles of multiple domains as in TV news program, we propose a Dynamic Dictionary Adaptation method. We adopted live resources of multilingual Named Entities and their translingual equivalences from Web sites of daily news, providing multilingual daily news in Chinese, English, Japanese and Korean. We also utilize Dynamic Domain Identification for automatic dictionary stacking. With these integrated approaches, we obtained average translation quality enhancement of 0.5 in Mean Opinion Score (MOS) for Korean-to-Chinese. We also had 0.5 and 0.1 average enhancement for Korean-English and Korean-Japanese, respectively. The average enhancement is 0.37, which means almost a third level up to the next higher MOS scale.

Keywords: Closed Caption, Multilingual Machine Translation, Dynamic Dictionary Adaptation, Named Entity Alignment, Dynamic Domain Identification.

1 Introduction

Closed Captions (CCs), which are hidden text in the video signal for deaf and hard of hearing and late deafened people, display the dialogue, narration and sound effects of a TV program. While closed captioning was originally developed for the hearing impaired, it can also be a great benefit for both foreign residents in Korea who do not understand Korean and Korean language learners.

All the terrestrial broadcasting stations in Korea are in a state of transition from analog to Digital Television (DTV) and to High-Definition TV (HDTV). As part of this transition, CCs must also be converted for service from analog to digital. Since the first broadcasting with analogue CCs by KBS (Korean Broadcasting System) in 1999, the rate of closed captioning TV program has

Y. Matsumoto et al. (Eds.): ICCPOL 2006, LNAI 4285, pp. 422–429, 2006.

been reached 31.7% (2005). But, closed captioning for DTV in Korea is not in progress yet and expecting 2007 or later for the DTV closed captioning service.

Closed captioning by Machine Translation (MT) system could be one of the most cost-effective choices for the multilingual closed captioning. There have been several approaches to translate analogue CC automatically with MT systems including *ALTo* (Simon Fraser Univ., Canada)[1], [2], *KANT* (CMU, USA)[3], and *CaptionEye* (ETRI, Korea)[4]. But, these systems resulted in rather hardly understandable translations so that they failed to reach practical systems. One of the major reasons for this is related to translating Named Entities (NEs) of proper names which are very popular in news and drama. NEs convey important information of news articles and drama scripts. So, correct translation of NEs is quite indispensable for effective comprehension of news and drama. But, it is impossible to enlist all NEs in the dictionary because we could encounter lots of new NEs of persons, locations and organizations everyday. In most case, we cannot predict the appearance of new entities. And that, the translation of the NEs cannot be achieved by simple combinations of each translation of the words in the NE expression. So, it is quite necessary to collect multilingual NEs and their certified multilingual translation equivalences automatically.

Another problem is related to translating domain-specific terminologies. Most MT systems support multiple domain dictionaries for better translation results. But they rely on the user for stacking multiple domain dictionaries and do not change the dictionaries while the translation is being performed. This can be a crucial weak point when we translate the TV news captions, in which lots of incidents belonging to different domains are reported.

In this paper, we propose a Dynamic Dictionary Adaptation methods for multilingual machine translation of CCs for DTV. To cope with frequent appearance of unregistered NEs and the articles of multiple domains as in TV news program, we adopted live Web resources of multilingual NEs and their translingual equivalences from Web sites of daily news, providing multilingual daily news in Chinese, English, Japanese and Korean. We also devised a Dynamic Domain Identifier (DDI) for news caption based on decision tree induction in order to activate and stack the multiple domain dictionaries dynamically [5].

In Sect. 2, we survey the related works on previous CC MT systems. In Sect. 3 we introduce Dynamic Dictionary Adaptation including NE Alignment for the translingual equivalences, Dynamic Domain Identifier, and Program Identification. In Sect. 4, we evaluate our system to show the overall performance enhancement. Concluding remarks will be found in Sect. 5.

2 Related Works

There have been several approaches for Closed Caption (CC) MT systems for EIA-608 analog CC. Simon Fraser Universities in Canada developed a fully automatic large-scale multilingual CC MT system, *ALTo*. In order to handle proper names, they used pattern matching and caching names in a name memory, where previously recognized names are stored, to recognize proper names [1], [2]. With 63 patterns, they reported a recall of 72.7% and a precision of 95.0%. Carnegie Mellon University (CMU) also briefly reported a real-time translation system of business news captions (analogue) from English to German based on their existing multi-engine MT systems [3]. Unknown words including human/company/place name are identified by Preprocessor. They only use in-house knowledge for recognizing proper names.

In Korea, *CaptionEye* systems had been newly developed by Electronics and Telecommunication Research Institute (ETRI) [4] from 1999 to 2000. *CaptionEye* system is multilingual MT systems among Korean-to-English, English-to-Korean, Korean-to-Japanese and Japanese-to-Korean language pairs. *CaptionEye* is essentially a kind of pattern-based system. The system did not pay many attentions to NEs of proper names. So, the system lacks of special module for handling proper names, although the target domain contains news captions. Simple Finite State Automata (FSA) based pattern matching for proper name recognition and gazetteers during morphological analysis are the only resources for handling NEs.

3 Dynamic Dictionary Adaptation

In this section, we present three components of Dynamic Dictionary Adaptation. The first one is an Automatic Named Entity Alignment for gathering multilingual NE translingual equivalences from the Web pages of daily news. The second is Dynamic Domain Identification for automatic dictionary stacking for news captions which has multiple domain articles. The last one is a Program Identification for activating program specific dictionary like dramas. By using **Electronic Program Guide (EPG)** information of TV programs, we could also exploit the very specific knowledge of a definite TV program when to translate the CCs of the program. With this dynamic adaptability of MT systems, we could make the translation quality higher.

3.1 Named Entity Alignment for Multilingual Named Entity Translation

Named Entities (NEs) of proper names are very popular in news and drama. NEs convey important information of news articles and drama scripts. So, correct translation of NEs is quite indispensable for effective comprehension of news and drama. But, it is impossible to enlist all NEs in the dictionary because we could encounter lots of new NEs of persons, locations and organizations everyday. And that, the translation of the NEs, cannot be achieved by simple combinations of each translation of the words in the NE expression. Incorrect translations of NEs make TY program viewers hard to understand the news. So, correct translation of NEs is quite indispensable for effective comprehension of news CCs. The same problems concerning to NEs can be found when to translate CCs of TV dramas.

We try to raise the translation quality higher to the commercial level by solving the translation problems of multilingual NEs with very practical and integrated techniques of machine translation and information extraction. Our solution is obtaining live translingual equivalences of NEs from Web sites of daily news, providing multilingual daily news in Chinese, English, Japanese and Korean. Most of the significant news articles can be found on the Web sites before we watch the same news on TV. So, we devised an intelligent Web crawler for extracting multilingual NEs from the Web sites and aligning their translingual equivalences from the non-parallel, content-aligned multilingual news articles. The aligned translingual NEs are lively updated in order to be used by the multilingual CC MT systems from Korean into Chinese/English/Japanese when the similar news on TV is translated.

The method for finding translingual equivalences between Korean and English NEs is basically based on the **Feature Cost Minimization Method** proposed by Huang *et al.* (2003) [6]. They proposed to extract NE translingual equivalences between Chinese and English based on the minimization of linearly combined multi-feature cost minimization. The costs include transliteration cost and translation cost, based on IBM model 1, and NE tagging cost by an NE identifier. They required NE Recognition on both the source side and the target side. They reported the NE translingual equivalence with 81translation score by 0.06 of NIST8 score (from 7.68 to 7.74). We adopt only two features: transliteration cost and translation cost. It is because that Korean and English NE Recognizers we developed are based on the SVM framework. SVM only output hard decision values for 2-class problem, zero or one. So, NE tagging cost used in [6] is meaningless in our model.

In case of the aligning between Korean and English NEs, the alignment cost, $C(K_{ne}, E_{ne})$ for a translingual NE pair of English and Korean (K_{ne}, E_{ne}) is their linear combination of the transliteration score, $C_{translit}(K_{ne}, E_{ne})$ and translation cost, $C_{translat}(K_{ne}, E_{ne})$:

$$C(K_{ne}, E_{ne}) = \lambda_1 C_{translit}(K_{ne}, E_{ne}) + \lambda_2 C_{translat}(K_{ne}, E_{ne}) \quad (1)$$

The transliteration score, $C_{translit}(K_{ne}, E_{ne})$ and translation cost $C_{translat}(K_{ne}, E_{ne})$ are adopted from [6] with modifications for Korean-to-English transliteration. Korean character (syllable) is almost independently transliterated into an English letter string through "Korean Syllable-to-Romanization" Table. Considering that mappings from Korean character to their English strings are mostly in a determinate way, ie., $P(e_i \mid k_i) \approx 1$. Given a *Hangeul* (Korean Alphabet) sequence ($k = "k_1 \dots k_l(k)"$) and a set of English letter sequence $E = \{e_1, \dots, e_n\}$. Here, $l(k)$ is the length of the string k. The goal is to find the most likely transliteration A* that maximize the transliteration likelihood as follows:

$$\begin{aligned} C_{translit}(K_{ne}, E_{ne}) &\equiv C_{translit}(K_{ne}, E_{ne} \mid A^*) \\ &= \sum_{(i,j)\in A^*} C_{translit}(K_i, E_j) \\ &= \sum_{(i,j)\in A^*} [argmin\{r_i \in E_{ki}\} - \log P(E_j \mid r_i)] \end{aligned} \quad (2)$$

Word translation probability P(e | k) can be estimated using the alignment models for statistical machine translation. Let K_{ne} denote a Korean NE and it is composed of i Korean words, $k_1, k_2, \ldots, k_i$. Let E_{ne} denote an English NE and it is composed of j English words, $e_1, e_2, \ldots, e_j$. The translation probability of a Korean and English NE pair $P_{translat}(K_{ne}, E_{ne})$ is computed as follows, which is known as the IBM model-1:

$$P_{translat}(K_{ne}, E_{ne}) = \frac{1}{L^j} \prod_{j=1}^{J} \sum_{i=1}^{L} p(e_j \mid k_i) \tag{3}$$

3.2 Dynamic Domain Identifier for Automatic Domain Dictionary Stacking

The base MT systems support 28 domain dictionaries for better translation results. But it relies on the user for stacking multiple domain dictionaries and does not allow the user to change the dictionaries while the translation is being performed. This can be a crucial weak point when we translate the news captions, because a news program reports lots of incidents belonging to different domains. So, we devised a **Dynamic Domain Identifier (DDI)** for news caption translation based on decision tree induction (C5.0) [5] in order to activate and stack the multiple domain dictionaries dynamically.

We identify the domain of an article with its lead sentence. Lead sentence is the first statement by the anchor, which highlights the topic of succeeding article. The DDI detects the shift of a new article by semi-structural analysis, analyzes the lead sentence and identifies the domain of subsequent article. The shift of a new article can be recognized at the very beginning of the news or just after the end of an article. The DDI activates top-one to top-three domain dictionaries with priorities. By using the proper stacking of domain-specific dictionaries, transfer ambiguities can be considerably resolved, and the quality of MT can be raised. We evaluate the DDI with lead sentences of MBC 9 Newsdesk caption corpus in January, 2005, which amounts to 1,060 sentences of 814 articles. The accuracy of first-ranked domain shows 65.7%. In the case of top-two, the accuracy increased up to 78.2%. The accuracy reaches 93% of precision with top one to three candidates. We use top-three domains for activating domain dictionaries.

3.3 Program Identifier for Program Specific Knowledge

Program System Information Protocol (PSIP) in ATSC DTV transport stream provides the program information which enables the DTV viewer to surf the world of DTV programs and easily choose a program. PSIP also provides metadata that may be used by the DTV receiver to display information about captioning (e.g. notifying the presence of CC) [7]. The program information is very crucial for activating program-specific dictionaries and program-specific components (e.g. domain identification for news caption) in CC translation system. TV Drama, especially soap opera, is a serialized program in which the same

Table 1. Evaluation Corpus

Evaluations	Subjective	Objective
Source	2005.8.1	2005.8.1-3
No.of sentences	458	1,106
No.of morphemes	10,703	33,546
Average morphemes per sentence	23.4	30.3

characters appear and the storyline follow the day-to-day lives of the characters. So, the NEs of the persons, locations and organizations) found in the current drama will surely be presented next time. We accumulate the program-specific NEs and provide their multilingual translations by live updating through smart update server. This program-specific translation knowledge is activated when the program name of current channel is identified.

4 Evaluation

We evaluate our multilingual CC translation system in two different ways: a subjective evaluation to measure the overall performance enhancement of multilingual CC translation system, and an objective one to quantify the effects of three components (NEA, DDI, and PI) we adopted to the base MT systems. The evaluation news corpus is part of the *MBC 9 Newsdesk* caption corpus in 2005. The corpus is unseen while the system has been developed. Table 1 shows the statistics of the evaluation news corpus.

4.1 Subjective Evaluation: MOS Test

As a subjective evaluation, we used Mean Opinion Score (MOS) test, which is one of the voice quality evaluation methods for Text-to-Speech (TTS) systems. The translation results were mapped to a full five point MOS scale, ranging from 1 to 5 based on the clearness of the translated sentence [8].

The empirical results of MOS evaluation are shown in Table 2. The columns except for those in the bottom line are the number of sentences rated 1 to 5. The average score is calculated by dividing the total score by the total number of sentences. With the integration of three components, we obtained less impressive enhancement of 0.1 for Korean-to-Japanese by (4.6-4.5). That's because of the similarity between two languages and the initial higher translation quality of the base Japanese-to-Korean MT system. But, the proposed Dynamic Dictionary Adulteration approach turn out more effective in Korean-to-Chinese and Korean-to-English as high as 0.5 and 0.5, respectively. The average enhancement is 0.37, which means almost a half level up to the next higher MOS scale.

4.2 Objective Evaluation: BLEU and NIST11 Score

In order to quantify the effects of the effects of three components (NE Alignment, Dynamic Domain Identification, and Program Identification); BLEU score

Table 2. The MOS test result of Korean-to-Chinese, Korean-to-English and Korean-to-Japanese CC translation

Lang. Pairs	Korean ⇒ Chinese		Korean ⇒ English		Korean⇒Japanese	
Environment	BMT	HMT	BMT	HMT	BMT	HMT
5(Perfect)	73	104	84	130	272	301
4(Good)	73	97	74	116	130	126
3(OK)	98	130	128	116	48	23
2(Poor)	174	113	130	71	5	5
1(Bad)	40	14	42	25	3	3
Average score	2.9	3.4	3.1	3.6	4.5	4.6

*BMT: Base MT, HMT:Hybrid MT (Base MT+ NEA, DDI, and PI)

by IBM, and NIST mteval V11 score (NIST11 score) by NIST [9]. Table 3 demonstrates the enhancement of CC translation with the integrated Dynamic Dictionary Adaptation components of NEA, DDI, and PI in case of Korean-to-English CC translation.

With the encouraging effects of NE Alignment, PI, and DDI, we could raise the overall translation accuracy. For news CCs, as we expected, the translingual NE recognition is the most effective. Those enhancements are meaningful to translation quality.

Table 3. Enhancement in Korean-to-English translation quality

Evaluation Metrics	BLEU Metrics	NIST11 Metrics
BASE	0.0513	3.1842
BASE + NEA	0.0613	3.3322
BASE + DDI	0.0551	3.2460
BASE + PI	0.0545	3.2288
BASE + ALL	0.0644	3.4076

5 Conclusions

In this paper, we present a preliminary multilingual CC translation system for Dital TV. In order to raise the translation quality at the practical level, we proposed Dynamic Dictionary Adaptation methods including multilingual Named Entity Aligning, Dynamic Domain Identification, and Program Identification. With the proposed integrated Dynamic Dictionary Adaptation approaches, we obtained average enhancement of 0.37 in MOS (Mean Opinion Score) for Korean-to-Chinese (2.9 to 3.4), Korean-English (3.1 to 3.6) and Korean-Japanese (4.5 to 4.6) in a news domain.The enhancement 0.37 means almost a third level up to the next higher MOS scale. The proposed methods is language independent. So, it is applicable to any language pairs.

References

1. Popowich, F., McFetridge, P., Nichoson, D., and Toole, J.: "Machine translation of closed captions", Machine Translation, Kluwer Academic Publishers, Vol. 15, (2000) 311–341
2. Turcato, D., Popowich, F., McFetridge, P., Nicholson, D. and Toole, J.: "Preprocessing closed captions for machine translation" Proc. 3rd Workshop on Interlinguas and Interlingual Approaches, Seattle, Washington, (2000) 38–45
3. Nyberg, E. and Mitamura, T.: "A real-time MT system for translating broadcast captions", Proc. Machine Translation Summit VI, San Diego, USA, Oct. 29-Nov.1, (1997) 51–57
4. Yang, S.-I., Kim, Y.-K., Seo, Y.-A., Choi, S.-K. and Park, S.-K.: "Korean to English TV caption translator: "CaptionEye/KE", Proc. 6th Natural Language Processing Pacific Rim Symposium (NLPRS), Tokyo, Japan, Nov. 27-30, (2001) 639–645
5. Kim, T.-W., Sim, C.-M., Yuh, S.-H., Jung, H.-M., Kim, Y.-K., Choi, S.-K., Park, D.-I., and Choi, K.-S.: "FromTo-CLIR: web-based natural language interface for cross-language information retrieval, int. Journal of Information Processing and Management, Vol.35, no.4, (1999) 559–586
6. Huang, F., Vogel, S., and Waibel, A.: "Automatic extraction of named entity translingual equivalences based on multi-feature cost minimization," Proc. 41st Annual Meeting of the Association for Computational Linguistics (ACL), Joint Workshop on Multilingual and Mixed-Language Named Entity Recognition: Combining Statistical and Symbolic Models, Sapporo, Japan, July 12, (2003) 9–16
7. Electronic Industries Alliance: Digital Television (DTV) Closed Captioning (EIA-708-B), Electronic Industries Alliance (EIA), Dec. (1999)
8. Yuh, S.-H., Lee, K.-J., and Seo, J.-Y.: "Multilingual closed caption translation system for digital television," IEICE Trans on Information and Systems, Vol.E89-D, no.6, (2006) 1885–1892
9. Lin, C.-Y. and Hovy, E.: "Automatic evaluation of summaries using N-gram cooccurrence statistics," Proc. 2003 Conference of the North American Chapter of the Association for Computational Linguistics on Human Language Technology, Edmonton, Canada, Vol.1, (2003) 71–78

Acquiring Concept Hierarchies of Adjectives from Corpora

Kyoko Kanzaki[1], Qing Ma[2], Eiko Yamamoto[1], Tamotsu Shirado[1], and Hitoshi Isahara[1]

[1] National Institute of Information and Communications Technology
3-5 Hikaridai, Seikacho, Sorakugun, Kyoto 619-0289, Japan
{kanzaki, eiko, shirado, isahara}@nict.go.jp
[2] Ryukoku University
Seta, Otsu 520-2194, Japan
qma@math.ryukoku.ac.jp

Abstract. We describe a method to acquire a distribution of the concepts of adjectives automatically by using a self-organizing map and a directional similarity measure. A means of evaluating concept hierarchies of adjectives extracted automatically from corpora is elucidated. We used Scheffe's method of paired comparison to test experimentally the validity of hierarchies thus obtained with human intuition and found that our method was effective for 43% of the hierarchies considered.

Keywords: thesaurus, hierarchies, abstract concept, adjectives.

1 Introduction

Since many inconsistencies are likely to exist in the thesauri compiled by human lexicographers, such thesauri should be revised or reexamined by comparing them to objective data extracted from corpora. There are two main approaches to extracting thesauri or an ontology from huge corpora automatically. One is categorization into semantic classes by calculating the distributions of words in the corpus using syntactic and/or surface patterns ([6],[8],[10],[16]). The other is the statistical extraction of semantic relationships among words, such as hypernym-hyponym relationships or part-whole relationships, from corpora using syntactic patterns ([1], [2], [7], [12]). A proper semantic class and its suitable label (or conceptual name) for a semantic class are important components in the organization of a thesaurus and ontology, as is the overall structure of concepts from the top to the bottom levels. We have been developing a means of automatically organizing the concepts of adjectives as part of a project aimed at enabling the automatic organization of the meanings of Japanese words. From corpora, we extract the concepts of adjectives – i.e., abstract nouns categorizing adjectives – semiautomatically, and then, we classify the concepts of adjectives by using Kohonen's self-organizing map and introducing a directional similarity measure to calculate an inclusion relationship. As a result, we obtain the similarity and hierarchical relationships of concepts of adjectives on the map.

In this paper, we explain how the self-organizing map is constructed, focusing especially on the hierarchical relationships among the concepts of adjectives.

Y. Matsumoto et al. (Eds.): ICCPOL 2006, LNAI 4285, pp. 430 – 441, 2006.

In Section 2 we explain the method to extract class names of adjectives from corpora. In Section 3, we explain the encoding of our input data for Kohonen's SOM and how hierarchies of the concepts of adjectives are constructed. In Section 4, we describe the process to select a feasible method for creating hierarchies in terms of surface and qualitative characteristics. We compare our created hierarchies with those of the EDR lexicon, a huge handcrafted Japanese lexicon, by using Scheffe's paired comparison method [15]. We conclude with Section 5.

2 Extracting Abstract Nouns to Categorize Adjectives from Corpora

Consider the Japanese syntactic structure, "Noun1 *wa* Noun2 *ga* Adj," where "Noun1 *wa*" refers to a topic and "Noun2 *ga*" refers to a subject. According to Takahashi [17], in the Japanese sentence, "*Yagi wa seishitsu ga otonashii* (The nature of a goat is gentle)," "*seishitsu* (nature)" (Noun2) is a superordinate concept of "*otonashii* (gentle)" (Adj), and the adjective "*otonashii* (gentle)" includes the meaning of an abstract noun "*seishitsu* (nature)." In this sentence, an abstract noun "*seishitsu* (nature)" can be omitted without changing the meaning of the sentence; i.e., the meanings of "*Yagi wa otonashii* (A goat is gentle)," and "*Yagi wa seishitsu ga otonashii* (The nature of a goat is gentle)" are the same. Our method is first to extract from the corpora all nouns preceded by the Japanese expression "*toiu,*" which is similar in meaning to "that" or "of" in English. "*Toiu* + noun (noun that/of ...)" is a typical Japanese expression, which introduces some information about the referent of the noun, such as an apposition. Therefore, we can elucidate the content of nouns found in this pattern by means of their modifiers. We then use syntactic patterns such as (1) "Noun1 *wa* Noun2 *ga* Adj" and (2) "Adj + Noun2_*no* + Noun1" to determine instance-category relationships among the extracted data. Noun1 is a concrete noun representing a topic or a subject, while Noun2 is a subject in pattern (1) and a noun that is elucidated by its modifier, an adjective, in pattern (2). "*No*" is a marker of adnominal noun usage. From the data, we manually select examples in which Noun2 can be omitted without changing the meaning of the original sentence or phrase. If Noun2 can be omitted, Noun2 may be an abstract concept of the modifying adjective. We have collected 365 abstract nouns from two years worth of articles from the Mainichi Shinbun, a Japanese newspaper, and extracted adjectives co-occurring with abstract nouns in the manner described above from 100 novels, 100 essays, and 42 years worth of newspaper articles.

3 Self-Organizing Map Using Directional Similarity Measures – To Find Similarity and Hierarchical Relationships of Concepts

A self-organizing map (SOM) can be visualized as a two-dimensional array of nodes on which a high-dimensional input vector can be mapped in an orderly manner through a learning process. After learning, a meaningful nonlinear coordinate system for different input features is created over a neural network. Such a learning process is competitive and unsupervised, and is called a self-organizing process [9]. In our previous work, similarity relationships between concepts were computed using feature vectors calculated from the Euclidian distance. In our current work, hierarchical relationships are computed as well as similarity relationships by introducing a directional similarity measure to the SOM. We used the complementary similarity measure (CSM) as the directional similarity measure to calculate an inclusion relationship for our data. The details of this are given in Section 3.1.2. An example of an extracted map is shown in Fig. 1. All concepts are distributed from the top concept "koto (thing)" to hyponyms on the map. The coordinates on the vertical and horizontal axes indicate the location of each concept based on the similarity between concepts calculated by the SOM using the CSM. The height of each concept above the plane in Fig. 1 corresponds to a CSM value of an inclusion relationship with "koto (thing)." In the next section, we explain the details regarding our input data for the SOM, the CSM, and the construction of hierarchies of all concepts using CSM values.

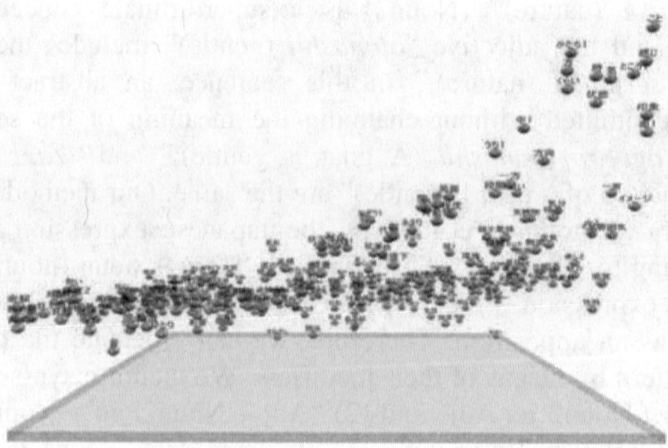

Fig. 1. Distribution of concepts on self-organizing map

3.1 Self-Organizing Map

The basic method of encoding the SOM follows Ma et al. [13].

Input Data

As explained in Section 2, we extracted abstract nouns categorizing adjectives and made a list like the one shown in Table 1 as learning data.

Table 1. List of abstract nouns and their adjectives

Abstract Noun (concept)	Co-occurring Adjectives (instances)
Kimochi (feeling)	*shiawase_na* (happy), *hokorashii* (proud), *kanashii* (sad), …
Joutai (status)	*aimai_na* (vague), *isogashii* (busy), …
Kanten (viewpoint)	*gakumonteki_na* (academic), *igakuteki_na* (medical), …
…	…

There were 365 abstract noun types, 10,525 adjective types, and 35,173 adjective tokens. The maximum number of co-occurring adjectives for a given abstract noun was 1,594.

Encoding

The semantic map of nouns was constructed by first defining each noun as a set of its modifiers (adjectives). From Table 1, for example, we can define "*kimochi* (feeling)" as a set of adjectives; i.e., "*kimochi*" = {"*shiawase_na* (happy)," "*hokorashii* (proud)," and "*kanashii* (sad),"…}. Suppose there is a set of nouns, w_i ($i = 1, \ldots, \omega$), that we plan to use for self-organizing. Any noun, w_i, can be defined by a set of its adjectives as $w_i = \{ a_1^{(i)}, a_2^{(i)}, \ldots, a_{\alpha i}^{(i)} \}$, where $a_j^{(i)}$ is the jth adjective of w_i and α_i is the number of adjectives of w_i. One method of encoding nouns so that they can be treated with an SOM is to use random coding, which is an ordinary method used to construct SOMs (see [9] for details). Through several preliminary computer experiments, however, we found that this method is not suitable for our task. Because in random coding each co-occurring word is represented by a random vector with fixed number of dimension (e.g., 100 dimensions), and each noun is represented by sum of vectors of all co-occurring words, when the number of co-occurring words with nouns become large (e.g., 10,525 adjectives co-occurred with nouns in our experiment), it is very difficult to encode nouns properly. Therefore, we used a new method as described below. Suppose we have a correlation matrix (Table 2) where d_{ij} is some metric of correlation (or distance) between nouns w_i and w_j. We can encode noun w_i from the correlation matrix as $V(w_i) = [d_{i1}, d_{i2}, \ldots, d_{i\omega}]^T$. We then use $V(w_i) \in \Re^{\omega}$ as the input to the SOM, i.e., $x = V(w_i)$ and $n = w$ ("x" is an input datum and "$\Re^n$" is an n-dimensional space.). T is the number of learning steps.

Table 2. Correlation matrix of nouns

	w_1 w_2 … w_ω
w_1	d_{11} d_{12} … $d_{1\omega}$
w_2	d_{21} d_{22} … $d_{2\omega}$
	⋮ ⋱
w_ω	$d_{\omega 1}$ $d_{\omega 2}$ … $d_{w\omega}$

Note that the individual d_{ij} of vector $V(w_i)$ only reflects the relationships between a pair of words when they are considered independently.

In this paper, d_{ij} is measured by the CSM, which calculates an inclusion relationship between words. This similarity measure was developed to enable recognition of degraded machine-printed text [5]. Yamamoto and Umemura [18] used the CSM to calculate a hypernym-hyponym relationship between a word pair.

$$\mathrm{CSM} = \frac{ad - bc}{\sqrt{(a+c)(b+d)}},$$

a indicates the number of times the two labels appear together; b indicates the number of times "label 1" occurs but "label 2" does not; c is the number of times "label 2" occurs but "label 1" does not; and d is the number of times neither label occurs. In our data, each abstract noun (w) is substituted into "label", a indicates the number of adjectives co-occurring with both abstract nouns, b and c indicate the number of adjectives co-occurring with either abstract noun ("label 1" and "label 2", respectively), and d indicates the number of adjectives co-occurring with neither abstract noun. Depending on the direction of the calculation, in other words, between the CSM value of w_i to w_j, and the CSM value of w_j to w_i, the numerical value substituted into b and c is reversed. That is, CSM calculates a similarity between words asymmetrically.

For a comparison with the CSM, we used another directional similarity measure, the overlap co-efficient (Ovlp). According to [11], "the overlap coefficient has the flavor of a measure of inclusion. It has a value of 1.0 if every dimension with a nonzero value for the first vector is also nonzero for the second vector or vice versa."

$$Ovlp\ (F, T) = \frac{|F \cap T|}{\min(|F|, |T|)} = \frac{a}{\min(a + b, a + c)}$$

The definitions of a, b, c, and d are the same as those of the CSM. F and T are the abstract nouns w_i and w_j in our data.

The learning step of an SOM consists of an ordering phase and a final phase. We used 30,000 learning steps in the ordering phase and 100,000 in the final phase. The map shown in Fig. 1 was the SOM of a 45 x 45 array where a neighborhood with a hexagonal topology was used. The initial radius of the neighborhood was set at 45 in the ordering phase and the last radius was set at 7 in the final phase.

3.2 Construction of Hierarchy Composed from Concepts When Using CSM

Using the CSM, we constructed a hierarchy composed of concepts of adjectives through the following process [19]. The hierarchical construction of concepts is a procedure that is independent from the process of distributing concepts on the two-dimensional map made by the SOM. After we obtained hierarchies, we plotted them on the map made by the SOM. Values of CSM were normalized, and we made a list of CSM values that we obtained. Examples of calculation results obtained using CSM are as follows.

Table 3. List of hypernym and hyponym relationship calculated by CSM

Word A	Word B	CSM value
Imeji (image)	*Inshou* (impression)	1.0
Koto (matter)	*Katachi* (form)	0.955
Inshou (impression)	*Kanji* (feeling)	0.936
Koto (matter)	*Kanji* (feeling)	0.925
…	….	…….

For example, in Table 3, "*Inshou* (impression)" and "*Kanji* (feeling)" is a hypernym-hyponym relationship.

1) For a hypernym A and a hyponym B, the initial hierarchy is A-B.
2) First, detect hyponyms deeper than B; i.e., hyponyms in the initial hierarchy A-B. Find the hyponym for B with the highest value in the list. This hyponym is denoted as X and connected behind B in the initial hierarchy. Next, find the hyponym of X with the highest value. This hyponym is denoted as Y and combined behind X. This process is repeated until such a hypernym/hyponym pair cannot be chosen. The hierarchy extended to hyponyms is A-B-X-Y.
3) Second, find superordinate nouns shallower than A, a hypernym in the initial hierarchy A- B. Find the hypernym of A with the highest value in the list. This hypernym is denoted as W and is connected in front of A. Next, find the hypernym of W with the highest value. This hypernym is denoted as V and is connected in front of W. This process is repeated until such a superordinate/subordinate pair cannot be chosen. The hierarchy extended to the hypernyms is V-W-A-B-X-Y.

In steps 2) and 3), the superordinate/subordinate relationship between two nouns must be retained. If the relationship is broken, we cannot connect the two nouns.

4) Last, insert "*koto* (matter)" at the top of all hierarchies. "*Koto* (matter)" is the topmost hypernym because all adjectives collocate with it. The hierarchy thus obtained is *koto*-V-W-A-B-X-Y.

4 Verification of Created Hierarchies

4.1 Comparison of CSM with Other Methods: Ovlp, and CSM Using Frequency Information

In this section, we compare hierarchies constructed automatically using several methods. We also determine which hierarchies provide a suitable point of comparison with the EDR concept hierarchy. The methods we compared were as follows.

1) CSM without frequency information (CSM)
2) Overlap coefficient without frequency information (Ovlp)
3) CSM with frequency information (Freq)

We set the normalized threshold value at 0.3 and 0.2 for CSM and Ovlp (CSM0.3, CSM0.2, Ovlp0.3, Ovlp0.2), and at 0.2 and 0.1 for Freq (Freq0.2, Freq0.1). Higher threshold values create hierarchies with a smaller number of nouns and lower threshold values lead to overly large and incoherent hierarchies. The thresholds that we set were found to enable the construction of hierarchies with a number of nouns that seemed suitable for our purposes. Among the hierarchies we created were some in which the same adjective was held by all concepts in a path from the bottom node to the top node as an instance of the concept. This type of hierarchy is considered "a hierarchy of an adjective" in this paper (see Section 4.2., Rules).

First, we counted the number of hierarchies of adjectives among the automatically extracted hierarchies to find the most plausible hierarchies. Then, we calculated the number of target adjectives covered by the automatically constructed hierarchies and calculated what percentage of the 365 abstract nouns appeared as elements in the hierarchies. Features of the hierarchies created using the different methods are shown in Table 4; that is, the percentage of hierarchies of adjectives among all automatically extracted hierarchies, the number of target adjectives covered by the automatically created hierarchies, and the percentage of the 365 abstract nouns that appeared in the hierarchies as elements. The top three scores for each category are circled in Table 4 (a bold circle indicates an especially good score). Poor scores are marked with an "*." We can see that CSM0.2 and Ovlp0.3 were suitable threshold values in terms of the three features we considered. CSM0.3 created the highest number of hierarchies, but these contained the lowest number of abstract nouns. In other words, CSM0.3 exhibited high precision but low recall. Ovlp0.2 created fewer hierarchies, but the created hierarchies covered almost all of the 365 abstract nouns (i.e., precision was low but recall was high). This means that Ovlp0.2 created overly large hierarchies. Freq0.1 exhibited a similar tendency, creating fewer hierarchies of adjectives that covered many abstract nouns. Freq0.2 created fewer hierarchies of adjectives, and the hierarchies typically covered a small number of abstract nouns. CSM0.3 exhibited a similar tendency. The above results indicate that CSM0.2 and Ovlp0.3 seem to create the most appropriate hierarchies, so these were the methods and threshold values we used in our evaluation. From the methods using frequency information, we used Freq0.2 in our evaluation because it created a greater number of hierarchies of adjectives than that created by Freq0.1.

Table 4. Features of hierarchies created by different methods

	Hierarchy depth	Percentage of hierarchy of adjectives	Number of adjectives in hierarchies	Percentage of abstract nouns in hierarchies
CSM0.3	13	90%	175 words	* 24%
CSM0.2	13	58%	101 words	90%
Ovlp0.3	9	63%	99 words	86%
Ovlp0.2	9	52%	53 words	92%
Freq0.2	8	53%	70 words	28%
Freq0.1	12	* 36%	67 words	* 82%

4.2 Comparison of Created Hierarchies with Existing Handcrafted Thesaurus

The EDR lexicon is a large Japanese lexicon for computers that was constructed by a great number people under a number of supervisors.

We performed an experiment to compare automatically generated hierarchies to EDR hierarchies. We used a method of psychological scaling, Scheffe's method of paired comparison [15], for this experiment.

Experimental Data

Based on the results in Section 4.1, we used the following as experimental data. For 30 adjectives, we obtained 30 hierarchies generated by CSM0.2, Ovlp0.3, and Freq0.2. We refer to these identical hierarchies generated by all three methods as "COMMON" hierarchies. We also obtained corresponding hierarchies from the EDR lexicon. For each adjective, we compared a pair of hierarchies that contained those adjectives consisting of a generated hierarchy and an EDR hierarchy (COMMON-EDR). We had participants in the experiment judge the plausibility of the paired hierarchies for the adjectives shown in Table 5.

Table 5. Adjectives in COMMON (CSM0.2, Ovlp0.3, Freq0.2)

Adj-ID: Adjective 1: *yuniku_na* (unique), 2: *atatakai* (warm), 3: *kyuu_na* (steep), 4: *furui* (old), 5: *shitashii* (close), 6: *zetsubouteki_na* (desperate), 7: *haya*i (early), 8: *hayai* (fast), 9: *osoi* (late), 10: *mendou_na* (troublesome), 11: *yasashii* (gentle), 12: *kitsui* (hard), 13: *sofuto_na* (soft), 14: *deriketo_na* (delicate), 15: *naibu_na* (naive), 16: *surudoi* (keen), 17: *kanbi_na* (sweet), 18: *ganko_na* (stubborn), 19: *ayaui* (dangerous), 20: *kiken'na* (dangerous), 21: kiraku_na (carefree), 22: *kouteiteki_na* (affirmative), 23: *takai* (high), 24: *kouki_na* (noble), 25: *jiyuu_na* (free), 26: *wakai* (young), 27: *juubun_na* (enough), 28: *yawarakai* (mild), 29: *juuyou_na* (important), 30: *dokusouteki_na* (original)

Participants

The 20 participants consisted of linguists, lexicographers, and people familiar with natural language processing (NLP).

Experimental Procedure

We showed the participants a pair of hierarchies for a certain adjective (a set consisting of an automatically generated hierarchy and an EDR hierarchy) and had them judge how valid each hierarchy was. They scored each hierarchy on a five point scale (-2, -1, 0, 1, 2) as illustrated below.

The test was conducted as follows, where a letter indicates the name of a concept. For example, A-B-C means that concepts A, B, and C are connected in a hierarchical relationship. In this case, the top concept is A and the bottom concept is C.

Adjective X	
A-B-C	P-Q-R

LEFT: hierarchy generated automatically
RIGHT: hierarchy from EDR lexicon

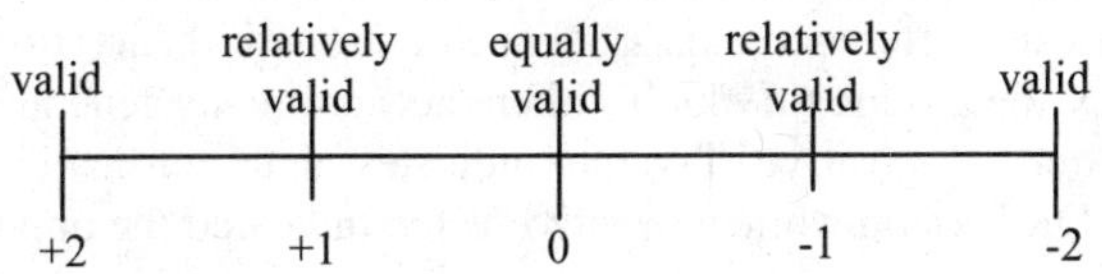

+2: The left expression is valid. +1: The left expression is more valid.
0: Both are equally valid.
-1: The right expression is more valid. -2: The right expression is valid.

A score of "0 (equally valid)" was given when both hierarchies appeared either valid or invalid. We provided two rules as criteria for judging the validity of a hierarchy.

Rule 1) If concepts are connected in a hierarchical relationship, an instance will appear at every level from the bottom to the top concept.

According to Lexicology [4], a system of hyponymic relationships constitutes a taxonomy or, in other words, a branching lexical hierarchy based on inclusiveness. For example, "creature" includes "bird" and "horse", "bird" includes "nuthatch" and "robin", and "nuthatch" includes "white-breasted nuthatch" and "red-breasted nuthatch" (p. 472). Conversely, an instance in a hyponymic concept is also an instance in a superordinate concept. For example, "breasted nuthatch" is an instance in "nuthatch," "bird," and "creature." In our data, "an instance" corresponds to "adjective X" for which a hierarchy was created automatically.

Rule 2) If concepts are connected in a hierarchical relationship, a hyponymic concept inherits features from a superordinate concept and a hyponymic concept unilaterally entails a superordinate concept.

According to Cruse [3], X will be said to be a hyponym of Y (and, by the same token, Y is a superordinate of X), if *A is f(X)* entails, but is not entailed by, *A is f(Y).*) For example, "This is a DOG" unilaterally entails "This is an ANIMAL" (pp. 88-89).

From our comparison of COMMON and EDR paired hierarchies (COMMON-EDR), we calculated a test statistic (T-value):

$$T = \frac{\overline{x_1} - \overline{x_2}}{\sqrt{\frac{s_1^2}{N_1} + \frac{s_2^2}{N_2}}},$$

where N is the number of samples, $\overline{x_1}$ and $\overline{x_2}$ indicate values averaged over the 20 participants for methods 1 and 2, respectively, and s_1^2 and s_2^2 indicate the unbiased variance over the participants for methods 1 and 2, respectively. Here, method 1 is COMMON and method 2 is EDR.

We investigated the difference between the performance of automatic generation and that of EDR at significance levels of 1%, 5%, and 25%. A significant difference means that an assessment of relative merit between two methods differs significantly. Conversely, no significant difference means that an assessment of relative merit between two methods does not differ significantly. At significance levels of 1% and 5%, we investigated whether a significant difference exists between automatic generation and EDR, and at significance levels of 5% and 25%, we investigated whether a significant difference exists between automatic generation and EDR. The T-value indicates the extent to which either the automatic generation or EDR was significantly better. A positive T-value indicates that automatic generation was superior to the EDR lexicon while a negative value indicated the opposite.

Experimental Results

The T-values from our comparison of the COMMON and EDR hierarchy pairs are shown in Fig. 2. These T-values were obtained through Scheffe's paired comparison method. (The numbers in the figure are the adjective ID numbers from Table 5 for each pair of hierarchies used in the experiment.)

A) Result 1 significance level: 5%, significant difference (T > 1.68)
At the 5% significance level, we found a significant difference for the following adjectives (T-values in parentheses).

(1) Adjective IDs for positive T-values: Total adjective IDs: 2/30
Adj ID: 22 (5.440), 26 (2.690)

(2) Adjective IDs for negative T-values: Total adjective IDs: 17/30
Adj ID: 1 (-5.575), 2 (-4.035), 4 (-5.379), 7 (-9.885), 8 (-6.016), 9 (-9.105), 11 (-2.818), 16 (-6.790), 19 (-6.892), 20 (-4.385), 21 (-4.884), 23 (-7.973), 24 (-3.952), 25 (-3.483), 27 (-3.921), 29 (-2.543), 30 (-3.353)

B) Result 2 significance level: 1%, significant difference: T > 2.4
At the 1% level, the result was the same as result 1.

C) Result 3 significance level: 5%, no significant difference: T < 1.68
At the 5% significance level, there was no significant difference for the following adjectives.

(1) Adjective IDs for positive T-values: Total adjective IDs: 5/30
Adj ID: 3 (0.565), 13 (0.802), 14 (1.457), 18 (0.256), 28 (0.960)

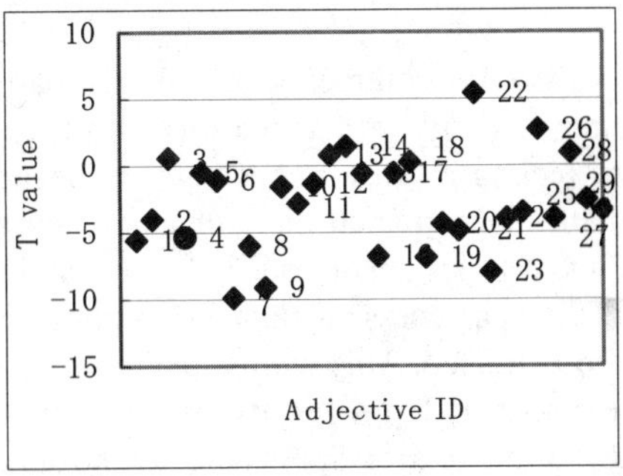

Fig. 2. T-values from assessment of paired hierarchies of 30 adjectives (COMMON-EDR)

(2) Adjective IDs for negative T-values: Total adjective IDs: 6/30
Adj ID: 5 (-0.488), 6 (-1.09), 10 (-1.550), 12 (-1.329), 15 (-0.56), 17 (-0.565)

D) Result 4 significance level: 25%, no significant difference: T < 0.68
At the 25% significance level, there was no significant difference for the following adjectives.

(1) Adjective IDs for positive T-values: Total adjective IDs: 2/30
Adj ID: 3 (0.565), 18 (0.256)

(2) Adjective IDs for negative T-values: Total adjective IDs: 3/30
Adj ID: 5 (-0.488), 15 (-0.56), 17 (-0.565)

Discussion

Comparing COMMON with EDR, the assessed validity of the EDR hierarchies was better for 17 pairs, whereas that of the COMMON hierarchies was better for 2 pairs (a significant difference was observed for these 19 pairs). The assessed validity for the remaining 11 pairs of the 30 test pairs did not significantly differ between EDR and COMMON. An example is shown below. Even in this raw comparison, our method generated results as good as those obtained from the EDR for 43% {(11+2)/30} of the hierarchies. As in other experiments, though we compared hierarchies generated only by CSM0.2 or Ovlp0.3 with those in EDR, EDR obviously differs more significantly than CSM0.2 and Ovlp0.3.

Adjective: Kouteitekina (Affirmative)

Automatically generated hierarchy

Koto/Men/Keikou/Mikata/Hyouka (thing/side/tendency/viewpoint/evaluation)

EDR

Gainen/Jishou/Koui/Taishoukoui /Mono wo taishou to suru koui /Jouhou no idou/Jouhou no jushin/Shiru/Ninchishutai to ninchitaishou to no ninchitekikyori genshou/Iken nado ni doui shiteiru sama

(concept/event/event (action, act)/action, act (action with thing or person as recipient of action)/movement (movement of information)/movement of information (reception of information)/reception of information (know of, get knowledge of, learn about, know by word of mouth)/a distance decrease between a person who recognizes a certain object and an object recognized by a person/of a reply, an attribute of being positive)

We asked examinees to write comments on the comparison between the EDR dictionary and our hierarchy. Many of their reasons for judging the EDR hierarchies as better were related to feeling a sense of incongruity because several nodes were missing from the automatically extracted hierarchies. When a lower concept in our hierarchy was directly linked to an upper concept, the examinees tended to consider the EDR better. Due to the sparseness of data in corpora, some necessary abstract nouns may not have been extracted. This can be overcome by using bigger corpora. Judging from these situations and the results of our raw comparison with EDR hierarchies, we believe our method is applicable to the automatic extraction of a thesaurus-like structure from huge corpora.

5 Conclusion

We have described the construction and evaluation of concept hierarchies of adjectives from corpora. First, using topological features of constructed hierarchies, we determined suitable methods and threshold values. We then conducted an evaluation to identify whether EDR hierarchies or our automatically generated hierarchies were objectively judged to be the best formed or most valid and to compare them from the viewpoint of content. From the topological features, we found that CSM0.2, Ovlp0.3, and Freq0.2 created better hierarchies, in that order. The evaluation results indicated that our automatic extraction was applicable to 43% {(11+2)/30} of the hierarchies. We will continue to develop and improve our methods of extracting the information needed to construct better hierarchies of adjectives.

References

1. Berland M. and Charniak E.: Finding Parts in Very Large Corpora. In Proceedings of 38th Annual Meetings of the Association for Computational Linguistics (ACL) (2000) 57-64.
2. Caraballo S. A.: Automatic construction of a hypernym-labeled noun hierarchy from text. In Proceedings of the 37th Annual Meeting of the Association for Computational Linguistics (ACL), (1999) 120-126.
3. Cruse D. A.: Lexical Semantics. Cambridge University Press, Cambridge (1986).
4. Cruse D. A., Hundsnurscher F., Job M., and Lutzeier P. R.: Lexicology, An international handbook on the nature and structure of words and vocabularies. Walter de Gruyter (2002).
5. Hagita N. and Sawaki M.: Robust Recognition of Degraded Machine-Printed Characters Using Complimentary Similarity Measure and Error-Correction Learning, In the Proceedings of the SPIE –The International Society for Optical Engineering (1995) 2442: 236-244.
6. Hatzivassiloglou V. and McKeown K.R.: Towards the Automatic Identification of Adjectival Scales: Clustering Adjectives According to Meaning. In Proceedings of the 31st Annual Meeting of the Association for Computational Linguistics (ACL) (1993) 172-182.
7. Hearst M. A.: Automatic Acquisition of Hyponyms from Large Text Corpora. In Proceedings of 14th International Conference on Computational Linguistics (COLING) (1992) 539-545.
8. Hindle D.: Noun Classification from Predicate-Argument Structures. In Proceedings of the 28th Annual Meeting of the Association for Computational Linguistics (ACL) (1990) 268-275.
9. Kohonen, T.: Self-Organizing Maps, Springer, Berlin (1995).
10. Lin D. and Pantel P.: Concept Discovery from Text. In Proceedings of 19th International Conference on Computational Linguistics (COLING) (2002) 768-774.
11. Manning C. D. and Shütze H.: Foundations of Statistical Natural Language Processing. The MIT Press, Cambridge (1999).
12. Pantel P. and Ravichandran D.: Automatically Labeling Semantic Classes. In Proceedings of Human Language Technology/North American Chapter of the Association for Computational Linguistics (HLT/NAACL) (2004) 321-328.
13. Ma Q., Kanzaki K., Zhang Y., Murata M., and Isahara H.: Self-organizing semantic maps and its application to word alignment in Japanese-Chinese parallel corpora. Journal of Neural Networks, The Official Journal of the International Neural Network Society, European Neural Network Society, and Japanese Neural Network Society, Elsevier (2004) 1241-1253.
14. Research Committee of Sensory Evaluation Union of Japanese Scientists and Engineers: Sensory Evaluation Handbook. JUSE Press Ltd. (in Japanese) (1973).
15. Scheffe H.: An analysis of variance for paired comparison. Journal of the American Statistical Association, 47 (1952) 381–400.
16. Schulte im Walde S. and Brew C.: Inducing German Semantic Verb Classes from Purely Syntactic Subcategorisation Information. In Proceedings of the 40th Annual Meeting of the Association for Computational Linguistics (ACL) (2002) 223-230.
17. Takahashi, T.: Various Phases Related to the Part-Whole Relationship Investigated in the Sentence, Studies in the Japanese language 103, The Society of Japanese Linguistics (in Japanese) (1975) 1-16.
18. Yamamoto E. and Umemura K.: A Similarity Measure for Estimation of One-to-Many Relationship in Corpus. Journal of Natural Language Processing (in Japanese) (2002) 45-75.
19. Yamamoto E., Kanzaki K., and Isahara H.: Extraction of Hierarchies Based on Inclusion of Co-occurring Words with Frequency Information. In Proceedings of 19th International Joint Conference on Artificial Intelligence (IJCAI) (2005) 1166-1172.

Pronunciation Similarity Estimation for Spoken Language Learning

Donghyun Kim and Dongsuk Yook

Speech Information Processing Laboratory
Department of Computer Science and Engineering
Korea University, Seoul, Korea
{kaizer, yook}@voice.korea.ac.kr

Abstract. This paper presents an approach for estimating pronunciation similarity between two speakers using the cepstral distance. General speech recognition systems have been used to find the matched words of a speaker, using the acoustical score of a speech signal and the grammatical score of a word sequence. In the case of learning a language, for a speaker with impaired hearing, it is not easy to estimate the pronunciation similarity using automatic speech recognition systems, as this requires more information of pronouncing characteristics, than information on word matching. This is a new challenge for computer aided pronunciation learning. The dynamic time warping algorithm is used for cepstral distance computation between two speech data with codebook distance subtracted to consider the characteristics of each speaker. The experiments evaluated on the Korean fundamental vowel set show that the similarity of two speaker's pronunciation can be efficiently computed using computers.

1 Introduction

The study of speech recognition using hidden Markov models (HMMs) [1] has been well understood for several decades. General speech recognition systems have been used to find the matched words of a speaker's utterance using the acostic score of the speech signal and the language model score of the word sequence. In the case of learning a foreign language and native language, for a student and a person with impaired hearing, respectively, it is important to find not only how well the meaning matches, but also how well the utterance is close to standard pronunciation. Therefore, it is not easy to estimate the pronunciation distance between two utterances and to determine the similarity using automatic speech recognition systems. In this paper, we propose to use a distance measure of the acoustic space and speaker space to automatically compute the pronunciation similarity.

The pronunciation of a speech segment can be characterized by acoustic features such as duration, pitch, accent, stress, formant, and so on [2][3]. These are typically used in the study of speech synthesis [4]. However, extracting those information from speech signals automatically is another difficult task. A rather simple approach using the distance between cepstral vector sequences is known to represent the

Y. Matsumoto et al. (Eds.): ICCPOL 2006, LNAI 4285, pp. 442 – 449, 2006.

characteristics of speech signals. This approach is motivated by utterance verification [5][6] and speaker verification [7], which deal with selecting the correct utterance and speaker, respectively, using confidence measures [8]. The likelihood ratio between the probabilistic distributions using correctly hypothesized model parameters, and incorrectly hypothesized model parameters, determines whether to accept or reject the utterances using a critical decision threshold. However, these methods are different from the proposed method in two aspects. First, the issue of this paper is to deal with the distance of two utterances, so it does not depend on the probabilistic distribution of the utterance model. Second, it analyzes the utterances not only in view of word matching, but also personal pronunciation characteristics. Computer-assisted language learning (CALL) has been studied in the research area of foreign language and pronunciation learning [9]. However, the main interest of those areas is learning the vocabulary or grammar.

This paper proposes a new approach to estimate the distance of two speaker's pronunciations and calculate a probabilistic score of this distance which can be considered as a confidence measure. To obtain the optimal confidence measure threshold, a support vector machine (SVM) is used. For the acoustic feature vector space, mel-frequency cepstral coefficients (MFCC) are used as feature vectors. In estimating the distance of two utterances, the dynamic time warping (DTW) method [10] and the distance between speakers using the codebook is used. The performance of the proposed method is evaluated using the seven fundamental Korean vowels from six speakers.

In section 2, the theory of the pronunciation similarity, the decision threshold finding procedure for a specific phone set, and a classification method are explained. Section 3 presents the experimental results. Some conclusions are discussed in section 4.

2 Pronunciation Similarity

Human speech is different for each speaker, since the vocal track of each speaker is different, even though the same word is spoken. In addition, as one speaker repeatedly utters the same word, speech is not always the same because the acoustic signal is generated through a very dynamic process inside the human body. Even if speech is different for each person, multiple utterances of the same word have very similar patterns which are typically analyzed with statistical methods such as HMMs. Therefore, the majority of automatic speech recognition systems have statistical patterns of speech in the form of parameters in the HMMs.

2.1 Estimate the Distance of Pronunciation Similarity

Inspired by utterance and speaker verification, the distance, D, of pronunciation dissimilarity can be defined as follows;

$$D = d_{\text{Speech}}(i, j) - d_{\text{Speaker}}(i, j), \tag{1}$$

where $d_{\text{Speech}}(\cdot)$ and $d_{\text{Speaker}}(\cdot)$ are methods to estimate statistical distance of speech signals and speakers, respectively, using speech data i and j. In the above criterion, the speech distance is subtracted by speaker distance since even though the speech distance for the same word is correctly estimated, some speakers are very different to each other by the underline characteristics of the personal vocal tract. If the reference speech is not sufficient to have enough statistics for HMMs, another method such as DTW may be needed to directly estimate the distance of acoustic feature vectors. Therefore, equation (1) can be changed to the following;

$$D = DTW - CB. \tag{2}$$

DTW is the utterance distance calculated using the DTW method and CB is the speaker distance estimated using codebooks. This paper uses two kinds of distance measure; the first is the utterance distance using a DTW method, the second is the codebook distance. If a small number of codewords are kept, the codebook can be trained using a k-means algorithm. To compensate for the unit difference, the above equation can be modified as;

$$D = w_1 \cdot DTW - w_2 \cdot CB \leq \tau, \tag{3}$$

where w_1 and w_2 are the compensating parameters or weights and τ is a decision threshold. This threshold determines the correctness of the pronunciation. In the next sector, we explain how to find these parameters.

2.2 Estimating Threshold Using a Support Vector Machine

The baseline of the pronunciation similarity is started with just a single phone pronunciation. To estimate the decision threshold of each phone for a reference speaker, the codebook statistic is first trained from the reference speaker's data. Then, for each speaker, the utterance distances and the codebook distance are computed using all training utterances. The codebook distance is calculated by averaging all codeword distances from the input data. The SVM [11][12] is used to find the optimal binary decision threshold. The SVM uses a linear kernel for the separating hyperplane. To describe this hyperplane, generally, the following form is used;

$$\sum_{i=1}^{2} w_i \cdot x_i - b = 0, \tag{4}$$

where x is the input vector (i.e., $x_1 = DTW$ and $x_2 = CB$) and b is the bias. Therefore, equation (4) is rewritten as the following equation;

$$DTW = \alpha \cdot CB + \beta. \tag{5}$$

where α and β, which come from w and b (i.e., $\alpha = -w_2/w_1$ and $\beta = b/w_1$). This equation can be used to easily show two dimensional space graphs. Fig. 1 shows the separating hyperplane for the distance between one vowel, "aa", of the reference speaker and seven vowels from other training speakers. In this figure, the utterance and speaker distances between the same phonemes (i.e., "aa" vs. "aa") are closer than distances between the different phonemes (i.e., "aa" vs. "ah" and "aa" vs. rest). The estimated values of the parameters are $\alpha = -0.87$ and $\beta = 26.91$. This hyperplane maximally separates the same vowels and the different vowels. The confidence measure is estimated using the distance from the hyperplane.

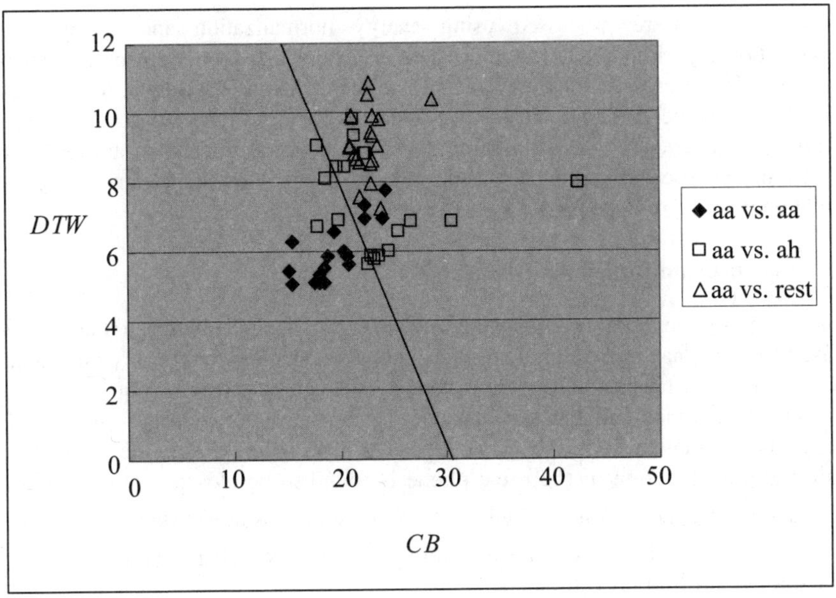

Fig. 1. Comparing "aa" to each vowel using *DTW* and *CB* distances and separating each group using SVM classifier

2.3 Confidence Measure

When a score of the pronunciation similarity between two utterances is requested, a probabilistic score can be provided using the confidence measure computed as in the previous section. Equation (5) which may be similar to the likelihood ratio test of utterance verification [6] provides a theoretical formulation to address the confidence measure problem. After the parameters are determined using the training procedure, the distance can be transformed to a confidence measure, y , using a sigmoid function;

$$y(i,j) = \frac{1}{1+e^{(DTW_{ij}-\alpha\cdot CB_{ij}-\beta)}}, \tag{6}$$

where $y(i,j)$ is limited within the probabilistic score. DTW_{ij} and CB_{ij} are estimated between utterances i and j.

3 Evaluation

To evaluate pronunciation similarity, automatic speech and speaker distance estimation and SVM classification for the decision threshold were implemented. The experiments were evaluated with seven fundamental Korean vowels from six male speakers. The utterances were recorded in 16 KHz sampling rate with a condenser microphone. Each vowel was pronounced ten times and data was extracted using speech detection algorithms. Data was then converted to a vector sequence of 13 dimensional MFCC features, which were processed using energy normalization and cepstral mean normalization algorithms. The utterance and codebook distance were calculated with other speakers for cross validation. To estimate the decision parameters, three speakers were trained. Namely, if one speaker was selected as a reference, two other speakers were used to generate the decision parameters. The decision threshold parameters were different in each speaker and in each of the seven vowels. After the training process, test evaluation with the rest speakers was performed.

3.1 Utterance and Codebook Distance

Table 1 shows the utterance distance and speaker distance between the reference speaker's phone "aa" and training speaker's phones, which include "aa", "ah", and the rest six vowels. After calculating the distances between phone pairs, equation (3) uses the average utterance and speaker distance, to show the compensated distance. The average is calculated from the same labeled vowel pairs. From the experimental result, the pronunciation of the same phone is found to be closer than other phones. The weight, w_2, like 1/4 is applied to show the compensated distance and estimate the decision threshold. Through the training process, two distances are obtained as input and the threshold is estimated.

Table 1. Average distance of vowels(DTW) and speaker(CB); training speakers pronunce each vowel to estimate the conpensated distance from "aa" with 1/4 weight of CB

	aa vs. aa	aa vs. ah	aa vs. rest
DTW	6.42	7.58	8.90
CB	18.84	22.18	22.33
$DTW-CB/4$	1.71	2.03	3.32
Pair	600	600	3600

3.2 Seven Vowels Test

To test the correct acceptance between phonemes, the number of vowel pairs is counted whenever both vowels are classed as matching vowels. To test correct rejection, it is also counted when both vowels are different. Each vowel of the reference speaker is compared to one of other speaker in the same vowel; a total of one hundred pairs were compared. The following experiments were evaluated with three reference and three testing speakers, and one vowel of the three reference speakers had a total of nine hundred distance pairs.

However, to estimate correct rejection, each vowel of the reference speaker was paired to six vowels of the testing speaker. The number of mismatched vowel pairs totaled five thousand four hundred. Table 2 shows the correct acceptance and correct rejection for pronunciation of seven Korean vowels. Using automatically selected threshold parameters, each pair was estimated and averaged. Since codebook size affects codebook distance, tables 3 and 4 show the results when using different codebook sizes.

Table 2. Average percent (%) of correctness for seven vowels with testing speakers using three codewords

CB_3	Correct Acceptance	Correct Rejection	Avg
aa	54	69	67
ah	53	83	79
ow	79	79	79
uw	73	81	80
uh	64	90	86
iy	64	91	87
ae	60	85	81
Average	**64**	**83**	**80**

Table 3. Average percent (%) of correctness for seven vowels with testing speakers using five codewords

CB_5	Correct Acceptance	Correct Rejection	Avg
aa	54	76	73
ah	41	87	80
ow	79	83	82
uw	73	84	83
uh	63	94	90
iy	64	92	88
ae	60	86	83
Average	**62**	**86**	**83**

Table 4. Average percent (%) of correctness for seven vowels with testing speakers using seven codewords

CB_7	Correct Acceptance	Correct Rejection	Avg
aa	55	75	72
ah	41	87	80
ow	79	82	82
uw	72	84	82
uh	63	94	90
iy	65	91	88
ae	60	86	82
Average	**62**	**86**	**82**

3.3 Pronunciation Similarity Score

The confidence measure driven from equation (6) is used as pronunciation similarity score for comparing a reference speaker to the testing speaker. Fig. 2 shows twenty pairs of the accumulated count probability graph where the compensated distance driven from equation (5) was converted as the similarity score, $y(i, j)$. The count of the matched vowel (i.e., "ow" vs. "ow") and mismatched vowel (i.e., "ow" vs. "uw") have different shape of probabilistic distributions. In the case of the matched vowel, the scores of 0.5 or more show pronunciation similarity.

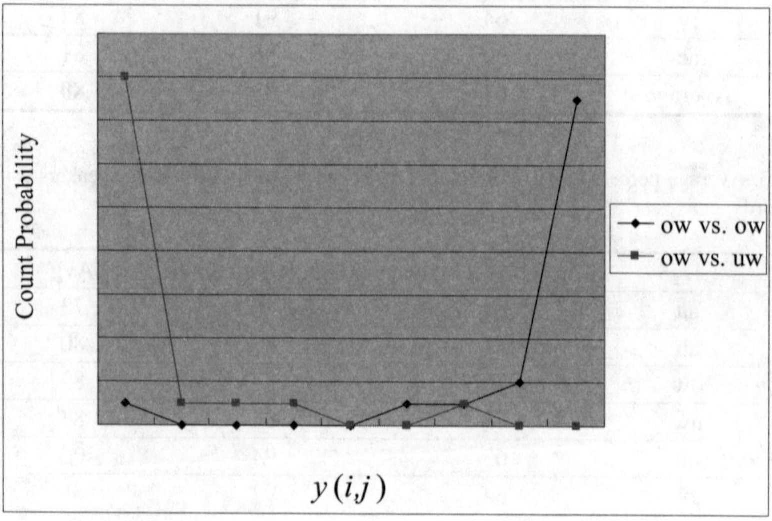

Fig. 2. Similarity scores for comparing "ow" to "uw"

4 Conclusions

This paper proposed a new approach to estimate pronunciation similarity using the distance between vector sequences of cepstral features for each speaker. This approach not only deals with speech word matching, but also characters of speaker pronunciation. In the results of the pronunciation similarity experiments on the seven fundamental Korean vowel data, it has been observed that the total average of the correctness showed the accuracy of 83%. Future work will consider speech features such as pitch, formant, duration, and accent using an automatic detection algorithm. In addition, speaker normalization methods such as vocal track length normalization may be incorporated since these methods can reduce inter speaker variation.

Acknowledgements

This work was supported by a grant (no. R01-2006-000-11162-0) from the Basic Research Program of the Korea Science & Engineering Foundation.

References

1. L. R. Rabiner, "A tutorial on hidden Markov models and selected applications in speech recognition", *Proceedings of the IEEE*, vol. 77, no. 2, pp. 257-286, 1989.
2. Q. Yan, S. Vaseghi, D. Rentzos, H. C. Ho, and E. Turajlic, "Analysis of acoustic correlates of Britich, Australian and American accents", *IEEE Workshop on Automatic Speech Recognition and Understanding*, pp. 345-350, 2003.
3. J. Humphries, *Accent modelling and adaptation in acoustic speech recognition*, Ph.D. thesis, Cambridge University, 1997.
4. Q. Yan and S. Vaseghi, "Analysis, modeling and synthesis of formants of British, American and Australian accents", in *Proceedings of International Conference on Acoustics, Speech, and Signal Processing*, pp. 712-715, 2003.
5. M. G. Rahim, C. H. Lee, and B. H. Juang, "Discriminative utterance verification for connected digits recognition", *IEEE Transactions on Speech and Audio Processing*, vol. 5, no. 3, pp. 266-277, 1997.
6. R. A. Sukkar, A. R. Setlur, C. H. Lee, and J. Jacob, "Verifying and correcting recognition string hypotheses using discriminative utterance verification", *Speech Communication*, vol. 22, pp. 333-342, 1997.
7. R. C. Rose, B. H. Juang, and C. H. Lee, "A training procedure for verifying string hypothesis in continuous speech recognition", in *Proceedings of International Conference on Acoustics, Speech, and Signal Processing*, pp. 281-284, 1995.
8. H. Jiang, "Confidence measure for speech recognition: A survey", *Speech Communication*, vol. 45, pp. 455-470, 2005.
9. S. M. Witt, *Use of Speech Recognition in Computer-assisted Language Learning*, Ph.D. thesis, Cambridge University, 1999.
10. C. Myers, L. Rabiner, and A. Rosenberg, "Performance tradeoffs in dynamic time warping algorithms for isolated word recognition", *IEEE Transactions on Acoustics, Speech, and Signal Processing*, vol. 28, no. 6, pp. 623-635, 1980.
11. V. N. Vapnik, "An overview of statistical learning theory", *IEEE Transactions on Neural Networks*, vol. 10, no. 5, pp. 998-999, 1999.
12. http://svmlight.joachims.org

A Novel Method for Rapid Speaker Adaptation Using Reference Support Speaker Selection

Jian Wang, Zhen Yang, Jianjun Lei, and Jun Guo

School of Information Engineering
Beijing University of Posts and Telecommunications
100876 Beijing, China
Wangjian200810@sohu.com

Abstract. In this paper, we propose a novel method for rapid speaker adaptation based on speaker selection, called reference support speaker selection (RSSS). The speakers, who are acoustically close to the test speaker, are selected from reference speakers using our proposed algorithm. Furthermore, a single-pass re-estimation procedure, conditioned on the selected speakers is shown. The proposed method can quickly obtain a more optimal reference speaker subset because the selection is dynamically determined according to reference support vectors. This adaptation strategy was evaluated in a large vocabulary speech recognition task. From the experiments, we confirm the effectiveness of proposed method.

1 Introduction

Typical speaker adaptation method includes the MAP family [1] and the linear transformation family (e.g., MLLR) [2]. These two families require significant amounts of adaptation date from the new speaker in order to perform better than a speaker independent (SI) system. Recently, a family of clustering and selection based speaker adaptation schemes (e.g., CAT, SST) has received much attention [3, 4]. This approach utilizes the correlations among different reference speakers and performs effectively in rapid speaker adaptation even if only one adaptation sentence has been used. Speaker selection training (SST) is a typical example of selection based speaker adaptation [3]. It trains a speaker dependent (SD)model for each of reference speakers and assumes that the adapted model for the test speaker must be a linear combination of the selected reference models. How to make a trade off between good coverage and small variance among the cohorts selected is still a very trick problems relied on the experiments. Dynamic instead of fixed number of close speaker selection seems to be a good alternative. Take support vector machine (SVM) as solution, we can choose an optimal set of reference models with a very limited amount of adaptation data. However, such selection based SVM is influenced by some critical factors, such as the kernel and its parameters, the complexity of the question and the noise data near the optimal hyperplane, etc.

Y. Matsumoto et al. (Eds.): ICCPOL 2006, LNAI 4285, pp. 450 – 456, 2006.

We can see clearly that the reference speakers are selected according to the support vectors (SV). In particular, it was observed that SVM solely trained on the SV set extracted by another machines with a test performance not worse than after training full dataset [5]. These support vectors lie close to the decision boundary between the two classes and carry all relevant information about the classification problem. This led to assume that it might be feasible to generate SV only without training SVM.

In this paper, agreeing with the assumption above, we proposed a heuristic method to extract the candidates of support vector for speaker selection and then re-estimate the speaker's model. In the next section, we discuss the basic idea of SVM based speaker selection. Our proposed method, reference support speaker selection (RSSS) is explained in section 3. Experimental results using our algorithm are shown in Section 4. Section 5 summarizes the paper and gives a conclusion.

2 SVM Based Speaker Selection

SVM is a promising machine learning technique developed from the theory of Structural Risk Minimization [6, 7]. It is typically constructed as a two class classifier. Fig.1 shows a typical two-class problem in which the examples are perfectly separable using a linear decision region. H1 and H2 define two hyperplanes. The closest in-class and out-of-class examples lying on these two hyperplanes are called the support vectors. For a separable data set, the system places a hyperplane in a high dimensional space so that the hyperplane has maximum margin. SVM used basically for binary (positive and negative) classification.

Given a training set $(x_i, y_i), i = 1,\ldots,n, x \in R^d, y \in (+1,-1)$, let $\phi : R^d \rightarrow H$ be a nonlinear map which transforms x from the input space into the feature space, $H = \{\phi(x) = z \mid x \in R^d\}$. Optimization on the input data in this case involves the use of a kernel-based transformation:

$$K(x_i, x_j) = \phi(x_i)^T \bullet \phi(x_j) \tag{1}$$

where, $\bullet$ denotes inner product. Different kernel functions form different SVM algorithms, here are three kernel functions are often used [8].

A kernel-based decision function allows dot product to be computed in a higher dimensional space without explicitly mapping the data into these spaces. It has the form:

$$f(x) = \sum_{i=1}^{N} \alpha_i y_i K(x, x_i) + b \tag{2}$$

The training set instances that lie closest to the hyper-plane are selected as the support vectors that always have nonzero α_i coefficients. Therefore, the SV set can fully describe the classification characteristics of the entire training set.

Regarding the reference speakers and the test speaker as two classes, we can use their feature vectors to train a SVM. The support vectors in reference speakers are approximately close to the class of test speaker, especially when these reference speakers distribute around the test equality. Then the reference speakers corresponding to these support vectors can be selected as a subset, called speakers support vector. Figure.2 illustrates the principle of this method.

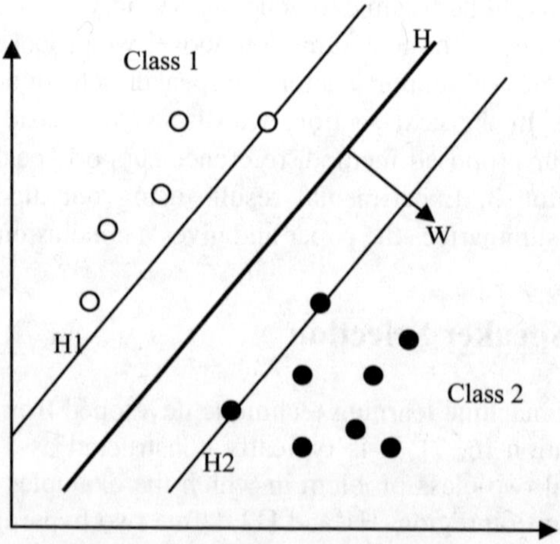

Fig. 1. H is the optimal hyperplane because it maximizes the margin the distance between the hyperplanes H1 and H2. Maximizing the margin indirectly results in better generalization.

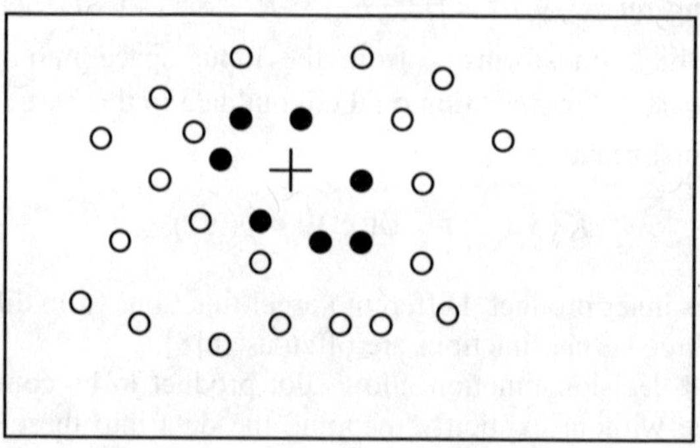

Fig. 2. Solid dot denote support vector speakers, hollow dot denote reference speakers, cross denotes test speaker

From above we can conclude the support vectors, which chosen as the representative points of the two classes and "support" the optimal hyperplane, are determined through maximizing the margin of the two classes. Hence, if we can find SV set directly without using SVM, the training time can be reduced greatly without much loss of classification precision.

3 The Proposed Algorithm

3.1 Main Procedure of Proposed Plan

The idea of the follow method is to find the points with the best probability become support vectors (called reference speaker support vectors). the two data-points of opposite class that are in distance closest to each other are the most probable support vectors and hence select them as the candidates support vectors. See the Figure .1. , given one point in positive examples, the point, which is closed in distance in the negative examples, is the most probable support vectors and hence select them as candidate speaker support vectors.

In the feature space, the distance between two points $z_i, z_j \in F$ can be computed by:

$$d_{ij} = \sqrt{k_{ii} - 2k_{ij} + k_{jj}} \tag{3}$$

where $k_{ij} = K(x_i, x_j)$ is kernel function.

For implementing reason, we extract the mean vectors of the GMM of M reference speaker to form supervectors $S_m (m = 1,2,\cdots M)$. We also apply MAP adaptation to the test utterance and extract the mean vectors of the Gaussians to form the supervector $\hat{S}_t$ of test speaker, every dimension of $\hat{S}_t$ has the order are the same as S_m.

Considering $\hat{S}_t$ as positive example and $S_m (m = 1,2,\cdots M)$ as negative examples, the distance between them can be calculated using equation 3. For verifying the speaker selection, a confidence measure of each speaker model can then be derived. The algorithm is described as follows. The algorithm is shown in Figure.3.

Input: given training set: $S_m \cup \hat{S}_t$.

Output: the selected candidate support vector.

Procedure :

for (m=1; M; m++) calculate d_{tm} (using formula 3);

find out the N(N can selected from 3 to 5)lest distances as measurement, then for every $S_m (m = 1,2,\cdots M)$,compute the confidence measure :

$$\delta(m) = \frac{\alpha(m)}{\beta(m)}$$

where $\alpha(m) = d_{tm} - \frac{1}{N}\sum_{j=1}^{N} d_{jt}$, $\beta(m) = \frac{1}{N}\sum_{j=1}^{N} d_{jt} - \frac{1}{M}\sum_{j=1}^{M} d_{jt}$

then we can selected the candidates SV through experiment threshold δ_T .

$$f(m) = \begin{cases} 0, \delta(m) < \delta_T \\ 1, \delta(m) \geq \delta_T \end{cases}$$

Fig. 3. The proposed algorithm

3.2 HMM Model Estimation

In this part, a single-pass re-estimation procedure, conditioned on the speaker-independent model, is adopted. In the adaptation procedure, there has no inherent structure's limitation of transformation-based adaptation schemes such as MLLR .A speaker adapted acoustic model is calculated from the HMM statistics of the selected speakers using a statistical calculation method.

The process of re-estimation would update the value of each parameter. The posteriori probability of occupying the m'th mixture component, $L_m^{i,r}(t)$, conditioned on the SI model, at time t for the r'th observation of the i'th cohort can be stored in advance .

The one-pass re-estimation formula may be expressed follows:

$$\tilde{\mu}_m = \sum_{i=1}^{N}\sum_{r=1}^{R_i}\sum_{t=1}^{T_r}(L_m^{i,r}(t)\cdot O^{i,r}(t)) \Big/ \sum_{i=1}^{N}\sum_{r=1}^{R_i}\sum_{t=1}^{T_r} L_m^{i,r}(t)$$
$$= \sum_{i=1}^{N} Q_m^i \Big/ \sum_{i=1}^{N} L_m^i \tag{4}$$

where :

$$L_m^i = \sum_{r=1}^{R_i}\sum_{t=1}^{T_r} L_m^{i,r}(t) \tag{5}$$

$$Q_m^i = \sum_{r=1}^{R_i}\sum_{t=1}^{T_r} L_m^{i,r}(t) Q^{i,r}(t) \tag{6}$$

Among them, $O^{i,r}(t)$ is the observation vector of the r'th observation of the i'th speaker at time t, $\tilde{\mu}_m$ is the estimated mean vector of the m'th mixture component of the target speaker. The variance matrix and the mixture weight of the m'th mixture component can also be estimated in a similar way.

4 Experiment Results

4.1 Experiment Setup

The database we used in these experiments is based on mandarin Chinese corpus provided by the 863 plan (China High-Tech Development Plan). About 60 hours of speech data from Chinese male speakers are used to train a gender-dependent SI model. We use 39 dimensional features consisting of 12 cepstral coefficients and log energy feature with the corresponding delta and acceleration coefficients. A five-state structure of a left-to-right HMM model with eight continuous density mixtures is trained. Then triphone-based HMM models are used in this continuous speech recognition.

The speakers ready for selection consist of 150 male speakers, with 250 utterances each. Typically one utterance, both in training and test set, lasts 3~5 seconds. Test set consists of 20 male speakers from the same accent with training set, 20 utterances each. 10 of them are used for selecting and adaptation. The other 10 are used for testing. It should be noted that we focus on very rapid adaptation of large-vocabulary system in this paper. All the adaptation methods in experiments are performed with only one adaptation sentence.

In order to evaluate the performance of the reference support vector, we perform different kernel methods and compare the results.

4.2 Experimental Results

Table 1 shows average recognition rates of RSSS (with 100 reference speakers) with different kernels. CSV represents the number of selected speakers. We take the SVM based speaker selection(SSVS) as comparison.

Table 1. Word accuracy percentage of different kernel

Accuracy %	RSSS			SSVS		
	Linear CSV=30	Poly CSV=10	RBF CSV=27	Linear SV=33	Poly SV=13	RBF SV=35
AWP	54.83	53.17	52.68	52.92	52.5	50.83

From Table 1 we can see that linear kernel based selection obtains the best recognition accuracy. As we known, although the polynomial kernel and RBF kernel can handle the case when the relation between class labels and attributes is nonlinear, but they use more parameters during training which influences the complexity of model selection. Then the support vectors obtained by linear kernel may bring a better performance in speaker selection.

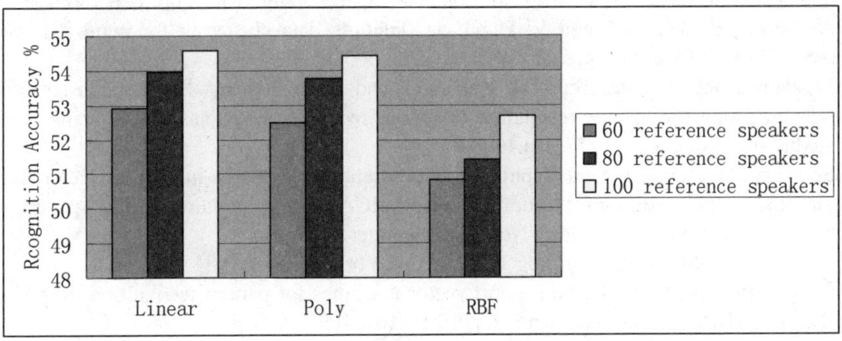

Fig. 4. Comparison with different number of reference speakers

From Figure.4, we observed an interesting result that heuristic procedure reduced the number of support vector, but the performance dose not decrease, we can also conclude from Figure.4 that as the number of reference speakers grows, the proposed method can select more accurate support vectors which are acoustically close to the test. So the performance will be improved. It is make out dynamically choosing the optimal cohort speakers for each target speaker is one of the key concerns in order to keep the balance between good coverage of phone context and acoustic similarity to the target speaker.

5 Conclusion and Discussion

Generally, performance of selection based speaker adaptation is very sensitive to the choice of initial models, the number of selected speakers is always fixed. A novel selection method based on SVM is proposed in this paper. It realizes dynamic speaker selection by finding the support vector and its corresponding speaker in the subset of reference speakers. Our experiments have shown the proposed scheme can improve the performance and robustness of adaptation, even few adaptation sentences is available. Further work will be focus on how to measure the relative contribution of each phone model of each specific speaker selected.

Acknowledgements

This research is partially supported by NSFC (National Natural Science Foundation of China) under Grant No.60475007, Key Project of Chinese Ministry of Education under Grant No.02029, and the Foundation of Chinese Ministry of Education for Century Spanning Talent.

References

1. J.L. Gauvain and C.H. Lee: Maximum a posterior estimation for multivariate Gaussian observations of Markov chains. IEEE Trans. Speech Audio Processing, vol. 2, pp. 291–298, Apr. 1994
2. C. J. Legetter and P. C. Woodland : Maximum likelihood linear regression for speaker adaptation of continuous density HMM's in Compute. Speech Lang., vol. 9, pp. 171–186, 1996
3. A. Sankar, F. Beaufays, and V. Digalakis: Training data clustering for improved speech recognition. in Proc. Eurospeech, 1995, pp. 502–505.4.
4. M. Padmanabhan, L. R. Bahl, D. Nahamoo, and M.A. Picheny: Speaker clustering and transformation for speaker adaptation in speech recognition systems. IEEE Trans. Speech Audio Processing, pp. 71–77, Jan.1998
5. Scholkopf B, Burges C, and Vapnik V. Incorporating invariances in support vector learing machines. In Artificial Neural Network–ICANN'96, volum 1112, pages 47-52,Berlin,1996.Springer Lecture Notes in Computer Science.
6. V. N. Vapnik: Statistical Learning Theory. New York: Wiley, 1998
7. C. J. C. Burges: A tutorial on support vector machines for pattern recognition. Knowledge Discovery Data Mining, vol. 2, no. 2, pp. 121–167, 1998
8. S. R. Gunn: Support vector machines for classification and regression. Technical Report Image Speech and Intelligent Systems Research Group, University of Southampton, 1997.

Using Latent Semantics for NE Translation

Boon Pang Lim and Richard W. Sproat

Dept of ECE, University of Illinois at Urbana Champaign, Urbana, IL 61801, USA

Abstract. This paper describes an algorithm that assists in the discovery of Named Entity (NE) translation pairs from large corpora. It is based on Latent Semantic Analysis (LSA) and Cross-Lingual Latent Semantic Indexing (CL-LSI), and is demonstrated to be able to automatically discover new translation pairs in a bootstrapping framework. Some experiments are performed to quantify the interaction between corpus size, features and algorithm parameters, in order to better understand the workings of the proposed approach.

1 Introduction

The problem of translating named entities (NE) across languages occupies an important place in machine translation. Such entities are in principle unlimited, making the exhaustive construction of translation dictionaries tedious if not impossible, and as such require special treatment and algorithms. The problem is compounded when we consider long names that span multiple words: some parts of which may be directly transliterated with only phonetic and orthographic knowledge, whilst others require some semantics.

In this paper we investigate an algorithm that uses contextual-semantic information to assist in the discovery of pairs of NE translations in two languages. The approach is generic enough to work for any sort of phrase translation, we will concentrate on features particularly salient to Named Entities, with particular attention to those found in abundant web news corpora. Some quantitative results are presented to illustrate the interaction of various factors such as corpus effects, feature sets and extraction, and tunable thresholds, and what impact these have on the performance of our algorithm.

2 Related Work

There is a large body of prior work that deals with the *transliteration* of Named Entities, in which phonemic and orthographic knowledge is chiefly used. Recent works such as [1], [2] and [3] tend to focus primarily on the use of such knowledge; less work has been done on exploiting semantic knowledge for transliteration.

Notable exceptions to this include [4], [5] and [6]. In [4], a simple context vector model was used in conjunction with a phonetic transliteration model, but while sophisticated statisical modeling for vector elements was applied, the method is still similar to a standard vector approach that employs a cosine-similarity measure. In [5], web news corpora was used to discover word translations (including NEs), by utilizing patterns in their temporal distribution across relevant news

Y. Matsumoto et al. (Eds.): ICCPOL 2006, LNAI 4285, pp. 457–465, 2006.

sources. The Canonical Correlation Analysis (CCA) technique studied in [6] is similar to LSA in that it extracts factors salient to translation.

Latent semantic analysis (LSA) is a purely statistical technique first pioneered by researchers in IR. It has been found to be robust in many different tasks and settings, and some recent work suggest that it may have some capacity to emulate human-like abilities at finding word associations. What we are proposing is a particular application of LSA to NE translation. It is *almost* identical to the cross-lingual latent semantic indexing (CL-LSI) algorithm first proposed in [7] and later further refined and studied in [8], [9] and other researchers in the field of Cross-Lingual Information Retrieval (CLIR). The basic idea of CS-LSI is to project feature vectors from either language into a joint space: this is accomplished by simply augmenting two document-term matrices together and performing a single SVD, side-stepping the need for a translation lexicon. Our algorithm improves on this by introducing an explicit step to map "true" pairs directly from one language's feature space to the other. This is possible since NEs tend to exhibit far less polysemy and have fewer paraphrases compared to common phrases, so that a nearly one-to-one correspondence exists across languages. Experiments in [8] demonstrate that CL-LSI works exceedingly well, with highly competitive performance against CLIR systems that use explicit translation, predicating some sort of implicit translation being performed.

These works collectively suggest that not only is the semantic knowledge that is embedded within contextual co-occurrence patterns relevant for NE translation, but an LSA approach can potentially work well. To our knowledge, our proposed method is one of the first to refine the use of LSA for NE translation, and is particularly powerful because it can potentially scale to diverse feature sets numbering in the hundreds of thousands on modest computing platforms.

3 Exploiting Semantic-Contextual Knowledge with LSA

Our algorithm is designed to function in the context of a larger bootstrapping framework that fuses information from multiple knowledge sources. This framework, illustrated in Figure 1, first uses some rudimentary method for Named Entity Recognition (NER) to produce a list of candidate NEs in each language. Modules which each score pairs of candidates according to some criterion criteria (be it phonetic, orthographic or contextual similarity), can be added to the framework at will. These scores then combined; top ranking pairs are added to a "correct" list that is used to further train and improve each module.

Our proposed algorithm has a two stage-training process as illustrated in Figure 2. First, we construct the semantic spaces for each language based on the list of "true" candidates. An explicit mapping function between either space is found. The next step is to produce a joint semantic space, and from this we can generate two linear projectors that will project context vectors from the feature space of either language onto the joint semantic space.

Let us denote the set $W = \{w_1, w_2, ...w_l\}$ containing l "true" pairs, where each pair $w_k \in W$ refers to a particular named entity. We have two sets of features

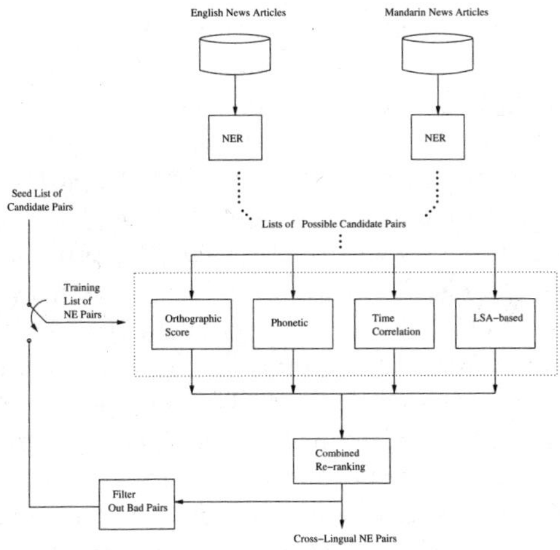

Fig. 1. A Bootstrapping Framework that employs Multiple Knowledge Sources

$F = \{f_1, f_2, ...f_n\}$ and $G = \{g_1, g_2, ...g_m\}$, where F contains n features from the first language, and similarly G for the second language. The pair-feature matrices T_1 and T_2, corresponding to feature sets F and G respectively, are constructed by counting the number of collocations between NE w_i and feature f_j (or g_j). The contribution of each feature is weighted using a TF-IDF score, so that each element in T_1 and T_2 is

$$T_{1,i,j} = n(w_i, f_j) log_2 \frac{N}{n(f_j)}, \text{ and } T_{2,i,j} = n(w_i, g_j) log_2 \frac{N}{n(g_j)}, \tag{1}$$

where N is the total number of features generated, $n(f_k)$ is the number of times feature f_k was generated over the entire corpus. The $log_2 \frac{N}{n(f_g)}$ term is analogous in function to the inverse document frequency (IDF) commonly used in information retrieval; however we must note that in our formulation, we are treating the contexts of transliteration pairs as documents and features as terms. Each NE-feature matrix is decomposed using Singular Value Decomposition (SVD),

$$T_1 = U_1 S_1 V_1^T, \text{ and } T_2 = U_2 S_2 V_2^T, \tag{2}$$

TRAINING I:	- Build Semantic Space for each Language from Seed pairs - Find an invertible mapping function between the two Semantic Spaces, - assuming that the seed pairs are point-to-point mappings.
TRAINING II:	- Build a Joint Semantic Space - Produce mapping functions from either language to the Joint Space
SCORING:	- Compute Joint Space vector for all candidates. - Compute a score between candidates with cosine similarity.

Fig. 2. Training and Scoring Algorithm for Contextual Features

such that Us and Vs are orthonormal matrices. Each row vector in T_1 corresponds to the feature vector in the language; each column vector corresponds to the context vector for a particular word. Since U_1 is orthonormal, defining a linear operator $P_1(x) = U_1^{-1} S_1^{-1} x$ projects the feature vector into a new space in which each the projected vectors are now spaced apart. Vector components that correspond to smaller singular values may be dropped altogether from the computation; this has been shown to empirically improve results on indexing tasks, by smoothing out noise in the data and ameliorating "overfitting".

Given the vectors x_k and y_k that correspond to the *kth* translation pair, we would like to find some invertible mapping $P : x_k \to y'_k$ such that the Euclidean distance between y_k and y'_k is minimized. This mapping associates the vectors within the space induced by the features of one language, to the space induced by the other language, and it is "optimal in some sense" with respect to correct translation pairs. If we take certain assumptions about the underlying probability distribution for components of context vectors (as in [10]), then the best projector is achieved by the linear operator

$$P_{1\to 2}(x) = U_2 S_2 V_2 V^T S^{-1} U^T x = T_2 T_1^* x, \tag{3}$$

where U's V's and S's are the results from the earlier singular vector decomposition, and T_1^* is the matrix pseudo inverse of T_1.

Let us denote $T_J = [T_1|T_2]$, applying the SVD gives $T_J = U_J S_J V_J^T$. The linear operator defined by $P_{1,2\to J}(x) = S_J^{-1} U_J^T x$ projects every context vector from T_J to an orthonormal set, i.e. $P_{1,2\to J}(T) = V_J$. Two linear projectors can be constructed that will simultaneously map a context vector from either language into the joint feature space and associate the cross-lingual LSA vectors for every seed pair. The projector from the first language to the joint space is given by

$$P_{1\to J}(x) = P_{1,2\to J} \begin{bmatrix} I \\ P_{1\to 2} \end{bmatrix} = S_J^{-1} U_J^{-1} \begin{bmatrix} I \\ U_2 S_2 V_2 V_1^{-1} S_1^{-1} U_1^{-1} \end{bmatrix} \tag{4}$$

This computation can be optimized by carefully ordering matrix multiplications; matrices U_J^{-1} and U_2 should be done first. The cosine similarity score between the joint space vectors is then computed and used as the module score.

4 Choosing Useful Features for NEs

Our choice of terminology (NE-features vs document-terms) reflects an understanding that the LSA approach can be extended beyond just simple word-occurrences and indeed to other levels of linguistic knowledge beyond semantics; features can be designed to capture syntactical, phonological and even morphological levels of knowledge, so long as they and the NEs exhibit some manifestation of the "distributional hypothesis".

A list of some features used in our experiments is shown in Table 1. Note that character unigrams, bigrams and trigrams are labelled as C1, C2 and C3 respectively. Complex feature spaces can be created by merging any combination

Table 1. Features and What They Capture

Feature	Description	Type of Knowledge
	Language independent	
H^*	Newspaper section-heading	topic information
Txx	Dates (Td), 5-day periods ($T5d$) and month (Tm)	temporal occurrence
p^*	Position of NE in document	Genre, writing style
sID, dID	Sentence or Document ID	parallel alignment
	English (word-segmented languages)	
dW	Words co-occuring with NE in the same document.	Topic of discourse
sW	co-occuring with NE in a short (± 5 words) window.	NE usage
Lx^*	letter n-grams occuring within NE.	Phonology, Orthography
	Written Chinese/written Chinese (Character, unsegmented)	
dCx	char n-grams co-occuring in same document.	Topic of discourse
sCx	char n-grams co-occuring in short (± 5) window.	general semantic use
ifC^*	chars occuring at initial/final within an NE.	possible NE type
neC^*	char n-grams occuring within NE.	Phonology, Orthography

of feature classes. The number of features in from each class can be limited by imposing threshold on feature's count IDF value.

The features listed here generally span higher levels of linguistic knowledge, notably semantics, topic of discourse and common usage (pragmatics). The obvious features based on word or n-gram character co-occurrences have common applications in topic-detection (TDT) tasks. The more unconventional features such as position within document, or temporal occurrence have the potential to capture some artifacts of writing style and the "herd-reporting" behaviour seen with news-worthy events. These may be particularly salient to NEs in the way they distribute in news corpora. The same can be said for section-headings or article level co-occurrence with other NEs as well.

5 Experiments

Effect of Explicit Association. The HKNews parallel corpus contains 44621 pairs of written Chinese and English news articles aligned at the sentence level. A list of 283 transliteration pairs were manually created from the corpus by examining the most frequently occurring bigram words and characters. Topical bias within the corpus resulted in several prominent semantic groups, such as place or street names (e.g. Tsim Sha Tsui), country names (e.g. US), organizations (e.g. United Nations), and named persons (e.g. Jiang Zemin).

This corpus was divided into three subsets: small, medium and large. These subsets are respectively one thirds, two-thirds and all of the corpus. Each sentence-level (or document-level) alignment is given a unique tag; these tags are used in-lieu of features. The projector from each language to the joint space is computed in two different ways: with and without explicit association of the seed pairs. Leaving out the association step is mathematically equivalent to replacing the long sequence of multiplications in Eq. 4 with the identity matrix.

Table 2. Mean Reciprocal Rank of Various Configurations (10-fold evaluation)

Without Pairwise mapping

Seeds	Corpus size		
	Small	Medium	Large
100	0.534	0.400	0.561
150	0.621	0.424	0.481
200	0.742	0.521	0.624

(a) Sent Ids as Feats

Seeds	Corpus size		
	Small	Medium	Large
100	0.717	0.655	0.751
150	0.804	0.760	0.817
200	0.869	0.852	0.881

(b) Doc Ids as Feats

With Pairwise mapping

Seeds	Corpus size		
	Small	Medium	Large
100	0.555	0.579	0.590
150	0.671	0.433	0.481
200	0.779	0.494	0.693

(c) Sent Ids as Feats

Seeds	Corpus size		
	Small	Medium	Large
100	0.717	0.707	0.783
150	0.832	0.818	0.844
200	0.881	0.870	0.893

(d) Doc Ids as Feats

A ten-fold cross validation was used. In each trial, the transliteration pairs were randomly divided into a seed/training and held-out/testing parts in varying proportions (100, 150, 200 seeds). The number of latent factors to retain was chosen per trial to maximize MRR. Averaged results are shown in Table 2.

As expected, we observe that either having a larger corpus or more seeds improves performance on the held-out test set. Furthermore, performing an explicit association consistently improves the performance over standard CL-LSI. Strangely, better results are not obtained sentence ids versus document ids. This could be explained by the fact that the algorithm effectively ignores the fact that there the ids themselves indicate alignment, in addition to where the boundaries of salient documents or sentences lie. Hence our algorithm has to do some extra work in automatically figuring out the alignment between documents (or sentences). Furthermore, the feature space of sentence ids could either be highly complex, so that our assumptions about the underlying joint distribution between features and NEs (Gaussian-ness) is no longer valid. This, in conjunction with having many numerous low-count features, gives us very poor feature space for LSA-based training methods. Another interesting anomaly observed here is that the results for medium sized corpora seem to perform a little poorer compared to the large and small sized corpora. More investigation is needed to look into this issue.

Number of Latent Factors to Retain. Empirical results have shown that dropping the latent factors with small eigenvalues can provide some sort of "smoothing effect" that reduces noise in the context vectors. The distribution of the "best" number of retained factors from our first experiment was analyzed to determine its dependence on the number of initial pairs and corpus size. Results are shown in Table 3 for the "best" number of retained factors averaged over 15 randomized seed-lists. It appears that as rule of thumb, discarding roughly a fifth of the latent factors should give reasonable results, but more empirical study is needed in order to figure out how many latent factors need to be discarded.

Table 3. Choosing Number of Latent Factors

Seeds	Small		Med		Large	
	Avg	Std. Dev	Avg	Std. Dev	Avg	Std. Dev
20	0.893	0.386	0.707	0.522	0.727	0.514
40	0.767	0.476	0.794	0.466	0.819	0.424
60	0.719	0.505	0.618	0.550	0.810	0.429
80	0.775	0.471	0.852	0.433	0.895	0.393
100	0.813	0.446	0.830	0.404	0.863	0.386
150	0.914	0.354	0.928	0.320	0.898	0.364
200	0.888	0.381	0.931	0.345	0.836	0.447

Combinations of Feature Spaces. The second experiment investigates the impact of using different features. The entire corpus was used in a ten-fold evaluation. In each trial a random set of 150 seeds were selected for training, and the remainder were held out for testing. The average MRR for selected pairs of features is shown in Table 4.

Table 4. MRR for different Feature Spaces (10-fold evaluation)

Written Chinese	English			
	sW	dW	sW+dW	sW+dW+T5d
sC1	0.362	0.264	0.371	0.395
sC2	0.292	0.192	0.340	0.384
sC1+sC2	0.354	0.275	0.408	0.422
dC1	0.404	0.403	0.462	0.522
dC2	0.291	0.699	0.376	0.440
dC1+dC2	0.374	0.538	0.512	0.556
sC1+dC1	0.397	0.349	0.450	0.492
dC1+T5d	0.368	0.491	0.459	0.643

(a) Basic Features

Written Chinese	English		
	Td	T5d	Tm
Td	0.831	0.718	0.586
T5d	0.684	0.861	0.724
Tm	0.582	0.717	0.861

(b) Temporal Features

The results indicate that the LSA approach to feature combination is sub-optimal. As yet, it is not possible to simply glob large feature spaces together and expect the algorithm to automatically sift out the best subset. This is evident when we observe that the document level features and temporal features perform best on their own, and that the overall result is worse when they are combined.

The experiments with the temporal features show that the best feature combination occurs when we use same sized time windows for both language corpus. This issue of scale is not consistent for the sentence and document level features. For instance the written Chinese `sC1` works best with English `sW+dW`, but written Chinese `dC2` and its combinations work better with English `dW`. This suggests that perhaps more study must be done on feature combination in order for it to be effective.

Bootstrapping. In the final experiment we demonstrate that our algorithm is viable in the bootstrapping setting. For this demonstration our best set of features along with optimal count and idf thresholds were chosen. All trials were run over our full set of news corpus. For each trial, a random list of 40, 60 or 100 initial seeds were chosen, new transliteration pairs were generated iteratively. During each iteration, the best scoring candidate pairs whose words are not in the original list are added to the list. The resulting list is then used to train the next iteration. The number of candidates to add is increased by one for every iteration. The micro-averaged recall and precision for 20 random trials for each configuration during each iteration shown in Table 5. Note that the Δ row simply records the net change in precision and recall after 10 bootstrapping iterations. The results show that with as little as 40 seed pairs, we can expect to obtain on average more correct pairs than wrong pairs with each iteration of the bootstrapping procedure.

Table 5. Prec and Recall for Bootstrapping Trials

Iteration	seeds					
	40		60		100	
	prec	rec	prec	rec	prec	rec
0	100.00	13.71	100.00	20.24	100.00	33.32
1	99.47	14.20	99.54	20.72	99.92	33.86
5	99.22	16.30	99.23	22.79	99.81	35.97
10	97.93	18.78	98.72	25.37	99.63	38.62
Δ	-2.06	+5.07	-1.27	+5.12	-0.36	+5.30

6 Conclusion

In this paper we have proposed an application of LSA towards Named Entity Translation. Experiments demonstrate its potential to combine information from large and diverse feature sets towards identifying NE pairs across languages. Furthermore, imposing an explicit pairwise mapping across pairs is helpful for Named Entity Translation. At the same time, less than stellar results suggest that it is unlikely that this approach by itself will yield spectacular results; rather it may be necessary to combine other measures and sources of information (as in Fig. 1) to get a practical system. More experimentation needs to be done to isolate variations due to the choice of seed pairs, possibly to semantic sub-clusters of translation pairs (e.g. place, person, org). Finally, some useful quantitative parameters were determined for our corpus, and we demonstrated the feasibility of this approach within a bootstrapping framework.

Future work include more in-depth experiments to compare this against analogous approaches based on support vector machines or artificial neural networks. In addition, we would like to extend the general approach to the case of multiple languages, and extract transliteration tuples from comparable corpora.

Acknowledgement

We would like to thank the reviewers for invaluable feedback on the initial draft; which have given us valuable insight for follow-up work, even though not all of their concerns could be adequately addressed with the final revision of this paper.

References

1. Al-Onaizan, Y., Knight., K.: Machine transliteration of names in Arabic text. In: Proc. of ACL Workshop on Computational Approaches to Semitic Languages. (2002.) 400–408
2. Oh, J.H., Choi, K.S.: An ensemble of grapheme and phoneme for machine transliteration. In: IJCNLP: Second International Joint Conference on Natural Language Processing, Jeju Island, Korea (2005)
3. Li, H., Zhang, M., Su, J.: A joint source-channel model for machine transliteration. In: Association for Computational Linguistics. (2004)
4. Huang, F., Vogel, S., Waibel, A.: Improving named entity translation combining phonetic and semantic similarities. In: HLT/NAACL. (2004)
5. Utsuro, T.: Translation knowledge acquisition from cross-linguistically relevant news articles (2004)
6. Cancedda, N., Dejean, H., Gaussier, E., Renders, J.M.: Report on CLEF-2003 experiments: two ways of extracting multilingual resources (2003)
7. Landauer, T.K., Littman, M.L.: A statistical method for language-independent representation of the topical context of text segments. In: Proceedings of the Sixth Annual Conference of the UW Centre for the New Oxford English Dictionary and Text Research. (1990) 31–38
8. Dumais, S., Letsche, T., Littman, M., Landauer, T.: Automatic cross-language retrieval using latent semantic indexing. In: American Association for Artificial Intelligence. (1997)
9. Mori, T., Kokubu, T., Tanaka, T.: Cross-lingual information retrieval based on LSI with multiple word spaces. In: Proceedings of the NTCIR Workshop 2 Meeting. (2001) 67–74
10. Yu-Seop Kim and Jeong-Ho Chang and Byoung-Tak Zhang: A comparative evaluation of data-driven models in translation selection of machine transliteration. In: COLING. (2002)

Chinese Chunking with Tri-training Learning

Wenliang Chen[1,2], Yujie Zhang[1], and Hitoshi Isahara[1]

[1] Computational Linguistics Group
National Institute of Information and Communications Technology
3-5 Hikari-dai, Seika-cho, Soraku-gun, Kyoto, Japan, 619-0289
[2] Natural Language Processing Lab
Northeastern University, Shenyang, China, 110004
{chenwl, yujie, isahara}@nict.go.jp

Abstract. This paper presents a practical tri-training method for Chinese chunking using a small amount of labeled training data and a much larger pool of unlabeled data. We propose a novel selection method for tri-training learning in which newly labeled sentences are selected by comparing the agreements of three classifiers. In detail, in each iteration, a new sample is selected for a classifier if the other two classifiers agree on the labels while itself disagrees. We compare the proposed tri-training learning approach with co-training learning approach on Upenn Chinese Treebank V4.0(CTB4). The experimental results show that the proposed approach can improve the performance significantly.

1 Introduction

Chunking identifies the non-recursive cores of various types of phrases in text, possibly as a precursor to full parsing or information extraction. Steven P. Abney was the first person to introduce chunks for parsing[1]. Ramshaw and Marcus[2] first represented base noun phrase recognition as a machine learning problem. In 2000, CoNLL-2000 introduced a shared task to tag many kinds of phrases besides noun phrases in English[3]. Much work has been done on Chinese chunking[4,5,6,7], of which supervised learning approaches are the most successful. However, to achieve good performance, the amount of manually labeled training data required by supervised learning methods is quite large.

Semi-supervised learning has recently become an active research area. It requires only a small amount of labeled training data and improves performance using unlabeled data. In this paper, we investigate the use of a semi-supervised learning approach, tri-training learning[8], on Chinese chunking. By considering that Chinese chunking is a sequence labeling problem, we propose a novel approach of selecting training samples for tri-training learning. The experimental results show that the proposed approach can improve the performance significantly using a large pool of unlabeled data.

The rest of this paper is as follows: Section 2 describes the definition of Chinese chunking. Section 3 describes the algorithm of tri-training for Chinese chunking. Section 4 introduces the related works. Section 5 explains the experimental results. Finally, in section 6 we draw the conclusions.

Y. Matsumoto et al. (Eds.): ICCPOL 2006, LNAI 4285, pp. 466–473, 2006.

2 Chinese Chunking

We defined Chinese chunks based on the Upenn Chinese Treebank V4.0 (CTB4)[1] as Chen et al. [9] did. And we use a tool[2] to generate the Chinese chunk dataset from CTB4.

2.1 Data Representation

To represent the chunks clearly, we represent the data with an IOB-based model as the CoNLL-2000 shared task did, in which every word is to be tagged with a chunk type label extended with I (inside a chunk), O (outside a chunk), and B (inside a chunk, but also the first word of the chunk).

With data representation, the problem of Chinese chunking can be regarded as a sequence labeling task. That is to say, given a sequence of tokens (words pairing with Part-of-Speech tags), $x = \{x_1, x_2, ..., x_n\}$, we need to generate a sequence of chunk tags, $y = \{y_1, y_2, ..., y_n\}$. In the following sections, we call an original sentence (x) as a data sequence, and a labeled sentence (y) as a labeled sequence.

2.2 Features

In this paper, we regard Chinese chunking as a sequence labeling problem. The observations are based on features that are able to represent the difference between two events. We utilize both lexical and Part-Of-Speech(POS) information as the features.

We use the lexical and POS information within a fixed window. The features are listed as follows:

- $W_i(i = -2, -1, 0, 1, 2)$
- $W_iW_{i+1}(i = -2, -1, 0, 1)$
- $P_i(i = -2, -1, 0, 1, 2)$
- $P_iP_{i+1}(i = -2, -1, 0, 1)$

Where W refers to a Chinese word while W_0 denotes the current word and $W_i(W_{-i})$ denotes the word i positions to the right (left) of the current word, and P refers to a POS tag while P_0 denotes the current POS and $P_i(P_{-i})$ denotes the POS i positions to the right (left) of the current POS.

3 Tri-training for Chinese Chunking

Tri-training was proposed by Zhou and Li [8], which was motivated from co-training. It designs three classifiers learning from unlabeled examples via an unlabeled example is labeled for a classifier if the other two classifiers agree on the labeling under certain conditions. This method can release the requirement of co-training with sufficient and redundant views. Additionally, tri-training learning considers the agreements of the classifiers while selecting new samples. We just need the labeled tags instead of the confident scores made by the classifiers. Thus we can use any classifier, which can be used in chunking, in tri-training learning.

---

[1] More detailed information at http://www.cis.upenn.edu/ chinese/

[2] Tool is available at http://www.nlplab.cn/chenwl/tools/chunklinkctb.txt

3.1 Algorithm of Tri-training

The algorithm of tri-training for chunking is presented in Table 1, and consists of three different classifiers. At each iteration, the unlabeled sentences are labeled by the current classifiers. Next, a subset of the sentences newly labeled is selected to be added to the training data. The general control flow of the algorithm is similar to the algorithm described by [8]. However, there are some differences in our algorithm. Firstly, we design a new measure to compute the agreement between two classifiers. Secondly, we propose a novel selection method based on the agreement measure.

Table 1. The pseudo-code for the Tri-training Algorithm

A, B, and C are three different classifiers.
M_A^i, M_B^i and M_C^i are the models for A, B and C at step i.
L is the original labeled data and U is the original unlabeled data.
U_A^i, U_B^i, and U_C^i are the unlabeled data at step i.
L_A^i, L_B^i, and L_C^i are the labeled training data for A, B and C at step i.

Initialise:
$L_A^0 = L_B^0 = L_C^0 = L$.
$U_A^0 = U_B^0 = U_C^0 = U$.
$M_A^0 \leftarrow Train(A, L_A^0)$.
$M_B^0 \leftarrow Train(B, L_B^0)$.
$M_C^0 \leftarrow Train(C, L_C^0)$.

Loop: {
M_A^i, M_B^i and M_C^i tag the sentences in U_A^i, U_B^i, and U_C^i.
Select newly labeled sentences L_A^{new}, L_B^{new} and L_C^{new} according to some selection method S.
$L_A^{i+1} = L_A^i + L_A^{new}$, $U_A^{i+1} = U_A^i - L_A^{new}$.
$L_B^{i+1} = L_B^i + L_B^{new}$, $U_B^{i+1} = U_B^i - L_B^{new}$.
$L_C^{i+1} = L_C^i + L_C^{new}$, $U_C^{i+1} = U_C^i - L_C^{new}$.
$M_A^{i+1} \leftarrow Train(A, L_A^{i+1})$.
$M_B^{i+1} \leftarrow Train(B, L_B^{i+1})$.
$M_C^{i+1} \leftarrow Train(C, L_C^{i+1})$.
if(all L_A^{new}, L_B^{new} and L_C^{new} are $\emptyset$) End Loop.
}

3.2 Select Training Samples

In selecting new training samples procedure, there are two steps: 1) compute the scores for all sentences; 2) select some sentences as new training samples according to the scores. We design a simple agreement measure to compute the scores for all sentences. Then based on the agreement measure, we propose a novel sample selection method, which is called Two Agree One Disagree method, to select newly labeled sentence as training samples.

Agreement Measure. Here, we describe how to compute the agreement between two classifiers for every sentence.

Suppose we have a data sequence $x = \{x_1, .., x_n\}$. Then use two classifiers to tag the sequence x, to get two labeled sequences $y_a = \{y_{1a}, .., y_{na}\}$ and $y_b = \{y_{1b}, .., y_{nb}\}$. Now, we compute the agreement A_g between y_a and y_b as follows:

$$A_g(y_a, y_b) = \frac{\sum_{1 \leq i \leq n} f(y_{ia}, y_{ib})}{n}. \qquad (1)$$

where n is the number of tokens in x, f is a binary function to tell whether y_{ia} and y_{ib} are the same label.

$$f(y_{ia}, y_{ib}) = \begin{cases} 1 & : \quad y_{ia} = y_{ib} \\ 0 & : \quad y_{ia} \neq y_{ib}. \end{cases} \tag{2}$$

A_g denotes the agreement between two labeled sequences or the agreement between two classifiers. The larger A_g is, the higher the agreement is.

Selection Methods. After computing the agreement, we should select new samples from the set of newly labeled sentences for next iteration. Suppose have three classifiers A, B, and C, we will select new samples for classifier A. As [10] suggested, we should select the sentences, which have high training utility. Thus we prefer to choose a sentence, which is correctly labeled by B or C and is not parsed correctly by the target classifier A, to be a new training sample.

We adopt two selecting principles: 1) If the higher agreement scores between the classifiers B and C at a sentence, the sentence is more likely correctly labeled. 2) If the classifier A disagree with the other classifiers (B and C) at a sentence, the sentence is not parsed correctly by A. To investigate how the selection criteria affect the learning process, we consider two selection methods.

We propose Two Agree method by applying the first principle. A newly labeled sentence is selected by:

- Two Agree Method (S_{2A}): Firstly, we tag U_A using B and C. Then we compute the agreements of all sentences by Equation 1. Finally, we rank the sentences by the agreements and select top m as new samples for A. The new samples are labeled by B.

By applying both two principles, we propose Two Agree One Disagree Method. A newly labeled sentence is selected by:

- Two Agree One Disagree Method (S_{2A1D}): Firstly, we tag U_A using A, B and C. Then we compute the agreements of A and B for all sentences, also compute the agreements of B and C for all sentences. Then we apply the intersection: the set of m percent highest-agreement scoring labeled sentences by B and C, and the set of m percent lowest-agreement scoring labeled sentences by A and B. The new samples are labeled by B.

Each selection method has a control parameter m, that determines the number of newly labeled sentences to add at each iteration. It also serves as an indirect control of the number of errors added to the training set[10]. In our experiments, we set m as 30% after tuning parameters.

4 Related Works

Semi-supervised learning is halfway between supervised and unsupervised learning. In addition to labeled data, the algorithm requires unlabeled data. The interest in semi-supervised learning increased recently for natural language processing[11,12,13].

There are some researchers who applied semi-supervised learning on chunking and parsing. Ando and Zhang[14] proposed a semi-supervised learning method that employs the structural learning for English chunking. Steedman et al.[15] learned the statistical parsers from small datasets using bootstrapping.

A prominent semi-supervised learning algorithm is co-training, which was first introduced in Blum and Mitchell[16] as a bootstrapping method. It has the similar control flow as tri-training. In this paper, we also investigate the application of co-training to Chinese chunking. The co-training algorithm used was presented in [11]. We do not use the confident scores. Instead, we select new samples, which have higher agreement score A_g of two classifiers.

5 Experiments

5.1 Experimental Setting

The UPENN Chinese Treebank-4(CTB4) consists of 838 files. In the experiments, we used the first 728 files (FID from chtb_001.fid to chtb_899.fid) as training data, and the other 110 files as testing data. In the training data, we used sizes of labeled sentences: 500 sentences. The other sentences (9,378) were used as unlabeled data for semi-supervised learning methods.

In the experiments, we used Support Vector Machines (SVMs), Conditional Random Fields (CRFs), and Memory-based Learning (MBL)[17] as the classifiers in tri-training. The first two models have achieved good performance in chunking[18][19]. Although the MBL model does not perform well as the first two models do, we hope that it can help to improve the performance of the first two models. And we used SVMs and CRFs as the classifiers in co-training. We used YamCha (V0.33)[3] to implement the SVMs model, CRF++ (V0.42)[4] to the CRFs model, and TiMBL[5] to the MBL model.

And we used all the default parameter settings of the package in our experiments. In learning procedures, we trained the models in 50 iteration rounds and selected 50 (maximum) newly labeled sentences as new training samples in each iteration. We evaluated the results as CONLL-2000 share-task did. In this paper, we report the results with F_1 score.

5.2 Experimental 1: Selection Methods of Tri-training

In this experiment, we investigated the performance of different selection methods: Two Agree method (S_{2A}) and Two Agree One Disagree method (S_{2A1D}). Figure 1 shows the experimental results, where CRF_S1 refers to the CRFs model with Two Agree method, CRF_S2 refers to the CRFs model with Two Agree One Disagree method, and the other strings have similar meanings.

From the figure, we found that Two Agree One Disagree method achieved better performance than Two Agree method. All three classifiers with Two Agree One Disagree

[3] YamCha is available at http://chasen.org/ taku/software/yamcha/

[4] CRF++ is available at http://chasen.org/ taku/software/CRF++/

[5] TiMBL is available at http://ilk.uvt.nl/timbl/

method provided better results. And the CRFs model provided the best results among three models. These results indicated that Two Agree One Disagree method can select new samples more efficiently via applying the second selecting principle. We also found that the MBL model can help to improve the performance of the CRFs and SVMs models, although it did not perform well as the CRFs and SVMs models did.

With Two Agree One Disagree method, tri-training boosts the performance of the MBL model from 83.32% to 86.43% (3.11% higher), the SVMs model from 85.92% to 87.06% (1.11% higher), and the CRFs model from 86.68% to 87.57% (0.89% higher).

5.3 Experimental 2: Tri-training vs. Co-training

In this experiment, we compared the performance between tri-training and co-training. Figure 2 shows the comparative results, where tri-training used Two Agree One Disagree method. We found that tri-training outperformed co-training. CRFs with tri-training achieved the best results with 87.57%, while CRFs with co-training got 87.16%.

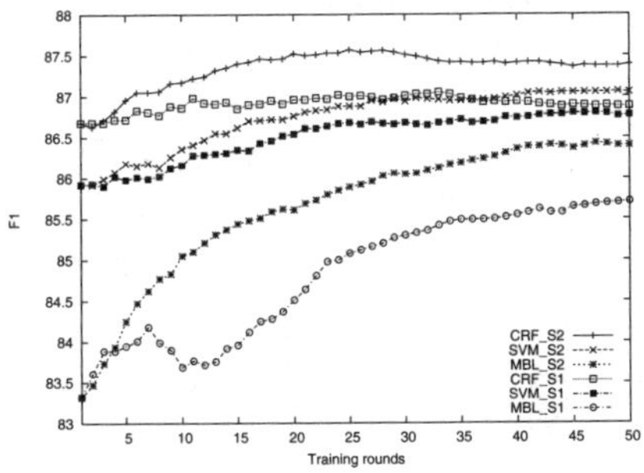

Fig. 1. Results of different selection methods for Tri-training

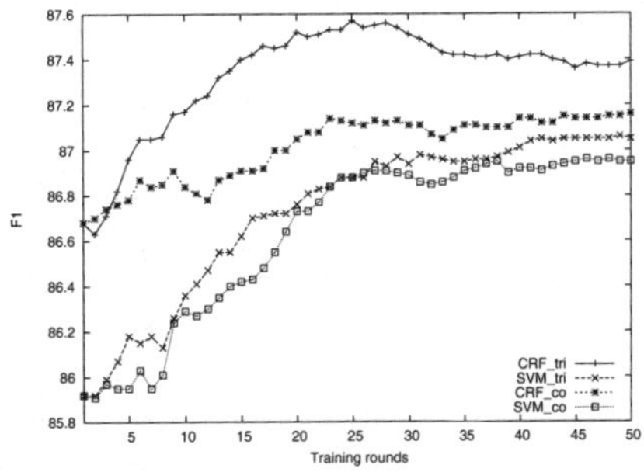

Fig. 2. Tri-training vs Co-training

5.4 Discussion

Our experimental results showed that tri-training learning approaches can exploit unlabeled data to improve the performance of Chinese chunking.

To test whether the improvements obtained by tri-training are significant, we performed one-tail paired t-test to compare the accuracy of tri-training against co-training and baseline systems. Here baseline systems refer to the models trained with seed data. For each token, if a classifier gives the correct tag, its score is 1, otherwise its score is 0. The p-values of one-tail paired t-test are shown in Table 2. From the table, we found that tri-training gave higher accuracy than the baseline at the level of significance 0.001 for both the CRFs and SVMs models. For the CRFs model, tri-training provided higher accuracy than co-training at the level of significance 0.01. In addition, for the CRFs model, co-training gave higher accuracy than the baseline at the level of significance 0.01.

Tri-training showed statistically significant improvement over the baseline, but the gap between them was not so big. We looked into the detail over all types of chunks.

Table 2. Summary of the p-value of one-tail paired t-test

nul hypothesis	t-test p-value
baseline vs. Co-training (CRFs)	0.0014
baseline vs. Tri-training (CRFs)	< 0.001
Co-training vs Tri-training(CRFs)	0.0012
baseline vs. Co-training (SVMs)	< 0.001
baseline vs. Tri-training (SVMs)	< 0.001
Co-training vs Tri-training(SVMs)	0.4623

Table 3 shows the results of all types of chunks generated by the CRFs model. Please note that the results of CLP and LST are 0.00 because there are not these two types of tags in 500 original labeled sentences. The results showed that the improvement was not observed uniformly on all types of chunks. Comparing tri-training with baseline systems, there were five types unchanged accuracy or changed very small, one type degraded, and six types changed from 0.2% to 3.35%. These indicated that we should try to improve the performance over all types to get better results.

Table 3. Results on all types of chunks (CRFs)

	baseline	Co-training	Tri-training
ADJP	78.70	79.26	80.09
ADVP	75.87	77.96	76.03
CLP	0.00	0.00	0.00
DNP	99.65	99.66	99.65
DP	99.67	99.70	99.70
DVP	84.16	60.11	86.98
LCP	99.85	99.80	99.82
LST	0.00	0.00	0.00
NP	86.16	86.51	87.36
PP	99.67	99.67	99.67
QP	94.66	94.77	95.89
VP	81.73	82.55	82.40
All	86.68	87.16	87.57

6 Conclusions

This paper presented an experimental study in which three classifiers were trained on labeled and unlabeled data using tri-training. We proposed a novel approach of selecting training samples for tri-training learning by considering the agreements of three classifiers. The experimental results showed that the proposed approach can improve the performance significantly.

In this paper, we trained the models based on word-based sentences with POS tags attached. However in real applications, the input is character-based sentences. Thus in future we will investigate the performance of tri-training learning method on character-based chunking.

References

1. Abney, S.P.: Parsing by chunks. In Berwick, R.C., Abney, S.P., Tenny, C., eds.: Principle-Based Parsing: Computation and Psycholinguistics. Kluwer, Dordrecht (1991) 257–278
2. Ramshaw, L., Marcus, M.: Text chunking using transformation-based learning. In Yarovsky, D., Church, K., eds.: Proceedings of the Third Workshop on Very Large Corpora, Somerset, New Jersey, Association for Computational Linguistics (1995) 82–94
3. Sang, E.F.T.K., Buchholz, S.: Introduction to the conll-2000 shared task: Chunking. In: Proceedings of CoNLL-2000 and LLL2000, Lisbin, Portugal (2000) 127–132
4. Li, H., Webster, J.J., Kit, C., Yao, T.: Transductive hmm based chinese text chunking. In: Proceedings of IEEE NLP-KE2003, Beijing, China (2003) 257–262
5. Tan, Y., Yao, T., Chen, Q., Zhu, J.: Applying conditional random fields to chinese shallow parsing. In: Proceedings of CICLing-2005, Mexico City, Mexico, Springer (2005) 167–176
6. Wu, S.H., Shih, C.W., Wu, C.W., Tsai, T.H., Hsu, W.L.: Applying maximum entropy to robust chinese shallow parsing. In: Proceedings of ROCLING2005. (2005)
7. Zhao, T., Yang, M., Liu, F., Yao, J., Yu, H.: Statistics based hybrid approach to chinese base phrase identification. In: Proceedings of Second Chinese Language Processing Workshop. (2000)
8. Zhou, Z.H., Li, M.: Tri-training: Exploiting unlabeled data using three classifiers. IEEE Transactions on Knowledge and Data Engineering **17** (2005) 1529–1541
9. Chen, W., Zhang, Y., Isahara, H.: An empirical study of chinese chunking. In: Proceedings of the COLING/ACL 2006 Main Conference Poster Sessions, Sydney, Australia, Association for Computational Linguistics (2006) 97–104
10. Steedman, M., Hwa, R., Clark, S., Osborne, M., Sarkar, A., Hockenmaier, J., Ruhlen, P., Baker, S., Crim, J.: Example selection for bootstrapping statistical parsers. Proceedings of the 2003 Conference of the North American Chapter of the Association for Computational Linguistics on Human Language Technology-Volume 1 (2003) 157–164
11. Pham, T., Ng, H., Lee, W.: Word sense disambiguation with semi-supervised learning. In: AAAI-05, The Twentieth National Conference on Artificial Intelligence. (2005)
12. Yarowsky, D.: Unsupervised word sense disambiguation rivaling supervised methods. In: In Proceedings of the 33rd Annual Meeting of the Association for Computational Linguistics (ACL). (1995)
13. Collins, M., Singer, Y.: Unsupervised models for named entity classification. Proceedings of the Joint SIGDAT Conference on Empirical Methods in Natural Language Processing and Very Large Corpora (1999) 100–110
14. Ando, R., Zhang, T.: A high-performance semi-supervised learning method for text chunking. Proceedings of the Annual Meeting of the Association for Computational Linguistics (ACL) (2005)
15. Steedman, M., Osborne, M., Sarkar, A., Clark, S., Hwa, R., Hockenmaier, J., Ruhlen, P., Baker, S., Crim, J.: Bootstrapping statistical parsers from small datasets. The Proceedings of the Annual Meeting of the European Chapter of the ACL (2003) 331–338
16. Blum, A., Mitchell, T.: Combining labeled and unlabeled data with co-training. Proceedings of the eleventh annual conference on Computational learning theory (1998) 92–100
17. Sang, E.F.T.K.: Memory-based shallow parsing. JMLR **2** (2002) 559–594
18. Sha, F., Pereira, F.: Shallow parsing with conditional random fields. In: Proceedings of HLT-NAACL03. (2003)
19. Kudo, T., Matsumoto, Y.: Chunking with support vector machines. In: In Proceedings of NAACL01. (2001)

Binarization Approaches to Email Categorization

Yunqing Xia and Kam-Fai Wong

Department of Systems Engineering and Engineering Management,
The Chinese University of Hong Kong, Shatin, Hong Kong
{yqxia, kfwong}@se.cuhk.edu.hk

Abstract. Email categorization becomes very popular today in personal information management. However, most n-way classification methods suffer from feature unevenness problem, namely, features learned from training samples distribute unevenly in various folders. We argue that the binarization approaches can handle this problem effectively. In this paper, three binarization techniques are implemented, i.e. one-against-rest, one-against-one and some-against-rest, using two assembling techniques, i.e. round robin and elimination. Experiments on email categorization prove that significant improvement has been achieved in these binarization approaches over an n-way baseline classifier.

Keywords: Binarization, assembling, email categorization, text classification.

1 Introduction

Email categorization becomes very popular today in personal information management. It seeks to assign a label to each incoming email and thus manage emails by folders. Text classification methods have already been applied to email categorization and anti-spam filtering with mixed success [3,9,11]. Nevertheless, as they intend to deal with all classes in one model, two problems are inevitable. One, n-way classification methods surfer from the *feature unevenness* problem, which is referred to as the fact that features learned from training samples distribute unevenly in classes. For example, observation on Enron email corpus [8] reveals that number of messages in folders ranges from 5 to 1,398. Features extracted from those folders might vary significantly. Two, number of classes in text classification is an important factor that influences classification quality. It is widely accepted that n-way classification methods perform less effective when considering more classes.

Many efforts have been contributed to tackle the two problems by designing novel classification methods or enhancing the ones in existence. We believe that quality of text classification can be improved via researching on multi-class basis on the one hand. But on the other hand, there exist other ways that are worth trying, the binarization approaches for instance.

The binarization approaches are investigated in this paper. The n-way classifier is first decomposed to multiple pairwise binary classifiers using strategies such as one-against-rest, one-against-one and some-against-rest. Then the binary classifiers are

Y. Matsumoto et al. (Eds.): ICCPOL 2006, LNAI 4285, pp. 474 – 481, 2006.

assembled using strategies such as round robin and elimination to achieve the objective of n-way classification. When elimination assembling strategy is adopted, quality of binary classifier used in elimination influences final prediction significantly. So some coupling strategies are adopted to help finding the optimal binary classifier. We implement the binarization approaches based on support vector machines (SVM) method. Substantial experiments are conducted on Enron email corpus [8] to justify our research efforts.

The remaining sections of this paper are organized as follows. In Section 2 and Section 3, three binarization techniques and two assembling techniques are presented, respectively. In Section 4, experiments on email categorization are presented as well as comparisons and discussions. We describe related works in Section 5 and conclude this paper in Section 6.

2 Binarization Techniques

The binarization techniques seek to decompose the n-way classification problem to multiple pairwise binary classification problems using binarization strategies such as one-against-rest, one-against-one and some-against-rest [10].

2.1 One-Against-Rest

One-against-rest (OAR) is a straightforward binarization strategy referred as unordered learning technique. It transforms an n-way classification problem into N two-class problems where N is total number of classes involved. OAR uses samples in class i as positive samples and ones in other classes $j(j=1, 2, \ldots, N; j\neq i)$ as negative samples. In most cases, N binary classifiers are not enough to make a correct prediction. Then the k-fold cross-validation method is used to produce $N*k$ base binary classifiers. Theoretically, bigger k will produce better results.

Compared to n-way classification, OAR is more accurate because two-class problem can be modeled more accurately than N classes using mathematical theory. Notably, more training samples can be used to produce accurate binary classifiers. Disadvantage of this method is that the feature unevenness problem still remains there as negative samples are usually more than that of positive samples. So the feature unevenness problem can not be well addressed in OAR.

2.2 One-Against-One

One-against-one (OAO) is another binarization technique designed to resolve the feature unevenness problem. For each possible pair of classes, a binary classifier is trained on samples from the two involved classes. Compared to OAR, OAO transforms the n-way problem into $C(N,2)$ two-class problems.

As classes compete against each other, the feature unevenness problem can therefore be limited to a lower degree in OAO. Quality of email categorization can be thus improved. The drawback of OAO is computational complexity in training. Besides, when training samples in some classes are very few, some extra techniques, say k-fold validation, are required to train the binary classifiers.

2.3 Some-Against-Rest

The feature unevenness problem remains serious if samples distribute very unevenly in folders. To make the binary decision process more flexible, the some-against-rest (SAR) binarization technique is proposed. The some-against-rest technique is a typical divide-and-conquer solution. Classes are first split into two groups, referred to as super classes. Training samples with class labels in the first group are considered as positives and samples from classes in the second group as negatives. Then binary classifiers are built with the re-labeled samples.

Advantage of SAR is that, not only is computational complexity saved, but equivalent quality of email categorization can be achieved provided that the grouping procedure is effective. The feature unevenness problem can be resolved more effectively than any other binarization approaches because pairwise coupling [6] is flexible in SAR. On the other hand, disadvantage lies in that computational complexity of SAR is higher than the other binarization techniques.

2.4 Comparison on a Real Case

The n-way classification and binarization techniques for a real classification problem are illustrated in Fig.1. For the 6-class problem shown in Fig.1(a), OAR learns 6 binary classifiers, in which the one for class "+" against the rest ones is shown in Fig.1(b). As a comparison, OAO learns $C(6,2)$ =6*5/2=15 binary classifiers, one for each pair of classes. Fig.1(c) shows the binary classifier <+, ^>. Obviously, in OAO case, the base classifier uses fewer examples than that in OAR case and thus has more freedom for fitting a decision boundary between the two classes. However, computational complexity in OAO training is increased from 6 classifiers to 15 ones.

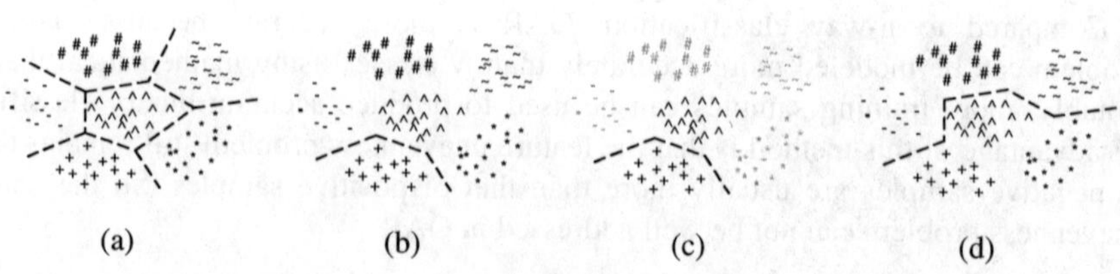

(a) (b) (c) (d)

Fig. 1. N-way classification and the binarization classification approaches: (a) n-way classify-cation; (b) one-against-rest (+, rest); (c) one-against-one (+,^); (d) some-against-rest

The SAR approach uses documents in two groups of classes as binary classification training samples. In this case, $C(6,1)+C(6,2)+C(6,3)$ =6+30+20=56 binary classifiers are learnt by the SAR. Fig.1(d) shows the binary classifier <(+, ^, *), (-, #, ~)>. In the SAR approaches, training samples can be arranged appropriately in terms of training sample number and content similarity.

3 Assembling Techniques

Assembling process aims to produce a final multi-class prediction based on predictions from the base binary classifiers generated in the binarization process. Two assembling strategies are discussed in this paper, i.e. round robin and elimination.

3.1 Round Robin Assembling

Round robin (RR) assembling technique is designed as a voting technique for OAO binarization technique [5], in which all the $N*(N-1)/2$ binary classifications are executed on given test samples. The label holding maximum votes from the base binary classifiers is output as final prediction. This concept can be revised and adopted in other binarization techniques.

RR is originally adopted to produce a vote between two base classes. Now we extend the concept to consider two groups of classes, namely two supper classes. For example, in OAR, we consider the "rest" classes as one super class. Now RR becomes applicable to OAR, and the OAR binary classifiers can be used to elect an optimal winner. For SAR, RR is made applicable by considering the "some" classes as one super class and the "rest" classes as another super class.

Another key issue in RR assembling is weight of the binary classifier generated by the binarization techniques. Intuitively, binary classifiers hold different prediction confidence. The confidence is in turn considered as weights of the binary classifiers when votes are counted for each class. The process to calculate weight of each binary classifier is referred to as pairwise coupling [6]. Typical metrics used in pairwise coupling are discriminativeness and evenness. Discriminativeness represents the degree that content of two classes is discriminative. Evenness represents the degree that training samples distribute evenly in two classes. Finally, the prediction confidence is product of the two degree values.

The ultimate goal of RR assembling is to produce a final prediction based on votes counted for each class. In the three binarization techniques, votes are counted in different ways. In OAR, we assign $\varpi \times 3$ to the "one" class and 0 to each of the "rest" classes if the "one" class wins; otherwise, the "one" class is assigned 0 and each of the "rest" classes is assigned $\varpi \times 3/(N\text{-}1)$, where ϖ is weight of this binary classifier. In OAO, we assign $\varpi \times 3$ to the winner class and 0 to the loser. In SAR, we assign $\varpi \times 3/K$ to the "some" classes and 0 to the "rest" classes if the "some" group wins the competition; otherwise 0 is assigned to the "some" classes and $\varpi \times 3/(N-K)$ to the "rest" classes, where K is number of classes in the "some" group.

3.2 Elimination Assembling

Elimination (EL) is another effective assembling technique in binarization approaches. It is often ignored by researchers because the elimination procedure is somehow error-prone. Intuitively, final prediction would be incorrect if the correct class is unfortunately eliminated in an earlier elimination round. However, we argue that such an incorrect elimination can be avoided when prediction confidence is incorporated in pairwise coupling [6].

The elimination process for OAO goes like this. A competition list is initialized to contain all classes and maintained after each run. In each run, two classes are randomly picked out of the list. The binary classifier trained with the two classes is applied to classify the test sample. When a prediction is made, the loser class is eliminated from the list and the winner is put back. Now size of the competition list is decreased by one. The elimination process is repeated until the competition list contains one class, which is considered as the optimal prediction that the test sample most likely belongs to. For OAR, the binary classifier with biggest weight is selected to classify the test sample. If the "one" class wins, this class will be output as prediction. Otherwise, further eliminations will be executed within the "rest" classes using OAR binarization technique until the final prediction is obtained. Elimination process in SAR is similar. If the "some" super class wins, classes in the "some" group remain in the competition list and classes in the "rest" group are eliminated. Then the SAR binarization technique is adopted for the "some" group and produces further eliminations until the final prediction is obtained.

Advantage of the EL assembling technique is the least computational complexity in searching for the final prediction. However, EL assembling technique is error-prone. When the binary classifier is selected inappropriately, the overall prediction will be put in danger. So prediction confidence is calculated to help finding the optimal binary classifiers. When the optimal classifiers are employed in elimination, some prediction errors can be avoided.

3.3 Comparison

We compare the two assembling techniques in two aspects, i.e. classification quality and computational complexity. Intuitively, the RR assembling is more reasonable than the EL because less errors occur in the former assembling technique. When prediction confidence is considered as weight for each voting binary classifier, quality of RR assembling can be further improved.

Elimination is advantageous in terms of computational complexity in predicting the optimal class. For example, with EL assembling, less than $N-1$ rounds of binary classification are required. But with RR assembling, $N*k$ rounds are required in OAO, $C(N,2)$ in OAO, and $C(N,1)+C(N,2)+....+\ C(N,[N/2])$ in SAR. Obviously, RR is much more computationally complex than EL.

4 Evaluations

4.1 Experiment Setup

Substantial experiments are conducted in this paper to evaluate the binarization techniques and assembling techniques in performing the task of email categorization. The training sets and test sets are extracted from the Enron email corpus [8] using the same data preparation method in [1]. We use the first 80% emails as training samples and the rest for test emails as test samples. Evaluation criteria are precision (p), recall (r) and F-1 measure (f) [12]. To produce overall results for each user, the micro-average method [12] is adopted.

We use SVM^{light} [7] as base binary classifier in our binarization implementations. SVM^{multiclass} [7] is considered as baseline method to evaluate the binarization approaches. Three binarization approaches are implemented, i.e. OAR, OAO and SAR, with two binarization and assembling techniques, i.e. RR and EL.

4.2 Results

We first run the baseline SVM n-way classification method. Then we run our approaches using various binarization and assembling techniques. Micro-averaged F-1 measure values achieved by the binarization approaches on seven users are presented in Table 1.

Table 1. Micro-averaged F-1 measure values for the seven users

Approach	beck	Farmer	kaminski	kitchen	lokey	sander	williams
SVM	0.512	0.656	0.553	0.532	0.598	0.706	0.939
OAR-EL	0.588	0.694	0.601	0.613	0.664	0.754	0.939
OAR-RR	0.629	0.710	0.623	0.642	0.687	0.791	0.942
OAO-EL	0.664	0.769	0.661	0.675	0.724	0.811	0.940
OAO-RR	0.669	0.770	0.663	0.680	0.726	0.817	0.941
SAR-EL	0.695	0.807	0.699	0.717	0.768	0.857	0.939
SAR-RR	0.718	0.838	0.717	0.731	0.791	0.878	0.943

4.3 Discussion I: N-Way vs. Binarization

The first comparison is made between the baseline n-way SVM method and the binarization approaches. Table 1 reveals that all the binarization approaches outperform SVM on all users. It thus proves that classification quality can be improved in the binarization approaches.

F-1 measure gain in SAR is around 0.18 over the baseline on average. The best performance is achieved on user *beck*, in which SAR-RR outperforms the baseline by 0.206 on F-1 measure. This can be addressed by observing training samples created by user *beck*. We find that user *beck* manages 101 folders and emails are distributed in folders very unevenly. It is thus not strange that SVM produces worst F-1 measure on user *beck*. Fortunately, overall classification quality equivalent to the binary classifiers can be achieved in the binarization approaches. Moreover, advantage of the binarization approaches gets more distinct when number of folders becomes bigger.

In contrast, F-1 measure improvement on user *williams* is very little, i.e. 0.004, over the baseline. We look into email collection maintained by user *williams* and find feature unevenness problem is extremely serious as 50.5% training samples come from folder *schedule_crawler*, 37% from *bill_williams_iii* and the remaining 12.5% from the other 14 folders. Classification problem in this case actually degrades to binary classification between folder *schedule_crawler* and *bill_williams_iii*. This is why F-1 measure on user *williams* is rather high and the binarization approaches do not improve much.

4.4 Discussion II: Binarization Approaches

Another general trend revealed in Table 1 is that SAR produces best quality among the three binarization approaches and OAR performs worst. This proves that the feature unevenness problem is resolved most effectively in SAR. However improvement varies on the seven users. This is because training environment is different for these users in terms of folder discriminativeness and evenness.

An interesting finding is that RR and EL in OAO produce satisfactory quality close to each other while RR outperforms El by 0.024 on average in OAR. This can be explained that elimination is most error-prone in OAR. When prediction confidence is incorporated in elimination assembling, optimal binary classifiers can be correctly selected in OAO. This is why RR outperforms EL by only 0.003 on average.

4.5 Discussion III: Round Robin vs. Elimination

According to Table 1, all binarization approaches using RR assembling outperform those using EL. This reveals that RR is more reasonable than EL since fewer errors occur in RR. Although prediction confidence can be calculated very accurately, quality of binarization approaches using EL never goes beyond those using RR.

Another finding is that RR outperforms EL by around 0.05 in SAR while by 0.024 in OAR and by 0.003 in OAO. It can be inferred that EL is more error-prone in OAR and SAR. This can be explained by methodological difference amongst the three binarization techniques. In SAR, fewest errors are made because the optimal binary classifier used in elimination is easily found. As comparison, OAO constructs binary classifiers with *two classes* in stead of *two groups of classes*. So the optimal binary classifiers found in OAO are not as effective as those found in SAR in resolving the feature unevenness problem. This proves that SAR is most feasible in adopting the elimination assembling technique while OAO is least.

5 Related Works

Decomposition and combination process in our binarization approaches are similar to error correcting output coding (ECOC) method discussed in [2]. Technically, ECOC is a general framework for the binarization methods to n-way classification problem and the binarization techniques discussed in this paper are special coding schemes for ECOC. However, we attempt to compare the binarization methods in email categorization, in which feature unevenness problem is serious.

The round robin assembling technique is described and evaluated in [5]. Our work is different. We focus on comparing round robin against other two binarization approaches on addressing the feature unevenness problem. Moreover, this is the first large-scale investigation on various binarization approaches to email categorization.

6 Conclusions

Classification quality is decreased by feature unevenness problem in email categorization. We propose to resolve this problem with binarization classification approaches. In this paper, three binarization techniques, i.e. one-against-rest, one-against-one and some-against-rest are implemented using two assembling techniques, i.e. round robin and elimination. Experiments show that the binarization approaches outperform the baseline n-way SVM classifier, in which some-against-rest improves most by 0.206. It is also revealed that round robin assembling is more accurate than elimination assembling in helping making optimal predictions, with the price of higher computational complexity.

References

1. Bekkerman, R., McCallum, A. and Huang, G.: Automatic Categorization of Email into Folders: Benchmark Experiments on Enron and SRI Corpora. UMass CIIR Technical Report IR-418 (2004).
2. Berger, A.: Error-correcting output coding for text classification. In IJCAI'99 Workshop on machine learning for information filtering (1999).
3. Cohen, W.: Learning Rules that Classify E-Mail. Proc. AAAI Spring Symposium on Machine Learning in Information Access, Stanford, California (1996).
4. Fisher, D.. and Moody, P.: Studies of Automated Collection of Email Records. University of California, Irvine, Technical Report UCI-ISR-02-4 (2001).
5. Furnkranz, J.: Round robin classification. Journal of Machine Learning Research (2002), 2:721-747.
6. Hastie, T. and Tibshirani, R.: Classification by pairwise coupling. In M. I. Jordan, M. J. Kearns, and S. A. Solla (eds.) *Advances in Neural Information Processing Systems 10* (NIPS-97), pages 507-513. MIT Press, 1998.
7. Joachims, T.: Learning to Classify Text Using Support Vector Machines, Methods, Theory, and Algorithms. Kluwer (2002).
8. Klimt, B. and Yang Y.: The Enron Corpus: A New Dataset for Email Classification Research. ECML04 (2004) pp:217-226.
9. Manco, G. Masciari, E. Rurolo, M. and Tagarelli, A.: Towards an adaptive mail classifier. In Proc AIIA'2002 (2002).
10. Schwenker, F.: Hierarchical support vector machines for multi-class pattern recognition. In Proc. IEEE KES2000 (2000), vol. 2, pp. 561–565.
11. Xia, Y., Dalli, A., Wilks, Y. and Guthrie, L.: FASiL Adaptive Email Categorization System. CICLing-05(2005): 723–734.
12. Yang, Y.: An evaluation of statistical approaches to text categorization. Journal IR (1999). 1(1/2):67-88.

Investigating Problems of Semi-supervised Learning for Word Sense Disambiguation

Anh-Cuong Le, Akira Shimazu, and Le-Minh Nguyen

School of Information Science
Japan Advanced Institute of Science and Technology
1-1 Asahidai, Nomi, Ishikawa, 923-1292, Japan

Abstract. Word Sense Disambiguation (WSD) is the problem of determining the right sense of a polysemous word in a given context. In this paper, we will investigate the use of unlabeled data for WSD within the framework of semi supervised learning, in which the original labeled dataset is iteratively extended by exploiting unlabeled data. This paper addresses two problems occurring in this approach: determining a subset of new labeled data at each extension and generating the final classifier. By giving solutions for these problems, we generate some variants of bootstrapping algorithms and apply to word sense disambiguation. The experiments were done on the datasets of four words: *interest*, *line*, *hard*, and *serve*; and on English lexical sample of Senseval-3.

1 Introduction

Word sense disambiguation involves the association of a given word in a text or discourse with a particular sense among numerous potential senses of that word. For this task, many supervised machine learning algorithms have been used for the WSD task, including Naïve Bayes, decision trees, an exemplar-based, support vector machines, maximum entropy, etc (see, e.g., [3]). However, supervised methods require large labeled data for training, which are expensive to obtain. Therefore, many researchers have recently concentrated their efforts on how to use unlabeled data to boost the performance of supervised learning for WSD, such as in [5,11,10,7]. The process of using both labeled and unlabeled data for training is called *semi-supervised learning* or *bootstrapping*.

In this paper, we focus on an approach that iteratively enlarges labeled data with new labeled examples which are obtained from unlabeled data. A common method for the extension of labeled data is to use the classifier trained on the current labeled dataset to detect labels for unlabeled examples. Among those new labeled examples, some highly accurate ones are selected and added to the current labeled dataset. This process is iteratively repeated until there is no unlabeled example left, or until the number of iterations reaches a pre-defined number. Two well-known methods of this approach are self-training [9] and co-training [1]. A general algorithm of this approach is sketched in Fig. 1 (it can be considered as the general self-training algorithm). We address two problems occurring in this algorithm as follows.

Y. Matsumoto et al. (Eds.): ICCPOL 2006, LNAI 4285, pp. 482–489, 2006.

Input:
L (labeled data); U (unlabeled data); h (a supervised learning algorithm)
$k = 0$; K is the maximum number of iteration
Output:
L_{out} is the final set of labeled data
or $h(L_{out})$ is the final classifier
Algorithm:
Repeat
1. $k \leftarrow k + 1$
2. use h and L to train a classifier $h(L)$
3. use $h(L)$ to label U, and obtain a labeled set U_L
4. get $L' \subset U_L$ consisting of high confident examples
5. $L \leftarrow L \cup L'$; $U \leftarrow U \setminus L'$

Until $U = \emptyset$ or $k > K$
6. $L_{out} \leftarrow L$

Fig. 1. A General Bootstrapping Algorithm

P_1: The first problem is how to determine a subset of new labeled examples at each extension of labeled data. To obtain "good" new labeled examples, we consider two criteria: the correctness of labeling and the number of new labeled examples. It is clear that adding a large number of misclassified examples into the labeled dataset will probably result in decreasing the accuracy of the classifier, and increasing confidence[1] of labeling may receive a small number of new labeled examples. Suppose that we have a new example which is assigned a label with a detection probability, previous studies normally use a threshold for this probability to decide whether a new labeled example will be selected or not, such as in [9,1]. However, choosing a high threshold will create difficulty in extending labeled data, and does not always result in correct classification. On the contrary, choosing a lower threshold may result in more misclassified examples. In this paper, we will increase confidence of new labeled examples by using one more supervised classifier for the detection process.

P_2: The second problem is that of how to generate the final classifier when the process of extending labeled data is completed. According to the framework in Fig. 1, this process will be stopped when the number of iterations reaches a pre-specified value, or until the unlabeled dataset becomes empty. Normally, the classifier, which is built on the labeled data obtained at the last iteration, is chosen as the final classifier. Some studies use a development dataset to find the most appropriate value for the number of iterations, such as in [7,5]. Others used an upper bound of error rate of training data as the condition for stopping this process, such as in [2]. However, the last classifier may be built based on new training data with some misclassified examples (related to problem P_1), so some advantages and some disadvantages are concurrently brought to the last classifier in comparison with the initial supervised classifier (i.e. the classifier

[1] The term "increasing confidence" of a classifier means using any method, by that we hope to increase the accuracy of label detection of this classifier.

which trained on the original labeled dataset). In our knowledge, this observation, which has not been observed in previous studies yet, may lead us to a solution of combining the advantages of both the initial and the last classifiers under classifier combination strategies.

The solutions for these two problems consequently generate variants of bootstrapping algorithms. They were evaluated through experiments on the four words *interest*, *line*, *hard*, *server*; and on English lexical sample of Senseval-3. Therefore, next section presents proposed solutions and new bootstrapping algorithms. Section 3 presents experiments with data preparation, results and discussion. Finally, conclusions are presented in section 4.

2 Problems and Solutions

2.1 Extending Labeled Data

This section presents problem P_1. An usual approach to this task is using a supervised learning algorithm to train a classifier based on the labeled dataset, and then using this classifier to detect labels for examples in a subset U' of the current unlabeled dataset U. Suppose h is the supervised classifier and $l(h,e)$ is the label of example e detected by h with probability $P(h,e)$. If $P(h,e)$ is greater than a threshold α, then example e with new label $l(h,e)$ will be considered to be added to L. As mentioned previously, using a classifier with threshold α for determining new labeled examples may cause a tradeoff problem between the extensibility and the accuracy of label detection. Furthermore, increasing α does not always result in an increase of accuracy in new labeled examples. To tackle this problem, we propose a solution that uses one more classifier, h'[2], as follows.

Suppose that at each extension, the maximum number of new labeled examples which are added to the current labeled dataset is N. We firstly use h and h' to detect labels for examples in U', and then select the examples e such that $P(h,e) \geq \alpha$. Among such new labeled examples, the examples which have agreements of labelling of h and h' are selected. If there exist more than N such examples, we will prefer the examples with high confidence of detection of h. In addition, if there are not enough agreements of labelling between h and h', we use h to select more new labeled examples, and also prefer the examples e with high $P(h,e)$. By this solution we can increase the confidence of new labeled examples, and also maintain the capability of extending labeled data.

Concerning this task, we also build a procedure of retaining class distribution. It is necessary because if a classifier is built based on the training data with a bias in some classes (i.e. some classes dominate others), then this bias will be increased at each extension of the labeled data. This problem is also considered in previous studies such as [1,8]. For a set of labeled examples, we divide it into the subsets such that the examples in each subset have the same label. We

[2] We assume that the two classifiers, h and h', are unbiased and the errors made by them would not fully overlap.

call them class-based subsets. So that, building a procedure of retaining class distribution for a set of new labeled examples means to resize its class-based subsets to keep class distribution of the original labeled dataset.

2.2 Generating Final Classifier

There is a fact that when extending labeled data, the feature space will be extended concurrently with adding some examples which are mislabeled. Therefore, the quality of labelling of the last classifier (trained on the last labeled dataset) will depend on each particular test example. In comparison with the initial supervised classifier, the final classifier may be better in detecting some test examples if these examples contain many new features covering by the new labeled examples. In the contrary case, if test examples can be well labeled by the initial supervised classifier, it is not necessary and risk to use the last classifier to label these examples. A natural way to utilize advantages of both these classifiers is to combine them when making decision of labelling. It then becomes a classifier combination problem. Based on OWA combination rules, which were also used for WSD, as presented in [4], we found that the *median* rule and *max* rule are intuitively applicable for this objective. These combination rules are recalled as follows.

For each example e, suppose $P_i(h_1, e),\ i = 1, \ldots, m$ is the probability distribution on class set $\{c_1, \ldots, c_m\}$ of using classifier h_1 to label e. A similar definition is applied to $P_i(h_2, e)$, for classifier h_2 and $i = 1, \ldots, m$.

Combining h_1 and h_2 using median and max rules, we obtain new probability distributions as follows:

$$P_i^{median}(h_1, h_2, e) = (P_i(h_1, e) + P_i(h_2, e))/2,\ i = 1, \ldots, m$$

$$P_i^{max}(h_1, h_2, e) = max\{P_i(h_1, e),\ P_i(h_2, e)\},\ i = 1, \ldots, m$$

Then the class c_k will be assigned to example e when using median rule (or max rule) iff:

$$P_k^{median}(h_1, h_2, e) \geq P_i^{median}(h_1, h_2, e), \forall i = 1, \ldots, m$$

$$P_k^{max}(h_1, h_2, e) \geq P_i^{max}(h_1, h_2, e), \forall i = 1, \ldots, m$$

2.3 A New Bootstrapping Algorithm

By the solutions as mentioned above, we generate a new bootstrapping algorithm as shown in Fig. 2. In this algorithm, $Resize(L_1, N)$ is the procedure for retaining class distribution, which returns new sizes for all class-based subsets of L_1 and satisfy that sum of all new sizes is less than or equal N; L is the labeled data; U is the unlabeled data; A and A' are two supervised learning algorithms, in which A is the primary one; α is a threshold; K is the maximum number of iteration; N is the maximum number of added labeled examples at each iteration; $C = \{c_1, \ldots, c_m\}$ is the set of classes; suppose S is a set of labeled examples, define $C_i(S) = \{e|e \in S,\ l(e) = c_i\}$, where $l(e)$ is the label of example e, and $i = 1, \ldots, m$.

Input: $L, U, A, A', \alpha, K, N$.
Output: classifier H
Algorithm:

initial: $k \leftarrow 0$; $L_0 \leftarrow L$; create a pool $U' \in U$ (randomly selected)
Repeat
1. $k \leftarrow k+1$; $L_1 \leftarrow \emptyset$; $L_2 \leftarrow \emptyset$
2. training A and A' on L and generate classifiers h and h', respectively
3. *for each* example $e \in U'$ *do*
   if $P(h,e) \geq \alpha$ *then* add e with new label $l(h,e)$ to L_1
4. $Resize(L_1, N)$
   obtaining n_i which is the new size of subset $C_i(L_1)$, for $i = 1, \ldots, m$
5. *for each* class $c_i \in C$ *do*
   5.1 sort examples of $C_i(L_1)$ in descending order, following the function $f(e)$:
   if $l(h,e) = l(h',e)$ *then* $f(e) \leftarrow 1 + P(h,e)$ *else* $f(e) \leftarrow P(h,e)$
   5.2. add first n_i examples of $C_i(L_1)$ to L_2
6. $L \leftarrow L \cup L_2$; rebuild U' from U and $L_1 \setminus L_2$

Until $k > K$ or $|L_2| = 0$
7. Let $A(L_0)$ and $A(L)$ be classifiers obtained by train A on L_0 and L, respectively then return H is the combination between $A(L_0)$ and $A(L)$

Fig. 2. A New Bootstrapping Algorithm

In fact, the new algorithm is an extension of the general self-training by providing solutions for problems P_1 and P_2. For experiment, we consider four variants of the bootstrapping algorithm as follows: A_1 is the general self-training algorithm; A_2 is the self-training algorithm with the solution for problem P_2 (we separate this by A_{2a} and A_{2b} respect to median and max rules for the combination step); A_3 is the self-training algorithm with the solution for problem P_1; A_4 is the self-training algorithm with the solutions for problem P_1 and P_2 (i.e. the new algorithm; we separate this by A_{4a} and A_{4b} respect to median and max rules for the combination step).

3 Experiment

3.1 Data

The first experiment was carried out on the datasets of the four words, namely *interest*, *line*, *serve*, and *hard*, which were obtained from Pedersen's homepage[3]. All examples in these datasets were tagged with the right senses. The sizes of

[3] http://www.d.umn.edu/~tpederse/data.html

these data are 2369, 4143, 4378, and 4342 for *interest*, *line*, *serve*, and *hard*, respectively. These datasets are large enough for dividing into labeled and unlabeled data sets. Furthermore, because we knew the tagged senses of examples in unlabeled dataset, we could evaluate the correctness of the new labeled examples (for problem P_1). We randomly extract 100 examples for labeled data, 300 examples for test data, and the remaining examples are treated as unlabeled examples.

The second test was carried out on English lexical sample from Senseval-3 data[4]. Unlabeled data in this experiment was collected from the British National Corpus (BNC) with about 3000 examples for each ambiguous word. Note that because the English lexical sample was also retrieved from BNC, so for a fair test, we removed all contexts from unlabeled dataset which also appear in the training or test datasets.

3.2 Feature Selection

Two of the most important kinds of information for determining the senses of an ambiguous word are the topic of the context and the relational information representing the structural relations between the target word and the surrounding words in a local context. A bag of unordered words in the context can represent the topic of the context, while collocation can represent grammatical information as well as the order relations between the target word and neighboring words. We also use more information about local context represented by words assigned with their positions, and their part-of-speech assigned with positions. These kinds of features are investigated in many WSD studies such as [6,5]. Particulary, all features used in our experiment fall in the kinds: a set of content words that include nouns, verbs, and adjectives, in a context window size 50; a set of collocations of words (we selected a set of collocations as presented in [3]); a set of words assigned with their positions in a window size 5; and a set of part-of-speech tags of these word also assigned with their positions, also in window size 5.

3.3 Parameters and Results

For the new algorithm, we chose naive Bayes (NB) as the primary supervised learning algorithm and support vector machines (SVM) as the second algorithm. That because SVM is a discriminative learning algorithm meanwhile NB is a generative one, so this difference will make SVM as an independent and confident classifier to verify the correctness of NB classifier's detection.

For the remaining parameters, we fix the maximum size of unlabeled dataset used at each iteration $|U'| = 800$ and the maximum size of new labeled examples which are added to the labeled dataset $N = 300$. The number of iteration K runs from 1 to 15 when testing on datasets of the four words, and then the best was used for testing on Senseval-3.

---

[4] http://www.senseval.org/senseval3

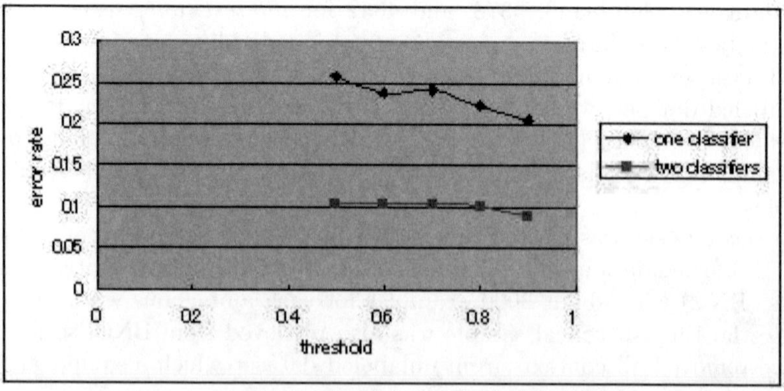

Fig. 3. Test problem P_1 using one classifier and two classifiers

Table 1. Test on Senseval-3

Algorithms		Average accuracy(%)
NB		70.13
SVM		69.45
A1 (self-training)		70.6
A2	median rule (A_{2a})	71.80
(self-training with P_2 solution)	max rule (A_{2b})	71.76
A3 (self-training with P_1 solution)		71.03
A4	median rule (A_{4a})	72.07
(self-training with P_1 and P_2 solutions)	max rule (A_{4b})	71.97

In the first experiment, we investigated the problem P_1 through various values of α, and the effectiveness of using one classifier (NB) and two classifiers (NB and SVM). The threshold α was tried the values $\{0.5, 0.6, 0.7, 0.8, \text{and } 0.9\}$. The result in Fig. 3 shows that using one more classifier much decrease error rate (from about 25% to 10%). In addition, this results also suggests the choice of $\alpha = 0.6$, which may ensure both the correctness and capability of the extension.

Table 1 shows a test on English lexical sample of Senseval-3[5] where we ran the bootstrapping algorithms 10 times and got the average. The algorithm A_3 is better than the algorithm A_1 (71.03% in comparison with 70.6%) shows the effectiveness of problem P_1. In addition, it should be noted that the algorithm with solution for problem P_2 still improves much the accuracy of self-training in the case not using solution for problem P_1 (the algorithm A_2). It can be explained that: permitting a high error rate of new labeled examples will expand more feature space which results in recognizing more new examples, and the proposed combination strategies keeps its advantage and decrease its disadvantage. Overall, combining the solutions for both problems P_1 and P_2 give the best

[5] We used the recall measure of fine-grained scoring for the evaluation.

result, and it increases about 1.9% of accuracy in comparison with supervised learning using NB.

4 Conclusion

In this paper, we have shown two problems of semi-supervised learning, particularly for self-training algorithm. They include the problem of extending labeled data and that of generating the final classifier. These problems have been investigated in WSD problem, and corresponding solutions were given. For the first problem, we used one more classifier to decrease error rate of new labeled examples. And for the second problem, we used two strategies of classifier combination including median and max rules to utilize both advantages of the last classifier (built based on the extended labeled data) and the initial supervised classifier. With those solutions, a new bootstrapping algorithm with several variants was generated. The experiments show that the proposed solutions are effective for improving semi-supervised learning. In addition, it also showed that unlabeled data significantly improve supervised learning in word sense disambiguation.

References

1. Blum A., and Mitchell T., 1998. Combining labeled and unlabeled data with co-training. In *Proc. COLT*, pages 92–100.
2. Goldman S. and Zhou Y., 2000. Enhancing supervised learning with unlabeled data. In *Proc. ICML*, pages 327–334, 2000.
3. Lee Y. K., and Ng H. T., 2002. An empirical evaluation of knowledge sources and learning algorithms for word sense disambiguation. In *Proc. EMNLP*, pages 41–48.
4. Le C. A., Huynh V-N, Dam H-C, Shimazu A., 2005. Combining Classifiers Based on OWA Operators with an Application to Word Sense Disambiguation. In *Proc. RSFDGrC*, Vol 1, pages 512–521.
5. Mihalcea R., 2004. Co-training and Self-training for Word Sense Disambiguation. In *Proc. CoNLL*, pages 33–40.
6. Ng H. T. and Lee H. B., 1996. Integrating multiple knowledge sources to Disambiguate Word Sense: An exemplar-based approach. In *Proc. ACL*, pages 40–47.
7. Pham T. P., Ng H. T., and Lee W. S., 2005. Word Sense Disambiguation with Semi-Supervised Learning. In *Proc. AAAI*, pages 1093–1098.
8. Pierce D., and Cardie C., 2001. Limitations of co-training for natural language learning from large datasets. In *Proc. EMNLP*, pages 1–9.
9. Yarowsky D., 1995. Unsupervised Word Sense Disambiguation Rivaling Supervised Methods. In *Proc. ACL*, pages 189–196.
10. Yu N. Z., Hong J. D., and Lim T. C., 2005. Word Sense Disambiguation Using Label Propagation Based Semi-supervised Learning Method. In *Proc. ACL*, pages. 395–402.
11. SU W., CARPUAT M., and WU D., 2004. Semi-Supervised Training of a Kernel PCA-Based Model for Word Sense Disambiguation. In *Proc. COLING*, pages 1298–1304.

Developing a Dialog System for New Idea Generation Support

Masahiro Shibata[1], Yoichi Tomiura[2], Hideki Matsumoto[3], Tomomi Nishiguchi[2], Kensei Yukino[2], and Akihiro Hino[4]

[1] Kyushu University Venture Business Laboratory, 6-10-1 Hakozaki Higashi-ku Fukuoka 812-8581 Japan

[2] Graduate School of Information Science and Electrical Engineering, Kyushu University, 744 Motooka Nishi-ku Fukuoka 819-0395 Japan

[3] Research and Development Center, Toshiba, 1 Komukai Toshiba-cho Saiwai-ku Kawasaki-shi 212-8582 Japan

[4] Level-5 Inc., Kyukan Daimyo Bldg. 2-6-28 Daimyo Chuo-ku Fukuoka 810-0041 Japan

Abstract. A knowledge-based dialog system gives correct answers; however, it is unsuitable for open-ended input. On the other hand, Eliza makes open-ended conversations, but it gives no new information to its user. We propose a new type of dialog system. Our system lies between the above two dialog systems, and it converses about various topics and gives information related to the user's utterances. This type of dialog is useful for generating new ideas especially when the user has an obscure desire to get information about his or her interest, but no concrete goal. Our system selects an appropriate sentence from a corpus to respond to a user's utterance. The most proper response will have surface cohesion and semantic coherence with the user's utterance. We made a trial system to converse about movies.

1 Introduction

Many practical dialog systems, such as online air travel planning [1] and telephone weather report announcing systems [2], are premised on the idea that users can make their requests clear enough and form their requests as lingual expressions. Under these conditions, a system can estimate the users' intentions using methods like pattern-matching, because the users' aims are definite and the input is restricted. Therefore, the system can give correct answers from prepared databases. In terms of answers, such a system needs fixed utterance patterns. This type of dialog system works well for specialized tasks, but is unsuitable for multiple tasks or open-ended input.

On the other hand, Eliza [3] is a well-known chat system for open-ended conversations. Eliza does not restrict the topic of conversation or the type of input. Eliza is a psychotherapeutic counseling system; therefore, it converts user utterances into interrogative sentences or makes nonsense responses, e.g., "Really?", "I see.", and "What happened?", to avoid answering to the user clearly and to promote introspection. Consequently, Eliza gives no new information to users.

Y. Matsumoto et al. (Eds.): ICCPOL 2006, LNAI 4285, pp. 490–497, 2006.

We propose a new dialog system that lies between these two types of dialog system. Our system converses about various topics and gives information related to the user's utterances, and it is useful for a new idea generation. When a user has an obscure desire to get information about his or her interest, but no concrete goal, he or she may generate some ideas through conversing with our system. Our system has three dialog modes. The information retrieval utterance mode uses a corpus as the source for the system's utterances. The proper answer to a user's utterance is selected from a corpus. To select the next system utterance, surface cohesion and lower semantic coherence are taken into account. In the study we report here, our system was made to converse about movies. Although the range of topics of our trial system is limited due to the limited corpus, the user can enter utterances freely.

2 Example of Dialog Using Our System

Below is a sample dialog in which our dialog system (S) talks about movies with a user (U). [1]

S0 : ĒGA NO HANASHI WO SHIMASHŌ. (Let's talk about movies.)
U1 : SAIKIN MITA ĒGA HA *CHĀRĪ TO CHOKORĒTO KŌJŌ* DESU. (The last film I saw was *Charlie and the Chocolate Factory*.)
S1 : OTOGI BANASHI TEKI YŌSO GA TSUYOKATTA NODE, "KODOMO NI ŌKU YOMARETEIRU SAKUHIN" TO IUNOGA WAKARU YŌNA KI GA SHIMASU. (It has many fairy-tale factors, so I am convinced many children have read this story.)
U2 : OTONA DEHA TANOSIMEMASENKA? (Can't adults enjoy the movie?)
S2 : CHĪSAI KODOMO HA MOCHIRON OTONA MO TANOSHIME- MASU! (Adults should enjoy it, as well as children!)
U3 : GEN NI WATASHI MO JŪBUN NI TANOSHIMIMASHITA. (Actually, I enjoyed it very much.)
S3 : SAISHO, ANO DOKUDOKUSISA TO TOPPI SUGIRU TENKAI NI HĪTEITANDESUGA, AMARINIMO SHŪRU NA SHĪN GA TUDUKU NODE ŌWARAI SHITE SHIMAIMASHITA. (The psychedelic and erratic development of the story spoiled my enjoyment at first. But so many surrealistic scenes occurred, I finally burst out laughing.)
U4 : UNPA=RUNPA WO ENJITA NOHA DARE DESHITAKANE? (Who plays the role of Oompa-Loompas?)
S4 : DHĪPU=ROI DESU. (Deep Roy plays them.)
U5 : SŌSŌ, DHĪPU=ROI NO ODORI NANKA SHŪRU DESHITANE. (Yeah, Deep Roy's dancing is surrealistic, I think.)
S5 : *CHĀRĪ TO CHOKORĒTO KŌJŌ* HA, AREDAKE BURAKKU DE SHŪRU NA KOTO WO SHITEOITE SAIGO NIHA KAZOKU AI NI MOTTEIKUNANTE... SHIKAMO IWAKAN NAKU. (I am surprised that *Charlie and the Chocolate Factory* ends up as a story of family love from such a black and surrealistic tale. However, this change is unforced.)

[1] In this paper, we use uppercase letters instead of Japanese charactors to describe Japanese words.

Table 1. Definition of terms in movie information database

Table name	Description	Attributes
Film	Information about each film title.	{film ID, film title (in Japanese), film title (in English), release year, country, running time, genre}
Person	Information about each person concerned with the movie.	{person ID, name (in Japanese), name (in English), occupation}
Role	Information about the work of person p in film f.	{film ID, person ID, works, character name (if p is an actor)}

3 System Overview

We suppose that the user's and the system's utterances are made alternately. Our dialog system has three modes of operation.

- **Accurate Understanding Utterance Mode (AUU mode)**
  When the user requests a correct answer to his or her question, our dialog system functions like a traditional knowledge-based dialog system, which uses a database and queries to get the correct answer. For our trial system, we used *PostgreSQL*, [2] a relational database management system, to create the movie information database. Table 1 lists the attributes of the relational tables we prepared. In the example in the previous section, the system's utterance S4 was generated with this mode.
- **Information Retrieval Utterance Mode (IRU mode)**
  To reply to a user's open-ended utterance, the system searches for the proper sentence from the corpus. In the example in the previous section, the system's utterances S1-S3, S5 were generated with this mode. This mode is explained in detail in the next section.
- **Nonsense Response Mode (NR mode)**
  When the system fails to get an answer using AUU mode and does not find a proper sentence using IRU mode, it responds as Eliza would. This response has no new information and may have no meaning to the user, but the system asks the user for the next utterance. This mode was not implemented in our trial system. Instead, the system's utters only "HĒ (Aha)" to signal that it selected NR mode to make the response.

4 Details of Information Retrieve Utterance Mode

In IRU mode, the system selects a sentence from a corpus to respond to the user's utterance. The user's utterance is only analyzed morphologically, because we could not prepare enough and reliable knowledge, such as case frames, to analyze a sentence syntactically for an open-ended conversation.

[2] http://www.postgresql.org/

We gathered sentences from web pages using the Japanese web search engine *goo*[3] with the keywords "α ĒGA" and constructed a movie corpus. "ĒGA" means "movie" in Japanese. The keyword "ĒGA" was used to exclude irrelevant pages. α is the expression which is to be the main theme of the dialog, that is to say, either a film title or the name of a person such as a director, an actor, or a writer of some film. Pages searched on the Internet included weblogs, diaries, and online shopping sites. Only a part of the page's content may be related to the main theme α or, even worse, only a few sentences may be relevant. Thus, the content of the gathered pages was hand-checked to remove sentences unrelated to α. Each sentence in our corpus was organized according to α, and $S(\alpha)$ denotes the set of sentences gathered by the keywords "α ĒGA".

Case elements are often abbreviated in Japanese. These invisible case elements are called zero pronouns. Anaphoric relations with zero or non-zero pronouns must be resolved in the original documents for our system to be able to use corpus sentences as candidates for the system's utterances. In our trial system, such anaphoric relationships were manually resolved in advance and each zero or non-zero pronoun in a corpus sentence was replaced with its correct antecedent.

In IRU mode, our system selects a proper response to a user's utterance from the corpus sentences through the three kinds of filtering and one ranking.We explain these filterings and the ranking respectively.

4.1 Filtering by Dialog Main Theme

We suppose that a dialog in our system will not be very long, and that the main theme of the dialog will be set at the beginning of the dialog and will not change. In our system, once a dialog starts, the system seeks the noun phrase that the first user's utterance most centrally concerns as the main theme of the dialog. We describe the centralness ranking later. Sentences from web pages whose main themes are the same as the dialog's main theme tend to be more appropriate for the system's utterances than ones on other web pages. Therefore, if the main theme of a dialog is α, our system restricts the candidates for its utterances to $S(\alpha)$. If $S(\alpha)$ is empty, the IRU mode fails to generate a sentence.

4.2 Filtering by Surface Cohesion

Our system selects utterance candidates in a way that the transition of the central concerns of the dialog will be natural. One theory that deals with such a transition in a discourse is the Centering Theory, and Walker, et al. applied this theory to Japanese[4]. When we create a transition model of central concerns according to the Centering Theory, the zero pronoun relationship must be resolved. Although anaphoric relationships have already been resolved in the corpus sentences, they must be resolved in the user's utterance, as well. Precise zero pronoun resolution requires a case frame dictionary, but constructing one

[3] http://www.goo.ne.jp/

entails a lot of effort. Moreover, there is the problem of word sense disambiguation using case frames[5,6]. Therefore, we suppose the following simple rule and our system selects candidates from $S(\alpha)$ according to this rule.

(1) The centralness of discourse entities in a sentence is ranked as follows:

(a zero pronoun) > (a noun phrase with the postposition "HA")
> (a noun phrase with the postposition "GA")
> (a noun phrase with the postposition "WO").

In Japanese, the case of a noun phrase is determined by the postposition appended to it. "GA" means the subjective case, and "WO" means the objective case. "HA" is not strictly a case marker; it indicates the topic of a sentence. We call this ranking the centralness ranking, in this paper. It expresses which noun phrase tends to be the antecedent of an anaphoric expression in the next sentence.

(2) If there is neither a noun phrase with a postposition "HA" nor one with a postposition "GA" in the user's utterance, we suppose that there is a zero pronoun with the case of "GA" in the utterance and that its antecedent is the entity which has the highest rank in the centralness ranking in the previous system's utterance. [4]

(3) We call what the system's utterance should most centrally concern the focus of the system's utterance. Let α be the dialog's main theme, and β be the entity which has the highest rank in the centralness ranking in the previous user's utterance. If $S(\alpha)$ has a sentence which includes a postpositonal phrase "β HA", the focus f is β and sentences which include a postpositonal phrase "β HA" are extracted from $S(\alpha)$. Otherwise, f is α and sentences which include a postpositonal phrase "α HA" are extracted from $S(\alpha)$. If f is β, the postpositional phrase "β HA" is removed from the extracted sentences.

The anaphoric relationships have already been resolved in the corpus sentences. Therefore, the sentences extracted in step (3) most centrally concern f. When f is set to the highest ranking entity in the centralness ranking of the previous user's utterance and therefore the postpositonal phrase "f HA" is abbreviated in the extracted sentences in step (3), we can determine that the extracted sentences have surface cohesion according to the Centering Theory. When f is set to the dialog's main theme, the local theme becomes the dialog's main theme, and this transition also seems natural.

4.3 Filtering by Predicate Coherence

Candidates are selected by the predicate coherence with the user's utterance. We show how to select proper candidates by predicate coherence.

[4] We do not check the selectional restriction.

Verb Coherence. A user's utterance is assumed to have a predicate verb. The predicate verb of the next utterance is restricted. If the system's utterance uses an unnatural predicate, the conversation will not be natural. Therefore, the system selects candidates to provide predicate verb coherence. In our system, verb coherence is calculated using mutual information given by

$$MI^V(v_i, v_j) = \log \frac{\hat{p}(v_i, v_j)}{\hat{p}(v_i)\,\hat{p}(v_j)} \ ,$$

where $\hat{p}(v)$ is the probability estimation for v and $\hat{p}(v_i, v_j)$ is the probability estimation for v_i and v_j to appear within two continuous sentences. To estimate the above probabilities, we gathered 5,875,515 sentences from web pages about movies. When the predicate verb of the user's utterance is v, candidates are selected on the condition that a sentence whose predicate is v_i are satisfied by $MI^V(v, v_i) \geq \theta$, where θ is a threshold. In our trial system, the threshold is set experientially as $\theta = 0.5$.

Adjective Cohesion. Discussing movies, a sentence whose main predicate is an adjective evaluates a quality of something such as movie a title or an actor. When a user makes an evaluation, he or she expects that the system will agree with his or her opinion. [5] The 1000 most frequent adjectives in the 5,875,515 sentences described in the above section were thus sorted into positive, negative, or other categories. When the user's utterance has a positive or negative adjective as its main predicate, the system's utterance should have same type as the predicate of user's utterance. When the predicate of the user's utterance is neither positive nor negative, the system selects any sentences with predicate adjectives regardless of their views. In our trial system, every adjective is sorted on manual. A study of collecting evaluating expressions from a corpus [7] may help us determine the adjective type automatically.

4.4 Ranking by Coherence of Nouns

At last, candidates are ranked according to noun coherence. The contents of the system's utterance should be related to the one of the user's. To evaluate this relationship, we focus on noun coherence between the user's utterance and each candidate for the system's utterance. We suppose that N is the set of all nouns in the user's utterance except for the current focus, and N' is a set of all nouns in each candidate for the system utterance except for the current focus [6]. According to N and N', we evaluate the noun coherence with mutual information of $(n, n') \in N \times N'$ as follows:

$$MI^N(n, n') = \log \frac{\hat{p}(n, n')}{\hat{p}(n)\,\hat{p}(n')} \ ,$$

[5] The system can object to an evaluation by providing reasons. However, this requires more complex processing than approval. Therefore, our trial system takes the strategy of agreement.

[6] In respect of the focus, its surface cohesion has been satisfied as we mentioned in Section 4.2. Thus, we exclude it from N and N' in evaluating noun coherence.

where $\hat{p}(n)$ and $\hat{p}(n')$ are the probability estimations of n and n' and $\hat{p}(n, n')$ is the joint probability estimation of n and n'. Here, n and n' are assumed to co-occur when a sentence having both n and n' exists in the learning data. The three highest pairs of $MI^N(\cdot)$s are selected and the sum is calculated as the cohesion of the user's utterance and each candidate. The noun cohesion is restricted to the top three pairs to reduce the influence on length of each candidate sentence.

Finally, the candidate sentence that has the highest noun cohesion with the user's utterance is selected from the corpus for the next system's utterance. In addition, this sentence is removed from the candidates in the corpus to prevent the system from selecting the same sentence in the same dialog.

5 Dialog Examples and Discussion

Below we discuss a few dialog examples. The following examples continue the dialog about "Charlie and the Chocolate Factory."

(Example 1)

U : KODOMO NI MISERU TO YOSASŌ DESUNE. (It seems good for children.)

S : KODOMO GA JŪBUN NI TANOSHIMERU SAKUHIN NI NATTEITA TO OMOIMASUSHI, DOKU MO JŪBUN NI HAITTEIRUN DESUGA, AKUMADE DŌWA TOSHITE WAKARIYASUI KYŌKUN WO FUKUNDEIRU NO GA SUBARASHĪ DESU. (I think that it is a very enjoyable movie for children. Moreover, it is splendid that it has an intelligible moral for a children's story, as well as rich black humor.)

This example is a success. It is a part of a series of utterances, and this user's utterance has a zero pronoun whose antecedent is "*Charlie and the Chocolate Factory*," and the antecedent of the noun phrase with "HA" in the system's utterance, which is abbriviated as a zero pronoun, is the same as the user's. In regard to the predicate, both "YOI igoodj" in the user's utterance and "SUBARASHĪ (splendid)" in the system's utterance are positive adjectives. In regard to the noun phrases, we show the top three pairs of noun phrases in $N \times N'$ as follows. The left-hand side of each pair is a noun phrase and the right-hand side is mutual information.

KODOMO (children) - KODOMO (children)	5.0388
KODOMO (children) - DŌWA(children's story)	1.5165
KODOMO (children) - DOKU (black humor)	0.8653.

(Example 2)

U : KŌHAN HA DAME DESHITAKA? (Do you dislike the latter half of the movie?)

S : *CHĀRĪ TO CHOKORĒTO KŌJŌ* HA, KŌHAN DANDAN TENKAI GA YOMETEKITA YŌNA KI MO SHIMASUGA, SONNA KOTO KI NI SHITECHA DAME DESU. (I can guess the plot of the latter half of *Charlie and the Chocolate Factory*, but that does not matter in this movie.)

This example is also a success. Focus cohesion is as good as in Example 1. The predicate adjective of the user is the same word "DAME DA" (bad). [7] Strictly

[7] In the translations, the word "bad" does not exist. It was rephrased.

speaking, "DAME DA" in the user's and the system's utterances are different. The subjective case of "DAME DA" in the user's utterance is the latter half of it, and the one in the system's utterance is the conversation partner (the user). To solve this problem, we need to resolve modification relationships.

6 Conclusion

We proposed a dialog system that makes a natural conversation about movies with users. The system selects a sentence from a corpus when the conditions of focus consistency and predicate cohesion are satisfied and noun cohesion is the highest. We had some persons use our system, and got the subjective evaluation that its responses seems very natural. We will evaluate the system quantitively in our future work.

The system considers only the local consistency at present. Therefore, the system may compliment a movie once but later denigrate it. In the future, we hope to achieve throughout a dialog. We also plan to review predicate cohesion. Our criterion is not necessarily the best for selecting the proper sentence. We must compare criterions from the other points of view and choose the best one.

References

1. Zue, V., Seneff, S., Polifroni, J., Phillips, M., Pao, C., Goddeau, D., Glass, J., Brill, E.: Pegasus: A spoken language interface for on-line air travel planning. Speech Communication **15** (1994) 331–340
2. V. Zue, S.S., Glass, J., Polifroni, J., Pao, C., Hanzen, T., Hetherington, L.: Jupiter: A telephone-based conversational interface for weather information. IEEE Trans. SAP **8** (2000) 100–112
3. Weizenbaum, J.: Eliza – a computer program for the study of natural language communication between man and machine. ACM **9** (1966) 36–45
4. Walker, M., Iida, M., Cote, S.: Japanese discourse and the process of centering. Computational Linguistics **20** (1994) 193–233
5. Fujii, A., Inui, K., Tokunaga, T., Tanaka, H.: Case contribution in example-based verb sense disambiguation. Journal of Natural Language Processing **4** (1997) 111–123
6. Kawahara, D., Kurohashi, S.: Improving japanese zero pronoun resolution by global word sense disambiguation. In: In Proc. of the 20th COLING. (2004) 343–349
7. Pang, B., Lee, L., Vaithyanathan, S.: Thumbs up? sentiment classification using machine learning techniques. In: Proc. of the 40th ACL. (2002) 417–424

The Incremental Use of Morphological Information and Lexicalization in Data-Driven Dependency Parsing

Gülşen Eryiğit[1], Joakim Nivre[2], and Kemal Oflazer[3]

[1] Department of Computer Engineering, Istanbul Technical Univ., 34469 Turkey
[2] School of Mathematics and Systems Engineering, Växjö Univ., 35195 Sweden
[3] Faculty of Engineering and Natural Sciences, Sabancı Univ., 34956 Turkey

Abstract. Typological diversity among the natural languages of the world poses interesting challenges for the models and algorithms used in syntactic parsing. In this paper, we apply a data-driven dependency parser to Turkish, a language characterized by rich morphology and flexible constituent order, and study the effect of employing varying amounts of morpholexical information on parsing performance. The investigations show that accuracy can be improved by using representations based on inflectional groups rather than word forms, confirming earlier studies. In addition, lexicalization and the use of rich morphological features are found to have a positive effect. By combining all these techniques, we obtain the highest reported accuracy for parsing the Turkish Treebank.

1 Introduction

An important issue in empirically minded research on natural language parsing is to what extent our models and algorithms are tailored to properties of specific languages or language groups. This issue is especially pertinent for data-driven approaches, where one of the claimed advantages is portability to new languages. The results so far mainly come from studies where a parser originally developed for English, such as the Collins parser [1], is applied to a new language, which often leads to a significant decrease in the measured accuracy [2,3,4,5,6]. However, it is often quite difficult to tease apart the influence of different features of the parsing methodology in the observed degradation of performance.

One topic that is prominent in the literature is the issue of lexicalization, i.e., to what extent the accuracy can be improved by incorporating features that refer to individual lexical items, as opposed to class-based features such as part-of-speech. Whereas the best performing parsers for English all make use of lexical information, the real benefits of lexicalization for English as well as other languages remains controversial [4,7,8].

Another aspect, which so far has received less attention, is the proper treatment of morphology in syntactic parsing, which becomes crucial when dealing with languages where the most important clues to syntactic functions are often found in the morphology rather than in word order patterns. Thus, for a language like Turkish, it has been shown that parsing accuracy can be improved by

Y. Matsumoto et al. (Eds.): ICCPOL 2006, LNAI 4285, pp. 498–508, 2006.

taking morphologically defined units rather than word forms as the basic units of syntactic structure [9].

In this paper, we study the role of lexicalization, morphological structure and morphological feature representations in data-driven dependency parsing of Turkish. More precisely, we compare representations based on the notion of *inflectional groups* proposed by Eryiğit and Oflazer [9] to a more traditional representation based on word forms. We experiment with different ways of representing morphological features in the input to the parser, and compare lexicalized and unlexicalized models to see how they interact with different representations of morphological structure and morphological features.

The parsing methodology is based on a deterministic parsing algorithm in combination with treebank-induced classifiers for predicting the next parsing action, an approach previously used for the analysis of Japanese [10], English [11,12], Swedish [13] and Czech [14]. Our study complements that of Eryiğit and Oflazer [9], which considers dependency parsing of Turkish in a probabilistic framework.

The rest of the paper is structured as follows. Section 2 describes some typologically prominent features of the Turkish language; section 3 introduces the framework of data-driven dependency parsing; and section 4 explains the experimental setup. In sections 5–7, we present experimental results on the impact of a morphologically defined tokenization (section 5), of rich inflectional features (section 6) and of lexicalization (section 7). Section 8 presents the results obtained with an optimized parsing model, in comparison to the state of the art, and section 9 contains our conclusions.

2 Turkish

Turkish is a flexible constituent order language. Even though in written texts, the constituent order of sentences generally conforms to the SOV or OSV structures, [15] the constituents may freely change their position depending on the requirements of the discourse context. From the point of view of dependency structure, Turkish is predominantly (but not exclusively) head final.

Turkish has a very rich agglutinative morphological structure. Nouns can give rise to hundreds of different forms and verbs to many more. Furthermore, Turkish words may be formed through productive derivations, and it is not uncommon to find up to five derivations from a simple root. Previous work on Turkish, [16,17,9] has represented the morphological structure of Turkish words by splitting them into inflectional groups (IG). The root and derived forms of a word are represented by different IGs separated from each other by derivational boundaries. Each IG is then annotated with its own part-of-speech and any inflectional features. Figure 1 shows the IGs in a simple sentence: "küçük odadayım" (*I'm in the small room*). The word "odadayım" is formed from two IGs; a verb is derived from an inflected noun "odada" (*in the room*).

Dependency relations in a sentence always hold between the final IG of the dependent word and some IG of the head word [17,9], so it is not sufficient to just

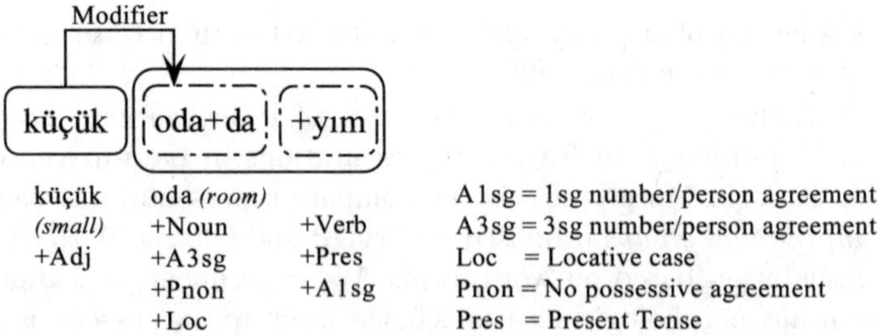

Fig. 1. Word and dependency representations

identify the words involved in a dependency relation, but the exact IGs. In the example, the adjective "küçük" (*small*) should be connected to the first IG of the second word. It is the word "oda" (*room*) which is modified by the adjective, not the derived verb form "odadayım" (*I'm in the room*). So both the correct head word and the correct IG in the head word should be determined by the parser.

3 Parsing Framework

A prominent approach to data-driven dependency parsing in recent years is based on the combination of three techniques:

1. Deterministic parsing algorithms for building dependency graphs [10,18]
2. History-based models for predicting the next parser action [19]
3. Discriminative classifiers to map histories to parser actions [10,20]

A system of this kind employs no grammar but relies completely on inductive learning from treebank data for the analysis of new sentences and on deterministic parsing for disambiguation. This combination of methods guarantees that the parser is robust, never failing to produce an analysis for an input sentence, and efficient, typically deriving this analysis in time that is linear or quadratic in the length of the sentence.

For the experiments in this paper we use the arc-standard variant of Nivre's parsing algorithm [18,21,22], a linear-time algorithm that derives a labeled dependency graph in one left-to-right pass over the input, using a stack to store partially processed tokens in a way similar to a shift-reduce parser.

The features of the history-based model can be defined in terms of different linguistic attributes of the input tokens, in particular the token on top of the stack, which we call the *top token*, and the first token of the remaining input, called the *next token*. The top token and the next token are referred to collectively as the *target tokens*, since they are the tokens considered as candidates for a dependency relation by the parsing algorithm. In addition to the target tokens, features can be based on neighboring tokens, both on the stack and in the remaining input, as well as dependents or heads of these tokens in the partially built dependency graph. The linguistic attributes available for a given token are the following: Lexical form (stem) (LEX), Part-of-speech category (POS), Inflectional features (INF), Dependency type (DEP).

To predict parser actions from histories, represented as feature vectors, we use support vector machines (SVM), which combine the maximum margin strategy introduced by Vapnik [23] with the use of kernel functions to map the original feature space to a higher-dimensional space. This type of classifier has been used successfully in deterministic parsing by Kudo and Matsumoto [10], Yamada and Matsumoto [11], and Sagae and Lavie [24], among others. To be more specific, we use the LIBSVM library for SVM learning [25], with a polynomial kernel of degree 2, with binarization of symbolic features, and with the one-versus-one strategy for multi-class classification.

4 Experimental Setup

The Turkish Treebank [16], created by METU and Sabancı University, has been used in the experiments. This treebank comprises 5635 sentences with gold standard morphological annotations and labeled dependencies between IGs. In the treebank, 7.2% of the sentences contain at least one dependency relation that is non-projective, not counting punctuation that is not connected to a head.[1] Each dependency link in the treebank starts from the final IG of the dependent word and ends in some IG of the head word.

Since the parsing algorithm can only construct projective dependency structures, we only use projective sentences for training but evaluate our models on the entire treebank.[2] More precisely, we use ten-fold cross-validation, where we randomly divide the treebank data into ten equal parts and in each iteration test the parser on one part, using the projective sentences of the remaining nine parts as training data.

The evaluation metrics used are the unlabeled (AS_U) and labeled (AS_L) attachment score, i.e., the proportion of tokens that are attached to the correct head (with the correct label for AS_L). A correct attachment implies that a dependent is not only attached to the correct head word but also to the correct IG within the head word. Where relevant, we also report the (unlabeled) word-to-word score (WW_U), which only measures whether a dependent is connected to (some IG in) the correct head word. Non-final IGs of a word are assumed to link to the next IG, but these links, referred to as *InnerWord* links, are not considered dependency relations and are excluded in evaluation scores. Results are reported as mean scores of the ten-fold cross-validation, with standard error, complemented if necessary by the mean difference between two models.

We use the following set of features in all the experiments described below:

- POS of the target tokens
- POS of the token immediately below the top token in the stack

[1] In the experiments reported in this paper, such dangling punctuation has been attached to the immediately following word in order to eliminate this uninteresting source of non-projectivity. Punctuation is also excluded in all evaluation scores.

[2] Our trial to use the pseudo-projective parsing strategy of Nivre and Nilsson [14] in order to process non-projective dependencies did not result in any improvement due to the limited amount of non-projective dependencies in the treebank.

Table 1. Summary table of experimental results

Section	Model	Unlexicalized		Lexicalized	
		AS_U	AS_L	AS_U	AS_L
5	Word-based	67.2±0.3	57.9±0.3	70.7±0.3	62.0±0.3
	IG-based	68.3±0.2	58.2±0.2	73.8±0.2	64.9±0.3
	IG-based deterministic	70.6±0.3	60.9±0.3	73.8±0.2	64.9±0.3
6	INF as single feature	71.6±0.2	62.0±0.3	74.4±0.2	65.6±0.3
	INF split	71.9±0.2	62.6±0.3	74.8±0.2	66.0±0.3
8	Optimized			76.0±0.2	67.0±0.3

- POS of the token immediately after the next token in the remaining input
- POS of the token immediately after the top token in the original input string
- DEP of the leftmost dependent of the top token
- DEP of the rightmost dependent of the top token
- DEP of the leftmost dependent of the next token

This is an unlexicalized feature model, involving only POS and DEP features, but we can get a lexicalized version by adding LEX features for the two target tokens. The value of each LEX feature is the stem of the relevant word or IG, rather than the full form. The reasoning behind this choice, which was corroborated in preliminary experiments, is that since the morphological information carried by the suffixes is also represented in the inflectional features, using the stem instead of the word form should not cause any loss of information and avoid data sparseness to a certain extent. The basic model, with and without lexicalization, is used as the starting point for our experiments. Additional features are explained in the respective subsections. An overview of the results can be found in table 1.

5 Inflectional Groups

In this set of experiments, we compare the use of IGs, as opposed to full word forms, as the basic tokens in parsing, which was found to improve parsing accuracy in the study of Eryiğit and Oflazer[9]. More precisely, we compare three different models:

- A word-based model, where the smallest units in parsing are words represented by the concatenation of their IGs.
- An IG-based model, where the smallest units are IGs and and where *InnerWord* relations are predicted by the SVM classifiers in the same way as real dependency relations.
- An IG-based model, where *InnerWord* relations are processed deterministically without consulting the SVM classifiers.

For these models, we use a reduced version of the inflectional features in the treebank, very similar to the reduced tagset used in the parser of Eryiğit and Oflazer [9]. For each IG, we use the part-of-speech of each IG and in addition

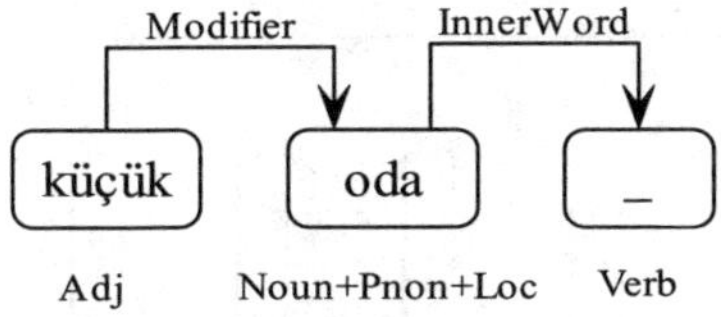

Fig. 2. IG-based representation with reduced inflectional features - LEX feature used in the parsing for each IG is shown within the rectangles and POS feature is given below the rectangles

include the case and possessive marker features if the IG is a nominal[3] or an adjective with present/past/future participle[4]. Using this approach, the POS feature of the word "odadayım" becomes +Noun+Pnon+Loc+Verb.

When lexicalizing the IG-based models, we use the stem for the first IG of a word but a null value ("_") for the remaining IGs of the same word. This representation, which is illustrated in figure 2 for the example given in figure 1, also facilitates the deterministic processing of *InnerWord* relations in the third model, since any top token can be directly linked to a next token with LEX="_", provided that the two tokens are adjacent.

In order to calculate the accuracy for the word-based models, we assume that the dependent is connected to the first IG of the head word. This assumption is based on the observation that in the treebank, 85.6% of the dependency links land on the first (and possibly the only) IG of the head word, while 14.4% of the dependency links land on an IG other than the first one.

The parsing accuracy obtained with the three models, with and without lexicalization, is shown in table 1. The results are compatible with the findings of Eryiğit and Oflazer [9], despite a different parsing methodology, in that the IG-based models generally give higher parsing accuracy than the word-based model, with an increase of three percentage points for the best models.

However, the results also show that, for the unlexicalized model, it is necessary to process *InnerWord* relations deterministically in order to get the full benefit of IG-based parsing, since the classifiers cannot correctly predict these relations without lexical information. For the lexicalized model, adding deterministic *InnerWord* processing has no impact at all on parsing accuracy, but it reduces training and parsing time by reducing the number of training instances for the SVM classifiers.

6 Inflectional Features

Instead of taking a subset of the inflectional features and using them together with the main part-of-speech in the POS field, we now explore their use as separate features for the target tokens. From now on, our POS tag set therefore consists only of the main part-of-speech tags found in the treebank.

[3] These are nouns, pronouns, and other derived forms that inflect with the same paradigm as nouns, including infinitives, past and future participles.

[4] These type of adjectives contain possessive agreement markers but no case markers.

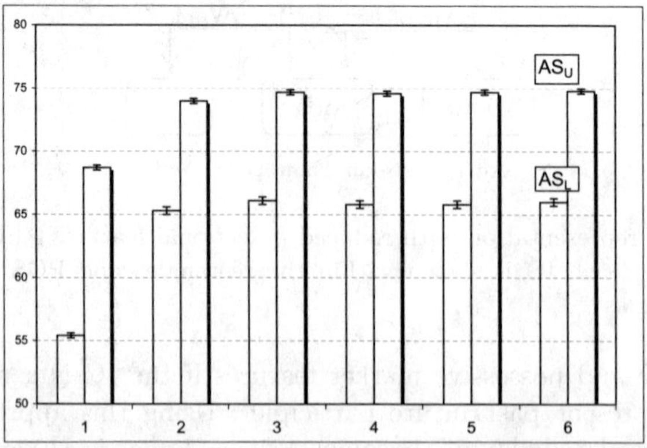

Fig. 3. Labeled and unlabeled accuracy for feature sets 1–6

As shown in earlier examples, the inflectional information available for a given token normally consists of a complex combination of atomic features such as +A3sg, +Pnon and +Loc. Thus, when adding inflectional features to the model, we can either add a single feature for each complex combination, or a single feature for each atomic component. As seen in table 1, both methods improve parsing accuracy by more than one percentage point across all metrics, but splitting features into their atomic components gives a slight advantage over the single feature approach. (The difference is quantitatively small but very consistent, with a mean difference of 0.4±0.1 for the labeled attachment score of the lexicalized models.)

Previous research has shown that using case and possessive features for nominals and possessive features for adjectives with participles improves parsing accuracy [9]. In order to get a more fine-grained picture of the influence of different inflectional features, we have tested six different sets, where each set includes the previous one and adds some more features. The following list describes each set:

1. No inflectional features at all
2. Case and possessive inflectional features for nominals and adjectives with participles
3. Set 2 + person/number agreement inflectional features for nominals and verbs
4. Set 3 + all inflectional features for nominals
5. Set 4 + all inflectional features for verbs
6. Set 5 + all inflectional features

The results, shown in figure 3, indicate that the parser does not suffer from sparse data even if we use the full set of inflectional features provided by the treebank. They also confirm the previous finding about the impact of case and possessive features. Besides these, the number/person agreement features available for nominals and verbs are also important inflectional features that give a significant increase in accuracy.

7 Lexicalization

Throughout the previous sections, we have seen that lexicalized models consistently give higher parsing accuracy than unlexicalized models. In order to get a more fine-grained view of the role of lexicalization, we have first studied the effect of lexicalizing IGs from individual part-of-speech categories and then from different combinations of them (see figure 4).[5] The results show that only the individual lexicalization of nouns and conjunctions provides a statistically significant improvement in AS_L and AS_U, compared to the totally unlexicalized model. Lexicalization of verbs also gives a noticeable increase in the labeled accuracy even though it is not statistically significant. A further investigation on the minor parts-of-speech of nouns[6] shows that only nouns with the minor part-of-speech "noun" has this positive effect, whereas the lexicalization of proper nouns does not improve accuracy. It can be seen from the chart of combinations that whereas lexicalization certainly improves parsing accuracy for Turkish, only the lexicalization of conjunctions and nouns has a substantial effect on the success.

a)

POS	AS_L	AS_U
Adjectives	62.7±0.3	72.3±0.2
Adverbs	62.8±0.3	72.4±0.2
Conjunctions	**63.5**±0.3	**73.4**±0.2
Determiners	62.7±0.3	72.2±0.2
Duplications	62.7±0.3	72.2±0.2
Interjections	62.6±0.3	72.2±0.2
Nouns	**64.0**±0.3	**73.1**±0.2
Numbers	62.6±0.3	72.2±0.2
PostPositions	62.7±0.3	72.2±0.2
Pronouns	62.7±0.3	72.2±0.2
Punctuations	62.8±0.3	72.3±0.2
Questions	62.6±0.3	72.2±0.2
Verbs	63.1±0.3	72.2±0.2

b)

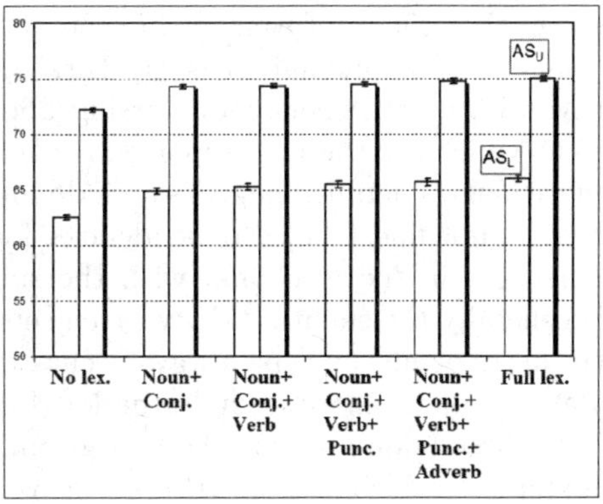

Fig. 4. Results of limited lexicalization: a)Individual b)Combination

Although the effect of lexicalization has been discussed in several studies recently [7,4,8], it is usually investigated as an all-or-nothing affair. The results for Turkish clearly show that the effect of lexicalization is not uniform across syntactic categories, and that a more fine-grained analysis is necessary to determine in what respects lexicalization may have a positive or negative influence. For the

[5] These results are not strictly comparable to those of other experiments, since the training data were divided into smaller sets (based on the POS of the next token), which reduced SVM training times without a significant decrease in accuracy.

[6] Nominals other than common nouns are tagged with an additional minor part-of-speech that indicates whether the nominal is a proper noun or a derived form – one of future participle, past participle, infinitive, or a form with zero derivation. The latter four do not contain lemma information.

models especially suffering from sparse data, it may even be a better choice to use some kind of limited lexicalization instead of full lexicalization. The results from the previous section suggests that the same is true for morphological information.

8 Optimized Parsing Model

After combining the results of the previous three sections, we performed a final optimization of the feature model. We found that using minor parts-of-speech instead of main parts-of-speech as the values of POS features and adding one more LEX feature for the token after the next token gave a best overall performance of AS_U=76.0±0.2 AS_L=67.0±0.3, and WW_U=82.7±0.5. We also tested our parser on two different subsets of the treebank. The first subset, which is used by Eryiğit and Oflazer [9] in order to evaluate their parser (giving AS_U=73.5±1.0 and WW_U=81.2±1.0), consists of the sentences only containing projective dependencies with the heads residing on the right side of the dependents. We obtained an AS_U=78.3±0.3, AS_L=68.9±0.2 and WW_U=85.5±1.0 on the same dataset again by using ten-fold cross-validation. Using the optimized model but omitting all lexical features resulted in AS_U=76.1±0.3, AS_L=65.9±0.4 and WW_U=82.8±1.2, which shows that the improvement in accuracy cannot be attributed to lexicalization alone. The second subset is the Turkish dataset of the CoNLL-X Shared Task on Multi-lingual Dependency Parsing [26]. We obtained an AS_U=75.82 and AS_L=65.68 which are the best reported accuracies on this dataset.

When we make a detailed analysis on individual dependency types, we see that the parser cannot find labeled dependencies for the types that have fewer than 100 occurrences in the treebank, with the single exception of RELATIVIZER, which is generally the enclitic "ki", written separately from the word it attaches to. Since this dependency type always occurs with the same particle, there is no sparse data problem. If we exclude the low-frequency types, we can divide the results into three main groups. The first group consists of determiners, particles and noun phrases which have an AS_U score over 79% and which are found within the closest distances. The second group mainly contains subjects, objects and different kinds of adjuncts, with a score in the range 55–79% and a distance of 1.8–4.6 IGs to their head. This is the group where inflectional features are most important for finding the correct dependency. The third group contains distant dependencies with a much lower accuracy. These are generally relations like sentence modifier, vocative, and apposition, which are hard to find for the parser because they cannot be differentiated from other nominals used as subjects, objects or normal modifiers. Another construction that is hard to parse correctly is coordination, which may require a special treatment.

9 Conclusion

Turkish is a language characterized by flexible constituent order and a very rich, agglutinative morphology. In this paper, we have shown that the accuracy achieved in parsing Turkish with a deterministic data-driven parser can be improved substantially by using inflectional groups as tokens, and by making extensive use of inflectional and lexical information in predicting the next parser action. Combining these techniques leads to the highest reported accuracy for parsing the Turkish Treebank.

However, besides showing that morpholexical information may improve parsing accuracy for languages with rich morphology and flexible word order, the experiments also reveal that the impact of both morphological and lexical information is not uniform across different linguistic categories. We believe that a more fine-grained analysis of the kind initiated in this paper may also throw light upon the apparently contradictory results reported in the literature, especially concerning the value of lexicalization for different languages.

Acknowledgments

This work is partially supported by a research grant from TÜBİTAK (The Scientific Technical Research Council of Turkey).

References

1. Collins, M.: Head-Driven Statistical Models for Natural Language Parsing. PhD thesis, University of Pennsylvania (1999)
2. Collins, M., Hajic, J., Ramshaw, L., Tillmann, C.: A statistical parser for Czech. In: Proc. of ACL-1999. (1999) 505–518
3. Bikel, D., Chiang, D.: Two statistical parsing models applied to the Chinese treebank. In: Proc. of the Second Chinese Language Processing Workshop. (2000) 1–6
4. Dubey, A., Keller, F.: Probabilistic parsing for German using sister-head dependencies. In: Proc. of ACL-2003. (2003) 96–103
5. Levy, R., Manning, C.: Is it harder to parse Chinese, or the Chinese treebank? In: Proc. of ACL-2003. (2003) 439–446
6. Corazza, A., Lavelli, A., Satta, G., Zanoli, R.: Analyzing an Italian treebank with state-of-the-art statistical parsers. In: Proc. of the Third Workshop on Treebanks and Linguistic Theories (TLT). (2004) 39–50
7. Klein, D., Manning, C.D.: Accurate unlexicalized parsing. In: Proc. of ACL-2003. (2003) 423–430
8. Arun, A., Keller, F.: Lexicalization in crosslinguistic probabilistic parsing: The case of French. In: Proc. of ACL-2005. (2005) 302–313
9. Eryiğit, G., Oflazer, K.: Statistical dependency parsing of Turkish. In: Proc. of EACL-2006. (2006) 89–96
10. Kudo, T., Matsumoto, Y.: Japanese dependency analysis using cascaded chunking. In: Proc. of Conll-2002. (2002) 63–69
11. Yamada, H., Matsumoto, Y.: Statistical dependency analysis with support vector machines. In: Proc. of IWPT-2003. (2003) 195–206
12. Nivre, J., Scholz, M.: Deterministic dependency parsing of English text. In: Proc. of COLING-2004. (2004) 64–70
13. Nivre, J., Hall, J., Nilsson, J.: Memory-based dependency parsing. In: Proc. of Conll-2004. (2004) 49–56

14. Nivre, J., Nilsson, J.: Pseudo-projective dependency parsing. In: Proc. of the ACL-2005. (2005) 99–106
15. Bozşahin, C.: Gapping and word order in Turkish. In: Proc. of the 10th International Conference on Turkish Linguistics. (2000)
16. Oflazer, K., Say, B., Hakkani-Tür, D.Z., Tür, G.: Building a Turkish treebank. In Abeille, A., ed.: Building and Exploiting Syntactically-annotated Corpora. Kluwer Academic Publishers (2003)
17. Oflazer, K.: Dependency parsing with an extended finite-state approach. Computational Linguistics **29**(4) (2003)
18. Nivre, J.: An efficient algorithm for projective dependency parsing. In: Proc. of IWPT 2003. (2003) 149–160
19. Black, E., Jelinek, F., Lafferty, J.D., Magerman, D.M., Mercer, R.L., Roukos, S.: Towards history-based grammars: Using richer models for probabilistic parsing. In: Proc. of the 5th DARPA Speech and Natural Language Workshop. (1992) 31–37
20. Veenstra, J., Daelemans, W.: A memory-based alternative for connectionist shift-reduce parsing. Technical Report ILK-0012, Tilburg University (2000)
21. Nivre, J.: Inductive Dependency Parsing. Springer (2006)
22. Nivre, J.: Incrementality in deterministic dependency parsing. In Keller, F., Clark, S., Crocker, M., Steedman, M., eds.: Proc. of the Workshop on Incremental Parsing: Bringing Engineering and Cognition Together (ACL). (2004) 50–57
23. Vapnik, V.N.: The Nature of Statistical Learning Theory. Springer (1995)
24. Sagae, K., Lavie, A.: A classifier-based parser with linear run-time complexity. In: Proc. of IWPT-2005. (2005) 125–132
25. Chang, C.C., Lin, C.J.: LIBSVM: A Library for Support Vector Machines. (2001) Software available at http://www.csie.ntu.edu.tw/~cjlin/libsvm.
26. Buchholz, S., Marsi, E., Krymolowski, Y., Dubey, A., eds.: Proc. of the CoNLL-X Shared Task: Multi-lingual Dependency Parsing, New York, SIGNLL (2006)

Pattern Dictionary Development Based on Non-compositional Language Model for Japanese Compound and Complex Sentences

Satoru Ikehara[1], Masato Tokuhisa[1], Jin'ichi Murakami[1], Masashi Saraki[2], Masahiro Miyazaki[3], and Naoshi Ikeda[4]

[1] Tottori University, Tottori-city, 680-8552 Japan
{ikehara, tokuhisa, murakami}@ike.tottori-u.ac.jp
[2] Nihon University, Tokyo, 101-0061 Japan
saraki@st.rim.or.jp
[3] Niigata University, Niigata-city, 950-2102 Japan
miyazaki@ie.niigata-u.ac.jp
[4] Gifu University, Gifu-city, 501-1112 Japan
ikeda@info.gifu-u.ac.jp

Abstract. A large-scale sentence pattern dictionary (SP-dictionary) for Japanese compound and complex sentences has been developed. The dictionary has been compiled based on the *non-compositional language model*. Sentences with 2 or 3 predicates are extracted from a Japanese-to-English parallel corpus of 1 million sentences, and the compositional constituents contained within them are generalized to produce a SP-dictionary containing a total of 215,000 pattern pairs. In evaluation tests, the SP-dictionary achieved a syntactic coverage of 92% and a semantic coverage of 70%.

Keywords: Pattern Dictionary, Machine Translation, Language Model.

1 Introduction

A wide variety of MT methods are being studied [1, 2, 3], including *pattern-based MT* [4, 5], *transfer methods*, and *example-based MT* [6, 7, 8], but it is proving to be difficult to obtain high-quality translations for disparate language groups such as English and Japanese. *Statistical MT* have been attracting some interest recently [9, 10, 11], but it is not easy to improve the quality of translations. Most practical systems still employ the *transfer method*, which is based on *compositional semantics*. A problem with this method is that it produces translations by separating the syntactic structure from the semantics and is thus liable to lose the meaning of the source text.

Better translation quality can be expected from *pattern-based MT* where the syntactic structure and semantics are handled together. However, this method requires immense pattern dictionaries which are difficult to develop, and so far this method has only been employed in hybrid systems [12, 13] where small-scale pattern dictionaries for specific fields are used to supplement a conventional *transfer method.*

Y. Matsumoto et al. (Eds.): ICCPOL 2006, LNAI 4285, pp. 509 – 519, 2006.

Example-based MT has been expected to resolve this problem. This method obtains translations by substituting semantically similar elements in structurally matching translation examples, hence there is no need to prepare a pattern dictionary. However, the substitutable elements depend on translation examples. This made it impossible to judge them at real time. This problem could be addressed by manually tagging each example beforehand, but the resulting method would be just another *pattern-based MT*.

This problem [14] has been partially resolved by a highly comprehensive valency pattern dictionary called *Goi Taikei* (A-Japanese-Lexicon) [15]. This dictionary contains 17,000 pattern pairs for the semantic analysis in the Japanese-to-English MT system ALT-J/E [16]. High quality translations with the accuracy of more than 90% has been performed for simple Japanese sentences, but there are still cases where a suitable translated sentence structure cannot necessarily be obtained. A valency pattern expresses the semantic relationship between independent words. The meaning of subordinate words (particles, auxiliary verbs, etc.) is dealt with separately, hence the original meaning is sometimes lost. Addressing this problem requires a mechanism that deals with the meaning of subordinate words within the sentence structure as a whole.

In order to realize such a mechanism, we propose a language model that focuses on the non-compositional expressions, and a method for creating patterns based on this model. This method obtains pattern pairs from parallel corpus by the semi-automatic generalization of compositional constituents.

2 Non-compositional Language Model

2.1 Compositional Constituents and Non-compositional Constituents

In the framework of expressions that come to mind during the process where a speaker is forming a concept, there are two types of constituents to consider. One is those that cause the overall meaning to be lost when they are substituted with other alternative constituents. And the other is those that do not cause the overall meaning to be lost. The former are referred to as *N-constituents (Non-compositional constituents)*, and the latter are referred to as *C-constituents (Compositional-constituents)*.

Definition 1: C- constituents* and *N-constituents
C-constituent is defined as a constituent which is interchangeable with other constituents without changing *the meaning of an expression structure*. All other constituents are *N-constituents*.

Definition 2: C-expressions* and *N-expression
C-expression (Compositional expression) is defined as an expression consisting of C-constituents, and N-expression (Non-compositional expression) is defined as an expression comprising one or more N-constituents.

Where a *constituent* is a part of an *expression* consisting of one or more words, one *constituent* can constitute one *expression*.

Before applying these definitions to actual linguistic expressions, *the meaning of an expression structure* is needed to be defined. Although a great deal of research has been made concerning the meaning of linguistic expressions, any statement is nothing more than a symbol as far as processing by a computer is concerned, and hence we just need to express meanings in a system of symbols that is free from semantic inconsistencies. In this study, considering applications to Japanese-to-English MT, *the meaning of expression structures* is defined in terms of an English expression.

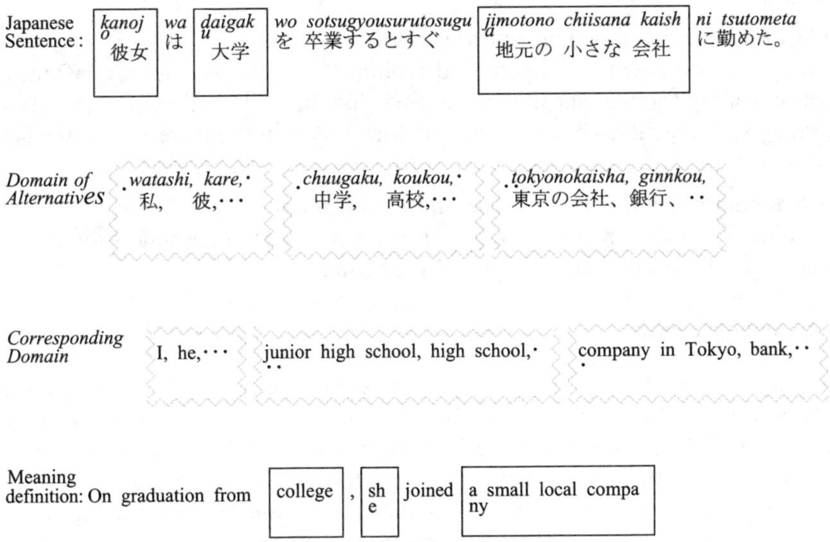

Fig. 1. Example of C-constituents

In Figure 1, the source sentence is a Japanese expression expressing a relationship between two events. *The meaning of the expression structure* is *Immediately after performing one action, somebody performed the other action.* This meaning is defined by using the English expression. For the constituent such as 彼女 *(she)*, 大学 *(college)* and 地元の小さな会社 *(small local company)*, there is a domain of substitutable constituents that doesn't change *the meaning of the expression structure*, therefore these are C-constituents.

2.2 Characteristics of C-*Constituents*

From the above definitions, it can be pointed out that a C-constituent possesses the following four important characteristics. From these characteristics, it is possible to obtain important guidelines for pattern-forming.

(1) Language pair dependence of C-constituent

Since one linguistic expression is used to define the meaning of another, the number and scope of C-constituents depends on the language pair. For languages that belong to the same group, the scope of C-constituents is large, while for disparate language groups it is expected to be smaller, as reflected in the different levels of difficulty of translating between the languages.

(2) Finite choice for alternative constituents

Although C-constituents can be substituted, that does not mean they can be substituted with anything at all. The range that can be substituted is limited both grammatically and semantically, thus this must be indicated in the pattern as the "domain" of the constituent.

(3) C-constituent dependent on constituent selection

The scope of constituents is determined arbitrarily. Hence whether a constituent is compositional or non-compositional depends on how the constituent is chosen. Accordingly, to obtain general-purpose patterns, it is better to increase the number of C-constituents.

(4) Simultaneity of a C-constituent and an N-expression

A so-called C-constituent is only compositional when seen in the context of the entire expression, and itself may actually be a N-expression.

2.3 Language Model

According to definition 1, a linguistic expression consists of C-constituents and N-constituents. According to characteristic (3), if we select a C-constituent from an expression with a meaningful range (e.g., word, phrase or clause), a C-constituent may itself also be an N-expression according to characteristic (4). Consequently a linguistic expression can generally be expressed with the language model shown in Fig. 2.

As this figure shows, when C-constituents are repeatedly extracted from N-expressions, the end result is an N-expressions that contains no C-constituents. Although the resulting N-expression may just be a single word, it could also be an idiomatic phrase that has no substitutable constituents. Thus, in this language model, linguistic expressions can be articulated into one or more N-expressions and zero or more N-constituents.

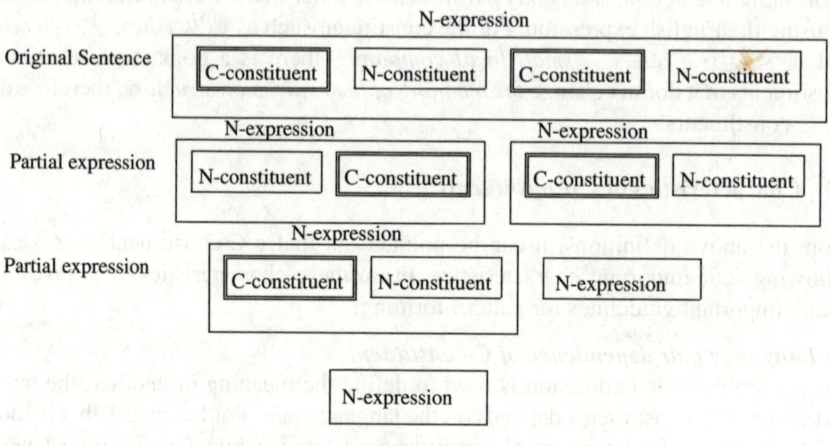

Fig. 2. Non-compositional language model

2.4 Patterns for N-Expressions

An important aspect of the language model is that the N-expressions that appear at each stage of the articulation are meaningful expression units. In this element decomposition process, loss of the original meaning can be avoided by using a semantic dictionary for N-expressions at each stage. For example, if linguistic expressions are classified into sentences, clauses and phrases, and semantic dictionaries are constructed for N-expressions at each of these levels, then this would constitute the bulk of a mechanism for assimilating the meaning of entire sentences.

It is thought that patterns are a suitable framework for expressing the syntactic structure of N-expressions, because:

(a) a N-constituent cannot be substituted with another constituent, thus a literal description is appropriate, and
(b) the order in which C- and N-constituents appear is often fixed, thus there is thought to be little scope for variation.

Therefore, in this study we will use a pattern-forming approach for meaningful N-expressions.

3 Development of SP (Sentence Pattern)-Dictionary

According to our language model, three kind of expression patterns (compound and complex sentence patterns, simple sentence patterns and phrase patterns) will be almost sufficient to cover Japanese expressions.

In this study, complex and compound sentences were targeted because the *Goi Taikei* [15] can gives good translations for most of simple sentences. But, complex and compound sentences are very difficult to obtain good translation results by the conventional MT systems. The number of predicates was limited to 2 or 3 because it is thought that complex and compound sentences with four or more predicates can often be interpreted by breaking them down into sentences with three predicates or fewer.

3.1 The Principles of Pattern-Forming

The Japanese-English parallel corpus is a typical example where the meaning of Japanese expressions is defined with English expressions. And when translation example is considered, the following two types of C-constituents can occur:

(1) cases where there is a constituent in the English expression that corresponds to a constituent in the Japanese expression, and
(2) cases where a constituent in the Japanese expression has no corresponding constituent in the English expression, but deleting this constituent from the Japanese expression does not cause any change in the corresponding English expression.

SP pairs were therefore produced by extracting components corresponding to these two cases from parallel corpus, and generalizing the results.

3.2 SP Generation Procedure

First, a parallel corpus was created by collecting together a sentence pair of 1 million basic Japanese sentences. From this corpus, 150,000 translation examples for compound and complex sentences with two or three predicates were extracted. Then, using resources such as Japanese-English word dictionaries, the semantic correspondence relationships between the constituents were extracted and converted into *variables*, *functions, symbols* in the following three stages to produce a SP-dictionary.

- ***Word-level generalization:*** compositional independent words (nouns, verbs, adjectives, etc.) are replaced by *variables*.
- ***Phrase-level generalization:*** compositional phrases (noun phrases, verb phrases, etc.) are replaced by *variables*.
- ***Clause-level generalization:*** compositional clauses (adnominal clauses and continuous clauses) are replaced by *variables*.

For C-constituents that can be semi-automatically recognized as such, the generalization is also performed semi-automatically.

3.3 Examples of SPs

An example of a SP is shown in Table 1. The meanings of the *variables*, *functions,* etc. used in this table are shown below.

***Word-level SPs*:** ①N1, N3, N4: *Noun variables.* ②V2, V5: *Verb variables.* Here, attached bracket represents semantic attribute numbers specifying semantic constraints on a variable. ③#1[...]: Omissible constituents. ④ /: Place of a constituent that need not appear. ⑤.*tekita*: Function for specifying a predicate suffix. ⑥AJ(V2): Adjectival form of the value of verb variable V2. ⑦N1ˆposs: Value of N1 transformed into possessive case.

***Phrase-level SPs*:** ①NP1: Noun phrase variable.

***Clause-level SPs*:** ①CL1: *Clause variable.* ②*so+that* (..., ...): *A sentence generation function* for *so that* sentence structure. ③subj(CL): Function that extracts the subject from the value of *a clause variable.*

3.4 The Number of Different SPs

Table 2 shows the number of SPs in the resulting SP-dictionary and the number of constituents replaced by *variables* at each level of generalization.

In Table 2, compared to the number of SPs of *word-level* and *phrase-level* SPs, the number of *clause level* SPs was particularly small. This indicates that most of the

Table 1. Examples of generated SPs

word-level **SP**	
Japanese SP	ha te wo ni #1[N1(G4)は]/V2(R3003)て/N3(G932)を/N4(G447)に/V5(R1809).*tekita*。
English SP	[N1|I] was so AJ(V2) as to V5 #1[N1^*poss*] N3 at N4.
Example	*ukkarisite teikikennwo ieni wasuretekita* うっかりして 定期券を 家に 忘れてきた。 I was so careless as to leave my season ticket at home.
phrase-level **SP**	
Japanese SP	ha ni nodakara NP1(G1022)は/V2(R1513).*ta*/N3(G2449)に/V4(R9100).*teiru*のだから/N5 (N1453).*dantei*。
English SP	NP1 is AJ(N5) in that it V4 on AJ(V2) N3.
Example	*sonoketsuronwa ayamattazenteini motozuite irunodakara ayamaridearu* その結論は 誤った前提に 基づいて いるのだから 誤りである。 The conclusion is wrong in that it is based on a false premise.
clause-level **SP**	
Japanese SP	node niatatteha CL1(G2492).*tearu*ので、N2(G2005)に当たっては/VP3(R3901).*gimu*
English SP	*so+that* (CL1, VP3.*must.passive* with subj (CL1)^*poss* N2)
Example	*sorewa kiwamete yuudokude arunode siyouniatattewa juunibunni* それは 極めて 有毒であるので、使用に当たっては 十二分に *chuuisinakerebanaranai* 注意しなくてはならない。 It is significantly toxic so that great caution must be taken with its use

Table 2. Number of different SPs and Ratio of C-constituents

Type of SPs	*word-level*	*phrase-level*	*clause-level*	Total
No. of pattern pairs	122,642 pairs	80,130 pairs	12,450 pairs	215,222 pairs
Ratio of C-constituent s	472,521/763,968 = 62 %	102,000/463,636 = 22 %	11,486/267,60 1 = 4.3 %	----

clauses in the parallel corpus are N-constituents which are impossible to generalize. The proportion of generalized C-constituents were 62% at the *word level* and 22% at the *phrase level*, but just 4.3% at the *clause level*.

For N-constituents, a semantically suitable translated result cannot be obtained when the constituent is extracted, translated and incorporated into the original sentence. Looking at the parallel corpus, most of the English translations of Japanese compound and complex sentences are simple sentences whose structures are very diverse. Regarding the results of Table 2, in the case of Japanese-to-English MT, high-quality translations cannot be achieved by conventional MT method based on *compositional semantics*.

4 The Coverage of the SP-Dictionary

4.1 Experimental Conditions

A *pattern parser* that compares input sentences against the SP-dictionary was used to evaluate the coverage of the SP-dictionary. The experiments were conducted by *cross-validation* manner and ten thousand input sentences were used. These were randomly selected from the example sentences used for creating the SPs. Since the input sentences will always match the SPs from which they were created, matches of this type were ignored and the evaluation was restricted matches to other SPs.

An input sentence many times matches to more than one SP and not all of them are necessarily correct. Therefore, the coverage was evaluated according to the following four parameters:

- ***Matched pattern ratio* (R):** The ratio of input sentences that are matched to at least one SP (*syntactic coverage*)
- ***Precision* (P1):** The ratio of matched SPs that are semantically correct
- ***Cumulative precision* (P2):** The ratio of matched SPs for which there is one or more semantically correct SP
- ***Semantic coverage* (C):** The ratio of input sentences for which there is one or more semantically correct SP (R×P2)

4.2 Saturation of *Matched Pattern Ratio*

Fig. 3 shows the relationship between the number of SPs and the *matched pattern ratio*. As you can see, there is a pronounced tendency for the *matched pattern ratio* to become saturated. When the SPs on the horizontal axis are rearranged in order of their frequency of appearance, the rate of saturation becomes about 5 times faster.

According to the previous study [17], the number of valency patterns required to more or less completely cover all simple sentences was estimated to be somewhere in the tens of thousands. We can say that the number of required SPs for complex and compound sentences is also expected to converge somewhere in the tens of thousands or thereabouts.

4.3 *Matched Pattern Ratio* and *Precision*

Table 3 shows the evaluation results. It was shown that 91.8% of the input sentences are covered syntactically by the whole dictionary. However, there were also many cases of

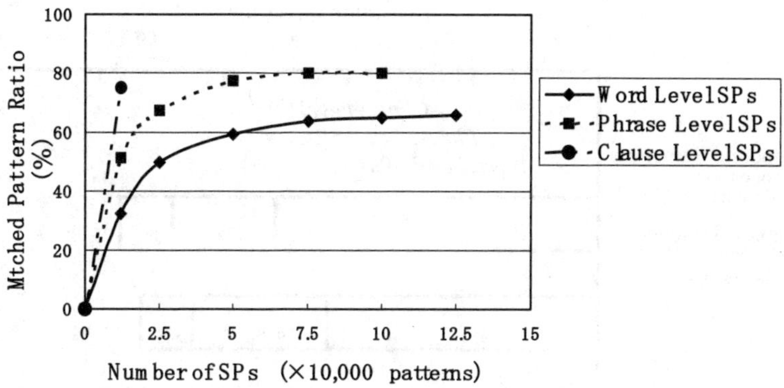

Fig. 3. Saturation of *Matched pattern ratio*

matches to semantically inappropriate SPs, and the semantic coverage decreased to 70% when these were eliminated. The number of *clause-level SPs* was just one tenth the number of *word-level SPs*, but had comparatively high coverage.

Table 3. Coverage of SP-dictionary

Type of SPs	R (*Matched pattern ratio*)	P1 (*Precision*)	P2 (*Cumulative precision*)	C=RxP2 (*Semantic coverage*)
Word Level	64.7 %	25 %	67 %	43.3 %
Phrase Level	80.0 %	29 %	69 %	55.2 %
Clause Level	73.7 %	13 %	68%	50.1 %
Total	91.8 %	——	——	70 %

4.4 *Semantic Coverage*

Since semantic ambiguity is small in the order of *word-level, phrase-level* and *clause-level SPs,* it is probably better to select and use the most semantically appropriate SP based on this sequence. Fig. 4 shows the ratio of SPs that are used when they are selected based on this sequence.

As Fig. 4 shows, about 3/4 of the meanings of Japanese compound and complex sentences are covered by the SP-dictionary. When MT is performed using the SP-dictionary, it is estimated that word-level SPs will be used for about half of the complex and compound sentences, while phrase-level and clause-level SPs will be applied to the other half.

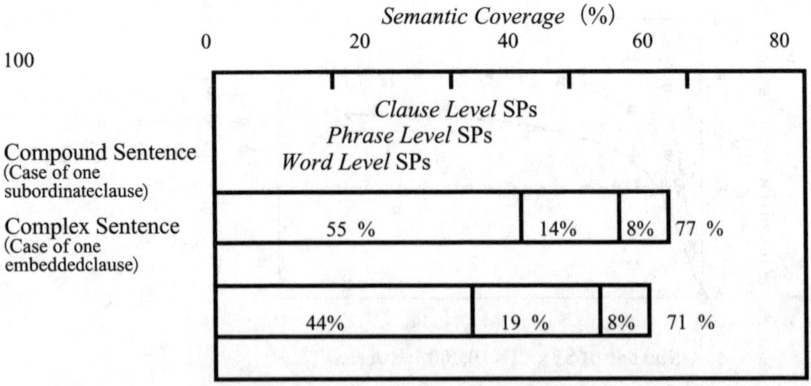

Fig. 4. *Semantic coverage* of SP-dictionary

5 Concluding Remarks

An Non-compositional language model was proposed and, based on this model, a sentence pattern dictionary was developed for Japanese compound and complex sentences. This dictionary contains 123,000 *word-level*, 80,000 *phrase-level* and 12,000 *clause-level* sentence pattern pairs (215,000 in total).

According to the results, the compositional constituents that could be generalized were 62% for independent words, 22% for phrases, whereas only 4.3% for clauses. This result shows that in Japanese-to-English MT hardly any Japanese compound and complex sentences can be translated into English as shown in a parallel corpus when they are translated by separating them into multiple simple sentences and then recombined.

Also, in evaluation tests of a SP-dictionary, the syntactic coverage was found to be 92%, while the semantic coverage was 70%. It is therefore proved that the SP-dictionary is very promising for Japanese to English MT.

Acknowledgements

This study was performed with the support of the *Core Research for Evolutional Science and Technology* (CREST) program of the *Japan Science and Technology Agency* (JST). Our sincere thanks go out to everyone concerned and to all the research group members.

References

1. Nagao. M.: Natural Language Processing, Iwanami Publisher (1996)
2. Ikehara. S.: Machine Translation, in Information Processing for Language, Iwanami Publisher (1998) 95-148

3. Tanaka, H. (Eds): Natural Language Processing - Fundamentals and Applications, Iwanami Publisher (1998)
4. Takeda, K.: Pattern-based Machine Translation, COLING, Vol. 2 (1996) 1155-1158
5. Watanabe, H. and Takeda, K.: A Pattern-based machine translation system extended by example based processing, COLING (1998) 1369-1373
6. Nagao, M.: A Framework of a Mechanical Translation between Japanese and English by Analogy Principle, in Artificial and Human Intelligence, North-Holland (1984) 173-180
7. Sato, S.: An example based translation and system, COLING (1992) 1259-1263
8. Brown, R. D.: Adding Linguistic Knowledge to a Lexical Example-Based Translation System, TMI 99 (1999) 22-32
9. Brown, P. F., John, C. S., Pietra, D., Jelinek, F. J., Lfferty, D. , Mercar, R. L. and Roossin, P. S.: A Statistical Approach to Machine Translation, Computational Linguistics, Vol. 16, No. 2 (1990) 79-85
10. Watanabe, T. and Sumita, E.: Bi-directional Decoding for Statistical Machine Translation, COLING (2002) 1075-1085
11. Vogel, S., Zhang, Y., Huang, F., Tribble, A., Venugopal, A., Zhao, B. and Waibel, A.: The CMU statistical machine translation system. MT Summit IX (2003) 402-409
12. Jung, H., Yuh, S., Kim, T., Park, S.: A Pattern-Based Approach Using Compound Unit Recognition and Its Hybridization with Rule-Based Translation, Computational Intelligence, Vol. 15, No. 2 (1999) 114-127
13. Uchino, H., Shirai, S., Yokoo, A., Ooyama, Y. and Furuse, K.: News Flash Translation System of ALTFLASH, IEICE Transactions, Vol. J84-D-II, No. 6 (2001) 1168-117
14. Ikehara, S.: Challenges to basic problems of NLP, J. of JSAI, Vol. 16, No. 3 (2001) 522-430
15. Ikehara, S., Miyazaki, M., Shirai, S., Yokoo, A., Nakaiwa, H., Ogura, K., Ooyama, Y. and Hayashi, Y.: *Goi Taikei* (A-Japanese Lexicon), Iwanami Publisher (1997)
16. Ikehara, S., Miyazaki, M., Shirai, S. and Hayashi, Y.: Speaker's conception and multi-level MT, J. of IPSJ, Vol. 28, No. 12 (1987) 1269-1279
17. Shirai, S., Ikehara, S., Yokoo, A. and Inoue, H.: The quantity of valency pattern pairs required for Japanese to English MT and their compilation. NLPRS '95, Vol. 1, (1995) 443-448.

A Case-Based Reasoning Approach to Zero Anaphora Resolution in Chinese Texts

Dian-Song Wu and Tyne Liang

Department of Computer Science
National Chiao Tung University, Hsinchu, Taiwan
{diansongwu, tliang}@cs.nctu.edu.tw

Abstract. Anaphora is a common phenomenon in discourses as well as an important research issue in the applications of natural language processing. In this paper, both intra-sentential and inter-sentential zero anaphora in Chinese texts are addressed. Unlike general rule-based approaches, our resolution method is embedded with a case-based reasoning mechanism which has the benefit of knowledge acquisition if the case size varies. In addition, the presented approach employs informative features with the help of two outer knowledge resources. Compared to rule-based approaches, our resolution to 1047 zero anaphora instances achieved 82% recall and 77% precision.

1 Introduction

Anaphoric relations include using noun phrases, pronouns, or zero anaphors to stand for previously mentioned nominal referents [1,2]. Therefore, anaphora resolution has become an indispensable part in message understanding as well as knowledge acquisition.

In the past literature, different strategies to identify antecedents of an anaphor have been presented by using syntactic, semantic and pragmatic clues [3,4,5]. In addition, a corpus-based approach is proposed by Dagan and Itai [6]. However, a large corpus is needed for acquiring sufficient co-occurring patterns and for dealing with data sparseness. Recently, outer resources like WordNet are applied for enhancing the semantic verification between anaphors and antecedents [7,8]. Nevertheless, using WordNet alone for acquiring semantic information is not sufficient for solving unknown words. To tackle this problem, a richer resource from the Web was exploited [9]. Anaphoric information is mined from Google search results at the expense of less precision.

Contrast to rich studies in English texts, efficient Chinese anaphora resolution has not been widely addressed. Recently, Wang et al. [10] presented an event-based partial parser and an efficient event-based reasoning mechanism to tackle anaphora and ellipsis appearing in primary school's mathematical problem description sentences. On the other hand, Wang et al. [11] presented a simple rule-based approach to handle nominal and pronominal anaphora in financial

Y. Matsumoto et al. (Eds.): ICCPOL 2006, LNAI 4285, pp. 520–531, 2006.

texts. However, any rule-based approach is essentially lacks of portability to other domains. Yeh and Chen [12] resolved zero anaphora by using centering theory and constraint rules. Though they had 65% F-score, yet they tackle the inter-sentential zero anaphors only. Contrast to the shallow parsing used by Yeh and Chen [12], Converse [13] used full parsing results from Penn Chinese Treebank for pronominal and zero anaphora resolution. Since less features were used at resolving intra-sentential zero anaphora, the overall resolution is not promising.

Xu [2] reported that zero anaphora is the most common phenomenon than pronominal and nominal anaphora in Chinese written texts. The zero anaphors can be in single sentence or in consecutive sentences. In this paper, the presented resolution is aimed to tackle both inter and intra sentential zero anaphora by exploiting more semantic information with the help of outer knowledge resources. The animate relation between nouns and verbs is acquired from the resources to improve the presented anaphora resolution. The kernel resolution module is embedded with a case-based reasoning (CBR) mechanism for its benefits in new knowledge acquisition [14]. Essentially such incremental learning approach is able to achieve optimal performance. In our design, the discovery of similar cases allows us to identify reference patterns and select possible antecedents. The experimental results on 1047 instances show that the presented approach can yield promising zero anaphora resolution in terms of 82% recall and 77% precision.

The remaining part of this paper is organized as follows. Section 2 introduces the Chinese sentence structures and zero anaphora. Section 3 describes the overall procedure of the proposed resolution in details. Section 4 gives the experimental results and analysis from different aspects. Section 5 is the final conclusions.

2 Chinese Sentence Structure and Zero Anaphora

In the survey of Xu [2], there are five basic types of simple sentence structures in Chinese. A simple sentence is defined to be a sentence bounded by punctuation marks like "，, 。, ；, ：, ！, ？" [15,16]. Several consecutive simple sentences form a complex sentence. In the following examples, we describe the sentence structures and the zero anaphors denoted by "ϕ". The intra-sentential cases are in Examples 2 and 3; inter-sentential cases are in other examples of complex sentences.

Example 1. SVO sentence: It can be "**subject** + transitive-verb + **object**" and "**subject** + intransitive-verb". Zero anaphors may occur in the position of **subject** or **object**.

李 先生 買 了 一些 蘋果，他 的 孩子們 都 吃完 ϕ 了。
Li xiansheng mai le yixie pingguo，ta de haizimen du chiwan ϕ le。
(Mr. Li bought some apples. His children ate (apples) up.)

Example 2. Serial verb sentence: A serial verb construction contains several verbs in a simple sentence, expressing simultaneous or consecutive actions. All verbs in the sentence have the same grammatical subject.

李四 參加 游泳 比賽 ϕ 贏得 冠軍。
Lisi canjia youyang bisai ϕ yingde guanjun。
(Lisi took part in the swimming contest, and (he) won the champion.)

Example 3. Pivotal sentence: A sentence is called a pivotal sentence if its predicate consists of two verb phrases and the object of the first verb is functioned as the subject of the second verb.

張三 通知 李四 ϕ 出席 明天 的 會議。
Zhangsan tongzhi Lisi chuxi mingtian de huiyi。
(Zhangsan informed Lisi to attend the meeting tomorrow.)

Example 4. Individual noun phrase sentence: The individual noun phrase is likely to be the antecedent of the succeeding zero anaphor.

總理 斯洛德，ϕ 宣布 德國 將 舉行 議會 選舉。
Zongli Siluode，ϕ xuanbu deguo jiang juxing yihui xuanju。
(The premier Schroeder declared that Germany will hold a councilor election.)

Example 5. Prepositional phrase sentences: When a prepositional phrase forms a simple sentence, a zero anaphor is likely to occur in front of the preposition.

人 的 生活 空間，ϕ 和 自然 環境 發生 了 對立。
Ren de shenghuo kongjian，ϕ he ziran huanjing fasheng liao duili。
(Human living space has conflict with natural environment.)

3 The Approach

Figure 1 illustrates the proposed resolution procedure including text preprocessing, zero anaphor (ZA) detection and antecedent (ANT) identification process, a case-based reasoning module (CBRM), a case base, and two lexicon resources. Each of these modules is described in the following subsections.

3.1 Text Preprocessing

The corpus we used in both training and testing phases is the well-known Academia Sinica Balancing Corpus (ASBC) [17]. Every text in the corpus is segmented into sentences, and every word is tagged with its part-of-speech. There are 46 kinds of POS tags used in ASBC Corpus. Using POS features, we construct a finite state machine chunker to chunk each base noun phrase as antecedent candidates.

In the design of the finite state machine, each state indicates a particular POS of a word. The arcs between states mean a word input from the sentence

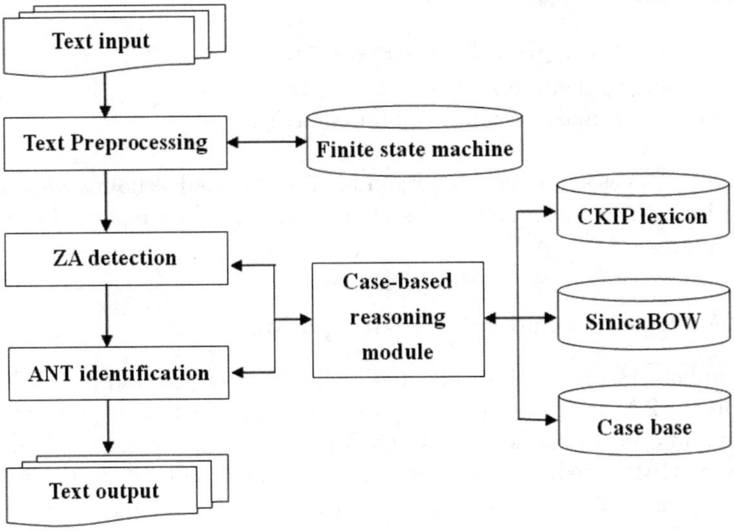

Fig. 1. The presented Chinese zero anaphora resolution procedure

sequentially. If a word sequence can be recognized from the initial state and ends in a final state, it is accepted as a base noun phrase with no recursion, otherwise rejected.

In addition, unlike western languages, Chinese grammar has little inflection information. Chinese verbs appear in the same form no matter whether they are used as nouns, adjectives, or adverbs [18]. The following examples show the usage for each of four cases by the word "放鬆" ("relax").

Example 6. Verb case:

時常(D)**放鬆**(VHC)自己(Nh)的(DE)情緒(Na)。
Shichang fangsong ziji de qingxu。
(A person must often relax his own emotion.)

Example 7. Noun case:

這(Nep)直接(VH)影響到(VJ)心情(Na)的(DE)**放鬆**(VHC)。
Zhe zhijie yingxiangdao xinqing de fangsong。
(This will influence the release of mood directly.)

Example 8. Adjective case:

在(P)人體(Na)**放鬆**(VHC)的(DE)狀態(Na)，
Zai renti fangsong de zhuantai，
(In the relaxed condition of body,)

Example 9. Adverb case:

環境衛生(Na)必須(D)毫不(D)放鬆(VHC)地(DE)改善(VC)。
Huanjingweisheng bixu haobu fangsong de gaishan。
(Sanitation must be improved unrelaxedly.)

Therefore, verbs as described in Example 7 are treated as nouns while performing chunking phase. Moreover, verbs in adjective cases are regarded as modifiers of noun phrases.

3.2 ZA Detection and ANT Identification

Zero anaphora resolution involves zero anaphor detection and antecedent identification. In ZA detection phase, verbs are examined to assign corresponding subjects and objects. If there is any omission, the ZA detection will submit the sentence to CBRM to decide whether there is a ZA or not. If yes, the ANT identification phase will be performed by using resolution template returned from CBRM.

In ZA detection phase, it must be noted that there are three conditions should be ignored while detecting ZA around verbs [19]. The conditions are described in the following examples. For instance, the object of verb "完成" (finish) in Example 10 is shifted to the position after the word "把" (Ba).

Example 10. "把" (Ba) sentence:

張三(Nb)已經(D)把(P)工作(Na)完成(VC)。
Zhangsan yijing ba gongzuo wancheng。
(Zhangsan finished the work.)

Example 11. "被" (Bei) sentence:

工作(Na)已經(D)被(P)張三(Nb)完成(VC)。
Gongzuo yijing bei Zhangsan wancheng。
(The work was finished by Zhangsan.)

Example 12. In adverb case: when the verb functions as a part of adverb as described in section 3.1, it would not be the related verb to a ZA.

3.3 Case-Based Reasoning Module

Case-based reasoning has been successful in the applications like legal case retrieval. It is an incremental and sustained learning since a new experience is retained each time a problem has been solved. In the application of the presented ZA resolution, a set of similar sentences measured by a given similarity function will be retrieved from the pre-constructed case base to detect whether an input sentence has a ZA or not. The most similar example will be reused if it is a ZA example. If necessary, this retrieved example will be revised to be a final version. Then the new version, its context and resolution will be retained in the case base. Figure 2 illustrates the procedure of the presented CBRM which is in charge of the communication between the ZA resolution and the outer resources. The module functions are summarized as follows:

1. Retrieve the most similar example w.r.t. the input sentence.
2. Reuse the resolution method of the most similar example on the basis of their feature and sentence pattern.
3. Revise the resolution process manually if there is any error.
4. Retain the refined example into case base.

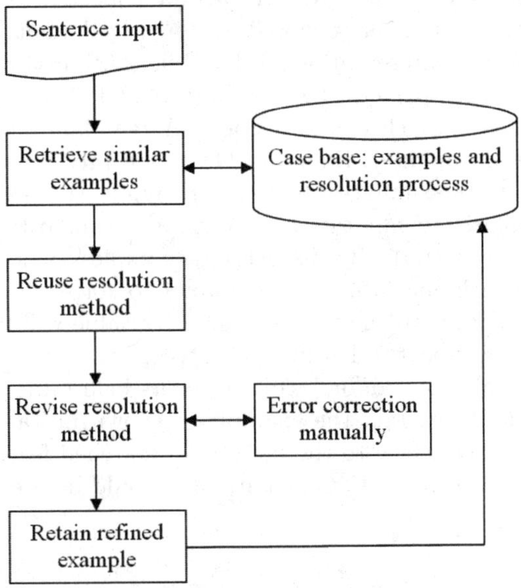

Fig. 2. The procedure of case-based reasoning module

Lexicon Resources. As mentioned above, two outer resources are used during ZA resolution to acquire informative features like animacy attribute of nouns and verbs. The resources are CKIP lexicon [20] and The Academia Sinica Bilingual WordNet (SinicaBOW), both of them were released from The Association for Computational Linguistics and Chinese Language Processing. CKIP lexicon contains 80,000 entries annotated with syntactic categories and semantic roles. For example, "作家" ("writer"), "司令"("commander"),"顧客"("customer") are regarded as animate nouns. SinicaBOW is a Mandarin-English bilingual database based on the frame of English WordNet and language usage in Taiwan. There are four kinds of verbs, as shown in Table 1, as animate verbs, namely, {cognition}, {communication}, {emotion}, and {social}.

Table 1. Examples of animate verbs

Unique Beginners	Verb Examples
{cognition}	思考 (think), 分析 (analyze,), 判斷 (judge)...
{communication}	告訴 (tell), 請求 (ask), 教導 (teach)...
{emotion}	感覺 (feel), 喜歡 (love), 害怕 (fear)...
{social}	參與 (participate), 使得 (make), 設立 (establish)...

Case Representation and the Case Base. In this paper, each case in the case base is represented in Tables 2 and 3 at different phases. At training phase, the case base contains 2738 examples which were collected from ASBC corpus and annotated with ZA marker (denoted as "ϕ") by human experts. Table 2 shows an input case at representation level and its ZA template which contains six features and sentence pattern described in Table 4. Table 3 shows one case stored in the case base which contains both ZA template and ANT template used as ZA resolution method. Equation (1) is the presented similarity function used to compute the similarity between the input case and the stored case examples. The similarity computation concerns the similarity of ZA template features and sentence pattern. By discovering examples analogous to the form of a problem, we can immediately use examples to solve the problem. Therefore, the ANT template with the highest similarity value will be retrieved from the case base and be used to identify the antecedent with respect to a given input sentence. For example, an omission ϕ occurs before the verb "宣佈" (announce) in Table 2. We extract the example as shown in Table 3 to infer the corresponding antecedent. Due to the number of matched features in the ANT template, we can decide that "主席" (chairman) should be the antecedent of ϕ in Table 2.

Table 2. Case representation at ZA detection

Representation Level at the ZA detection	主席(Na)評估(VE)意見(Na)後(Ng)，ϕ宣佈(VE)下(Nes)個(Nf)月(Na)舉行(VC)協調會(Na)。 (After the chairman evaluated the opinions, (he) announced that the reconciliation meeting will be held next month.)
Implementation Level (ZA template)	STR: 宣布 POS: VE ROLE: subject Anim_Verb: Y ORDER: 2 First_Verb: Y PATTERN: VNVN

Table 3. Case representation in the case base

Representation Level at ANT identification	中央(Nc)銀行(Nc)主席(Na)再度(D)調整(VC)政策(Na)，ϕ宣布(VE)下(Nes)季(Nd)降低(VHC)存款(Na)準備率(Na)。 (The chairman of the central bank adjusted the policy again; (he) announced that the preparation rate on deposit will be reduced next quarter.)
Implementation Level (ZA template)	STR: 宣布 ROLE: subject POS: VE Anim_Verb: Y ORDER: 2 First_Verb: Y PATTERN: VNVN
Implementation Level (ANT template)	ROLE: subject POS of Head: Na NUM: single Anim_Noun: Y DEFINITE: N ORDER: 1 REPEAT: N

Table 4. Description of template features

	Feature	Description
ZA template	STR	Word of related verb
	ROLE	Grammatical role of current position: subject, object, or other
	POS	Part-of-Speech of related verb
	Anim_Verb	Y if related verb is an animate verb; else N
	ORDER	ZA occurs in the i-th sentence
	First_Verb	Y if related verb is the first one; else N
	PATTERN	POS sequence that consider only nouns (N), verbs (V), and prepositions (P). For given two sentences S_I and S_C with sentence pattern P_I and P_C respectively, If pattern P_I equal to pattern P_C, then PATTERN(S_I,S_C)=1; Else if pattern P_I is a subsequence of pattern P_C, then PATTERN(S_I,S_C)=0.5; Else PATTERN(S_I,S_C)=0;
ANT template	ROLE	Grammatical role of candidate: subject, object, or other
	POS_Head	Part-of-Speech of candidate head noun
	NUM	Single, plural
	Anim_Noun	Y if candidate is an animate noun; else N
	DEFINITE	Y if candidate is a definite noun phrase; else N
	ORDER	Candidate occurs in the i-th sentence;
	REPEAT	Y if candidate repeat more than once; else N

$$SIM(S_I, S_C) = \frac{MATCH(S_I, S_C) \times \alpha}{number\ of\ features} + PATTERN(S_I, S_C) \times \beta \quad (1)$$

S_I: input sentence
S_C: case sentence
α, β: weighting factors, where $\alpha + \beta = 1$
$MATCH(S_I, S_C)$: number of matched features in case S_I and S_C
$PATTERN(S_I, S_C)$: the value of PATTERN in case S_I and S_C

4 Experiments and Analysis

The presented resolution is justified with narrative report texts selected from the ASBC corpus. Table 5 lists the statistical data of both training and testing corpus. Table 6 shows the performance results in terms of precision and recall at various matching thresholds. It is observed that the optimal performance (in terms of F-score) is achieved when α and β value are 0.7 and 0.3 respectively.

Table 5. Statistical information of training and testing data

	Training data	Testing data
Articles	192	100
Sentences	6,625	3,532
Words	63,682	30,826
Zero Anaphors	2,738	1,047

Table 6. Performance at various thresholds

Threshold α	0.1	0.2	0.3	0.4	0.5	0.6	0.7	0.8	0.9	1.0
Recall	0.42	0.54	0.58	0.61	0.72	0.80	**0.82**	0.80	0.76	0.73
Precision	0.39	0.46	0.47	0.52	0.64	0.76	**0.77**	0.76	0.74	0.74

In order to verify feature impact, a baseline model is built in such a way that only grammatical role feature is used in ANT identification. Figure 3 shows that the highest F-score is obtained when all the ZA template features indicated as 'ALL' are concerned and the baseline yields the worst performance at comparison. It is also noticed that the resolution performance can be enhanced significantly after applying animate feature (denoted as 'A') and sentence pattern mapping (denoted as 'B'). On the other hand, we also verify the sensitivity of the training case size in our presented CBR approach to ZA resolution. It is found from Figure 4 that the feasible performance results can be obtained when training corpus size is close to the testing one. If the training case size is half of the testing case size, the performance may drop 20%.

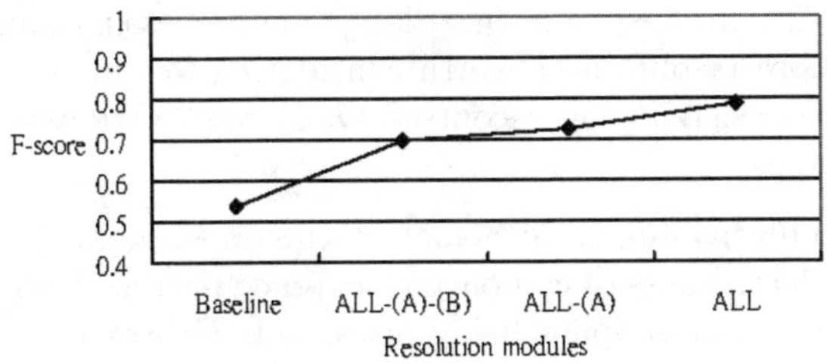

Fig. 3. F-score after applying resolution modules

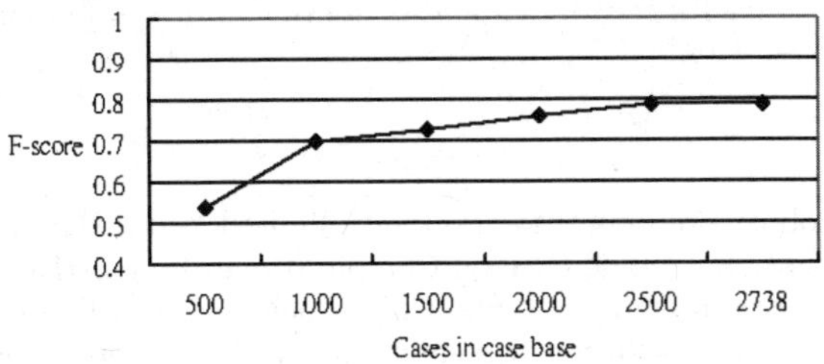

Fig. 4. F-score over different case base scale

While analyzing the ZA resolution, we classified three main sources of errors as follows:

(1) Preprocessing error: there are 6% ZA errors attributed to base noun phrase chunker.
(2) Animacy ambiguity: there are some verbs that can be both animate and inanimate such as "使得" (make), making the ZA resolution a wrong animate feature selection.
(3) Inappropriate template: there exists some exception sentence structure given by the authors, making the ZA identification unresolvable. There are 14% errors attributed to the inapplicability of the stored template method.

5 Conclusions

In this paper, we presented a case-based reasoning approach to Chinese zero anaphora resolution. Compared to other rule-based resolution methods , the presented approach turns out to be promising to deal with both intra-sentential and inter-sentential zero anaphora. In addition, we introduced two new features, namely animate relation acquired from outer resources and sentence patterns, in both ZA detection and ANT identification. Experimental results show that they can contribute 9% improvement to overall resolution performance. The drawback with this approach is the construction of the case base in advance. However, our experimental analysis shows that feasible performance results can be obtained when training corpus size is close to the testing one.

With the growing interest in natural language processing and its various applications, anaphora resolution is worth considering for further message understanding and the consistency of discourses. Our future work will be directed into the following studies:

(1) Modifying similarity function: Semantic classes of nouns may be employed to enhance the effectiveness of comparing sentence pattern similarity.
(2) Extending the set of anaphor being processed: This analysis aims at identifying instances (such as definite and pronominal anaphor) that could be useful in anaphora resolution.
(3) Exploiting web resource: The web resource can be utilized to identify sentence patterns and other useful features such as gender and number of entities.

References

1. Mitkov, R.: Robust pronoun resolution with limited knowledge. In Proceedings of the 18th International Conference on Computational Linguistics. (1998) 869–875
2. Xu, J.J.: Anaphora in Chinese Texts. China social science, Beijing. (2003)
3. Hobbs, J.: Pronoun Resolution. Research Report 76-1, Department of Computer Sciences, City College, City University of New York. (1976)
4. Lappin, S., Leass, H.: An Algorithm for Pronominal Anaphora Resolution. Computational Linguistics. **20** (1994) 535–561
5. Kennedy, C., Boguraev, B.: Anaphora for everyone: Pronominal anaphora resolution without a parser. In Proceedings of the 16th International Conference on Computational Linguis-tics. (1996) 113–118
6. Dagan, I., Itai, A.: Automatic processing of large corpora for the resolution of anaphora ref-erences. In Proceedings of the 13th International Conference on Computational Linguistics. (1990) 330–332
7. Mitkov, R., Richard, E., Orasan, C.: A new, fully automatic version of Mitkov's knowledge-poor pronoun resolution method. In Proceedings of the 3rd International Conference on Computational Linguistics and Intelligent Text Processing. (2002) 168–186
8. Liang, T., Wu, D.S.: Automatic Pronominal Anaphora Resolution in English Texts. International journal of Computational Linguistics and Chinese Language Processing. **9** (2004) 1–20
9. Modjeska, N.N., Markert, K., Nissim, M.: Using the Web in Machine Learning for Other-Anaphora Resolution. In Proceedings of the 2003 Conference on Empirical Methods in Natural Language Processing. (2003) 176–183
10. Wang, Y.K., Chen, Y.S., Hsu, W.L.: Empirical Study of Mandarin Chinese Discourse Analysis: An event-based approach. In Proceedings of the 10th IEEE International Conference on Tools with AI. (1998) 466–473
11. Wang, N., Yuan, C.F., Wang, K.F., Li, W.J.: Anaphora Resolution in Chinese Financial News for Information Extraction. In Proceedings of the 4th World Congress on Intelligent Control and Automation. (2002) 2422–2426
12. Yeh, C.L., Chen, Y.C.: Zero anaphora resolution in Chinese with shallow parsing. Journal of Chinese Language and Computing. (2005) (to appear)
13. Converse, S.P.: Resolving Pronominal References in Chinese with the Hobbs Algorithm. In Proceedings of the 4th SIGHAN Workshop on Chinese Language Processing. (2005) 116–122

14. Aamodt, A., Plaza, E.: Case-Based Reasoning: Foundational Issues, Methodological Variations, and System Approaches. AI Communications, IOS Press. **7** (1994) 39–59
15. Chen, K.J., Liu, S.H.: Word identification for Mandarin Chinese sentences. In Proceedings of the 14th Conference on Computational linguistics. (1992) 101–107
16. Wang, H.F., Mei, Z.: Robust Pronominal Resolution within Chinese Text. Journal of Software. **16** (2005) 700–707
17. CKIP.: A study of Chinese Word Boundaries and Segmentation Standard for Information processing. Technical Report, Taiwan, Taipei, Academia Sinica. (1996)
18. Ding, B.G., Huang, C.N., Huang, D.G.: Chinese Main Verb Identification: From Specification to Realization. International journal of Computational Linguistics and Chinese Lan-guage Processing. **10** (2005) 53–94
19. Liu, Y.H., Pan, W.Y., and Gu, W.: Shiyong Xiandai Hanyu Yufa (Practical Modern Chinese Grammar). The Commercial Press. (2002)
20. CKIP.: The content and illustration of Sinica corpus of Academia Sinica. Technical Report no. 95–02, Institute of Information Science, Academia Sinica. (1995)

Building a Collocation Net

GuoDong Zhou[1,2], Min Zhang[2], and GuoHong Fu[3]

[1] School of Computer Science and Technology, Suzhou University, China 215006
gdzhou@suda.edu.cn
[2] Institute for Infocomm Research, Singapore 119613
mzhang@i2r.a-star.edu.sg
[3] Department of Linguistics, The University of Hong Kong, Hong Kong
ghfu@hotmail.com

Abstract. This paper presents an approach to build a novel two-level collocation net, which enables calculation of the collocation relationship between any two words, from a large raw corpus. The first level consists of atomic classes (each atomic class consists of one word and feature bigram), which are clustered into the second level class set. Each class in both levels is represented by its collocation candidate distribution, extracted from the linguistic analysis of the raw training corpus, over possible collocation relation types. In this way, all the information extracted from the linguistic analysis is kept in the collocation net. Our approach applies to both frequently and less-frequently occurring words by providing a clustering mechanism resolve the data sparseness problem through the collocation net. Experimentation shows that the collocation net is efficient and effective in solving the data sparseness problem and determining the collocation relationship between any two words.

1 Introduction

In any natural language, there always exist many highly associated relationships between words, e.g. "strong tea" and "powerful computer". Although “strong” and “powerful” have similar syntax and semantics, there exist contexts where one is much more appropriate than the other (Halliday 1966). For example, we always say “strong tea” instead of “strong computer” and “powerful computer” instead of “powerful tea”. Psychological experiments in Meyer et al (1975) also indicated that human’s reaction to a highly associated word pair was stronger and faster than that to a poorly associated one. Lexicographers use the terms “collocation” and “co-occurrence” to describe various constraints on pairs of words. Here, we restrict “collocation” in the narrower sense between grammatically bound words, e.g. “strong” and “tea”, which occur in a particular grammatical order, and “co-occurrence” for the more general phenomenon of relationships between words, e.g. “doctor” and “nurse”, which are likely to be used in the same context (Manning et al 1999). This paper will concentrate on “collocation” rather than “co-occurrence” although there is much overlap between these two terms.

Collocations are important for a number of applications such as natural language generation, computational lexicography, parsing, proper noun discovery, corpus linguistic research, machine translation, information retrieval/extraction, etc. As an

Y. Matsumoto et al. (Eds.): ICCPOL 2006, LNAI 4285, pp. 532 – 541, 2006.

example, Hindle et al (1993) showed how collocation statistics can be used to improve the performance of a parser where lexical preferences are crucial to resolving the ambiguity of prepositional phrase attachment.

Currently, there are two categories of approaches used to discover collocations and co-occurrences: statistics-based approaches and parsing-based approaches. On the one hand, the statistics-based approaches are widely used to extract co-occurrences, where two words are likely to co-occur in the same context, from a large raw corpus, by using a statistical criterion, such as frequency (Justeson et al 1995; Ross et al 1975; Kupiec et al 1995; Zhao et al 1999), mean and variance (Smadja 1993), t-test (Church et al 1989; Church et al 1993), chi-square test (Church et al 1991; Snedecor et al 1989), likelihood ratio (Dunning 1993) and mutual information (Rosenfeld 1994; Denniz 1998; Zhou et al 1998; Zhou et al 1999). On the other hand, the parsing-based approaches rely on linguistic analysis and extract collocations, which differentiate between different types of linguistic relations, from the parsed trees of a large corpus. Normally such methods are combined with the frequency-based method to reject the ones whose frequencies are below a predefined threshold (Yang 1999).

Generally, both the statistics and parsing-based approaches are only effective on frequently occurring words and not effective on less frequently occurring words due to the data sparseness problem. Moreover, the extracted collocations or co-occurrences are always stored in a dictionary, which only contains a limited number of entries with limited information for each one. Finally, the collocation dictionary normally does not differentiate the strength degree among various collocations.

This paper combines the parsing-based approach and the statistics-based approach, and proposes a novel structure of collocation net. Through the collocation net, the data sparseness problem is resolved by providing a clustering mechanism and the collocation relationship between any two words can be easily determined and measured from the collocation net. Here, the collocation relationship is calculated using novel estimated pair-wise mutual information (EPMI) and estimated average mutual information (EAMI). Moreover, all the information extracted from the linguistic analysis is kept in the collocation net. Compared with the traditional collocation dictionary, the collocation net provides much more powerful facility since it can determine and measure the collocation relationship between any two words quantitatively.

The layout of this paper is as follows: Section 2 describes the novel structure of collocation net. Section 3 describes estimated pair-wise mutual information (EPMI) and estimated average mutual information (EAMI) to determine and measure the collocation relationship between any two words while Section 4 presents a method for automatically building a collocation net given a large law corpus. Experimentation is given in Section 5. Finally, some conclusions are drawn in Section 6.

2 Collocation Net

The collocation net is a kind of two-level structure, which stores rich information about the collocation candidates and others extracted from the linguistic analysis of a large raw corpus. The first level consists of word and feature bigrams[1] while the second level consists of classes that are clustered from the word and feature bigrams in the first level. For convenience, each word and feature bigram in the first level is also regarded as a class (atomic class). That is to say, each first level atomic class contains only one word and feature bigram while each second level class contains one or more word and feature bigrams clustered from first level atomic classes.

Meanwhile, each class in both levels of the collocation net is represented by its related collocation candidate distribution, extracted from the linguistic analysis. In this paper, a collocation candidate is represented as a 3tuple: a left side, a right side and a collocation relation type, which represents the collocation relationship between the left side and the right side. Both the left and right sides can be either a word and feature bigram or a class of word and feature bigrams. For example, a collocation candidate can be either $wf_i - CR_k - wf_j$ or $C_{hi} - CR_k - C_{gj}$, where wf_i is a word and feature bigram; C_{hi} is the i-th class in the h-th level and CR_k is a relation type.

Briefly, the collocation net is defined as follows:

$$CoNET = \{wf, CR, L1, L2, P_{h->g}\} \tag{1}$$

- wf stores possible word and feature bigrams
- CR stores possible collocation relation types
- $L1$ and $L2$ are the first and second levels in the collocation net, respectively;

$$L_h = \{< C_{hi}, FDCC_{C_{hi}} >\}, \tag{2}$$

where $C_h = \{C_{hi}, 1 \le i \le |C_h|\}$ is the class set in L_h (Obviously, $C_1 = wf$); C_{hi} is the i-th class in C_h; $|C_h|$ is the number of the classes in C_h and $FDCC_{C_{hi}} (1 \le i \le |C_h|)$ is the frequency distribution of collocation candidates related with C_{hi}. That is, each word and feature bigram or class in the collocation net is represented by the distribution of its related collocation candidates. In this way, all the information extracted via the linguistic analysis is stored in the collocation net.

- $P_{h->g}$ defines the class transition probability from a class to another class in the collocation net:

$$P_{h->g} = \{P(C_{hi} \to C_{gj}), 1 \le i \le |C_h|, 1 \le j \le |C_g|\}, \tag{3}$$

where $P(C_{hi} \to C_{gj})$ is the class transition probability from C_{hi} to C_{gj}. In this paper, $P(C_{hi} \to C_{gj})$ is defined using a kernel function $k(C_{gj}, C_{hi})$, which measures the similarity between C_{gj} and C_{hi} by calculating their shared probability distribution of collocation candidates:

[1] The reason to use the word and feature bigram is to distinguish the same word with different features, which can be "word sense", "part-of-speech", etc. In this paper, "part-of-speech" is used as the feature.

$$P(C_{hi} \to C_{gj}) = k(C_{hi}, C_{gj})$$
$$= \sum_{k=1}^{|CR|} \sum_{m=1}^{|wf|} \min(P(C_{hi} - CR_k - wf_m \mid C_{hi}), P(C_{gj} - CR_k - wf_m \mid C_{gj}))$$
$$+ \sum_{k=1}^{|CR|} \sum_{m=1}^{|wf|} \min(P(wf_m - CR_k - C_{hi} \mid C_{hi}), P(wf_m - CR_k - C_{gj} \mid C_{gj})) \quad (4)$$

Through the class transition probability facility $P_{h->g}$, the collocation net provides a word-clustering mechanism to resolve the data sparseness problem and becomes effective in determining and measuring the collocation relationship between any two words, whether they are frequently occurring or not. We will discuss this in more details in Section 3.

3 Measuring a Collocation Candidate

For a collocation candidate, e.g. $wf_i - CR_k - wf_j$, we propose two quantitative measurements to calculate the collocation relationship between the two word and feature bigrams wf_i and wf_j given the collocation relation type CR_k: estimated pair-wise mutual information (EPMI) and estimated average mutual information (EAMI). Moreover, we also extend the EPMI and EAMI to determine and measure the collocation relationship between any two words. In this way, we can not only determine the most possible collocation relationship between any two words but also measure the strength of the collocation relationship between them.

3.1 EAMI: Estimated Average Mutual Information

Traditionally in information theory, average mutual information (AMI) measures the co-occurrence relationship between two words as follows:

$$AMI(w_i, w_j) = P(w_i, w_j) \cdot \log \frac{P(w_i, w_j)}{P(w_i) \cdot P(w_j)} \quad (5)$$

For a collocation candidate, e.g. $wf_i - CR_k - wf_j$, we can extend the above notion and measure the average mutual information between the two word and feature bigrams wf_i and wf_j given the collocation relation type CR_k as follows:

$$AMI(wf_i - CR_k - wf_j) = P(wf_i - CR_k - wf_j \mid CR_k) \cdot \log \frac{P(wf_i - CR_k - wf_j \mid CR_k)}{P(wf_i - CR_k - * \mid CR_k) \cdot P(* - CR_k - wf_j \mid CR_k)} \quad (6)$$

Here, we use "*" to indicate all the possibilities on the corresponding part. The problem with the above equation is that it only works on frequently occurring word and feature bigrams and is not reliable on less-frequently occurring word and feature bigrams (e.g. frequency < 100). In order to resolve this problem, we propose a modified version of AMI, called estimated average mutual information (EAMI), to measure the collocation relationship of a collocation candidate when one or two word and feature bigrams do not occur frequently. This is done by finding two optimal classes in the collocation net and mapping the less-frequently occurring word and feature bigrams to them through the word-clustering mechanism provided in the collocation net as follows:

$$
\begin{aligned}
EAMI(wf_i - CR_k - wf_j) &= EAMI(C_{1i} - CR_k - C_{1j}) \\
&= \max_{\substack{C_{1i}\to C_{hm},\\ C_{1j}\to C_{gn}}} \{P(C_{hm} - CR_k - C_{gn} \mid CR_k)\cdot P(C_{1i} \to C_{hm})\cdot P(C_{1j} \to C_{gn}) \\
&\quad \cdot \log\frac{P(C_{hm} - CR_k - C_{gn} \mid CR_k)}{P(C_{hm} - CR_k - * \mid CR_k)\cdot P(* - CR_k - C_{gn} \mid CR_k)}\} \\
&= \max_{\substack{C_{1i}\to C_{hm},\\ C_{1j}\to C_{gn}}} \{P(C_{1i} \to C_{hm})\cdot P(C_{1j} \to C_{gn})\cdot AMI(C_{hm} - LR_k - C_{gn})\}
\end{aligned}
\tag{7}
$$

where $P(C_{hm} - CR_k - C_{gn} \mid CR_k)\cdot P(C_{1i} \to C_{hm})\cdot P(C_{1j} \to C_{gn})$ is the estimated joint probability of the collocation candidate $C_{1i} - CR_k - C_{1j}$ given the class transitions $C_{1i} \to C_{hm}$ and $C_{1j} \to C_{gn}$. Here, C_{hm} can be either C_{1i} itself or any class in $L2$ while C_{gn} can be either C_{1j} itself or any class in $L2$. That is, C_{1i} / C_{1j} can be either mapped to itself when the word and feature bigram occurs frequently or mapped to any class in $L2$ when the word and feature bigram does not occur frequently.

3.2 EPMI: Estimated Pair-Wise Mutual Information

Similarly in information theory, pair-wise mutual information (PMI) measures the change of information between two words as follows:

$$
PMI(w_i, w_j) = \log\frac{P(w_i, w_j)}{P(w_i)\cdot P(w_j)}
\tag{8}
$$

For an collocation candidate, e.g. $wf_i - CR_k - wf_j$, we can also extend above notion to measure the PMI of the collocation candidate as follows:

$$
PMI(wf_i - CR_k - wf_j) = \log\frac{P(wf_i - CR_k - wf_j \mid CR_k)}{P(wf_i - CR_k - * \mid CR_k)\cdot P(* - CR_k - wf_j \mid CR_k)}
\tag{9}
$$

Similar to AMI, the problem with the above equation is that it only works on frequently occurring word and feature bigrams. In order to resolve this problem, we also propose a modified version of PMI, called estimated pair-wise mutual information (EPMI), to calculate the information change of a collocation candidate when one or two word and feature bigrams does not occur frequently. This is done by using the two optimal classes found in calculating EAMI as follows:

$$
\begin{aligned}
\underset{C_{hm}, C_{gn}}{EPMI}(wf_i - CR_k - wf_j) &= \underset{C_{hm}, C_{gn}}{EPMI}(C_{1i} - LR_k - C_{1j}) \\
&= P(C_{1i} \to C_{hm})\cdot P(C_{1j} \to C_{gn}) \\
&\quad \cdot \log\frac{P(C_{hm} - CR_k - C_{gn} \mid CR_k)}{P(C_{hm} - CR_k - * \mid CR_k)\cdot P(* - CR_k - C_{gn} \mid CR_k)} \\
&= P(C_{1i} \to C_{hm})\cdot P(C_{1j} \to C_{gn})\cdot PMI(C_{hm} - LR_k - C_{gn})
\end{aligned}
\tag{10}
$$

Equation (10) measures the pair-wise mutual information using the collocation candidate between the two optimal classes and takes the class transitions $C_{1i} \rightarrow C_{hm}$ and $C_{1j} \rightarrow C_{gn}$ into consideration. In this paper, EAMI is used not only as a quantitative measurement for a collocation candidate but also as a selection criteria to determine the two optimal classes in calculating EPMI since EAMI takes the joint probability into consideration, while EPMI is used to measure the strength degree of a collocation candidate. For example, parse tree re-ranking can be performed by considering EPMI of the included collocation candidates in parse trees.

3.3 Collocation Relationship Between Any Two Words

Given any two words w_i and w_j, the EPMI and EAMI between them are defined as the EPMI and EAMI of the optimal collocation candidate related with the two words. Here, the optimal collocation candidate is determined by maximizing the EPMI among all the related collocation candidates over all possible word and feature bigrams and all the possible collocation relation types:

$$EPMI(w_i, w_j) = \max_{all\ \ CR_k,\ wf_i,\ and\ \ wf_j} \{EPMI(wf_i - CR_k - wf_j)\} \tag{11}$$

$$EAMI(w_i, w_j) = EAMI(wf_i - CR_k - wf_j) \tag{12}$$

Here in Equation (12), the collocation candidate $wf_i - CR_k - wf_j$ is determined through maximizing $EPMI(w_i, w_j)$ in Equation (11).

4 Building a Collocation Net

Given a large raw corpus and a general-purpose full parser, a collocation net can be built iteratively as follows:

1) Extract collocation candidates via linguistic analysis: First of all, all the sentences in the large raw corpus are parsed into parsed trees using a general-purpose full parser. For every sentence, the N-best (e.g. N=20) parsed tree hypotheses (PTHs) are kept and their relative probabilities are computed. Then, all the possible collocation candidates are extracted from the PTHs and their frequencies are accumulated. Assume T_i the set of the N-best PTHs for the i-th sentence in the corpus and T_{ij} the j-th PTH in T_i, the frequency of a collocation candidate in T_{ij} is equal to the relative probability of T_{ij} in T_i. $f(wf_i - CR_k - wf_j)$, the frequency of the collocation candidate $wf_i - CR_k - wf_j$, is summed over all the PTHs where $wf_i - CR_k - wf_j$ occurs. In this way, we have a large set of collocation candidates with their frequencies.
2) Build the collocation net based on the extracted collocation candidates: Given the previously extracted collocation candidates, the collocation net is built by first building the first level through the statistics of the collocation candidates

and then clustering similar classes in the first level to construct the second level using the k-means clustering algorithm.

3) Examine whether the collocation net is to be re-built. For example, whether the average probability ratio between the best parsed tree hypothesis and the second best parsed tree hypothesis for each sentence converges or begins to drop. If yes, exit the building process. If no, re-build the collocation net. In this paper, the threshold for the average probability ratio is set to 0.99.

5 Experimentation

The experimentation has been done on the Reuters corpus, which contains 21578 news documents of 2.7 million words in the XML format. In this paper, the Collins' parser is applied and all the collocations are extracted between the head and one modifier of a phrase. In our experimentation, only six most frequently occurring collocation relation types are considered. Table 1 shows them with their occurrence frequencies in the Reuters corpus.

Table 1. Six most frequently occurring collocation relation types

Collocation Relation Type	Remark	Freq
VERB-SUB	The right noun is the subject of the left verb	37547
VERB-OBJ	The right noun is the object of the left verb.	59124
VERB-PREP	The right preposition modifies the left verb	80493
NOUN-PREP	The right preposition modifies the left noun	19808
NOUN-NOUN	The right noun modifies the left noun	109795
NOUN-ADJ	The right adjective modifies the left noun	139712

To demonstrate the performance of the collocation net, the N-best collocations are extracted from the collocation net. This can be easily done through computing the EAMI and EPMI of all the collocation candidates extracted from the corpus, as described in Section 3. Then all the collocation candidates whose EPMIs are larger than a threshold (e.g. 0) are kept as collocations and sorted according to their EPMIs. As a result, 31965 collocations are extracted from the Reuters corpus. Table 2 gives some of examples. It shows that our method can not only extract the collocations that occur frequently in the corpus but also extract the collocations that seldom occur in the corpus. Another advantage is that our method can determine the collocation relationship between any two words and measure its strength degree. In this way, our method can even extract collocations that never occur in the corpus. Table 3 gives some of them. For example, the collocation candidate NOUN(abatement)_NOUN-ADJ_ADJ(eligible) can be measured as a collocation with EAMI of 1.01517e-05 and EPMI of 1.174579 although this collocation candidate doesn't exist in the corpus. The main reason is that the collocation net provides a word-clustering mechanism to resolve the problem of data sparseness. This is done by using the word-clustering mechanism in the collocation net as shown in Section 3. Table 4 shows an example class "finance/tax" in the second level of the collocation net.

Table 2. Examples of N-best collocations

NO.	Left Side	Relation Type	Right Side	EPMI	EAMI	Freq
1	NOUN(complex)	NOUN-ADJ	ADJ(aidsrelated)	10.8	0.00023	3
2	NOUN(fraction)	NOUN-ADJ	ADJ(tiny)	10.7	0.00023	3
3	NOUN(politician)	NOUN-ADJ	ADJ(veteran)	10.5	0.00029	3
1001	NOUN(publishing)	NOUN-ADJ	ADJ(desktop)	6.22	0.00045	8
1002	VERB(start)	VERB-SUB	NOUN(talk)	6.22	0.00049	2
1003	NOUN(science)	NOUN-ADJ	ADJ(political)	6.21	.000040	9
5001	VERB(give)	VERB-OBJ	NOUN(breakdown)	3.94	0.00073	11
5002	VERB(introduce)	VERB-OBJ	NOUN(tax)	3.94	0.00018	3
5003	NOUN(fund)	NOUN-NOUN	NOUN(trust)	3.94	0.00051	11
10001	VERB(cut)	VERB-OBJ	NOUN(cost)	2.69	0.00170	11
10002	NOUN(session)	NOUN-NOUN	NOUN(house)	2.69	0.00007	3
10003	NOUN(challenge)	NOUN-PREP	PREP(of)	2.68	0.00147	6
15001	NOUN(factor)	NOUN-ADJ	ADJ(seasonal)	1.85	0.00009	16
15002	NOUN(report)	NOUN-NOUN	NOUN(acreage)	1.85	0.00009	3
15003	NOUN(menu)	NOUN-PREP	PREP(of)	1.85	0.00024	5
20001	NOUN(investor)	NOUN-ADJ	ADJ(Norwegian)	1.20	0.00007	2
20002	NOUN(conflict)	NOUN-ADJ	ADJ(serious)	1.20	0.00001	5
20003	NOUN(country)	NOUN-PREP	PREP(in)	1.20	0.00198	50
31963	NOUN(infusion)	NOUN-PREP	PREP(into)	5^{e}-4	8.27e-9	3
31964	VERB(ask)	VERB-OBJ	NOUN(leader)	4^{e}-4	1.70e-8	3
31965	VERB(land)	VERB-SUB	NOUN(plane)	2e-4	3.19e-8	4

Table 3. Examples of collocations not existing in the corpus

Left Side	Collocation Relation Type	Right Side	EPMI	EAMI
NOUN(accountan)	NOUN-ADJ	ADJ(associate)	3.22	8.68e-06
NOUN(worker)	NOUN-NOUN	NOUN(professional)	2.89	1.12e-04
VERB(regain)	VERB-SUB	NOUN(ability)	2.21	8.41e-05
NOUN(stock)	NOUN-ADJ	ADJ(borrowed)	2.12	8.61e-05
NOUN(share)	NOUN-NOUN	NOUN(Malaysia)	1.89	8.65e-05
NOUN(activity)	NOUN-NOUN	NOUN(buying)	1.27	8.71e-05
NOUN(business)	NOUN-NOUN	NOUN(customer)	1.22	9.66e-05
NOUN(abatement)	NOUN-ADJ	ADJ(eligible)	1.17	1.02e-05
VERB(transfer)	VERB-SUB	NOUN(business)	1.06	5.18e-05

Table 4. An example class ("finance/tax") in the second level of the collocation net

NOUN(asbestos) NOUN(abatement) NOUN(abba) NOUN(market) NOUN(share) NOUN(stock) NOUN(tax) NOUN(currency) NOUN(contract) NOUN(income) NOUN(reserve) NOUN(investment) NOUN(bid) NOUN(trade) ......

In order to further evaluate the usefulness of the collocation net, we have used it in parse tree re-ranking using the standard PARSEVAL metrics. Here, Collins' parser is used with the standard setting (sections 2-21 as training data, section 24 as development data and section 23 as testing data) while 20-best parse trees for each sentence are considered in re-ranking. This is done by building a collocation net on the golden parse trees in the training data and adjusting the probability of each parse tree candidate using the collocation net to achieve parse tree re-ranking. For each parse tree candidate, e.g. T_{ij} with the original probability $P_{T_{ij}}$, the probability of the parse tree candidate can be adjusted by considering the contribution of its included collocation candidates:

$$\log P_{T_{ij}} = \log P_{T_{ij}} + \sum_{i=1}^{|CC|} EPMI(CC_i) \qquad (13)$$

where $CC = \{CC_i, 1 \le i \le |CC|\}$ includes all the collocation candidates extracted from T_{ij} ; $|CC|$ is the number of collocation candidates extracted from T_{ij} and $EPMI(CC_i)$ is the estimated pair-wise mutual information, which measures the change of information when the collocation candidate CC_i is collocated. Then, all the parse tree candidates for a sentence can be re-ranked according to their adjusted probabilities as calculated in Formula (2).

Table 5 shows the effect of parse tree re-ranking using the collocation net. It shows that the use of the collocation net can increase the F-measure by 1.6 in F-measure.

Table 5. Application of the collocation net in parse tree re-ranking

	P(%)	R(%)	F1
Before re-ranking	88.26	88.05	88.15
After re-ranking	89.82	89.66	89.74

6 Conclusion

This paper proposes a novel structure of two-level collocation net and a method capable of automatically building the collocation net given a large raw corpus. Through the collection net, the collocation relationship between any two words can be calculated quantitatively using novel estimated average mutual information (EAMI) as the selection criterion and estimated pair-wise mutual information (EPMI) as the strength degree. Obviously, the two-level collocation net can be easily extended to more levels through cascading such a two-level structure.

Future works include systematic evaluation of the collocation net on a much larger corpus, its application to other languages such as Chinese and in a general-purpose parser for adaptation to a new domain/application, and development of a more-level collocation net.

References

1. Church K.W. and Patrick H. (1989). Word Association Norms, Mutural Information and Lexicography. *ACL'1989* :76-83.
2. Church K.W. and William A.G. (1991). A Comparison of the Enhanced Good Turing and Deleted Estimation Methods for Estimating Probabilities of English Bigrams. *Computer, Speech and Language*. 5(1) :19-54.
3. Church K.W. and Robert L.M. (1993). Introduction to Special Issue on Computational Linguistics Using Large Corpora. *Computational Linguistics*. 19(1) :1-24.
4. Dunning T. (1993). Accurate Methods for the Statistics of Surprise and Coincidence. *Computational Linguistics*. 19(1) :61-74.
5. Halliday M. (1966). Lexis as a linguistic level. *In memory of J.R.Firth,* edited by C.Bazell, J.Catford, M.Halliday and R.Robins, Longman.
6. Hindle D. and Rooth M. (1993). Structural Ambiguity and Lexical Relations. *Computational Linguistics*. 19(1) :102-119.
7. Justeson J.S. and Katz S.M. (1995). Technical Terminology: Some Linguistic Properties and an Algorithm for Identification in Text. *Natural Language Engineering*. 1(1):9-27.
8. Julian K., Pederson J. and Chen F. (1995). A Trainable Document Summarizer. *SIGIR'1995*: 68-73.
9. Manning C.D. and Schutze H. (1999). *Fundations of Statistical Natural Language Processing*. pp.185. The MIT Press.
10. Meyer D. et al. (1975). Loci of Contextual Effects on Visual Word Recognition. *In Attention and Performance V*, edited by P.Rabbitt and S.Dornie. Acdemic Press, pp. 98-116.
11. Ross I.C. and Tukey J.W. (1975). Introduction to these Volumes.. In John Wilder Tukey(ed.), *Index to Statistics amd Probability*, pp.Iv-x. Los Altos, CA: R&D Press.
12. Rosenfeld R. (1994). Adaptive Statistical Language Modeling: A Maximum Entropy Approach. *Ph.D. Thesis,* Carneige Mellon University.
13. Smadja F. (1993). Retrieving Collocations from Text: Xtract. *Computational Linguistics*. 19(1):143-177.
14. Snedecor G.W. and William G.C. (1989). Statistical Methods. Ames: Iowa State University Press. pp.127.
15. Yang J. (1999). Towards the automatic Acquisition of Lexical Selection Rules. *MT Summit VII* :397-403. Singapore.
16. Yuret D. (1998). Discovery of Linguistic Relations Using Lexical Attraction". *Ph.D thesis. cmp-lg/9805009*. MIT.
17. Zhao J. and Huang C.N. (1998). Aquasi-Dependency Model for the Structural Analysis of Chinese BaseNPs. *COLING-ACL'1998:*1-7. Univ. de Montreal. Canada.
18. Zhou G.D. and Lua K.T. (1998). Word Association and MI-Trigger-based Language Modeling. *COLING-ACL'1998:*1465-1471. Univ. of Montreal,Canada.
19. Zhou G.D. and Lua K.T. (1999). Interpolation of N-gram and MI-based Trigger Pair Language Modeling in Mandarin Speech Recognition. *Computer, Speech and Language*, Vol. 13(2) : 123-135.

Author Index

Printing: Mercedes-Druck, Berlin
Binding: Stein+Lehmann, Berlin

Lecture Notes in Artificial Intelligence (LNAI)

Vol. 4333: U. Reimer, D. Karagiannis (Eds.), Practial Aspects of Knowledge Management. XII, 338 pages. 2006.

Vol. 4304: A. Sattar, B.-H. Kang (Eds.), AI 2006: Advances in Artificial Intelligence. XXVII, 1303 pages. 2006.

Vol. 4293: A. Gelbukh, C.A. Reyes-Garcia (Eds.), MICAI 2006: Advances in Artificial Intelligence. XXVIII, 1232 pages. 2006.

Vol. 4289: M. Ackermann, B. Berendt, M. Grobelnik, A. Hotho, D. Mladenič, G. Semeraro, M. Spiliopoulou, G. Stumme, V. Svatek, M. van Someren (Eds.), Semantics, Web and Mining. X, 197 pages. 2006.

Vol. 4285: Y. Matsumoto, R. Sproat, K.-F. Wong, M. Zhang (Eds.), Computer Processing of Oriental Languages. XVII, 544 pages. 2006.

Vol. 4274: Q. Huo, B. Ma, E.-S. Chng, H. Li (Eds.), Chinese Spoken Language Processing. XXIV, 805 pages. 2006.

Vol. 4265: L. Todorovski, N. Lavrač, K.P. Jantke (Eds.), Discovery Science. XIV, 384 pages. 2006.

Vol. 4264: J.L. Balcázar, P.M. Long, F. Stephan (Eds.), Algorithmic Learning Theory. XIII, 393 pages. 2006.

Vol. 4259: S. Greco, Y. Hata, S. Hirano, M. Inuiguchi, S. Miyamoto, H.S. Nguyen, R. Słowiński (Eds.), Rough Sets and Current Trends in Computing. XXII, 951 pages. 2006.

Vol. 4253: B. Gabrys, R.J. Howlett, L.C. Jain (Eds.), Knowledge-Based Intelligent Information and Engineering Systems, Part III. XXXII, 1301 pages. 2006.

Vol. 4252: B. Gabrys, R.J. Howlett, L.C. Jain (Eds.), Knowledge-Based Intelligent Information and Engineering Systems, Part II. XXXIII, 1335 pages. 2006.

Vol. 4251: B. Gabrys, R.J. Howlett, L.C. Jain (Eds.), Knowledge-Based Intelligent Information and Engineering Systems, Part I. LXVI, 1297 pages. 2006.

Vol. 4248: S. Staab, V. Svátek (Eds.), Managing Knowledge in a World of Networks. XIV, 400 pages. 2006.

Vol. 4246: M. Hermann, A. Voronkov (Eds.), Logic for Programming, Artificial Intelligence, and Reasoning. XIII, 588 pages. 2006.

Vol. 4223: L. Wang, L. Jiao, G. Shi, X. Li, J. Liu (Eds.), Fuzzy Systems and Knowledge Discovery. XXVIII, 1335 pages. 2006.

Vol. 4213: J. Fürnkranz, T. Scheffer, M. Spiliopoulou (Eds.), Knowledge Discovery in Databases: PKDD 2006. XXII, 660 pages. 2006.

Vol. 4212: J. Fürnkranz, T. Scheffer, M. Spiliopoulou (Eds.), Machine Learning: ECML 2006. XXIII, 851 pages. 2006.

Vol. 4211: P. Vogt, Y. Sugita, E. Tuci, C.L. Nehaniv (Eds.), Symbol Grounding and Beyond. VIII, 237 pages. 2006.

Vol. 4203: F. Esposito, Z.W. Raś, D. Malerba, G. Semeraro (Eds.), Foundations of Intelligent Systems. XVIII, 767 pages. 2006.

Vol. 4201: Y. Sakakibara, S. Kobayashi, K. Sato, T. Nishino, E. Tomita (Eds.), Grammatical Inference: Algorithms and Applications. XII, 359 pages. 2006.

Vol. 4200: I.F.C. Smith (Ed.), Intelligent Computing in Engineering and Architecture. XIII, 692 pages. 2006.

Vol. 4198: O. Nasraoui, O. Zaïane, M. Spiliopoulou, B. Mobasher, B. Masand, P.S. Yu (Eds.), Advances in Web Mining and Web Usage Analysis. IX, 177 pages. 2006.

Vol. 4196: K. Fischer, I.J. Timm, E. André, N. Zhong (Eds.), Multiagent System Technologies. X, 185 pages. 2006.

Vol. 4188: P. Sojka, I. Kopeček, K. Pala (Eds.), Text, Speech and Dialogue. XV, 721 pages. 2006.

Vol. 4183: J. Euzenat, J. Domingue (Eds.), Artificial Intelligence: Methodology, Systems, and Applications. XIII, 291 pages. 2006.

Vol. 4180: M. Kohlhase, OMDoc – An Open Markup Format for Mathematical Documents [version 1.2]. XIX, 428 pages. 2006.

Vol. 4177: R. Marín, E. Onaindía, A. Bugarín, J. Santos (Eds.), Current Topics in Artificial Intelligence. XV, 482 pages. 2006.

Vol. 4160: M. Fisher, W. van der Hoek, B. Konev, A. Lisitsa (Eds.), Logics in Artificial Intelligence. XII, 516 pages. 2006.

Vol. 4155: O. Stock, M. Schaerf (Eds.), Reasoning, Action and Interaction in AI Theories and Systems. XVIII, 343 pages. 2006.

Vol. 4149: M. Klusch, M. Rovatsos, T.R. Payne (Eds.), Cooperative Information Agents X. XII, 477 pages. 2006.

Vol. 4140: J.S. Sichman, H. Coelho, S.O. Rezende (Eds.), Advances in Artificial Intelligence - IBERAMIA-SBIA 2006. XXIII, 635 pages. 2006.

Vol. 4139: T. Salakoski, F. Ginter, S. Pyysalo, T. Pahikkala (Eds.), Advances in Natural Language Processing. XVI, 771 pages. 2006.

Vol. 4133: J. Gratch, M. Young, R. Aylett, D. Ballin, P. Olivier (Eds.), Intelligent Virtual Agents. XIV, 472 pages. 2006.

Vol. 4130: U. Furbach, N. Shankar (Eds.), Automated Reasoning. XV, 680 pages. 2006.

Vol. 4120: J. Calmet, T. Ida, D. Wang (Eds.), Artificial Intelligence and Symbolic Computation. XIII, 269 pages. 2006.

Vol. 4114: D.-S. Huang, K. Li, G.W. Irwin (Eds.), Computational Intelligence, Part II. XXVII, 1337 pages. 2006.

Vol. 4108: J.M. Borwein, W.M. Farmer (Eds.), Mathematical Knowledge Management. VIII, 295 pages. 2006.

Vol. 4106: T.R. Roth-Berghofer, M.H. Göker, H.A. Güvenir (Eds.), Advances in Case-Based Reasoning. XIV, 566 pages. 2006.

Vol. 4099: Q. Yang, G. Webb (Eds.), PRICAI 2006: Trends in Artificial Intelligence. XXVIII, 1263 pages. 2006.

Vol. 4095: S. Nolfi, G. Baldassarre, R. Calabretta, J.C.T. Hallam, D. Marocco, J.-A. Meyer, O. Miglino, D. Parisi (Eds.), From Animals to Animats 9. XV, 869 pages. 2006.

Vol. 4093: X. Li, O.R. Zaïane, Z. Li (Eds.), Advanced Data Mining and Applications. XXI, 1110 pages. 2006.

Vol. 4092: J. Lang, F. Lin, J. Wang (Eds.), Knowledge Science, Engineering and Management. XV, 664 pages. 2006.

Vol. 4088: Z.-Z. Shi, R. Sadananda (Eds.), Agent Computing and Multi-Agent Systems. XVII, 827 pages. 2006.

Vol. 4087: F. Schwenker, S. Marinai (Eds.), Artificial Neural Networks in Pattern Recognition. IX, 299 pages. 2006.

Vol. 4068: H. Schärfe, P. Hitzler, P. Øhrstrøm (Eds.), Conceptual Structures: Inspiration and Application. XI, 455 pages. 2006.

Vol. 4065: P. Perner (Ed.), Advances in Data Mining. XI, 592 pages. 2006.

Vol. 4062: G. Wang, J.F. Peters, A. Skowron, Y. Yao (Eds.), Rough Sets and Knowledge Technology. XX, 810 pages. 2006.

Vol. 4049: S. Parsons, N. Maudet, P. Moraitis, I. Rahwan (Eds.), Argumentation in Multi-Agent Systems. XIV, 313 pages. 2006.

Vol. 4048: L. Goble, J.-J.C.. Meyer (Eds.), Deontic Logic and Artificial Normative Systems. X, 273 pages. 2006.

Vol. 4045: D. Barker-Plummer, R. Cox, N. Swoboda (Eds.), Diagrammatic Representation and Inference. XII, 301 pages. 2006.

Vol. 4031: M. Ali, R. Dapoigny (Eds.), Advances in Applied Artificial Intelligence. XXIII, 1353 pages. 2006.

Vol. 4029: L. Rutkowski, R. Tadeusiewicz, L.A. Zadeh, J.M. Zurada (Eds.), Artificial Intelligence and Soft Computing – ICAISC 2006. XXI, 1235 pages. 2006.

Vol. 4027: H.L. Larsen, G. Pasi, D. Ortiz-Arroyo, T. Andreasen, H. Christiansen (Eds.), Flexible Query Answering Systems. XVIII, 714 pages. 2006.

Vol. 4021: E. André, L. Dybkjær, W. Minker, H. Neumann, M. Weber (Eds.), Perception and Interactive Technologies. XI, 217 pages. 2006.

Vol. 4020: A. Bredenfeld, A. Jacoff, I. Noda, Y. Takahashi (Eds.), RoboCup 2005: Robot Soccer World Cup IX. XVII, 727 pages. 2006.

Vol. 4013: L. Lamontagne, M. Marchand (Eds.), Advances in Artificial Intelligence. XIII, 564 pages. 2006.

Vol. 4012: T. Washio, A. Sakurai, K. Nakajima, H. Takeda, S. Tojo, M. Yokoo (Eds.), New Frontiers in Artificial Intelligence. XIII, 484 pages. 2006.

Vol. 4008: J.C. Augusto, C.D. Nugent (Eds.), Designing Smart Homes. XI, 183 pages. 2006.

Vol. 4005: G. Lugosi, H.U. Simon (Eds.), Learning Theory. XI, 656 pages. 2006.

Vol. 4002: A. Yli-Jyrä, L. Karttunen, J. Karhumäki (Eds.), Finite-State Methods and Natural Language Processing. XIV, 312 pages. 2006.

Vol. 3978: B. Hnich, M. Carlsson, F. Fages, F. Rossi (Eds.), Recent Advances in Constraints. VIII, 179 pages. 2006.

Vol. 3963: O. Dikenelli, M.-P. Gleizes, A. Ricci (Eds.), Engineering Societies in the Agents World VI. XII, 303 pages. 2006.

Vol. 3960: R. Vieira, P. Quaresma, M.d.G.V. Nunes, N.J. Mamede, C. Oliveira, M.C. Dias (Eds.), Computational Processing of the Portuguese Language. XII, 274 pages. 2006.

Vol. 3955: G. Antoniou, G. Potamias, C. Spyropoulos, D. Plexousakis (Eds.), Advances in Artificial Intelligence. XVII, 611 pages. 2006.

Vol. 3949: F.A. Savacı (Ed.), Artificial Intelligence and Neural Networks. IX, 227 pages. 2006.

Vol. 3946: T.R. Roth-Berghofer, S. Schulz, D.B. Leake (Eds.), Modeling and Retrieval of Context. XI, 149 pages. 2006.

Vol. 3944: J. Quiñonero-Candela, I. Dagan, B. Magnini, F. d'Alché-Buc (Eds.), Machine Learning Challenges. XIII, 462 pages. 2006.

Vol. 3937: H. La Poutré, N.M. Sadeh, S. Janson (Eds.), Agent-Mediated Electronic Commerce. X, 227 pages. 2006.

Vol. 3932: B. Mobasher, O. Nasraoui, B. Liu, B. Masand (Eds.), Advances in Web Mining and Web Usage Analysis. X, 189 pages. 2006.

Vol. 3930: D.S. Yeung, Z.-Q. Liu, X.-Z. Wang, H. Yan (Eds.), Advances in Machine Learning and Cybernetics. XXI, 1110 pages. 2006.

Vol. 3918: W.-K. Ng, M. Kitsuregawa, J. Li, K. Chang (Eds.), Advances in Knowledge Discovery and Data Mining. XXIV, 879 pages. 2006.

Vol. 3913: O. Boissier, J. Padget, V. Dignum, G. Lindemann, E. Matson, S. Ossowski, J.S. Sichman, J. Vázquez-Salceda (Eds.), Coordination, Organizations, Institutions, and Norms in Multi-Agent Systems. XII, 259 pages. 2006.

Vol. 3910: S.A. Brueckner, G.D.M. Serugendo, D. Hales, F. Zambonelli (Eds.), Engineering Self-Organising Systems. XII, 245 pages. 2006.

Vol. 3904: M. Baldoni, U. Endriss, A. Omicini, P. Torroni (Eds.), Declarative Agent Languages and Technologies III. XII, 245 pages. 2006.

Vol. 3900: F. Toni, P. Torroni (Eds.), Computational Logic in Multi-Agent Systems. XVII, 427 pages. 2006.